Design of
Machine Elements

PRENTICE-HALL INTERNATIONAL, INC., *London*
PRENTICE-HALL OF AUSTRALIA, PTY. LTD., *Sydney*
PRENTICE-HALL OF CANADA, LTD., *Toronto*
PRENTICE-HALL OF INDIA PRIVATE LIMITED, *New Delhi*
PRENTICE-HALL OF JAPAN, INC., *Tokyo*

M. F. SPOTTS

Mechanical Engineering Department
The Technological Institute
Northwestern University

Design of

Machine Elements

FOURTH EDITION

PRENTICE-HALL, INC., Englewood Cliffs, N. J.

13-200550-6

Library of Congress Card Catalog Number: 78-162274

Current printing (last digit):

10 9 8 7 6

Throughout the wide fields of engineering, the most fundamental requirement is a sound knowledge of the first principles of mechanics coupled with an intimate understanding of the properties of materials. More problems arise from weaknesses in mechanical design than from any other cause, impressing upon us the necessity of being well versed in the basic principles of mechanical engineering.

SIR GEORGE H. NELSON,
President,
Institution of Electrical Engineers

Preface to Fourth Edition

The essential features of the previous editions have been retained in the fourth edition. As before, the object of the book is to present, at a professional level, a comprehensive survey of analytical design methods which can be applied in the broad and ever-widening field of mechanical equipment.

The chapters are largely independent of each other, and may be studied in any order. Chapter 1 has been retained for reference, but can be easily omitted by students already familiar with the principles of strength of materials. Mathematical derivations are presented in full whenever possible, and are valuable as indicating the analytical basis for the resulting design equations. The contents of the book has been gathered from a wide variety of sources, but has been simplified and coordinated and a unified system of notation used. Information on material selection and dimensioning and tolerancing has been retained.

An effort has been made to have a book that is as self-explanatory as possible therefore reducing the work load on the instructor. In addition, this feature should prove of assistance to engineers in industry who are attempting to gain an understanding of the principles of design by means of home study. The many numerical examples will illustrate the application of the various formulas.

Although design decisions are influenced to a large extent by judgment and experience, the engineer must also be familiar with results as obtained by calculation. In fact, it is only after the principles of elementary analysis have become familiar working tools that the designer can turn his attention to the broader aspects of the problem as a whole. This statement applies in general to all engineering, and is not restricted solely to those who expect to become professional designers.

One must of necessity make use of empirical methods and test results when analytical solutions are not yet available. However, anyone who is thoroughly familiar with the fundamental theories presented in this book will be able to proceed with greater confidence and assurance into more indefinite areas.

The presentation has been simplified and updated to bring it in line with current design practice. Sufficient details have been given so that the usual design problem can be handled. The book can thus be retained by the student, or the engineer in industry, as a reference and source book. Footnotes provide readily available sources where additional design information can be obtained.

The author again wishes to express appreciation for the many helpful suggestions received from users of the previous editions.

M. F. SPOTTS

Evanston, Illinois

Introduction

1. Machine design. Machine design is the art of planning or devising new or improved machines to accomplish specific purposes. In general, a machine will consist of a combination of several different mechanical elements properly designed and arranged to work together, as a whole. During the initial planning of a machine, fundamental decisions must be made concerning loading, type of kinematic elements to be used, and correct utilization of the properties of engineering materials. Economic considerations are usually of prime importance when the design of new machinery is undertaken. In general, the lowest over-all cost is desired. Consideration should be given not only to the cost of design, manufacture, sale, and installation, but also to the cost of servicing. The machine should of course incorporate the necessary safety features and be of pleasing external appearance. The objective is to produce a machine which is not only sufficiently rugged to function properly for a reasonable life, but is at the same time cheap enough to be economically feasible.

The engineer in charge of the design of a machine should not only have adequate technical training, but must be a man of sound judgment and wide experience, qualities which are usually acquired only after considerable time has been spent in actual professional work. A start in this direction can be made with a good teacher while the student is yet at the university. However, the would-be designer must expect to get a substantial portion of his training after leaving school through further reading and study, and especially by being associated in his work with competent engineers.

2. Design of machine elements. This book, as the title indicates, will not deal with the broader aspects of the design of complete machines, but will attempt to explain the fundamental principles required for the correct design of the separate elements which compose the machine.

The principles of design are of course universal. The same theory or equations may be applied to a very small part, as in an instrument, or to a larger but similar part used in a piece of heavy equipment. In no case, however, should mathematical calculations be looked upon as absolute and final. They are all subject to the accuracy of the various assumptions which must necessarily be made in engineering work. Sometimes only a portion of the total number of parts in a machine are designed on the basis of analytic calculations. The form and size of the remaining parts are then usually determined by practical considerations. On the other hand, if the machine is very expensive, or if weight is a factor, as in airplanes, design computations may then be made for almost all the parts.

The purpose of the design calculations is of course to attempt to predict the stress or deformation in the part in order that it may safely carry the loads which will be imposed upon it, and that it may last for the expected life of the machine. All calculations are, of course, dependent on the physical properties of the construction materials as determined by laboratory tests. A rational method of design attempts to take the results of relatively simple and fundamental tests such as tension, compression, torsion, and fatigue and apply them to all the complicated and involved situations encountered in present-day machinery.

In addition, it has been amply proved that such details as surface condition, fillets, notches, manufacturing tolerances, and heat treatment have a marked effect on the strength and useful life of a machine part. The design and drafting departments must specify completely all such particulars, and thus exercise the necessary close control over the finished product.

Training in rapid and accurate numerical work is invaluable to the designer. The designer should keep an accurate notebook, as it is frequently necessary for him to refer to work which he has done in the past. A sketch, carefully drawn to scale, is also a necessity, and provides a convenient place for putting down a portion of the data used in connection with the problem. It goes without saying that all data, assumptions, equations, and calculations should be written down in full in order to be intelligible when referred to at a later date. The student should start forming such habits, and it is recommended that the problems in this book be worked out and preserved as reference material.

Contents

7. Welded and Riveted Connections   281

8. Lubrication   315

9. Ball and Roller Bearings   352

10. Spur Gears   375

Design of
Machine Elements

1 Fundamental Principles

Design methods for the various machine elements are founded on the theories of mechanics and strength of materials. The scope of such theories is very extensive, and the purpose of this chapter is to present, for review and ready reference, those topics that are generally used by designers, and that will be referred to throughout the book. These theories are more or less simplified approximations, and attention should be directed to the limitations imposed by the assumptions that had to be made in arriving at working formulas.

A thorough grounding in these fundamentals will prove of great value in attacking new and unfamiliar problems. In fact, only after such theories have become working tools is it possible to achieve the broad perspective and balanced judgment that must be expressed by the really competent machine designer.

A, area
b, width of cross section parallel to neutral axis
c, maximum distance, neutral axis to edge of cross section
d, distance, diameter of cirle

E, modulus of elasticity for normal stress
FS, factor of safety
G, modulus of elasticity for shear stress

1

h, height of cross section perpendicular to neutral axis
i, radius of gyration
I, moment of inertia
l, length
M, bending moment
P, load
r, radius of circle, radius to
 center of curvature of deflected beam
s, s_n, normal stress
 s_s, shear stress
 s_x, normal stress, x-direction
 s_y, normal stress, y-direction
 s_{xy}, shear stress, x- and y-directions
 s_1, maximum normal stress
 s_2, minimum normal stress
s_{smax}, maximum shear stress

s_{yp}, yield point stress
 v, distance on cross section of
 beam perpendicular to neutral axis
V, total shear force on cross
 section
w, distributed load, pounds per
 unit length
y, deflection of beam
γ, (gamma), shearing deformation, weight per cubic inch
 of a material
δ, (delta), axial deformation
ϵ, (epsilon), strain or elongation
μ, (mu), Poisson's ratio. For
 metals, μ is usually taken as
 0.3.

1. Statical Equilibrium

When a body is at rest, or in motion with constant velocity, the external forces acting upon it are in equilibrium. This statement applies to the body as a whole or to any portion of it. When a force analysis is to be made, it is sometimes advantageous to consider only a portion of the body, which can be obtained by assuming that cutting planes are passed through the body at the desired locations. The internal forces that were acting at the locations of the cuts must then be represented as a system of external forces properly distributed to maintain equilibrium of the separate parts and to preserve the original state of stress in the material. When a problem is analyzed in this manner the loading will consist entirely of external forces and moments. It is not necessary to consider the internal stresses.

Statical equilibrium means that both forces and moments are in balance. When a body is in equilibrium, the sum of the components of the forces in any given direction must be equal to zero. Likewise the moments about any given line as an axis must be equal to zero. If the body is undergoing acceleration, the inertia forces must be included in the equilibrium equations.

2. Engineering Materials

The mathematical equations used in designing are derived for an idealized material, which is assumed to have the following properties.

(a) *Perfect Elasticity.* Loads or forces acting on a body cause changes in its shape and dimensions. A perfectly elastic material is one that returns to its original form immediately upon removal of the loads. The equations used in designing are nearly always derived on the assumption of perfect elasticity. If the material is such that this assumption cannot be made, the mathematical complications, in many cases, become too great for practical calculations. It should never be forgotten, however, that in some cases there may be a considerable variation between the actual stresses in the body and the stresses obtained from the equations for an idealized substance. A material may exhibit a high degree of elasticity for small loads, but may retain a permanent deformation when the loads become sufficiently great.

(b) *Homogeneity*. A homogeneous body is one that has the same properties throughout its entire extent.

(c) *Isotropy*. An isotropic material is one whose elastic properties are the same in all directions.

Actually, a metal is not a homogeneous substance. It consists of an aggregate of very small crystals whose strength depends upon their orientation with respect to the applied force. When the minute crystals have a random orientation, the location in the body, or the inclination at which a test specimen is taken, has no effect on the results of the test. The assumption that the material is homogeneous and isotropic, for all practical purposes, is fulfilled. This is true for cast, hot-rolled, or annealed metals. In contrast, materials that have been cold rolled or drawn may have a preferred orientation of crystals, and may exhibit a definite grain effect with a variation in strength depending on the direction of the applied load. The assumption cannot be made that such materials are homogeneous and isotropic.

3. Tension and Compression Stress

The eyebar, in Fig. 1-1(a), which supports the load P, is said to be in *tension*, or to have an internal force of tension. Such a force causes an increase

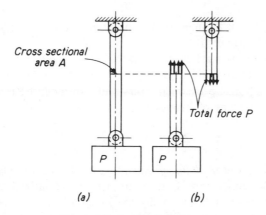

Figure 1-1 Eye-bar loaded in tension.

in the length of the bar. A solution for the stress can be effected by means of cutting planes as described in Section 1.

If the bar, shown in Fig. 1-1(a), is cut normal to its axis, as shown in Fig. 1-1(b), equal and opposite tension forces, uniformly distributed, must be applied to the cut surfaces. In magnitude, each must be equivalent to the load P. The average stress s, or force per unit of cross-sectional area A, is then equal to P divided by A. Hence

$$s = \frac{P}{A} \tag{1}$$

Thus the magnitude of the external forces on the cross sections in Fig. 1-1(b) constitutes the measure of the internal force in the bar shown in Fig. 1-1(a). Forces are usually expressed in pounds and areas in square inches.

It is, of course, obvious that this solution is correct only if the assumption regarding the distribution of stress on the cut surfaces is correct. Had the bar been cut near one end where the shape is no longer a prism, it is apparent that the situation would be more complicated, and the stress system would no longer be simple tension uniformly distributed over the cross section.

Since the assumption relative to homogeneity of the material is never exactly fulfilled, the stresses on the cross section will not be entirely uniform but will be subject to small local variations. Equation (1) does, however, give the average value of the tensile stress over the whole cross section.

A *compressive stress* is one that causes a decrease in the length of the body in the direction of the force.

The total change of length in a uniform body caused by an axial load is called the *deformation*, δ. If the deformation is divided by the original length, l, of the body, the result is the deformation per unit length, and is called *elongation* or *strain*, ϵ. It can be represented mathematically by the equation

$$\epsilon = \frac{\delta}{l} \tag{2}$$

Although elongation ϵ is a dimensionless number, it is customary to speak of it in terms of inches per inch.

For most materials used in engineering, stress and strain are directly proportional; when this condition exists, the material is said to follow *Hooke's law*. The linear relationship between stress and strain can be represented by an equation if a constant of proportionality is introduced as follows:

$$s = \epsilon E \qquad \text{or} \qquad \epsilon = \frac{s}{E} \tag{3}$$

Constant E is called the *modulus of elasticity*, or *Young's modulus*, for the material. It has the dimension of stress and can be visualized as the tensile stress that would cause a body to double in length, $\epsilon = 1$, provided the material would remain elastic with such excessive loading. Values of E for common materials used in engineering are given in Table 2-1 of the following chapter.

Substitution of Eqs. (1) and (3) into Eq. (2) gives the important relationship

$$\delta = \frac{Pl}{AE} \tag{4}$$

Equations (3) and (4) are valid either for tension or compression. Tension stress and increase of length are considered positive, whereas compressive stress and decrease in length are considered negative.

Example 1. In Fig. 1-1, let load P be equal to 5,000 lb, and let the bar be 3 in. wide and 0.5 in. thick. The uniform portion of the bar is 60 in. long. The material is steel.

    (a) Find the stress in the uniform portion of the bar.

    (b) Find the deformation of the uniform portion of the bar.

Solution. Cross-sectional area: $A = 3 \times 0.5 = 1.5$ in.²

Stress, by Eq. (1): $s = \dfrac{5,000}{1.5} = 3,333$ psi

Deformation, by Eq. (4): $\delta = \dfrac{5,000 \times 60}{1.5 \times 30,000,000} = 0.00667$ in.

For statical equilibrium, the summation of the forces in any direction must equal zero, and the summation of the moments about any axis also must equal zero. The following equations must therefore be fulfilled.

$$\Sigma F = 0, \qquad \Sigma M = 0$$

4. Statically Indeterminate Problems in Tension and Compression

Machine parts are sometimes arranged in a manner whereby the axial forces cannot be determined by the equations of statics alone. Such force systems are said to be statically indeterminate. They are characterized by the presence of more supports or members than the minimum required for the equilibrium of the structure. For such situations, the deformations of the parts must be taken into consideration. The following example will illustrate a typical method for solving such problems.

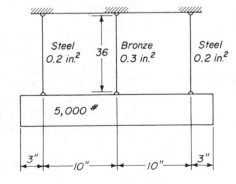

Example 2. Find the force in each of the vertical bars in Fig. 1-2. The weight can be assumed to be rigid and to maintain the connections for the three vertical bars in a straight line. Assume the support at the top to be rigid also.

Solution. Because both geometry and loading are

Figure 1-2 Example 2.

symmetrical, the forces will be equal in the two outer bars. From statical equilibrium, the following equation can be written:

$$2F_1 + F_2 = 5,000 \tag{a}$$

Since two unknowns are present, it is necessary to obtain another equation to effect a solution. This can be done by considering the deformations of the bars. From the given data, all bars will have the same amount of deformation. Hence

$$\delta_1 = \delta_2, \quad \text{or} \quad \frac{F_1 l_1}{A_1 E_1} = \frac{F_2 l_2}{A_2 E_2}$$

Numerical values should be substituted into the equation above.

$$\frac{F_1 36}{0.2 \times 30,000,000} = \frac{F_2 36}{0.3 \times 15,000,000} \quad \text{or} \quad 3F_1 = 4F_2 \tag{b}$$

Equations (a) and (b) should now be solved simultaneously to give

$$F_1 = 1,818 \text{ lb} \quad \text{and} \quad F_2 = 1,364 \text{ lb}$$

The distribution of the forces in an indeterminate structure is sensitive to small variations in the dimensions of the parts. A small undetected error in machining a dimension can cause a large change in the distribution of the loads. Any calculations made by the designer will not be valid unless the fit of the parts can be rigidly maintained as originally planned.

A variation in temperature may change the values of the forces in an indeterminate system. If temperature causes a relative change between the lengths of the various parts of the assembly, the effect can be similar to that of an error in machining. The designer must therefore consider temperature variations as well as dimensional errors in his calculations for an indeterminate structure.

In the solution of indeterminate problems, it is usually necessary to assume that certain members or supports are rigid. Since complete rigidity cannot be achieved, any deformation of such assumed rigid elements will cause changes in the calculated values of the forces.

5. Center of Gravity

An equation for finding the center of gravity of an area can be derived from Fig. 1-3. The distance from the x-axis to the center of gravity is called $\bar{y}$, and the distance to an element dA parallel to the x-axis is called y. The moment arm for dA about a horizontal axis through the center of gravity C is equal to $(y - \bar{y})$. Mathematically, center of gravity means that the sum of the moments of the areas about the axis through C must be equal to zero as indicated by the following equation:

$$\int (y - \bar{y}) \, dA = 0 \quad \text{or} \quad \bar{y} = \frac{\int y \, dA}{\int dA} = \frac{\int y \, dA}{A} \tag{5}$$

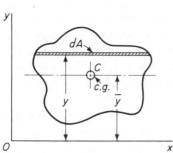

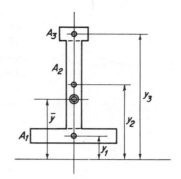

Figure 1-3 Center of gravity.

Figure 1-4 Center of gravity of composite section.

A similar equation can be written for the moments about the y-axis, which permits $\bar{x}$ to be found.

A composite figure can usually be divided into simple areas, and Eq. (5) can be applied by making the numerator equal to the sum of the $\int y\, dA$ terms of the separate parts. The denominator is the total area. If the location of the center of gravity for each of the separate areas is known, it is not necessary to perform mathematical integrations. Each $\int y\, dA$ term can be replaced by the product of the area times the distance from the axis to its center of gravity as is shown in Fig. 1-4. Thus

$$\bar{y} = \frac{A_1\bar{y}_1 + A_2\bar{y}_2 + \cdots}{A_1 + A_2 + \cdots} \tag{6}$$

Example 3. Find the location of the center of gravity of the T-shaped cross section in Fig. 1-21(b).

Solution. Divide the area into two parts of 6 in.² and 4 in.² by extending the vertical sides of the stem to the bottom of the cross section. Let the axis of moments be taken along the bottom of the T.

In Eq. (6): $\bar{y} = \dfrac{6 \times 3 + 4 \times 0.5}{6 + 4} = 2$ in.

The center of gravity is located 2 in. up from the bottom.

6. Bending of Beams

Suppose a long, thin, straight beam is bent into a curve by moments M applied at the ends, as shown in Fig. 1-5(a). The beam and moments lie in the xy-plane with the origin at the left end and the y-axis positive downward. At distance x from the left end, the deflection of the beam is given by distance y, as shown. Figure 1-5(b) shows, enlarged, a slice AB of differential length dx cut from the beam at location x.

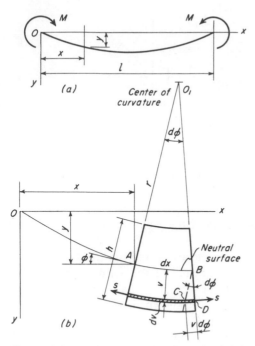

Figure 1-5 Bending stress for beam loaded by moments at ends.

The planes cutting the right and left end surfaces of AB are taken perpendicular to the longitudinal axis of the originally straight beam. It is customary to assume that these cross sections will remain plane and perpendicular to the longitudinal elements of the beam after moments M are applied. Laboratory experiments have in general verified this assumption. After bending, some of the elements have been lengthened, some have been shortened, and at one location, called the *neutral surface*, no change in length has taken place.

The loading of Fig. 1-5 is called *pure bending*. No shear or tangential stress will exist on the end surfaces of AB, and the only stress will be s, acting normally to the surfaces as shown. An equation can be derived for giving the value of this bending stress at any desired distance v from the neutral surface. Let O_1 be the center of curvature for slice AB of the deflected beam. Let $d\varphi$ be the small angle included between the cutting planes, and let r be the radius of curvature. Consider a horizontal element located a distance v below the neutral surface. Draw line BC parallel to O_1A. Angle AO_1B is equal to angle CBD and the following proportion can be written.

$$\frac{v}{r} = \frac{v\,d\varphi}{dx} = \epsilon \tag{a}$$

Since the total deformation of the element $v\,d\varphi$ divided by the original length dx is the unit deformation or strain ϵ, Eq. (a) indicates that the elongation of the element varies directly with the distance v from the neutral surface. Let it be assumed that the material of the beam follows Hooke's law. Substitution of $\epsilon = s/E$ into Eq. (a) gives

$$\frac{s}{E} = \frac{v}{r} \qquad \text{or} \qquad s = \frac{E}{r}v \tag{7}$$

Thus the stress also varies directly with the distance from the neutral surface, becoming larger as v increases. Equation (7) is of course valid only for stresses in the elastic range of the material. Above the neutral surface, for negative values of v, the stress is compressive and increases uniformly with the distance from the neutral surface. Equation (7), obtained from the geometry of the deformed beam, gives only a portion of the solution. It is now necessary to consider the equilibrium of the beam.

Figure 1-6(a) shows the beam after the left-hand portion has been

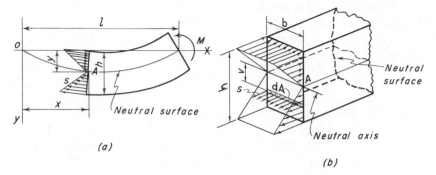

(a)

(b)

Figure 1-6 Stress caused by bending moment.

removed by passing a cutting plane at A. It will be observed that the stresses on the left-hand cross section are distributed in accordance with Eq. (7), and that the given moment M is acting at the right-hand end. A perspective of the stress system is shown in Fig. 1-6(b). The intersection of the neutral surface with the cross section upon which the stresses are acting is called the *neutral axis*.

The portion of the beam in Fig. 1-6(a) must be in equilibrium with respect to both forces and moments. Since the given loading M is a pure moment, it is necessary that the resultant horizontal force on the left end surface be zero. The force on an element of area dA, shown in Fig. 1-6(b), is equal to $s\,dA$. When this force is added up for the entire area, $\int s\,dA$, the result must be equal to zero. The value of s from Eq. (7) should now be substituted as follows:

$$\int s\,dA = \int \frac{E}{r} v\,dA = \frac{E}{r} \int v\,dA = 0 \qquad (b)$$

or

$$\int v\,dA = \bar{v}A = 0 \qquad (8)$$

The integral of Eq. (8) represents the total moment of area about the neutral axis. As was shown in Section 5, it can be set equal to the product of the area A of the cross section times the distance $\bar{v}$ of its center of gravity from the neutral axis. The only way in which this product can be zero is for $\bar{v}$ to have a zero value. It must therefore be concluded that the neutral axis passes through the center of gravity of the cross section.

Since the beam in Fig. 1-6(a) is in equilibrium, the moment of the stresses of the left-hand end surface must be equal to the applied moment load M. The force on an element of area, $s\,dA$, when multiplied by the distance v to the neutral axis, and integrated over the entire area, is equal to the moment M. Hence

$$M = \int sv\,dA = \int \frac{E}{r} v^2\,dA = \frac{E}{r} \int v^2\,dA = \frac{EI}{r} \qquad (9)$$

The integral $\int v^2\, dA$ is customarily called the *moment of inertia of the area* and is represented by the letter I. This substitution has been made in the last form of Eq. (9). If the radius of curvature r is eliminated between Eqs. (7) and (9), the following important equation results.

$$s = \frac{Mv}{I} \tag{10}$$

This equation gives the value of the bending stress at any distance v from the neutral axis. The greatest stress is found at the location on the cross section where v is the largest. This maximum value of v is usually denoted by c, and the equation for the maximum bending stress then becomes

$$s = \frac{Mc}{I} \tag{11}$$

It should be noted that the magnitude of the stress s given by Eq. (11) is independent of the kind of material composing the beam. The ratio I/c is called the *section modulus of the cross section*.

Although Fig. 1-6 illustrates a rectangular beam, the foregoing theory is valid for any shape of cross section. The maximum stress is located on the boundary at the point farthest removed from the neutral axis.

According to the original assumption, this theory should be applied only to long, thin beams loaded in pure bending. However, under most conditions, the equations give satisfactory results for bending stress when the bending moment is caused by transverse forces applied to the beam. Transverse forces also cause compressive stress between the elements in the neighborhood of the loads.

If the material in the beam does not follow Hooke's law, the magnitude of the stress is no longer proportional to the distance from the neutral axis. If Eq. (11) is applied to such beams, the results may be only approximate.

7. Moment of Inertia

The integral $\int v^2\, dA$ appeared in connection with the bending theory of beams. For convenience in writing, this integral, as was mentioned in the foregoing section, is usually replaced by the symbol I and is called the *moment of inertia*. Expressions for I for the commonly used geometric shapes of cross sections can be found in the mechanical engineering handbooks. See also Fig. 1-7.

If the integral is computed for a rectangle of width b and depth h about an axis parallel to side b through the center of gravity, as shown in Fig. 1-7, it is found to be equal to

$$I = \frac{bh^3}{12} \tag{12}$$

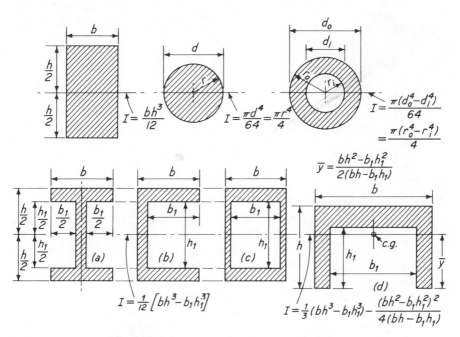

Figure 1-7 Moments of inertia of various sections.

Width b is parallel to the axis about which the moment tries to rotate the cross section.

Example 4. Let the beam in Fig. 1-5 be 2 in. wide and 3 in. deep. Let the bending moment M at each end be equal to 40,000 in. lb. Find the value of the bending stress.

Solution. By Eq. (12): $I = \dfrac{bh^3}{12} = \dfrac{2 \times 3^3}{12} = 4.5$ in.^{4}

By Eq. (11): $s = \dfrac{Mc}{I} = \dfrac{40{,}000 \times 1.5}{4.5} = 13{,}330$ psi

Example 5. A steel bandsaw blade is 0.028 in. thick. Find the value of the bending stress when the blade is passing around a pulley of 18-in. diam. $E = 30{,}000{,}000$ psi.

Solution. By Eq. (7): $s = \dfrac{Ev}{r} = \dfrac{30{,}000{,}000 \times 0.014}{9} = 46{,}670$ psi

Another expression for the bending stress of a beam of rectangular cross section can be secured by substituting Eq. (12) into Eq. (11).

$$s = \frac{6M}{bh^2} \tag{13}$$

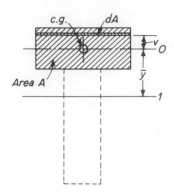

For a circle of diameter d or radius r about a diameter through the center, the value of the integral $\int v^2 \, dA$ becomes

$$I = \frac{\pi d^4}{64} = \frac{\pi r^4}{4} \tag{14}$$

When Eq. (14) is substituted into Eq. (11), the following expression for the bending stress of a beam of circular cross section is obtained.

$$s = \frac{32M}{\pi d^3} \tag{15}$$

Figure 1-8 Moment of inertia about parallel axis.

Suppose that the moment of inertia for the shaded rectangle in Fig. 1-8 is desired about axis 1. Axis 0 should be drawn through the center of gravity parallel to axis 1 about which the moment of inertia is desired. The integral of distance squared times differential area then becomes

$$I_1 = \int (v + \bar{y})^2 \, dA = \int (v^2 + 2\bar{y}v + \bar{y}^2) \, dA \tag{a}$$

The integration can be made term by term and is to extend over the shaded area. The first integral, $\int v^2 \, dA$, is equal to the moment of inertia about axis 0 and may be written I_0. The second term, $2\bar{y} \int v \, dA$, is equal to zero since it represents merely the moment of the area about an axis through the center of gravity. The third term is seen to be equal to $A\bar{y}^2$, where A is the area. The moment of inertia for area A about axis 1 is then equal to

$$I_1 = I_0 + A\bar{y}^2 \tag{16}$$

This equation is known as the *parallel axis theorem*. The moment of inertia for a composite cross section can be found by dividing it into elementary parts, finding the moment of inertia about the desired axis for each by Eq. (16), and then adding the results together. Axis 0, however, must be taken through the center of gravity of the area under consideration and must be parallel to the axis about which the total value of I is desired.

The equation for moment of inertia is sometimes written

$$I = i^2 A \tag{b}$$

where i is called the *radius of gyration* for the area. It is hypothetical distance at which the entire area A would have to be concentrated in order to give the same value for I as determined by the integral.

Example 6. Find the value of the moment of inertia of the T-shaped cross section of Fig. 1-21(b).

Solution. By Example 3, the center of gravity is located 2 in. up from the bottom.

By Eq. (16), for the stem: $I = \dfrac{1 \times 6^3}{12} + 6 \times 1^2 = 24$ in.⁴

By Eq. (16), for the T: $I = \dfrac{4 \times 1^3}{12} + 4 \times 1.5^2 = 9.33$ in.⁴

Total: $I = 24 + 9.33 = 33.33$ in.⁴

8. Principle of Superposition

Stresses and deformations are produced in a body by the forces that are exerted upon it. It is natural to assume that the resultant effect at any chosen point is the sum of the effects of the various loads. In general, experiments have shown that this is so. The idea that the resultant effect is the sum of the separate effects is known as the *principle of superposition*. In general, it is valid for cases of loading only where the magnitude of the stress and deflection is directly proportional to the load.

Example 7. Calculate and plot the distribution of stress over a cross section of the offset link shown in Fig. 1-9(a). The main body of the link is straight and is $\frac{3}{4}$ in. thick.

It is obvious that the loading of Fig. 1-9(a) produces both direct tension and bending stress on the cross section. The principle of superposition applies, and each stress can be computed separately. The two stresses can then be added algebraically to obtain the resultant.

At this point, good use may be made of a principle from statics whereby a given force may be resolved into a parallel force and a couple. By doing so,

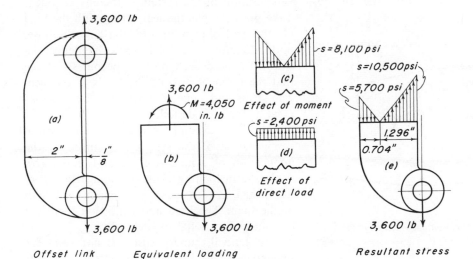

Figure 1-9 Offset link with tension load.

the equivalent loading on the cross section shown by Fig. 1-9(b) is secured. The moment arm is equal to the distance from the line of action of the force to the center of gravity of the cross section.

Solution. The computations are as follows.

$$A = \frac{3}{4} \times 2 = 1.5 \text{ in.}^2$$

$$I = \frac{bh^3}{12} = \frac{3}{4} \times 8 \times \frac{1}{12} = 0.5 \text{ in.}^4$$

In Eq. (1): $s = \dfrac{P}{A} = \dfrac{3,600}{1.5} = 2,400$ psi direct stress

The moment arm is the distance from the line of action of the force to the center of the cross-sectional area.

In Eq. (11): $s = \dfrac{Mc}{I} = \dfrac{3,600 \times 1.125 \times 1}{0.5} = 8,100$ psi bending stress

Superposition applies and the resultant stress is given by

$$s = \frac{P}{A} + \frac{Mc}{I} \tag{17}$$

On the inside edge: $s = 2,400 + 8,100 = 10,500$ psi tension
On the outside edge: $s = 2,400 - 8,100 = -5,700$ psi compression

Views (c) and (d) show the effect of moment and direct stress separately; the resultant stress on a cross section through the main body of the link is given by (e). Note that views (b) and (e) are in static equilibrium. Also note that the line of zero stress no longer passes through the center of gravity of the cross section.

Superposition cannot be applied if the loads produce deflections that are so great that the basic configuration of the system is thereby changed.

(a) Beam with large deflection

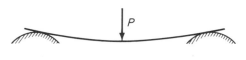

(b) Deflection causes change in span length

Figure 1-10 Examples where deflection changes original geometry; superposition does not apply.

Take, for example, the leaf spring with the load on the end shown in Fig. 1-10(a). Suppose, upon doubling the load, that the deflection, shown by the dashed outline, becomes so large that the moment arm of the loads is reduced. The stress and deflection will not be twice as great as for a single load.

Another example where the fundamental configuration of the system is changed by the application of a load is given in Fig. 1-10(b). A change in loading causes a change in deflection, which in turn causes a change in the length of the span. The load and deflection therefore are not proportional to each other.

In general, superposition is not valid for slender members loaded in compression. After the load reaches a value known as the *critical* or *buckling load*, a large lateral deflection of the member results from a small additional increase in the axial load.

9. Additional Beam Equations

For most beams used in engineering, the slope φ, shown in Fig. 1-5, has a very small value. Therefore, the tangent dy/dx can be considered as having a value very close to the angle φ. Then

$$\varphi = \frac{dy}{dx} \quad \text{and} \quad \frac{d\varphi}{dx} = \frac{d^2y}{dx^2}$$

The slope φ is reduced in Fig. 1-5 in passing from A to B, and the increment $d\varphi$ of the angle is thus a negative quantity. From the figure, $dx = -r\,d\varphi$. Then

$$\frac{d\varphi}{dx} = -\frac{1}{r} = \frac{d^2y}{dx^2} \tag{18}$$

The negative sign is inserted because both dx and r are positive quantities. Substitution of Eq. (9) gives

$$\frac{d^2y}{dx^2} = -\frac{M}{EI} \tag{19}$$

Equation (19) is the fundamental equation of beam theory. For the general case of loading, moment M is a function of x and not a constant as in Fig. 1-5. If the beam has a large deflection, as, for example, a leaf spring in some types of service, the approximation $\varphi = dy/dx$ cannot be used, and the mathematics of the solution becomes very complicated.

When the bending moment of the beam is not constant, but varies with x, the loading for slice AB is more complicated than in the case of pure bending previously considered. Shearing forces, as well as moments, exist on the end surfaces of the slice. In general the top surface of the beam is acted upon by transverse loads.

A slice from such a beam is shown in Fig. 1-11. The rate at which the bending moment is changing in value is equal to dM/dx, and the distance over which this change takes place in passing from A to B is dx. The change in value for the moment then is $(dM/dx)\,dx$. The moment on the right end is equal to the moment M of the left end plus the increment or change. Similarly, the shear force, as a function of x, changes from a value V on the left to the amount on the right, as shown. Should either the moment or shear be decreasing with increasing x, the corresponding derivative would be negative, and would thus effect

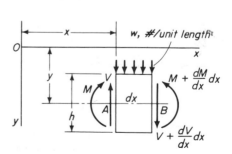

Figure 1-11 Element loaded by forces and moments.

a reduction for the right-hand value. All shears and moments in Fig. 1-11 have positive directions. Note, however, that the arrows have opposite directions on ends A and B.

The moments on the A–B element must be in equilibrium. The equation for moments about point A, for example, is

$$M + \frac{dM}{dx}\,dx - \left(V + \frac{dV}{dx}\,dx\right)dx - w\,dx \times \frac{dx}{2} - M = 0$$

When differentials of higher order are neglected, this equation reduces to

$$\frac{dM}{dx} = V \tag{20}$$

The location of the point of maximum bending moment is found by differentiating M with respect to x and setting the result equal to zero. However, dM/dx is equal to V by Eq. (20). Therefore, a maximum (or a minimum) value of the bending moment occurs at those points for which the shear V is equal to zero. The load w on the top surface of the beam may be uniform, may vary with x, or may be equal to zero except at the points where concentrated forces or moments are acting.

10. Deflection of Beams

The foregoing equations can be used for deriving equations for the deflections of beams. If the expression for the bending moment, as a function of x, is substituted in Eq. (19), and if two integrations are performed and the constants of integration are evaluated, the equation for the deflection y at location x is obtained. The process is illustrated by the following example.

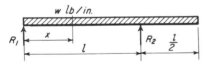

Figure 1-12 Example 8.

Example 8. Derive the equation for the deflection y for values of x located between the supports for the beam shown in Fig. 1-12.

Solution. By taking moments, the values of the reactions are found to be $R_1 = 3wl/8$ and $R_2 = 9wl/8$.

At location x:
$$M = \frac{3}{8}wlx - \frac{wx^2}{2}$$

In Eq. (19):
$$EI\frac{d^2y}{dx^2} = -M = -\frac{3}{8}wlx + \frac{wx^2}{2}$$

Integrate:
$$EI\frac{dy}{dx} = -\frac{3wlx^2}{16} + \frac{wx^3}{6} + C_1$$

Since the slope dy/dx is not known for any point on the beam, constant C_1 must be retained and evaluated later.

Integrate:
$$EIy = -\frac{3wlx^3}{48} + \frac{wx^4}{24} + C_1x + C_2$$

At the left end, $x = 0$ and $y = 0$. When these are substituted, $C_2 = 0$. At the right end, $x = l$ and $y = 0$. When these are substituted,

$$C_1 = \frac{wl^3}{48}$$

The equation for the deflection thus is

$$y = \frac{w}{48EI}(2x^4 - 3lx^3 + l^3x)$$

A transverse load on a beam causes a deflection or change in elevation at the point of application. However, a reaction generally remains fixed in elevation even though it carries a force.

Figures 1-13 to 1-16 give the shears, moments, and deflections for a number of beams of various types of loads and conditions of support. In these figures, a simple support is assumed to offer no resistance to lateral motion or to rotation in the plane of the moments.

Example 9. Suppose it is specified that the deflection from its own weight at the center of a simply supported steel shaft should not exceed 0.010 in. per foot of span.

  (a) Find the maximum permissible length for a 3-in.-diam shaft.

  (b) Find the stress caused by the weight of the shaft. For steel, $\gamma = 0.283$ lb/in.^{3}, $E = 30,000,000$ psi.

Solution. (a) By the conditions of the problem the deflection at the center is $0.010l/12$, where l is the length in inches. Hence by No. 8 of Fig. 1-14,

$$\frac{5wl^4}{384EI} = \frac{0.010l}{12} \quad \text{or} \quad l^3 = \frac{0.010 \times 384EI}{12 \times 5w}$$

By Eq. (14):

$$I = \frac{\pi d^4}{64} = \frac{\pi 3^4}{64} = 3.976 \text{ in.}^4$$

$$w = \gamma A = 0.283\frac{\pi 3^2}{4} = 2.00 \text{ lb/in.}$$

Substitution of these values gives

$$l^3 = 3,817,000 \quad \text{or} \quad l = 156.3 \text{ in.} = 13.02 \text{ ft}$$

(b)

$$M = \frac{wl^2}{8} = \frac{2 \times 156.3^2}{8} = 6,106 \text{ in. lb}$$

$$s = \frac{Mc}{I} = \frac{6,106 \times 1.5}{3.976} = 2,300 \text{ psi}$$

Example 10. Show that the spring constants, load for a unit deflection, for the two beams of Fig. 1-17 are the same.

Solution. The spring constant can be derived from the deflection equation for the beam by setting the load P equal to k when the deflection y becomes equal to unity.

By No. 5 of Fig. 1-13:

$$1 = \frac{kl^3}{3EI} \quad \text{or} \quad k = \frac{3EI}{l^3} = \frac{3E}{l^3}\frac{bh^3}{12} = \frac{Ebh^3}{4l^3}$$

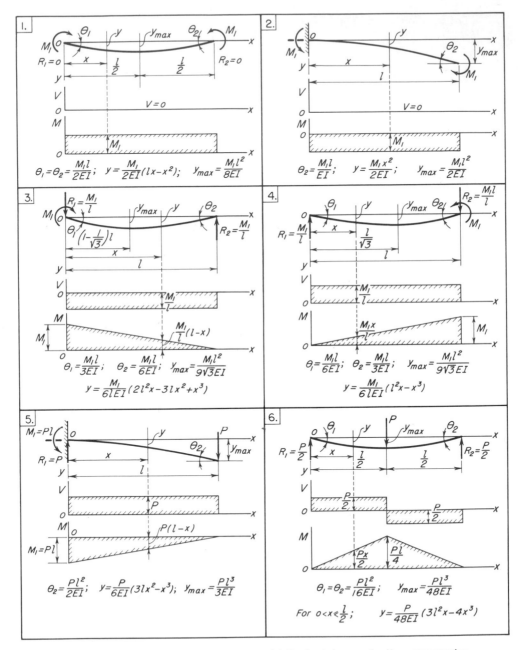

Figure 1-13 Shear, moment, and deflection in beams of uniform cross section.

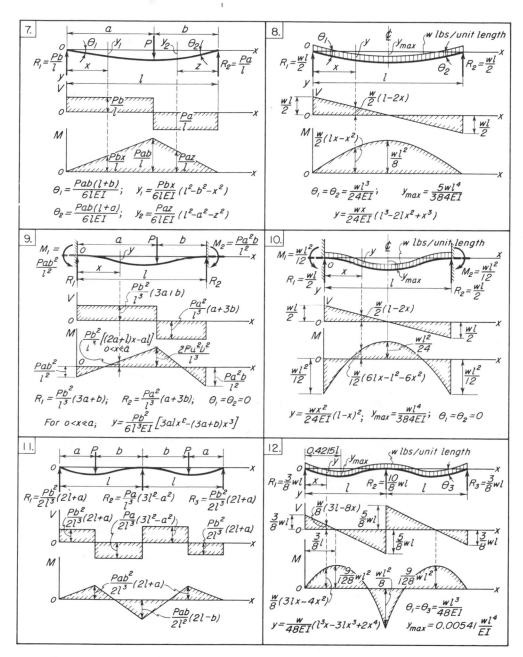

Figure 1-14 Shear, moment, and deflection in beams of uniform cross section.

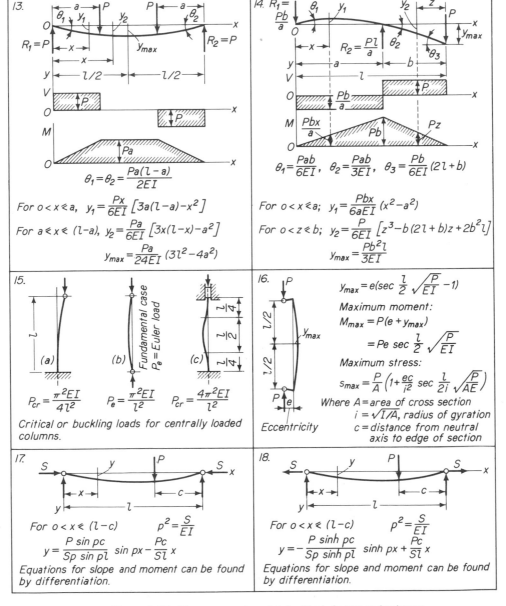

13.

$$\theta_1 = \theta_2 = \frac{Pa(l-a)}{2EI}$$

For $0 < x \leqslant a$, $\quad y_1 = \frac{Px}{6EI}\left[3a(l-a) - x^2\right]$

For $a \leqslant x \leqslant (l-a)$, $\quad y_2 = \frac{Pa}{6EI}\left[3x(l-x) - a^2\right]$

$$y_{max} = \frac{Pa}{24EI}(3l^2 - 4a^2)$$

14.

$$\theta_1 = \frac{Pab}{6EI}, \quad \theta_2 = \frac{Pab}{3EI}, \quad \theta_3 = \frac{Pb}{6EI}(2l+b)$$

For $0 < x \leqslant a$; $\quad y_1 = \frac{Pbx}{6aEI}(x^2 - a^2)$

For $0 < z \leqslant b$; $\quad y_2 = \frac{P}{6EI}\left[z^3 - b(2l+b)z + 2b^2 l\right]$

$$y_{max} = \frac{Pb^2 l}{3EI}$$

15.

$$P_{cr} = \frac{\pi^2 EI}{4l^2} \qquad P_e = \frac{\pi^2 EI}{l^2} \qquad P_{cr} = \frac{4\pi^2 EI}{l^2}$$

Critical or buckling loads for centrally loaded columns.

16.

$$y_{max} = e\left(\sec\frac{l}{2}\sqrt{\frac{P}{EI}} - 1\right)$$

Maximum moment:

$$M_{max} = P(e + y_{max})$$

$$= Pe\,\sec\frac{l}{2}\sqrt{\frac{P}{EI}}$$

Maximum stress:

$$s_{max} = \frac{P}{A}\left(1 + \frac{ec}{i^2}\sec\frac{l}{2i}\sqrt{\frac{P}{AE}}\right)$$

Where A = *area of cross section*
　　$i = \sqrt{I/A}$, *radius of gyration*
　　c = *distance from neutral axis to edge of section*

17.

For $0 < x \leqslant (l-c)$ $\qquad p^2 = \frac{S}{EI}$

$$y = \frac{P\sin pc}{Sp\,\sin pl}\sin px - \frac{Pc}{Sl}x$$

Equations for slope and moment can be found by differentiation.

18.

For $0 < x \leqslant (l-c)$ $\qquad p^2 = \frac{S}{EI}$

$$y = -\frac{P\sinh pc}{Sp\,\sinh pl}\sinh px + \frac{Pc}{Sl}x$$

Equations for slope and moment can be found by differentiation.

Figure 1-15. Shear, moment, and deflection in beams and columns.

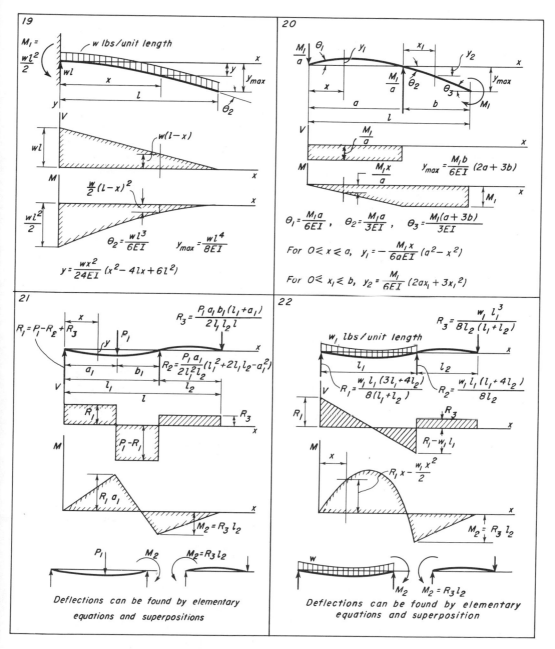

Figure 1-16 Shear, moment, and deflection in beams of uniform cross section.

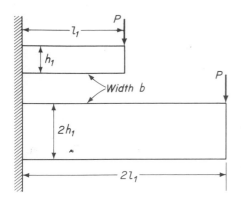

Figure 1-17 Example 10.

For sketch (a): $k = \dfrac{Ebh_1^3}{4l_1^3}$

For sketch (b): $k = \dfrac{Eb(2h_1)^3}{4(2l_1)^3} = \dfrac{Ebh_1^3}{4l_1^3}$

If the beam has a cross-section width 8 or 10 or more times as great as the thickness, the beam is stiffer, and the deflection is less than that indicated by the equation for a narrow beam. The large cross-section width prevents the lateral expansion and contraction of the material, and the deflection is thereby reduced. A better value for the deflection of a "wide" beam is obtained by multiplying the result given in the equation for a narrow beam by $(1 - \mu^2)$ where μ is Poisson's ratio. Poisson's ratio is the ratio of the unit lateral contraction to the unit axial elongation experienced by a bar in tension.

Data for the stress and deflection of flat plates are given in Chapter 12.

11. Effect of Ribs on Castings

Ribs are sometimes added to the webs of castings to give greater strength and rigidity. It is possible, however, that the addition of a shallow rib to a body loaded in bending may cause an increase in stress rather than a decrease. The low rib gives a small increase in the moment of inertia, but the distance from the neutral axis to the edge of the cross section becomes relatively greater, and the stress is accordingly increased. The situation is illustrated by the following example.

Example 11. Figure 1-18 represents the cross section through a simply supported beam 60 in. long that carries a 200-lb load at the center. $E = 15,000,000$ psi.

(a) Find the value of the bending stress and the deflection at the center if the ribs were omitted.

(b) Find the value of the bending stress and the deflection at the center when the ribs are present.

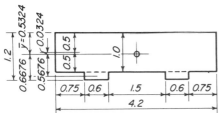

Figure 1-18 Cross section for ribbed beam, Example 11.

Solution.

(a) $I = \dfrac{bh^3}{12} = \dfrac{4.2 \times 1^3}{12} = 0.35 \text{ in.}^4$

$M = \dfrac{Pl}{4} = \dfrac{200 \times 60}{4} = 3{,}000 \text{ in. lb}$

$s = \dfrac{Mc}{I} = \dfrac{3{,}000 \times 0.5}{0.35} = 4{,}290 \text{ psi}$

$$y = \frac{Pl^3}{48EI} = \frac{200 \times 60^3}{48 \times 15,000,000 \times 0.35} = 0.1714 \text{ in.}$$

(b) Area: $A = 4.2 \times 1 + 2 \times 0.2 \times 0.6 = 4.44$ in.[2]

To find $\bar{y}$, take the axis along top of the cross section.

$$4.44\bar{y} = 4.2 \times 0.5 + 0.24 \times 1.1 = 2.364$$

$$\bar{y} = 0.5324 \text{ in.}$$

For the main area: $\quad I = \dfrac{bh^3}{12} + A\bar{y}^2 = 0.35 + 4.2 \times 0.0324^2 = 0.3544$ in.[4]

For the ribs: $\quad I = \dfrac{1.2 \times 0.2^3}{12} + 0.24 \times 0.5676^2 = 0.0781$ in.[4]

Total: $\quad I = 0.3544 + 0.0781 = 0.4357$ in.[4]

$$s = \frac{Mc}{I} = \frac{3,000 \times 0.6676}{0.4325} = 4,630 \text{ psi}$$

$$y = \frac{Pl^3}{48EI} = \frac{200 \times 60^3}{48 \times 15,000,000 \times 0.4325} = 0.1387 \text{ in.}$$

As shown above, shallow ribs cause an increase in the bending stress. The deflection, however, has been decreased. If the ribs were made somewhat larger, the stress would be decreased, and the beam would be stronger.[1]

Ribbed structures are frequently made of a brittle material, such as cast iron, which is weak in tension. If possible, the ribs should be in compression. When a cast-iron body with parallel ribs is bent and the ribs are in tension, care must be exercised to make certain that all ribs are of the same height, or they may fail progressively beginning with the highest, and the full strength of the body could not be realized.

12. Shearing Stress

Suppose an element is loaded by shearing stresses acting tangentially to its sides as shown in perspective in Fig. 1-19(a) or in plan in Fig. 1-19(b) and (c). Such loading causes no change in the length of the sides of the element, but merely produces a distortion or change in the value of the 90° angles in the corners.

Shearing stresses are usually denoted by double subscripts. The first subscript indicates the direction of the normal to the plane under consideration, and the second subscript indicates the direction of the stress. Hence stress s_{xy} lies in a plane whose normal is in the x-direction, while the stress acts in the y-direction. For similar reasons s_{yx} indicates that the stress is in a plane perpendicular to the y-axis, and is parallel to the x-axis. Since the ele-

[1]See Marin, Joseph, "Stiffness of Ribbed Plates" *Machine Design*, **19**, May, 1947, p. 145, and Radich, E. A., "Strength and Stiffness of Ribbed Plates," *Machine Design*, **21**, Sept., 1949, p. 149.

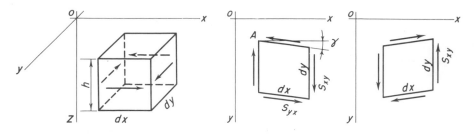

(a) Element loaded in shear   (b) Positive shear   (c) Negative shear

Figure 1-19 Element loaded by shearing or tangential stress.

ment is in equilibrium, the moments of the forces about a point, say A, must add up to zero. The stress should be multiplied by the area and then by the moment arm to give

$$s_{xy}\, dx\, dy\, h - s_{yx}\, dx\, dy\, h = 0$$

or

$$s_{xy} = s_{yx} \tag{21}$$

Equation (21) shows that the shearing stresses in two perpendicular directions at a point are equal. Usually no distinction in notation is made, and both would be represented by the same symbol.

It should be noted, however, that four arrows are necessary to specify a state of shear for an element, and for equilibrium these arrows must be arranged either as in Fig. 1-19(b) for positive shear, or as in Fig. 1-19(c) for negative shear. Thus if the direction of one arrow is reversed, all four must be reversed. In other words, if shearing stress exists on one side of an element, then, in general, shearing stress must exist on all four sides, as shown in Fig. 1-19.

The shearing strain or angular deformation γ is proportional to the shearing stress for values within the elastic range, and Hooke's law for shear becomes

$$s_{xy} = \gamma G \tag{22}$$

The constant of proportionality, G, is called the *modulus of elasticity in shear*. It has dimensions of psi. In magnitude it would be equal to the stress that would cause the angular deformation to become equal to one radian, provided Hooke's law is valid for such imaginary loading. The mathematical relationship between the three elastic constants E, G, and μ is given by

$$G = \frac{E}{2(1 + \mu)} \tag{23}$$

where μ is Poisson's ratio.

13. Transverse Shearing Stress in Beams

In addition to the bending stresses, the loads on a beam may also cause shearing stresses between the elements. The designer is interested in the magnitude of these stresses, since machine parts made of ductile materials are usually designed on the basis of shearing stress.

If a vertical load is supported by a stack of laminated strips, the shearing effect is as shown in Fig. 1-20(a). In a solid beam the elements do not slide on each other, but the shearing stress tending to make them do so is present.

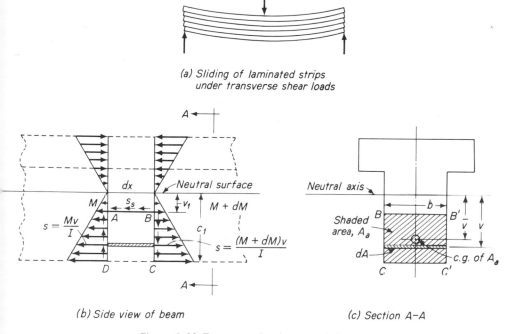

(a) Sliding of laminated strips
under transverse shear loads

(b) Side view of beam

(c) Section A-A

Figure 1-20 Transverse shearing stress in beam.

Figure 1-20(b) shows a portion $ABCD$ cut from a beam of uniform cross section by two adjacent vertical planes and a horizontal plane, located at a distance v_1 below the neutral surface. Moment M is acting on the cross section of the left end of the slice, while moment $M + dM$ is acting on the right.

To maintain equilibrium in the horizontal direction, shearing stress s_s must act toward the left on surface AB. This shear is necessary because the normal stresses from the bending moment are assumed to be larger on surface BC on the right than on surface AD on the left.

An element of area dA on surface AD has the normal force $s\,dA$ or $Mv\,dA/I$ acting as shown in Fig. 1-20(b). The total force on the left end extending from A to D is then

$$\int_{v_1}^{c_1} \frac{Mv}{I} dA \tag{a}$$

Similarly, the total force on the right end BC is

$$\int_{v_1}^{c_1} \frac{(M + dM)v}{I} dA \tag{b}$$

The shearing force on the horizontal surface AB is $s_s b\,dx$, where b is the width of the beam at the location where the shear stress s_s is acting.

The equilibrium equation for horizontal forces for $ABCD$ is then

$$s_s b\,dx + \int_{v_1}^{c_1} \frac{Mv\,dA}{I} = \int_{v_1}^{c_1} \frac{(M + dM)v\,dA}{I} \tag{c}$$

or

$$s_s = \frac{1}{b} \int_{v_1}^{c_1} \frac{dM}{dx} \frac{v\,dA}{I} = \frac{V}{Ib} \int_{v_1}^{c_1} v\,dA \tag{24}$$

In the last form of Eq. (24), shear V has been substituted for dM/dx. In Eq. (24), $v\,dA$ represents the moment of the area of the element about the neutral axis. This is integrated over the entire surface from v_1, the location where the shearing stress s_s is desired, to the outer edge. This integral can also be written $\bar{v}A_a$, where A_a is the shaded area of view A-A, and $\bar{v}$ is the distance from its center of gravity to the neutral axis. Equation (24) can then be written

$$s_s = \frac{V}{Ib} \bar{v} A_a \tag{25}$$

As we proved in the preceding section, the shearing stress on the vertical end surfaces at distance v_1 from the neutral axis is also equal to the horizontal shear stress s_s as determined by Eqs. (24) or (25).

For composite cross sections it is convenient to divide area A_a into several parts, find $\bar{v}A_a$ for each of them, and then add together for the final result. For such beams, Eq. (25) is written

$$s_s = \frac{V}{Ib} \sum \bar{v} A_a \tag{26}$$

The total shear force on the cross section is represented by V. The distance from the neutral axis to the point where the shearing stress is desired is given by v_1.

Example 12. Find the transverse shear in the material 3 in. from the top surface for the beam of Fig. 1-21(a). Also find the value of the transverse shearing stress at the neutral axis.

Solution. As is shown in Example 4, the center of gravity of the cross section in Fig. 1-21(b) is found to be 2 in. up from the bottom. As shown in Example

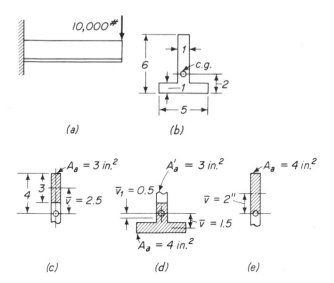

Figure 1-21 Example 12.

6, the moment of inertia about the horizontal axis through the center of gravity is found to be 33.33 in.[4]

Referring to Fig. 1-21(c), it is seen for location 3 in. from the top that $\bar{v} = 2.5$ in. and $A_a = 3$ in.[2] Substitution in Eq. (25) gives

$$s_s - \frac{10,000}{33.33 \times 1} \times 2.5 \times 3 - 2,250 \text{ psi}$$

It is of course immaterial whether A_a is taken above or below the location at which the stress is desired. If taken below, the situation is as shown in Fig. 1-21(d). Equation (26) gives

$$s_s = \frac{10,000}{33.33 \times 1} (1.5 \times 4 + 0.5 \times 3) = 2,250 \text{ psi}$$

At the neutral surface, values of $\bar{v}$ and A_a shown in Fig. 1-21(e) are used and the value of the shearing stress becomes

$$s_s = \frac{10,000}{33.33 \times 1} \times 4 \times 2 = 2,400 \text{ psi}$$

When Eq. (24) is applied to rectangular cross section, $dA = b\, dv$ and $c_1 = h/2$. After making these substitutions and integrating, the following result is obtained.

$$s_s = \frac{V}{2I} \left(\frac{h^2}{4} - v_1^2 \right) = \frac{3V}{2A} \left(1 - \frac{4v_1^2}{h^2} \right) \tag{27}$$

The transverse shearing stress in a rectangular beam thus varies as a second-degree parabola in v_1. Its value is proportional to the length of the arrows in Fig. 1-22(a). The maximum value occurs at the neutral axis, where $v_1 = 0$,

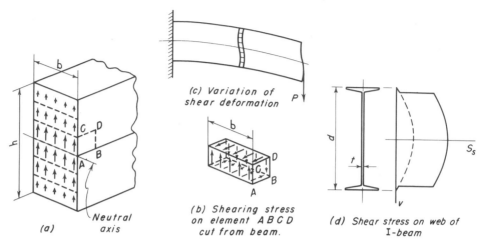

Figure 1-22 Distribution of shearing stress over cross section of beam.

and is equal to

$$s_s = \frac{3V}{2A} \tag{28}$$

The shear decreases both above and below the axis in accordance with Eq. (27) until at the upper and lower edges it becomes zero.[2] The end surface of the beam in Fig. 1-22(a) is also acted upon by the system of normal stresses caused by the bending moment, as shown by Fig. 1-6(b). In Fig. 1-22(a) these have been omitted for greater clarity.

Within reasonable limits, the presence of the shearing stress has no effect on the value of the bending stress, and vice versa.

The shearing deformation also varies over the surface of a cross section; it is maximum at the neutral axis and zero at the top and bottom. The shearing stress causes a warping of the cross sections, which were originally plane and perpendicular to the longitudinal elements of the beam.[3] For a cantilever with a load on the end, the situation is as shown in Fig. 1-22(c).

For a solid circular cross section, Eq. (24) gives the following value for the maximum transverse shear, which occurs at the neutral axis.[4]

$$s_{smax} = \frac{4}{3} \cdot \frac{V}{A} \tag{29}$$

For a circular tube with very thin walls, the maximum transverse shear stress at the neutral axis is given by

---

[2] See p. 116 of Reference 3, end of chapter.

[3] For deflection of beams caused by shear, see p. 170 of Reference 3, end of chapter.

[4] A more exact analysis gives values of $1.38P/A$ and $1.23P/A$ for the shearing stress at the center and ends, respectively, of the neutral axis.

$$s_{smax} = 2\frac{V}{A} \tag{30}$$

The distribution of transverse shear stress for an I-beam is shown in Fig. 1-22(d). The stress is practically uniform except in the regions near the top and bottom. A good approximate value for the stress is obtained by dividing the total shearing force V by the area of the web td with the web considered as extending the entire depth of the beam.

14. Shear and Bending Moment Diagrams

The effects of the forces and moments that act upon the different parts of a machine are of primary importance to the machine designer. Forces arise from a variety of causes. A force may be due to weight or to inertia if a body is being accelerated. A force may be transmitted to the body by another member of the machine at the point where the two parts are in contact. It is common practice to represent forces by means of arrows on sketches in solving force problems. It is very important, except for gravity and inertia, to keep in mind that the body must have contact with the rest of the structure at the point where a force, represented by an arrow, is considered as acting. As an example, consider the shaft of Fig. 1-23(a). The entire loading occurs at point A, where the bracket is attached to the shaft. The equivalent loading diagram is shown in Fig. 1-23(b). The diagram of Fig. 1-23(c) is incorrect because forces P are not in contact with the shaft.

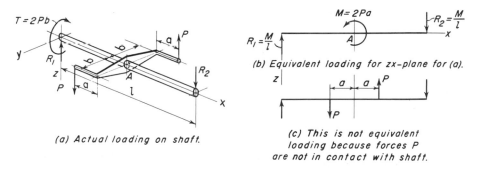

(a) Actual loading on shaft.

(b) Equivalent loading for zx-plane for (a).

(c) This is not equivalent loading because forces P are not in contact with shaft.

Figure 1-23 Actual and equivalent load diagram for shaft.

Moment loads, such as those described in Fig. 1-23(b), frequently occur in machine parts. Thus in the tank shown in Fig. 1-24(a) the floor beam not only resists the vertical pressure of the fluid, but has moments applied to it at the ends by the uprights, which are acted upon by the horizontal pressure of the fluid. The equivalent loading is shown in Fig. 1-24(b). In Fig. 1-24(c)

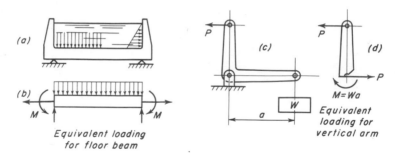

Figure 1-24 Examples of beams with moment loads.

the vertical bar is loaded at the base by the moment Wa as well as by the force P. The equivalent loading is shown in Fig. 1-24(d).

In solving most stress problems it is first necessary to find the reactions that the given loading produces. It is customary to make use of the following equations from statics.

$$\sum F_x = 0, \qquad \sum F_y = 0, \qquad \sum F_z = 0$$
$$\sum M_x = 0, \qquad \sum M_y = 0, \qquad \sum M_z = 0$$

According to these equations, the components of the forces, as well as the moments, in each of the three coordinate directions must add to zero. For static equilibrium, each of these equations must be satisfied. It should be noted that each is satisfied independently of the others. So far as the reactions are concerned, a force can be considered to be acting anywhere along its line of action, and a moment load can be considered to be acting anywhere in its plane. The state of stress in the material, however, is determined by the location, on the line of action, at which the force is considered to be applied.

The shearing force and bending moment, acting internally on the cross section of a beam at any point, are equal to the force and moment required for the equilibrium of each portion of the beam after it has been cut in two at the given point. For example, consider the simply supported beam, shown in Fig. 1-25(a), with load P and reactions R_1 and R_2. In (b) and (c), after cutting at distance x from the left end, shear V_1 and moment M_1 are required to maintain equilibrium of each portion. Reactions R_1 and R_2 in (b) and (c) are the same as in (a). Shear V_1 is found by summation of vertical forces. Moment M_1 is found by making a summation of moments, with the location of the cut taken as the moment center. The shear and moment diagrams shown in (d) and (e) are constructed by taking different values of x and finding V and M for each until a sufficient number of values has been found to plot the diagrams. It is generally assumed that a reaction does not change in elevation. Other transverse loads deflect the beam, and the elevations of the loads are thereby changed.

Moments and shears can, of course, be either positive or negative,

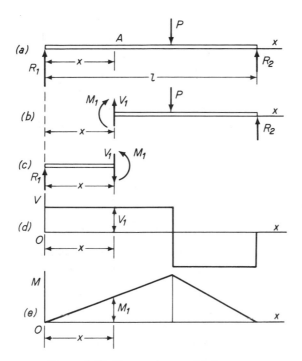

Figure 1-25 Shear and moment in beam.

depending on the direction in which they act. The shear and moment at point A for the beam shown in Fig. 1-25 are both positive. Positive directions for the arrows representing these quantities are accordingly shown in the figures. Even though both shear and moment are positive at A, note that a reversal in direction for both the V_1 and M_1 arrows occurs in (b) and (c). The direction of the arrow thus depends on whether the portion of the beam to the right of the cut, as in (b), is considered, or whether the portion to the left, as in (c), is considered. Note that both (b) and (c) are in agreement with Fig. 1-19(b) with respect to direction of positive shear. Although the y-axis is taken positive downward, both the V- and M-axes are taken positive upward.

Depending on the loading and method of supporting the beam, the shear at a cross section may have one sign, and the moment may have the opposite sign.

Another method for determining the sign of the moment is shown in Fig. 1-26. Here a moment tending to bow the beam concavely upward is considered positive, whereas a moment tending to make the beam concave downward is negative.

Example 13. A simply supported beam is loaded by a concentrated moment of 25,000 in. lb as shown in Fig. 1-27. Find values of reactions R_1 and R_2, and draw and dimension the bending moment diagram.

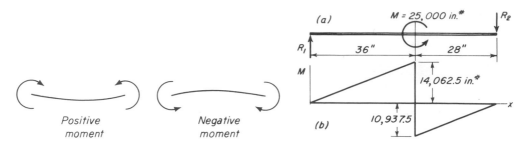

Figure 1-26 Sign convention for moments.   **Figure 1-27** Example 13.

Solution. Reaction R_1 can be found by writing a moment equation for the entire beam with center of moments taken at R_2. The resulting equation is

$$64R_1 = 25,000; \qquad R_1 = 390.6 \text{ lb}$$

Reaction R_2 can be found by a moment equation with R_1 taken as the center.

$$64R_2 = 25,000; \qquad R_2 = 390.6 \text{ lb}$$

The moment diagram consists of segments of straight lines. Let it be assumed that the beam is cut just to the left of the point of application of the moment load. The bending moment is positive and is equal to

$$M = 36R_1 = 36 \times 390.6 = 14,062.5 \text{ in. lb}$$

Again let it be assumed that the beam is cut just to the right of the point of application of the load. The moment is negative.

$$M = 28R_2 = 28 \times 390.6 = 10,937.5 \text{ in. lb}$$

15. Slender Compression Members or Columns

When a short block is loaded in compression as in Fig. 1-28(a), the average compressive stress in the material is found by dividing the load by the cross-sectional area. However, when the member is long and slender, as that shown in Fig. 1-28(b), the situation is complicated by the possibility of lateral buckling. Buckling does not occur when the bar is straight and a load, smaller than the critical value, is centrally applied. Such a column is stable; the bar, if given a lateral deflection, returns to its originally straight condition upon removal of the lateral force.

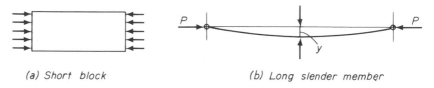

(a) Short block   *(b) Long slender member*

Figure 1-28 Bodies under compressive loading.

Force P in Fig. 1-28(b) can be increased until the straight form of the bar becomes unstable, and the axial loads will then maintain the bar in a curved form. The smallest load capable of maintaining the bar in a slightly bent form is called the *critical* or *buckling load*. A load only slightly greater can cause a relatively large lateral deflection. After buckling the stress increases rapidly; in addition to the direct compression, a bending stress from moment Py is present.

Hence the phenomenon of stability or buckling is quite different from the phenomenon of bending. A beam with lateral load starts to deflect as soon as any load is present, and the deflection is directly proportional to the magnitude of the load. A slender member in compression, in contrast, exhibits no lateral deflection until the critical or buckling load has been reached. Any increase of load then causes a large increase of deflection with accompanying danger of failure. It is obvious that the principle of superposition does not apply to columns.

For the straight column with hinged ends loaded by centrally applied forces, it can be shown that the critical buckling load is given by

$$P_e = \frac{\pi^2 EI}{l^2} \tag{31}$$

This is commonly known as the Euler equation. See No. 15 of Fig. 1-15. Moment of inertia I should be about the axis of rotation of the cross section.

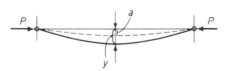

Sometimes a column may not be perfectly straight, but has some initial crookedness a, as shown in Fig. 1-29. The bending moment then will be larger, and the carrying capacity will be reduced. It can be shown[5] that the working load P for such a column with factor of safety FS is given by

Figure 1-29 Column with initial crookedness a.

$$P^2 - \left[s_{yp}A + \left(1 + \frac{ac}{i^2}\right)P_e\right]\frac{P}{FS} + \frac{s_{yp}AP_e}{(FS)^2} = 0 \tag{32}$$

where s_{yp} = yield point stress for the material,

$\quad A$ = area of cross section,

$\quad c$ = distance from neutral axis to edge of cross section, and

$\quad i = \sqrt{I/A}.$ (33)

Here i is called the *radius of gyration* of the cross section.[6]

As for other machine elements, a factor of safety must be used in column design. Because the stress is not proportional to the load, the factor of safety FS is applied to the load rather than to the stress. Let P_{yp} be the load on the

[5]See p. 31 of Reference 5, end of chapter; also p. 13 of the author's *Mechanical Design Analysis*, Englewood Cliffs, N.J.: Prentice-Hall, Inc., 1964.

[6]Some books represent the radius of gyration by r. When this is done, care must be taken not to confuse the radius of gyration with the radius r of a round cross section.

column that causes the maximum stress to be equal to the yield point value s_{yp}. Let the working load P be equal to P_{yp} divided by the factor of safety.

$$P = \frac{P_{yp}}{FS} \tag{34}$$

When this equation is substituted into Eq. (32), the result is

$$P_{yp}^2 - \left[s_{yp}A + \left(1 + \frac{ac}{i^2}\right)P_e\right]P_{yp} + s_{yp}AP_e = 0 \tag{35}$$

It can also be shown that the maximum stress in the material is given by

$$s = \frac{P}{A}\left(1 + \frac{ac}{i^2} \times \frac{P_e}{P_e - P}\right) \tag{36}$$

If P in this equation is increased to the yield point value P_{yp}, then stress s becomes equal to the yield point stress s_{yp}.

Example 14. A solid circular steel column with hinged ends is 36 in. long and $2\frac{5}{8}$ in. in diameter. Yield point stress for the material is 50,000 psi. Initial crookedness a is $\frac{1}{16}$ in. Find the load capacity for an FS of 4.

Solution.   Area: $A = \dfrac{\pi d^2}{4} = \dfrac{\pi 2.625^2}{4} = 5.412$ in.^{2}

Moment of inertia: $I = \dfrac{\pi d^4}{64} = \dfrac{\pi 2.625^4}{64} = 2.331$ in.^{4}

Radius of gyration: $i^2 = \dfrac{I}{A} = \dfrac{d^2}{16} = \dfrac{2.625^2}{16} = 0.431$ in.^{2}

Euler load: $P_e = \dfrac{\pi^2 EI}{l^2} = \dfrac{\pi^2 E 2.331}{36^2} = 532,500$ lb

In Eq. (32):

$$P^2 - \left[50,000 \times 5.412 + \left(1 + \frac{0.0625 \times 1.3125}{0.431}\right)532,500\right]\frac{P}{4}$$

$$+ \frac{50,000 \times 5.412 \times 532,500}{4^2} = 0$$

$$P^2 - 226,100P + 9,005,400,000 = 0$$

Upon solution, this quadratic gives

$$P = 51,600 \text{ lb,} \qquad \text{working load}$$

For hinged or pin-connected ends, the assumption is made that there is no restraint against rotation at the ends of the column. For square or fixed ends, all rotation is prevented as shown by No. 15(c) of Fig. 1-15. The deflection curve consists of two quarter-waves and one half-wave. Bending moments are present at the ends, but no moments act at the inflection points. Calculations are usually made by taking one-half the actual length for l in the equations for columns with hinged ends. For many practical cases, the end conditions are intermediate between being completely hinged and completely restrained.

The well-known secant equation for the maximum stress in a column with an eccentrically applied load is given in No. 16 of Fig. 1-15. For small values of e and a, the secant equation gives practically the same result as the equation for the column with initial crookedness. The secant equation, however, is very inconvenient for design calculations.

An actual column under load may behave differently from an ideal column, and the design of this element therefore presents many difficulties. Uncertainties such as amount of restraint at the ends, eccentricity of the load, initial crookedness, nonhomogeneity of the material, and deflection caused by the load can produce large variations in the behavior of a column. The choice of a suitable value for the factor of safety also presents great difficulties. Empirical equations are frequently used for column designing. The possibility of buckling about both of the principal axes of the cross section should be investigated.

Sometimes only a portion of a stressed body is in compression, as, for example, the compression flange of a beam. The danger of buckling may be present here if sufficient lateral support is lacking. Such action has been the cause of failures and should be carefully guarded against by the designer.[7]

If the load P approaches the critical load P_e, Eq. (36) indicates that the stress becomes infinite. Although there can be no such stress, it is characteristic of column equations to indicate the buckling phenomenon in this manner.

16. Stresses in Any Given Direction

The stresses in a body, as found by the equations of this chapter, have definite directions. It is sometimes necessary to have the stresses at directions other than those given by the equations.

Figure 1-30(a) shows an element of a plate with the vertical surfaces subjected to the general two-dimensional state of stress. The element has been cut from a larger plate so that stresses s_x, s_y, and s_{xy} represent the effect of the surrounding material on the element. A plan view of the element is shown in Fig. 1-30(b). Suppose stresses s_x, s_y, and s_{xy} are known, and that it is necessary to find the values of the stresses on an inclined surface whose normal makes an angle φ with the x-axis as shown in Fig. 1-30(c). Angle φ is an arbitrarily chosen angle and determines the directions of the n- and n'-axes.

Assume that stress s_r must be applied to the cut surface in order to maintain equilibrium of the remaining portion of the plate. Resultant stress s_r can be resolved into the components of normal stress s_n and shear stress s_s, as shown.

If the area of the inclined surface is A, then the area of the horizontal side of the body will be $A \sin \varphi$, and the area of the vertical side, $A \cos \varphi$. Since the plate of Fig. 1-30(c) is in equilibrium, the projections of the forces

[7]A complete treatment of this subject may be found in the reference works at the end of the chapter.

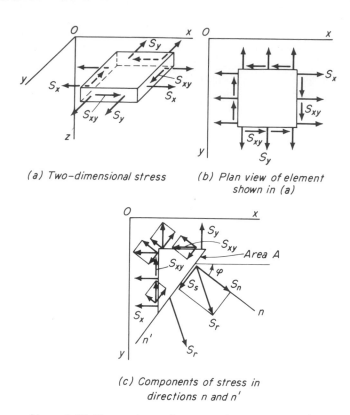

(a) Two-dimensional stress

(b) Plan view of element
shown in (a)

(c) Components of stress in
directions n and n'

Figure 1-30 Shear and normal stress on element at any angle φ.

on the perpendicular to the inclined surface must be in equilibrium. Multiplication of stress by area and then by the appropriate trigonometric function gives the following equation for s_n.

$$s_n = 2s_{xy} \sin \varphi \cos \varphi + s_x \cos^2 \varphi + s_y \sin^2 \varphi$$

The trigonometric terms should be changed by the substitution of the equations involving the double angles. Then,

$$s_n = \frac{s_x + s_y}{2} + \frac{s_x - s_y}{2} \cos 2\varphi + s_{xy} \sin 2\varphi \tag{37}$$

If the element in Fig. 1-30 is cut at 90° to the direction in sketch (c), summation of the forces will give the equation for the normal stress in the n' direction.

$$s_n' = \frac{s_x + s_y}{2} - \frac{s_x - s_y}{2} \cos 2\varphi - s_{xy} \sin 2\varphi \tag{38}$$

Thus the normal stresses in the material at any desired angle φ can be found by use of the above equations. Should the equation give a negative result, the corresponding stress is compressive.

In a similar manner, s_s can be found by making the sum of the projections of all forces parallel to the cut surface equal to zero. Hence,

$$s_s = s_{xy}(\cos^2 \varphi - \sin^2 \varphi) - (s_x - s_y) \sin \varphi \cos \varphi$$

or

$$s_s = s_{xy} \cos 2\varphi - \frac{s_x - s_y}{2} \sin 2\varphi \tag{39}$$

The shear stress s_s at any desired angle φ can thus be found by Eq. (39). A positive result for s_s means that the stress is directed as in Fig. 1-30(c), and a negative result means that the stresses is directed oppositely.

Angle φ is positive when taken clockwise from the x-axis.

17. The Mohr Circle

A graphical solution to the combined stress problem, known as the Mohr circle, will now be given. Use of this method rather than the previously derived equations usually effects a considerable saving in time. However, certain conventions regarding signs and directions must be understood and carefully followed.

Figure 1-31 shows the perpendicular axes s_n and s_s. Normal stresses, regardless of the inclination of the surface on which they act, are plotted horizontally—positive, or tension, to the right of the origin, and negative, or compression, to the left. Shear stresses are plotted vertically upward or downward on the diagram. The normal and shear stresses at a point in the body thus become the coordinates of a point on the circle.

Stresses s_x and s_{xy} acting on the right and left edges of the plate in Fig. 1-30(b) locate point A in Fig. 1-31. Tension s_x is plotted to the right in accordance with the above-mentioned rule. Since shear stress s_{xy} tends to rotate the element in a clockwise direction, it is plotted upward. Stresses s_y and s_{xy} of the upper and lower edges of the plate shown in Fig. 1-30(b) locate point B in Fig. 1-31. Tension s_y is plotted to the right. Since shear stress s_{xy} on these surfaces tends to produce counterclockwise rotation, it is plotted downward. The Mohr circle is drawn with line AB as a diameter. Greater facility in the determination of angles will be obtained if radii AC and BC are marked x-axis and y-axis, respectively.

To find the stresses on an element oriented at angle φ, as shown in Fig. 1-31(c), the angle 2φ is laid off from CA, the x-axis of the circle, in the same direction as angle φ is turned in the body. Diameter DE is thus located.

The horizontal projection of CD has the value shown in the figure. When this is added to OC, the result is the value of s_n as given by Eq. (37). The vertical projection of CD has the value shown on the figure. This is equal to s_s as given by Eq. (39). It is plain that the coordinates of point D of the circle are equal to the normal and shear stresses as found by the combined stress equations.

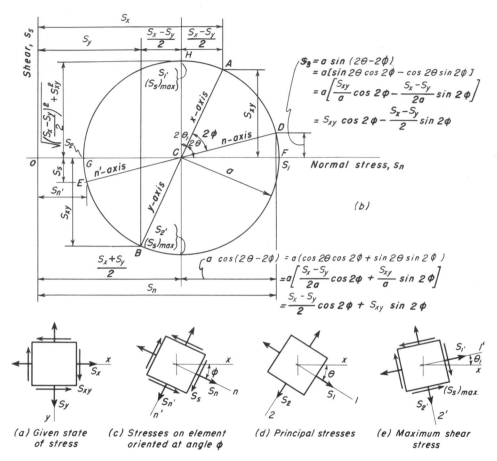

Figure 1-31 Mohr circle for two-dimensional stress.

Stresses $s_{n'}$ and s_s for the surface, in Fig. 1-31(c), whose normal lies at angle $(90° + \varphi)$ from the x-axis, are given by the coordinates of point E.

A clockwise angle φ on the body corresponds to a clockwise angle of 2φ on the circle, and vice versa.

Values of stresses s_n, $s_{n'}$, and s_s change as angle φ is changed. The maximum and minimum values of the normal stresses are called the principal stresses, and are designated s_1 and s_2, respectively. Their values can be found from the abscissas for points F and G in Fig. 1-31(b). The element for the principal stresses is oriented at angle θ to the x-axis as shown in Fig. 1-31(d). As shown by the circle, the value of θ can be found by the following equation.

$$\tan 2\theta = \frac{2s_{xy}}{s_x - s_y}, \qquad \text{for principal stresses} \tag{40}$$

The radius of the circle has the value shown. The equations for s_1 and s_2 are as follows.

$$s_1 = \frac{s_x + s_y}{2} + \sqrt{\left(\frac{s_x - s_y}{2}\right)^2 + s_{xy}^2} \tag{41}$$

$$s_2 = \frac{s_x + s_y}{2} - \sqrt{\left(\frac{s_x - s_y}{2}\right)^2 + s_{xy}^2} \tag{42}$$

It should be noted that the sides of the element for principal stresses are free from shearing stress. If shear stress s_{xy} should be equal to zero, stresses s_x and s_y would become the principal stresses.

The maximum shearing stress to which the material is subjected has a value equal to the radius of the circle. On the circle, point H is located $90°$ from points F and G for principal stresses. In the body, the surfaces for maximum shear stress are thus inclined $45°$ to the surfaces for the principal stresses. The element of maximum shearing stress, as shown in Fig. 1-31(e), is inclined at θ_1 to the x-axis. As shown by the circle, the value of θ_1 can be found by the following equation.

$$\tan 2\theta_1 = -\frac{s_x - s_y}{2s_{xy}}, \qquad \text{for maximum shear stress} \tag{43}$$

The value of the maximum shearing stress is

$$s_{smax} = \sqrt{\left(\frac{s_x - s_y}{2}\right)^2 + s_{xy}^2} \tag{44}$$

Let the axes for maximum shear stress be called $1'$ and $2'$. In Fig. 1-31, the element for maximum shear stress has normal stresses on the sides of value

$$s_{1'} = s_{2'} = \frac{s_x + s_y}{2} \tag{45}$$

Another useful equation for maximum shear stress is obtained by subtracting s_2 from s_1.

$$s_{smax} = \tfrac{1}{2}(s_1 - s_2) \tag{46}$$

Example 15. Let the state of stress at some point in a body be defined as follows.
$$s_x = 20{,}000 \text{ psi}, \qquad s_y = -4{,}000 \text{ psi}, \qquad s_{xy} = 5{,}000 \text{ psi}$$

(a) Draw the view of the element for the given state of stress and mark values thereon.

(b) Draw the Mohr circle for the given state of stress and mark completely.

(c) Draw the element oriented $30°$ clockwise from the x-axis and show values of all stresses.

(d) Draw the element correctly oriented for principal stresses and show values.

(e) Draw the element for maximum shear stress and mark values of all stresses.

Solution (a), (b). The given state of stress and the Mohr circle are shown in Figs. 1-32(a) and (b), respectively.

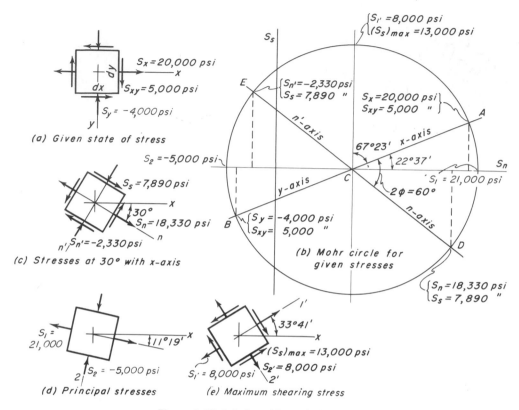

Figure 1-32 Solution of Example 15 by Mohr circle.

(c) Diameter *ECD* should be drawn at 60° clockwise to the *x*-axis of the circle, and stresses s_n and $s_{n'}$ scaled and placed on the element of Fig. 1-32(c). Since point *D* lies below the s_n-axis, shear stress s_s crosses the *n*-axis of Fig. 1-32(c) in the direction that causes a counterclockwise moment on the element. Likewise, since *E* lies above the s_n-axis, stress s_s crosses the *n'*-axis in sketch (c) in the direction that causes a clockwise moment on the element.

(d) Principal stresses s_1 and s_2, together with their angle of inclination, are scaled directly from the circle, and are shown acting on an element properly oriented in Figs. 1-32(d).

(e) The maximum shear stress s_{smax} and the corresponding normal stress $s_{1'}$ are shown on the element of Fig. 1-32(e). The arrows are directed in accordance with the previously explained rules.

The advantages of the graphical method for solving combined stress problems should now be apparent. Not only is the method more rapid, but the state of stress for any direction can be scaled directly from the circle. When the equations are used for solving a problem, a separate computation must be made for each desired direction. The Mohr circle also aids in forming a mental picture of the state of stress in the body. In working problems, care must be exercised that all necessary information is placed on the drawing for the circle as well as on the views of the various elements.

The reader should check all values shown in Figs. 1-32(c), (d), and (e) by using the appropriate equations. Note that for $\varphi = 30°$, Eq. (39) gives a negative result for s_s. This result checks with the circle and indicates that the shear stress for this direction is acting oppositely to that shown in Fig. 1-30(c).

18. Stresses and Deformations in Two Directions

The elongation ϵ_x in the x-direction caused by the tensile stress s_x is accompanied by a decrease in the width of the body at right angles to the stress, as shown in Fig. 1-33. This decrease of width is a definite proportion of the increase of length and is expressed by the equation

$$\epsilon_y = -\mu \frac{s_x}{E} \text{ in./in.} \tag{47}$$

where the factor μ is known as *Poisson's ratio*. In a similar manner, a tensile stress in the y-direction causes a decrease of the length in the x-direction equal to

$$\epsilon_x = -\mu \frac{s_y}{E} \tag{48}$$

Conversely, a compressive stress causes an increase in the width at right angles to the stress. This result is confirmed by Eqs. (47) and (48), since compressive or negative stress values are substituted in the equations. For most metals in engineering service, μ has a value between 0.25 and 0.30.

Superposition of the stresses shown in Figs. 1-33 and 1-34 gives those shown in Fig. 1-35. Stress s_x causes an increase in length in the x-direction, whereas in this same direction a shortening occurs because of s_y. The net effect is given by

$$\epsilon_x = \frac{1}{E}(s_x - \mu s_y) \text{ in./in.} \tag{49}$$

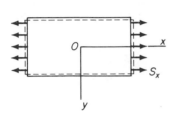

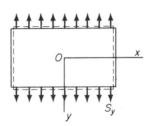

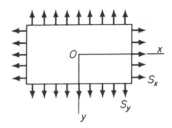

Figure 1-33 Deflection due to stress s_x.

Figure 1-34 Deflection due to stress s_y.

Figure 1-35 Tension in two directions.

In the y-direction a similar equation applies.

$$\epsilon_y = \frac{1}{E}(s_y - \mu s_x) \tag{50}$$

Equations (49) and (50) represent Hooke's law for two-dimensional stress. These equations can be solved simultaneously for the stresses to give

$$s_x = \frac{E}{1-\mu^2}(\epsilon_x + \mu\epsilon_y) \tag{51}$$

$$s_y = \frac{E}{1-\mu^2}(\epsilon_y + \mu\epsilon_x) \tag{52}$$

19. Deflection of Beam from Shearing Stress

In addition to the bending, warping of the cross sections, shown in Fig. 1-22(c), from the shearing stress causes additional deflection of a beam. The total deflection at the center for a simply supported beam carrying a uniform load w per unit length is given by

$$y = \frac{5wl^4}{384EI}\left(1 + \frac{25\alpha I}{l^2 A}\right) \tag{53}$$

The shear increment is represented by the second term of the parentheses. Here α represents the ratio of the maximum shearing stress to the average shearing stress, V/A, on the cross section. It therefore has the value 3/2 for rectangular cross sections and 4/3 for circular cross sections. For other shapes the value must be computed. In Eq. (53) it is assumed that there is no restraint to the free warping of the cross sections.

For a simply supported beam carrying a concentrated load P at the center, the equation is

$$y = \frac{Pl^3}{48EI}\left(1 + \frac{31.2\alpha I}{l^2 A}\right) \tag{54}$$

Example 16. Find the per cent of increase in deflection for a simply supported 8-in. I-beam with load P at the center. Area A is 7.09 in.^{2}. The thickness of the web is 0.24 in. The moment of inertia I is 83.4 in.^{4}. Length of the span is equal to 10 times the depth of the beam.

Solution. In accord with Section 13, the maximum shearing stress will be taken as the shear force V divided by the area of the web.

$$Max\ s_s = \frac{V}{0.24 \times 8} = \frac{V}{1.92}$$

The average shearing stress: $Av\ s_s = \frac{V}{A} = \frac{V}{7.09}$

Then: $\alpha = \frac{7.09}{1.92} = 3.69$

By Eq. (54): $y = \dfrac{Pl^3}{48EI}\left(1 + \dfrac{31.2 \times 3.69 \times 83.4}{80^2 \times 7.09}\right) = \dfrac{Pl^3}{48EI}(1 + 0.212)$

The deflection due to shear is thus 21.2 per cent as great as that due to bending.[8]

20. Principle of St. Venant

It was pointed out in Section 3 that equation $s = P/A$ was not applicable in the region close to the eye of the eyebar. At some distance away, however, the stress distribution becomes simple tension uniformly distributed, and the use of the equation for finding the value of the stress is justified.

A similar situation prevails for the use of equation $s = Mc/I$ for finding bending stress. The moments M in Fig. 1-5 could be applied to the beam in a number of ways as shown in Fig. 1-36. It is obvious that the state of stress

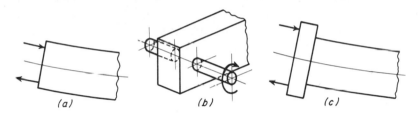

(a)   *(b)*   *(c)*

Figure 1-36 Methods for applying moment to end of beam.

near the end of the beam in Fig. 1-36(a) is not the same as that in Fig. 1-6(b), which represents the distribution for $s = Mc/I$. The same is true for Fig. 1-36(b). Equation (11) might be applied all the way to the end of the beam for sketch (c) because the basic assumption that the cross sections are planes is substantially fulfilled. At locations sufficiently far from the ends in Fig. 1-36(a) and (b), the use of Eq. (11) for finding the bending stress is satisfactory.

The remarks made above can be applied to the stress situation in the neighborhood of a concentrated force (either load or reaction). In Fig. 1-37 the stress situation for cross section OO' at the load is complicated by the vertical compressive stresses between the elements from the load. For cross section NN' located a distance equal to or greater than the depth of the beam from OO', the stress situation is approximately the same as that represented by Fig. 1-6(b) for bending stress and Fig. 1-22(b) for shear stress.

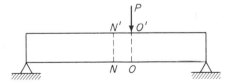

Figure 1-37 Beam with concentrated load.

[8]For derivation of Eqs. (53) and (54) as well as for a more precise method for computing α, see Chapter 5, Reference **3**, end of chapter.

The phenomenon that the stress system tends to become regular at distances removed from the disturbance is known as the principle of St. Venant. This principle states that if the forces acting on a small region of the body are replaced by a different, but statically equivalent system, no change in the stress or deformation will be experienced at points sufficiently far removed from the loads. It is valid for both normal and shear stress.

A sudden change in shape or form, arising from notches, holes, or fillets, causes the stresses to be increased beyond the values indicated by the equations of this chapter. This increase in stress is usually local in nature, and occurs in the immediate neighborhood of the discontinuity. It is taken care of in the computations by multiplying the stress, as given in the usual equation, by a stress concentration factor, which will be explained in the following chapter.

When a metal is stressed beyond the elastic limit in tension, it will subsequently be found that the elastic limit in compression is reduced. Conversely, the elastic limit in tension is reduced for a material strained beyond the elastic limit in compression. This phenomenon is known as the *Bauschinger effect*.

It must always be kept in mind that the equations in this book are derived on the assumption that the material is perfectly elastic, and that at all times the stress is proportional to the strain.

REFERENCES

1. Faupel, J. H., *Engineering Design*, New York: John Wiley & Sons, Inc., 1964.

2. Roark, R. J., *Formulas of Stress and Strain*, 3d ed., New York: McGraw-Hill Book Company, 1954.

3. Timoshenko, S., *Strength of Materials*, 2d ed., Vol. 1, New York: Van Nostrand Reinhold Company, 1940.

4. Timoshenko, S., *Strength of Materials*, 2d ed., Vol. 2, New York: Van Nostrand Reinhold Company, 1941.

5. Timoshenko, S., and J. M. Gere, *Theory of Elastic Stability*, 2nd ed., New York: McGraw-Hill Book Company, 1961.

PROBLEMS

See Table 2-1 of Chapter 2 for mechanical properties of engineering materials.

1. The lower ends of the two hangers in Fig. 1-38 were at the same elevation before the loads were applied. The horizontal member is of uniform cross section. Find the value of length l if the horizontal member (assumed to be rigid) is horizontal after all loads are acting.   *Ans.* $l = 38.3$ in.

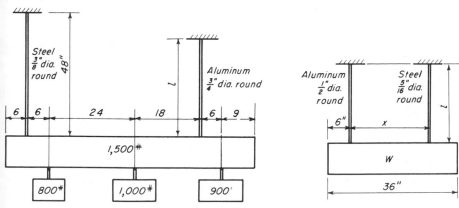

Figure 1-38 Problem 1.   **Figure 1-39** Problem 2.

2. The bottom member in Fig. 1-39 is of uniform cross section and can be assumed to be rigid. Find the value of the distance x if the lower member is to be horizontal.   *Ans.* $x = 22.2$ in.

3. The bottom member in Fig. 1-40 is of uniform cross section and can be assumed to be rigid. Its hinge is frictionless. Find the number of degress of rotation of the lower member.   *Ans.* $\varphi = 0.137°$.

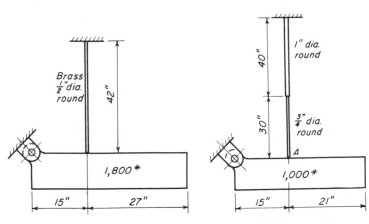

Figure 1-40 Problem 3.   **Figure 1-41** Problem 4.

4. The bottom member in Fig. 1-41 is of uniform cross section. Its hinge is frictionless. The rods are of steel. Find the distance point A drops upon attachment of the weight.   *Ans.* $\delta = 0.00475$ in.

5. In Fig. 1-42, find the drop of the 500-lb weight.   *Ans.* $\delta = 0.0126$ in.

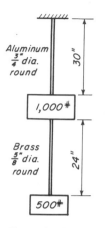

Figure 1-42 Problem 5.

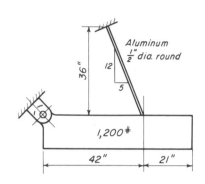

Figure 1-43 Problem 6.

6. In Fig. 1-43 the lower member is of uniform cross section and can be assumed to be rigid. Find the angular rotation of the lower member in degrees.
Ans. $\varphi = 0.0286°$.

7. In Fig. 1-44 the lower member is of uniform cross section and can be assumed to be rigid. Find the change in elevation of the left end because of the stretch of the rods.   *Ans.*   Drop $= 0.0352$ in.

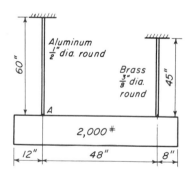

Figure 1-44 Problem 7.

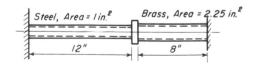

Figure 1-45 Problem 8.

8. The members in Fig. 1-45 have a neat fit at the time of assembly. Find the force caused by an increase in temperature of 100°F. Supports are immovable.
Ans. $F = 25{,}050$ lb.

9. After being drawn up snug, the nut in Fig. 1-46 is given one-quarter additional turn. Find the force in the pipe and bolt.   *Ans.* $F = 13,750$ lb.

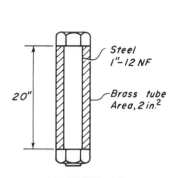

Figure 1-46 Problem 9.

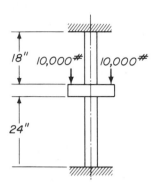

Figure 1-47 Problem 10.

10. The bars in Fig. 1-47 are fitted top and bottom to immovable supports. The bars are of the same material and have the same cross section. Find the force in each bar.   *Ans.*   Top, 11,430 lb; bottom, 8,570 lb.

11. In Fig. 1-48 the outer bars are symmetrically placed with respect to the center bar. The top member is rigid and located symmetrically on the supports. Find the load carried by each of the supports.

Ans.   Center, 8,312 lb; outer, 5,844 lb.

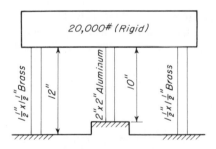

Figure 1-48 Problem 11.

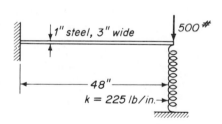

Figure 1-49 Problem 12.

12. In Fig. 1-49, find the deflection at the load.   *Ans.* $y = 1.167$ in.

13. Because of an error in machining, the center strut in Fig. 1-50 was made 0.010 in. shorter than the other two. The members on top and bottom can be considered rigid. Bars are made of the same material and have equal cross sections. Find the load carried by each bar. Bars are 2 × 2-in. steel.

Ans.   Outer, 173,330 lb; inner, 53,330 lb.

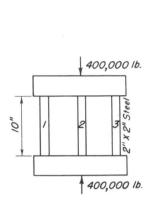

Figure 1-50 Problem 13.

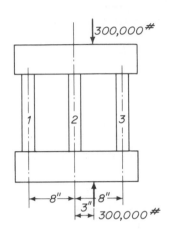

Figure 1-51 Problem 14.

14. In Fig. 1-51, the struts are of the same material and have equal cross-sectional areas. Members on top and bottom can be considered rigid. Find the force in each bar.   *Ans.*   43,750 lb; 100,000 lb; 156,250 lb.

15. Find the distance x in Fig. 1-52 that causes the 1,000-lb weight to remain level if the lower ends of the hanger are at the same elevation before the weight is applied.   *Ans.* $x = 21.2$ in.

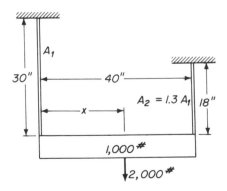

Figure 1-52 Problem 15.

16. Find the force in each bar in Fig. 1-53. $E_1 = 2E_2$. The 5,000-lb weight can be considered rigid.   *Ans.* $F_1 = 1,600$ lb; $F_2 = 1,440$ lb.

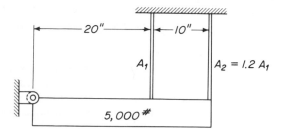

Figure 1-53 Problem 16.

17. The rigid beam in Fig. 1-54 was level before the load was applied. Find the force in each hanger.   *Ans.*   Steel, 10,210 lb; aluminum, 7,660 lb.

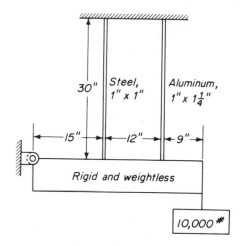

Figure 1-54 Problem 17.

18. The bars in Fig. 1-55 are of the same material and have equal cross-sectional areas. There is no stress in the bars before the load is applied. Find the load carried by each bar.   *Ans.*   Outer, 4,390 lb; inner, 7,470 lb.

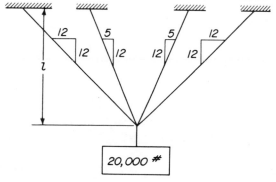

Figure 1-55 Problem 18.

19. The bars in Fig. 1-56 have the same cross-sectional area. There is no stress in the bars before the load is applied. Each bar is 0.5 in. square.

(a) Find the force in each bar.

(b) Find the force in each bar if the temperature drops 100°F.

Ans. (a) Steel, 8,340 lb; brass, 5,560 lb.

(b) Steel, 8,100 lb; brass, 5,970 lb.

Draw the shear and bending moment diagrams for the beams shown in the following figures and find the values of the maximum bending stress.

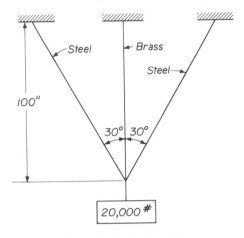

Figure 1-56 Problem 19.

20. Figure 1-57.   *Ans.* $s = 1,000$ psi.

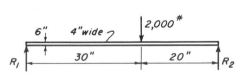

Figure 1-57 Problem 20.

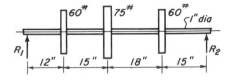

Figure 1-58 Problem 21.

21. Figure 1-58.   *Ans.* $s = 19,500$ psi.

22. Figure 1-59.   *Ans.* $s = 12,220$ psi.

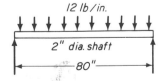

Figure 1-59 Problem 22.

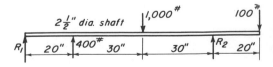

Figure 1-60 Problem 23.

23. Figure 1-60.   *Ans.* $s = 9,450$ psi.

24. A steel saw blade 0.05 in. thick is bent into an arc of a circle of 2-ft radius. Find the bending stress.   *Ans.* $s = 31{,}250$ psi.

25. Find the reactions and also the value of the bending stress at a point 5 ft from the left end for the beam of Fig. 1-61.   *Ans.* $s = 18{,}090$ psi.

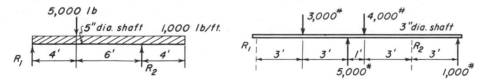

Figure 1-61 Problem 25. Figure 1-62 Problem 26.

26. Find the reactions and also the value of the bending stress at a point 6.5 ft from the left end for the beam in Fig. 1-62.   *Ans.* $s = 10{,}880$ psi.

27. Figure 1-63 illustrates the front side rod of a steam locomotive. Find the bending stress from inertia and dead weight for a train speed of 90 mph. Weight of steel equals 0.283 lb/in.[3] Assume the rod to be a simply supported beam of uniform cross section *A-A*.   *Ans.* $s = 12{,}430$ psi.

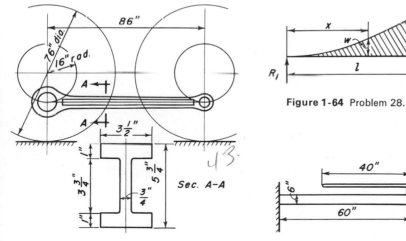

Figure 1-63 Problem 27.   Figure 1-64 Problem 28.

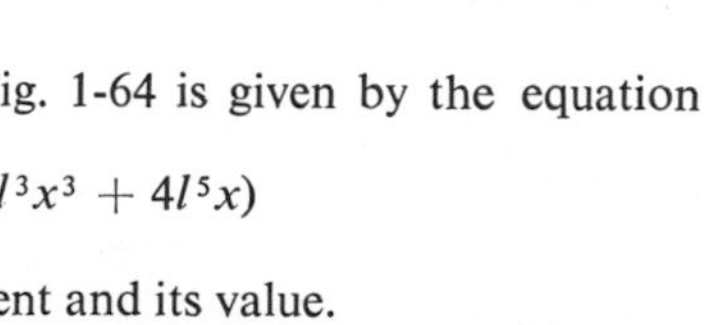

Figure 1-65 Problem 29.

28. The deflection of the beam shown in Fig. 1-64 is given by the equation

$$y = \frac{w_1}{360EIl^2}(x^6 - 5l^3x^3 + 4l^5x)$$

Find the location of the maximum bending moment and its value.

Ans.   Max $M = 0.039w_1l^2$.

29. Compute the change of elevation of the end of the light pointer attached to the beam of Fig. 1-65 upon the application of the load. The beam is 4 in. wide. $E = 1{,}500{,}000$ psi.

30. Moments M_1 and M_2 are applied to the ends of the simply supported beam shown in Fig. 1-66. Light pointers of lengths $2l/3$ and $l/3$ are attached to the ends as shown and are directed along the axis of the beam before applying the moments. If the distance a is measured, and if E, I, and l are known, show that moment M_1 is equal to $6EIa/l^2$.

The idea of this problem can be used for finding the value of the unknown end moments acting on a beam. The pointers are clamped in position, and distance a is measured. Length l must be free from transverse loads.

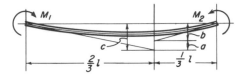

Figure 1-66 Problem 30.

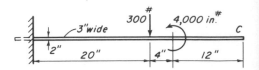

Figure 1-67 Problem 31.

31. Find the deflection at the end C of the beam in Fig. 1-67. $E = 1,600,000$ psi.
Ans. $y_c = 0.17$ in. up.

32. Find the deflection of the end A of the beam in Fig. 1-68. $E = 1,500,000$ psi.
Ans. $y_a = 0.126$ in. down.

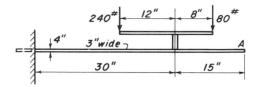

Figure 1-68 Problem 32.

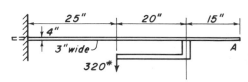

Figure 1-69 Problem 33.

33. Find the deflection of the end A of the beam of Fig. 1-69. $E = 1,500,000$.
Ans. $y_a = 0.158$ in. down.

34. Find the deflection at A in Fig. 1-70. Steel.   Ans. $y = 0.518$ in.

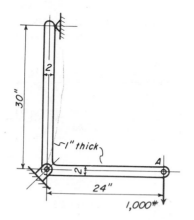

Figure 1-70 Problem 34.

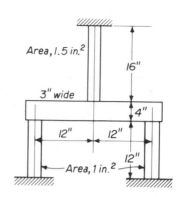

Figure 1-71 Problem 35.

35. The supports top and bottom in Fig. 1-71 can be considered immovable. The upper strut was found to be $\frac{1}{32}$ in. too long to be assembled without stress. If this member is driven into place, what force will be induced therein? All parts are of steel.   *Ans.* $F = 27,040$ lb.

36. The steel bar in Fig. 1-72 was welded into place with a neat fit at time of assembly. Find the force in the bar if the temperature drops 120°F. Supports can be considered rigid.   *Ans.* $F = 9,510$ lb.

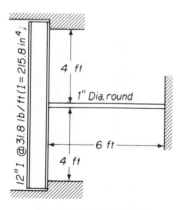

Figure 1-72 Problem 36.

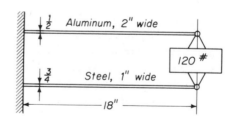

Figure 1-73 Problem 37.

37. Find the deflection of the weight in Fig. 1-73.   *Ans.* $\delta = 0.185$ in.

38. The steel beams in Fig. 1-74 are $1\frac{1}{2}$ in. wide and 2 in. deep. Find the deflection of each beam.   *Ans.*   Top, 0.160 in.; bottom, 0.813 in.

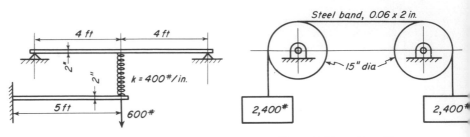

Figure 1-74 Problem 38.   **Figure 1-75** Problem 39.

39. Find the value of the stress in the steel band when it is passing around a pulley in Fig. 1-75. Pulley bearings are frictionless.   *Ans.* $s = 140,000$ psi.

40. Compute the values of the transverse shear at points 1, 2, 3, and 4 in. below the top surface of the beam in Fig. 1-76 for cross sections to left of the load. The beam is 6 in. wide and 8 in. deep.   *Ans.* $s_s = 328, 563, 703,$ and 750 psi.

Find the value of the maximum transverse shear for the beams of the following figures:

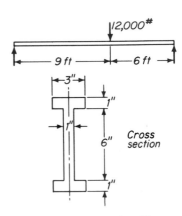

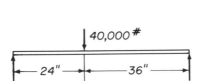

Figure 1-76 Problem 40.   **Figure 1-77** Problem 41.

41. Figure 1-77.   *Ans.* $s_s = 1,170$ psi.

42. Figure 1-78.   *Ans.* $s_s = 2,400$ psi.

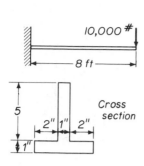

Figure 1-78 Problem 42.

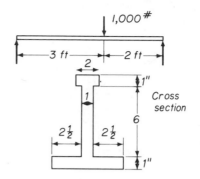

Figure 1-79 Problem 43.

43. Figure 1-79.   *Ans.* $s_s = 99$ psi.

44. Find the thickness of the web required to make the value of the maximum transverse shearing stress equal to 600 psi for the beam in Fig. 1-80.

Ans. $t = 0.33$ in.

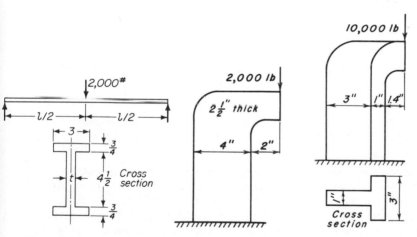

Figure 1-80 Problem 44.   **Figure 1-81** Problem 48.   **Figure 1-82** Problem 49.

45. Utilizing Eq. (24), prove Eq. (29).

46. Find the maximum value of the transverse shear for a hollow circular shaft having outer and inner radii of r_1 and r_2, respectively.

Ans. $s_s = V(r_1^3 - r_2^3)/3I(r_1 - r_2)$.

47. Prove Eq. (30).

48. Find the values and plot the distribution of stress over the cross section of the upright of Fig. 1-81. Locate the point of zero stress.   *Ans.* $s_c = 1,400$ psi.

49. Find the values and plot the distribution of stress over the cross section of the upright of Fig. 1-82. Locate the point of zero stress.   *Ans.* $s_t = 6,860$ psi.

50. Determine the width d of the offset link of Fig. 1-83 if the permissible value of the working stress is 8,000 psi. Carry out the work in a manner similar to Example 7.   *Ans.* $d = 2.73$ in.

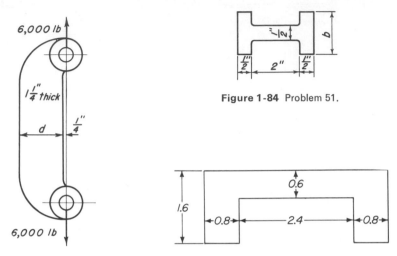

Figure 1-83 Problem 50.   **Figure 1-84** Problem 51.

          Figure 1-85 Problem 52.

51. The loading and general arrangement of a link are the same as for Problem 50, but the cross section is shown in Fig. 1-84. Find width of flange b.
                                    Ans. $b = 1.55$ in.

52. Use the equations of Fig. 1-7 to calculate the moment of inertia of the cross section of Fig. 1-85 about the horizontal axis through the center of gravity.
                                    Ans. $I = 0.8197$ in.^{4}

53. Make a sketch showing the equivalent loading of axial force and moments about the x- and y-axes for the bar shown in Fig. 1-86. Compute and show on isometric sketches the distribution of stress on cross section $ABCD$ caused by each of the separate loads. Also make a sketch showing resultant distribution of stress and mark the value at each corner. Dimension the location of the neutral axis for the cross section.   *Ans.*   Max $s = 600$ psi.

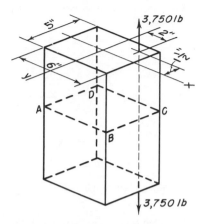

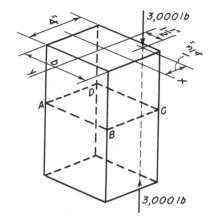

Figure 1-86 Problem 53.   **Figure 1-87** Problem 54.

54. Make a sketch for the equivalent loading for the bar of Fig. 1-87. What must be the dimension b in order that the maximum stress on a typical cross section, such as $ABCD$, will be equal to 1,000 psi? Make sketches and mark values for the stresses due to the separate loads. Also make the sketch for the resultant stresses and mark values at corners. Dimension the location of the neutral axis.

$$Ans. \quad b = 4.29 \text{ in.}$$

55. The inclined load shown in view M-M, Fig. 1-88, is applied to the pin at the end of the beam. Resolve load into horizontal and vertical components and find the bending stress due to each. Add algebraically to get the stress at points A, B, C, and D. $Ans.$ At A, $s = 1,562.5$ psi.

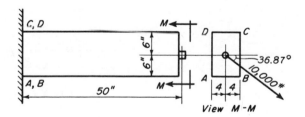

Figure 1-88 Problem 55.

56. The moment of inertia about axis 1-1 in Fig. 1-89 is equal to 120 in.⁴ The area is equal to 20 in.² Find the moment of inertia about axis 2-2.

Ans. $I_2 = 220$ in.⁴

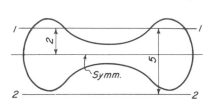

Figure 1-89 Problem 56.

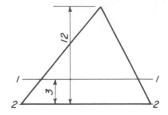

Figure 1-90 Problem 57.

57. The moment of inertia about axis 1-1 in Fig. 1-90 is equal to 540 in.⁴ Area is equal to 60 in.² Find the moment of inertia about axis 2-2.

Ans. $I_2 = 1,440$ in.⁴

58. Find the value of the moment of inertia about the left edge in Fig. 1-91.

Ans. $I = 75.78$ in.⁴

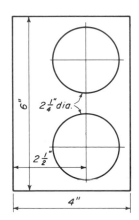

Figure 1-91 Problem 58.

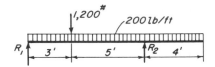

Figure 1-92 Problem 59.

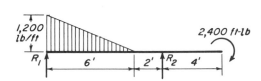

Figure 1-93 Problem 60.

59. Find the reactions, draw shear and moment diagram, and dimension significant points for the beam shown in Fig. 1-92.   *Ans.* $R_1 = 1,350$ lb.

60. Repeat Problem 59 using the beam of Fig. 1-93. Does the curve for shear have a horizontal tangent at a point 6 ft from the left end? Why? Also find the value of the maximum bending moment and its location.

Ans.   Max $M = 2,770$ ft lb at 2.54 ft from left end.

61. Draw and dimension the bending moment diagram for the member shown in Fig. 1-94. *Ans.* Max $M = 28,800$ in. lb.

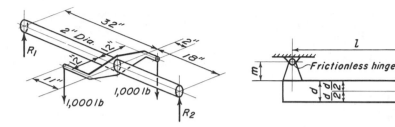

Figure 1-94 Problem 61.   Figure 1-95 Problem 62.

62. Repeat Problem 61 for Fig. 1-95.

63. Cut the *AB* portion of the beam of Fig. 1-96 free from the balance of the beam and place the shears and moments on the end surfaces which were acting before cutting. Is this portion of the beam now in equilibrium?

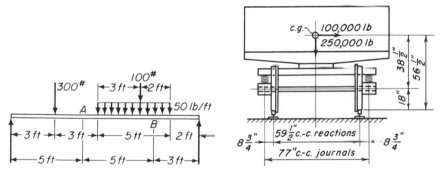

Figure 1-96 Problem 63.   Figure 1-97 Problem 64.

64. Figure 1-97 shows the loads and general dimensions of a freight car. The horizontal load is due to centrifugal effects in passing around a curve. If the car has four axles all equally loaded, make a force diagram showing the equilibrium of a unit consisting of one axle and two wheels. The centrifugal force is assumed to be applied to the axle at the inner bearing only, and to be resisted by the flange of the outer wheel only. Also draw and dimension the bending moment diagram for the axle. *Ans.* Max $M = 832,800$ in. lb.

Draw and dimension the load and reaction diagram, the shear diagram, and the bending moment diagram for the beams shown in the following figures:

65. Figure 1-98.

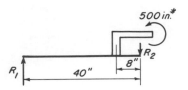

Figure 1-98 Problem 65.

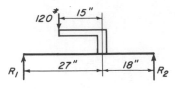

Figure 1-99 Problem 66.

66. Figure 1-99.

67. Figure 1-100.

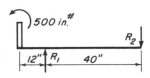

Figure 1-100   Problem 67.

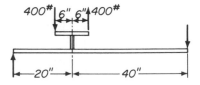

Figure 1-101 Problem 68.

68. Figure 1-101.

69. Figure 1-102.

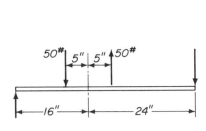

Figure 1-102 Problem 69.

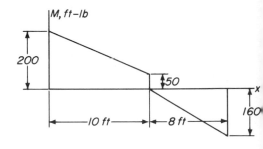

Figure 1-103 Problem 70.

70. Draw and dimension the shear and load diagram for the beam whose moment diagram is given in Fig. 1-103.

71. If the values in Fig. 1-104 represent pounds of shear, draw and dimension the corresponding load and moment diagrams for this beam.

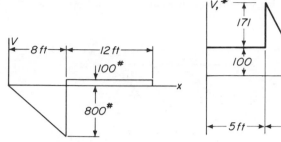

Figure 1-104 Problem 71.

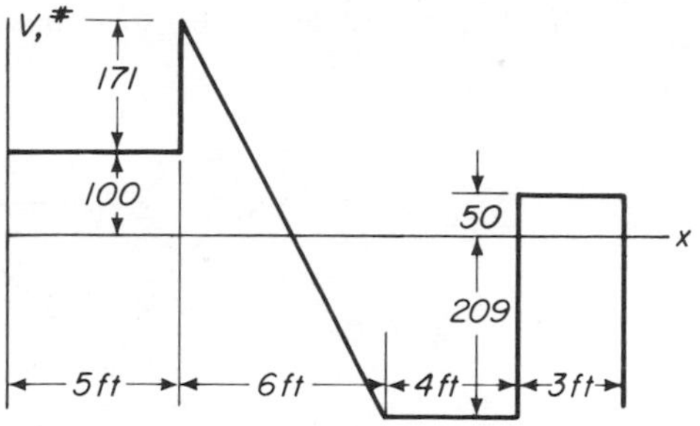

Figure 1-105 Problem 73.

72. If, however, the values in Fig. 1-104 represent foot-pounds of moment, draw and dimension the corresponding load and shear diagrams.

73. Repeat Problem 71 using Fig. 1-105.

74. Repeat Problem 72 using Fig. 1-105.

75. Draw and dimension the load and moment diagrams for the beam whose shear diagram is represented by Fig. 1-106.

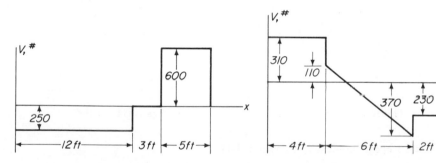

Figure 1-106 Problem 75.

Figure 1-107 Problem 76.

76. Repeat Problem 75 for the beam whose shear diagram is given in Fig. 1-107.

77. Draw and dimension the load and shear diagrams for the beam whose moment diagram is represented by Fig. 1-108.

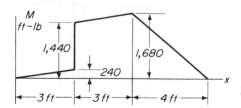

Figure 1-108 Problem 77.

78. Repeat Problem 77 for the moment diagram shown in Fig. 1-109.

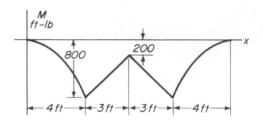

Figure 1-109 Problem 78.

79. Make an exploded isometric view of the three portions of the beam of Fig. 1-110 and show all forces and moments necessary for equilibrium.

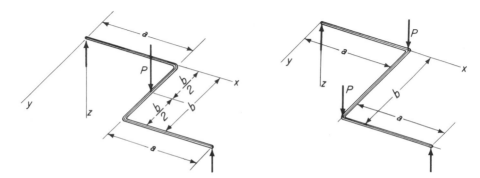

Figure 1-110 Problem 79.   **Figure 1-111** Problem 80.

80. Repeat Problem 79 using Fig. 1-111.

81. Draw and dimension the bending moment diagram for the beam shown in Fig. 1-112. Find the location where the moment is maximum and find its value.

Ans.   Max $M = 4{,}900$ in. lb.

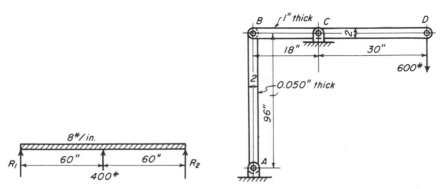

Figure 1-112 Problem 81.   **Figure 1-113** Problem 82.

82. Find the deflection of point D in Fig. 1-113. All parts are steel.

Ans.   0.485 in. at D.

83. A simply supported beam is 60 in. long and carries a 5,000-lb load at the center.

(a) If the cross section is a hollow square 6 in. high on the outside and 5 in. high on the inside, find the value of the maximum bending stress.

(b) Suppose the cross section is a hollow circle of the same area as for part (a) with wall thickness equal to 0.5 in. Find the value of the maximum bending stress.   *Ans.*   (a) $s = 4,020$ psi.

(b) $s = 4,090$ psi.

84. A simply supported cast-iron beam is 36 in. long and carries a 300-lb load at the center. Note that the three cross sections in Fig. 1-114 have areas equal to one another. Find the stress and deflection at the center for each beam. $E = 15,000,000$.

Ans.   (b) $s = 4,460$ psi; $y = 0.0472$ in.

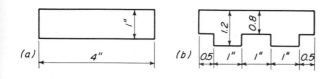

Figure 1-114 Problem 84.

85. Three beams have the equal cross-sectional areas shown in Fig. 1-115. Find the stress in each beam for a bending moment of 30,000 in. lb.

Ans.   (b) $s = 8,820$ psi.

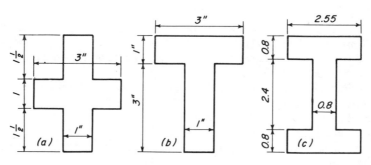

Figure 1-115 Problem 85.

86. A 2 × 2-in. steel column with hinged ends is 60 in. long. Yield strength of the material is 40,000 psi. $FS = 2.5$. Initial crookedness $a = 0.05$ in. Find the load capacity.   *Ans.* $P = 35,600$ lb.

87. A column is made from a piece of steel pipe 144 in. long. OD = 4.50 in. ID = 4.30 in. Yield strength of the material is 36,000 psi. $FS = 10$. Initial crookedness $a = 0.2$ in. Find the load capacity if the column can be considered as having hinged ends.   *Ans.* $P = 6,990$ lb.

88. The material in a body is subjected to the following stresses: $s_x = -10,000$ psi, $s_y = -4,000$ psi, and $s_{xy} = 4,000$ psi. Draw the Mohr circle and mark all significant points. Draw a view of the element, properly oriented, for maximum normal stress and show values of all stresses. Do the same for the element with maximum shearing stress.

Ans. $s_1 = -2,000$ psi; $s_2 = -12,000$ psi; $s_{s\,max} = 5,000$ psi.

89. A uniformly distributed normal stress, either tension or compression, is applied to the edges of the plate of Fig. 1-116 in the x-direction, and a uniform tension or compression is also applied in the y-direction. The dimensions of the deformed plate are 15.010 and 9.996 in. Find the stresses s_x and s_y. $E = 30,000,000$ psi. Poisson's ratio $\mu = 0.3$.

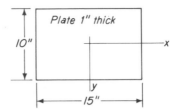

Figure 1-116 Problem 89.

90. Draw the Mohr circle for a prismatic bar loaded in tension and prove the rule: "The maximum shearing stress is equal to one-half the axial stress and is located at 45° to the direction of the axial stress."

91. Draw the Mohr circle for an element loaded on pure shear, and determine the value of the maximum normal stress and its direction.

92. An element is acted upon by the following stresses: $s_x = 7,000$ psi, $s_y = -2,000$ psi, and $s_{xy} = 2,000$ psi.

(a) By means of the equations compute the stresses on the sides of an element oriented 30° clockwise with the x-axis.

(b) Find the value of φ for maximum and minimum normal stress and compute the values of these stresses.

(c) Repeat (b) for maximum shear.

(d) Make a view showing the given state of stress, and draw the corresponding Mohr circle. Scale the stresses for the element oriented 30° from the x-axis, and compare them with the results secured by use of the equations. Draw a view of this element with the stresses placed thereon.

(e) Scale the values of the maximum and minimum normal stress and represent them by arrows on an element oriented at the proper angle.

(f) Repeat for the element that is subjected to the maximum shearing stress.

Ans.   Max $s = 7,420$ psi; max $s_s = 4,920$ psi.

93. Repeat Problem 92 using given stresses of $s_x = 1,500$ psi, $s_y = 22,500$ psi, and $s_{xy} = -10,000$ psi.   *Ans.*   Max $s = 26,500$ psi; max $s_s = 14,500$ psi.

94. The total normal force uniformly distributed over edge AB of the plate of Fig. 1-117 is 15,000 lb, and the total shear force is 12,000 lb. For edge BC the normal and shear forces are 135,000 and 30,000 lb, respectively. Draw a view showing an element with the given state of stress and also draw the corresponding Mohr circle. Draw an element with sides parallel to the x- and y-axes and show the stresses acting on it. *Ans.* $s_x = -2,710$ psi.

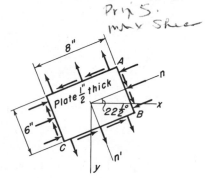

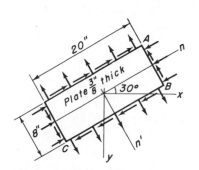

Figure 1-117 Problem 94.   **Figure 1-118** Problem 95.

95. The uniformly distributed normal and shear forces for edge AB of the plate of Fig. 1-118 arc 15,000 and 24,000 lb, respectively. For edge BC the normal and shear forces are 100,000 and 32,000 lb, respectively. Draw an clement for the given state of stress; also draw the Mohr circle. Draw the element with sides parallel to the x- and y-axes and show the stresses that are acting on it.

Ans. $s_n = 5,050$ psi.

96. For the loading shown in Fig. 1-119, draw the element located at 45° from the x-axis and place the stresses thereon that are acting on it.

Ans. $s_s = 4,000$ psi.

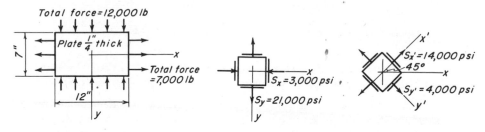

Figure 1-119 Problem 96.   **Figure 1-120** Problem 97.

97. In Fig. 1-120 the elements show the normal stresses that are acting at the same point in a body. Determine the values of the shearing stresses for both elements and mark them with arrows properly directed.

98. Make a drawing for the element at A of the beam in Fig. 1-121 with horizontal and vertical sides, and show the stresses acting on it. Construct the corresponding Mohr circle. Draw the element for the principal stresses correctly oriented and show the stresses acting on it. Do the same for the element of maximum shear stress.   *Ans.* Min $s = -2,940$ psi; max $s_s = 1,570$ psi.

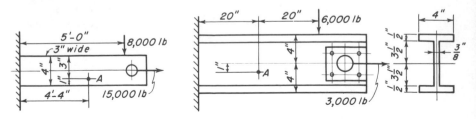

Figure 1-121 Problem 98.   Figure 1-122 Problem 99.

99. Repeat Problem 98 for the element at A of Fig. 1-122.
          Ans. Min $s = -3,060$ psi; max $s_s = 2,390$ psi.

100. Draw a view of the element at A in Fig. 1-123 with horizontal and vertical sides and show all stresses acting on it.   *Ans.* $s = 62.5$ psi.

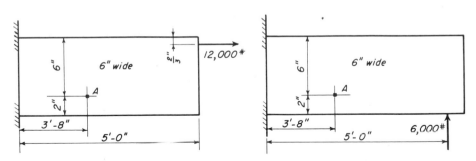

Figure 1-123 Problem 100.   Figure 1-124 Problem 101.

101. Draw a view of the element at A in Fig. 1-124 with horizontal and vertical sides and show all stresses acting on it.   *Ans.* $s = 750$ psi; $s_s = 140.6$ psi.

102. Draw a view of the element at A in Fig. 1-125 with horizontal and vertical sides and show all stresses acting on it. Draw the Mohr circle for this element and determine the value of the maximum normal stress and the maximum shear stress.
          Ans. $s_{max} = 707$ psi; $s_{smax} = 382$ psi.

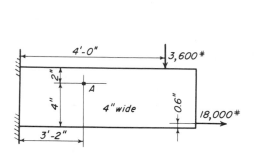

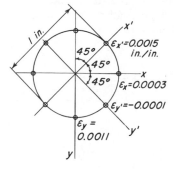

Figure 1-125 Problem 102.   Figure 1-126 Problem 103.

103. Strain gage measurements at a certain point on a body give the elongations shown by Fig. 1-126. Draw a view of the element with horizontal and vertical sides and show all normal and shear stresses acting on it. Do the same for the element whose sides are inclined 45° with the x-axis. Check all results with the Mohr circle. $E = 30{,}000{,}000$ psi; $\mu = 0.3$.

Ans. $s_x = 20{,}770$ psi; $s_y = 39{,}230$ psi; $s_{xy} = 18{,}460$ psi.

104. Repeat Problem 103 for the strain gage readings shown by Fig. 1-127.

Ans. $s_{x'} = 10{,}880$ psi; $s_{y'} = 6{,}260$ psi; $s_{x'y'} = 11{,}540$ psi.

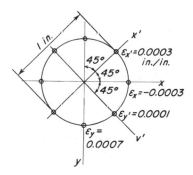

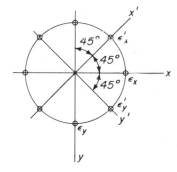

Figure 1-127 Problem 104.   Figure 1-128 Problem 105.

105. Prove that the strain measurements when taken as in Fig. 1-128 must fulfill the equation

$$\epsilon_x + \epsilon_y = \epsilon_{x'} + \epsilon_{y'}$$

106. When free of loads, a steel plate is 15 in. long in the x-direction and 12 in. long in the y-direction. After the loads are applied the 12-in. length becomes 12.006 in. Find the value of s_y if $s_x = 20{,}000$ psi.   Ans. $s_y = 21{,}000$ psi.

107. After loads in the x- and y-directions are applied to the steel plate of Fig. 1-129 the length in the y-direction is increased by 0.01 in. Find the change of length in the x-direction.   *Ans.* $\delta_x = -0.0184$ in.

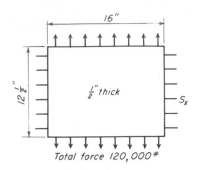

Figure 1-129 Problem 107.

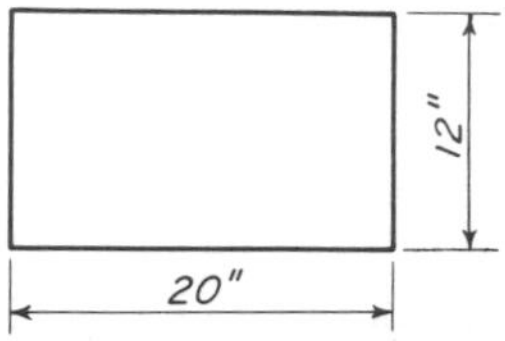

Figure 1-130 Problem 108.

108. A uniformly distributed stress, either tension or compression, is applied to the right and left edges of the plate shown in Fig. 1-130. A uniform tension or compression is also applied to the top and bottom edges. After the loads are acting, the final dimensions of the plate are 11.988 and 20.008 in.

(a) Find the values of s_x and s_y if the material is 1045 steel in the as-rolled condition.

(b) Find s_x and s_y if the material is Class 25 cast iron.

$$\textit{Ans.} \quad \text{(a) } s_x = 3,300 \text{ psi}; s_y = -29,010 \text{ psi.}$$
$$\text{(b) } s_x = 1,560 \text{ psi}; s_y = -13,730 \text{ psi.}$$

109. The moment of inertia about the 1-1 axis in Fig. 1-131 is equal to 4,320 in.[4] The area is equal to 120 in.[2] Find the moment of inertia about axis 2-2.

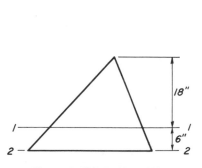

Figure 1-131 Problem 109.

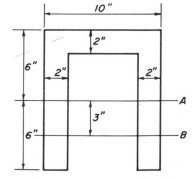

Figure 1-132 Problem 110.

110. The moment of inertia about axis A in Fig. 1-132 is equal to 880 in.[4] Find the moment of inertia about axis B.

111. The load shown on the beam of Fig. 1-133 is removed and replaced by a new load w uniformly distributed over the entire beam. The maximum bending moments for both loadings are the same. Find the value of w.

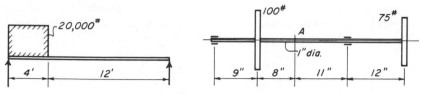

Figure 1-133 Problem 111.

Figure 1-134 Problem 112.

112. In Fig. 1-134 let the bearings be considered simple supports. The shaft is steel. Find the deflection at point A.

113. The pins in Fig. 1-135 are all at the same elevation. The bar is of spring steel. If the bar was originally straight, find the value of the maximum bending stress.

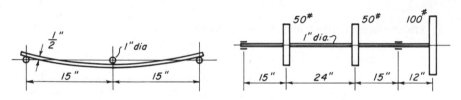

Figure 1-135 Problem 113.

Figure 1-136 Problem 115.

114. Let the cross section through an industrial car be similar to that of the car in Fig. 1-97. The loading for a single axle is 4,000 lb vertical and 1,000 lb horizontal. Rails are 36 in. center to center, and journals are 44 in. center to center. Wheels are 16 in. in diameter. The center of gravity of load is 19 in. above the center of the axle. Draw and dimension the bending moment diagram for the axle.

115. Find the deflection at the midpoint between the bearings in Fig. 1-136. The shaft is steel.

116. In Fig. 1-137 assume that the forces in the oil films are symmetrically disposed about the center line of the bearing. Determine the value of the eccentricity e so that the slope assumed by the bearing is the same as the slope of the shaft at A. The shaft is steel. The bearing is Class 30 cast iron.

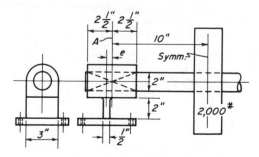

Figure 1-137 Problem 116.

117. A 12-in. standard I-beam of 31.8 lb per ft has $I_x = 215.8$ in.⁴ and $I_y = 9.5$ in.⁴ The width of the flange is 5 in. Find the percentage increase in bending stress caused by changing a vertical load to one inclined at 1° to the vertical.

118. The casting shown in section in Fig. 1-138 has its center of gravity located as shown. Assume that all loads can be considered as acting at the center of the 3-in. bearing. Find the value of the bending stress for this point.

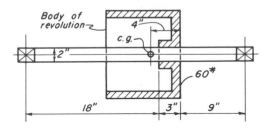

Figure 1-138 Problem 118.

119. An element in two-dimensional stress has $s_x = 12,000$ psi, $s_y = -4,000$ psi, and $s_{xy} = 3,900$ psi. Draw and dimension the Mohr circle. Draw and dimension the elements for the principal stress and for the maximum shearing stress.

120. Strain gage readings for a point in a stressed body show $\epsilon_x = -0.0006$ in./in., $\epsilon_y = .0026$ in./in., $\epsilon'_x = 0.0022$ in./in., and $\epsilon'_y = -0.0002$ in./in. Axes x and x' are at 45° to each other. $E = 30,000,000$ psi, $\mu = \frac{1}{3}$.

(a) Draw the view of an element with horizontal and vertical sides and show values of the stresses acting on it.

(b) Do the same for an element that is inclined 45°.

Check all results by making a Mohr circle.

121. The valve push rod for an overhead valve engine is $\frac{1}{4}$ in. in diameter and 14 in. long. Find the critical load when the rod is considered as a column with round ends. Assume the rod is perfectly straight.

122. A solid circular steel column with hinged ends is 1.96 in. in diameter and 54.75 in. long. Yield strength of the material is 36,000 psi. Initial crookedness $a = 0.049$ in. Find the yield point loading for this column.

Ans. $P_{yp} = 56,200$ lb.

123. An 8 × 5.25-in. I-beam at 17 lb/ft is used as a column with hinged ends. Length is 96 in. Area of cross section is 5 in.² Moment of inertia (in weak direction) is 6.16 in.⁴ Yield strength of the material is 33,000 psi. Initial crookedness $a = 0.50$ in. $FS = 4$. Find the value of the working load.   *Ans.* $P = 16,000$ lb.

124. A solid circular steel column with hinged ends is 36 in. long and carries a load of 50,000 lb. Initial crookedness a is equal to $\frac{1}{40}$ of the diameter d. $FS = 4$. Yield strength of the material is 50,000 psi. Find the required diameter.

2

Working Stresses

The problem of mechanical strength is one of the most important features of the design of machine parts. Stress equations in general are applicable to idealized homogeneous materials subjected to steady loads. This chapter will extend the theories of designing to include cases of fluctuating loads, where the fatigue strength of the material has an important influence on the success of the design. Machine parts should be shaped so that stress concentrations at points of high loading are avoided as much as possible. Suitable adjustments must also be made when the material carries loads in two directions. A distinction must be made between ductile and brittle materials. Lack of knowledge or appreciation of the behavior of engineering materials under actual service conditions has been the cause of many expensive failures.

FS, factor of safety

K, stress concentration factor

s_{av}, average stress

s_e, endurance limit stress for reversed bending

s_1, s_2, principal stresses

s_r, range stress

s_s, shear stress

s_{yp}, yield point stress, tension

s_{syp}, yield point stress, shear

s_{uc}, ultimate stress, compression

s_{ut}, ultimate stress, tension

μ, Poisson's ratio. For metals, μ is usually taken as 0.3.

1. Stress Strain Diagrams

Much useful information concerning the behavior of materials and their suitability for engineering purposes can be obtained by making tensile tests and plotting a graph for the relationship between stress and strain. The characteristic shape of the stress–strain diagram for low-carbon steel is shown in Fig. 2-1(a). It should be noted that the material followed Hooke's law until

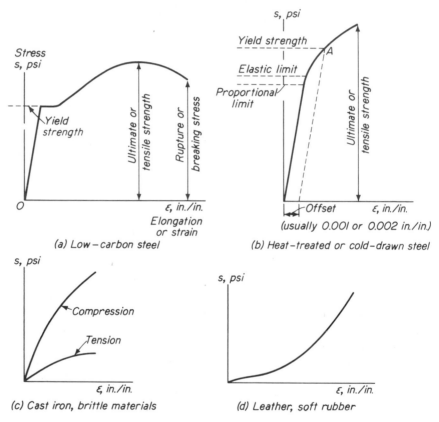

Figure 2-1 Stress-strain diagrams for various kinds of materials.

the loading became a little more than one-half of the *ultimate strength*. This material has a well-defined *yield point* or stress at which a marked increase in elongation occurs without increase in load. The *proportional limit* marks the maximum value of the stress for which Hooke's law holds. The modulus of elasticity of the material can be found from the slope, s/ϵ, of the straight-line portion of the curve, or from Eq. (3) of Chapter 1, $E = s/\epsilon$.

After the ultimate stress is reached, soft steel specimens undergo a marked reduction in diameter, called *necking*, at some point in the stressed material. It is customary, however, to construct the diagram on the basis of stresses computed by using the original cross-sectional area. The ratio of the loss of cross-sectional area at failure to the original area of the specimen is called the *reduction of area*. This quantity, together with the elongation at failure, gives useful information concerning the *ductility* of the material. The speed at which the load ,is applied affects the shape of the diagram. The yield point and ultimate stresses become higher as the speed of loading increases.

Many steels, especially those that have been heat treated, do not have a well-defined elastic limit, but yield gradually after passing the proportional limit, as shown in Fig. 2-1(b). If the loading were stopped at point *A* at a higher stress than the elastic limit, and if the specimen were then unloaded and readings taken, the curve would follow the dashed line, and a permanent set, or plastic deformation, would exist. For such materials, the stress corresponding to some given permanent set (usually 0.001 or 0.002 in./in.) is called the *yield strength*, and is taken as the limit of the engineering usefulness of the material.

Most materials do not exhibit a permanent set if loaded slightly beyond the proportional limit. The maximum value of such stress is known as the *elastic limit*, which is usually difficult to determine experimentally. *Proof loads* or *proof stresses* refer to loading that the material or part must sustain while fulfilling specified conditions relative to failure or deformation.

For ductile materials, the value of the yield strength in shear is equal to about 0.5–0.6 of the yield strength in tension.

Table 2-1 AVERAGE VALUES FOR MECHANICAL PROPERTIES
OF ENGINEERING MATERIALS

Material	Modulus of Elasticity		γ, Wt lb/in.^{3}	α, Coefficient of Linear Expansion in./in./°F
	Tension, psi E	Shear, psi G		
Cast iron	See Table 14-13		0.256	0.000 0056
Steel	30,000,000	11,500,000	0.283	0.000 0065
Stainless steel, 18-8	28,000,000	10,000,000	0.295	0.000 0096
Brass, bronze	15,000,000	5,300,000	0.30–0.32	0.000 0102
Aluminum	10,000,000	3,850,000	0.100	0.000 0128
Magnesium	6,500,000	2,400,000	0.065	0.000 0145

Poisson's ratio, $\mu = 0.3$.

Variations in values shown in Table 2-1 are possible, depending on composition and method of manufacture.

Nonductile or brittle materials, such as cast iron and ceramics, do not follow Hooke's law to any noticeable degree. The characteristic stress–strain diagram for either tension or compression is shown in Fig. 2-1(c). Leather and rubber have diagrams similar to that in Fig. 2-1(d).

Mechanical properties for a number of widely used engineering materials are given in Table 2-1.

Elementary elastic theories as discussed in Chapter 1 apply, in general, to bodies of uniform cross section and are unable to take account of the effect of a change in shape on the resulting stresses.

2. Stress Concentration Caused by Sudden Change in Form

Only rarely does the failure of a machine part occur because of the sudden application of a single heavy load. Breakage, in the great majority of cases, is caused by repeated or fatigue loading, and takes place at a point of stress concentration where an abrupt change in the form of the part occurs. Such failures can occur without warning or plastic deformation. The average stress for the cross section may be below the elastic limit for the material.

Consider, for example, the state of stress in the tension member of two widths illustrated in Fig. 2-2. Near each end of the bar the internal force

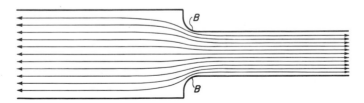

Figure 2-2 Stress concentration caused by sudden change in cross section.

is uniformly distributed over the cross sections. The nominal stress in the right portion can be found by dividing the total load by the smaller cross-sectional area; the stress in the left portion can be found by dividing by the larger area. However, in the region where the width is changing, a redistribution of the force within the bar must take place. In this portion the load is no longer uniform at all points on a cross section, but the material in the neighborhood of points B in Fig. 2-2 is stressed considerably higher than the average value. The stress situation is thus more complicated, and the elementary equation P/A is no longer valid. The maximum stress occurs at some point on the fillet, as at B, and is directed parallel to the boundary at that point.

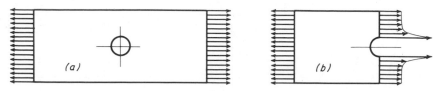

Figure 2-3 Stress concentration for bar with hole loaded in tension.

Another example is a bar in tension with a circular hole as shown in Fig. 2-3(a). If the bar is cut on the cross section of the hole, the tension stresses will be as shown in Fig. 2-3(b). The stress distribution along the cut surface is practically uniform until the neighborhood of the hole is reached, where it suddenly increases. High stresses such as these cause a fatigue crack to start under fluctuating loading.

This irregularity in the stress distribution caused by abrupt changes of form is called *stress concentration*. It occurs for all kinds of stress, axial, bending, or shear, in the presence of fillets, holes, notches, keyways, splines, tool marks, or accidental scratches. Inclusions and flaws in the material or on the surface also serve as stress raisers. The maximum value of the stress at such points is found by multiplying the nominal stress as given by the elementary equation by a stress concentration factor K, which is defined as follows.

$$K = \frac{\text{highest value of actual stress on fillet, notch, hole, etc.}}{\text{nominal stress as given by elementary equation for minimum cross section}} \quad (1)$$

Values of stress concentration factors can be found experimentally by photoelastic analysis or direct strain gage measurement. For a number of cases, solutions have been obtained by mathematical analysis.

3. Stress Concentration Factors

Stress concentration factors have been determined for a wide variety of geometric shapes and types of loading. The factors[1] for rectangular bars of two widths in tension or compression are given by the curves of Fig. 2-4. When such a bar is loaded in pure bending, the stress concentration factors can be obtained from Fig. 2-6. Corresponding curves for shafts are given by Figs. 2-5 and 2-7. The use of such curves is illustrated by the following example.

[1]Such curves, by R. E. Peterson, appeared in *Machine Design*, **23**, 1951 in the February, March, May, June, and July issues. These have now been collected into an enlarged book, Reference 7, end of chapter.

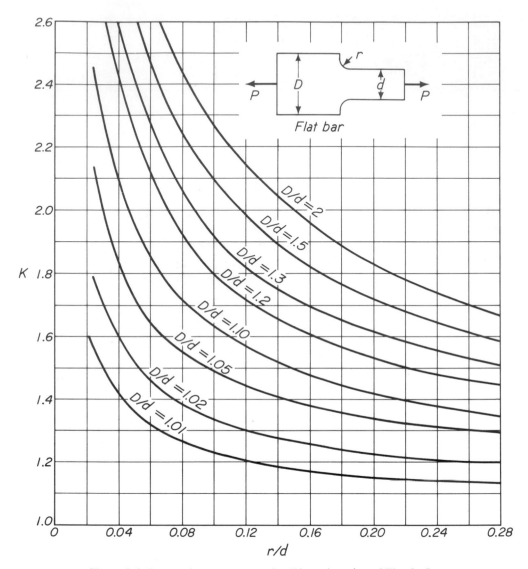

Figure 2-4 Factors of stress concentration K for various sizes of fillets for flat bar in tension or compression to be applied to the stress in the section of width d.

Example 1. Let the minor width of the flat bar in Fig. 2-2 be 1.25 in., the major width be 2.25 in., and the radius of the fillet be 0.25 in.

(a) Find the value of the stress concentration factor when the bar is loaded in tension.

(b) Find the value of the stress concentration factor if the loading is a pure moment instead of tension.

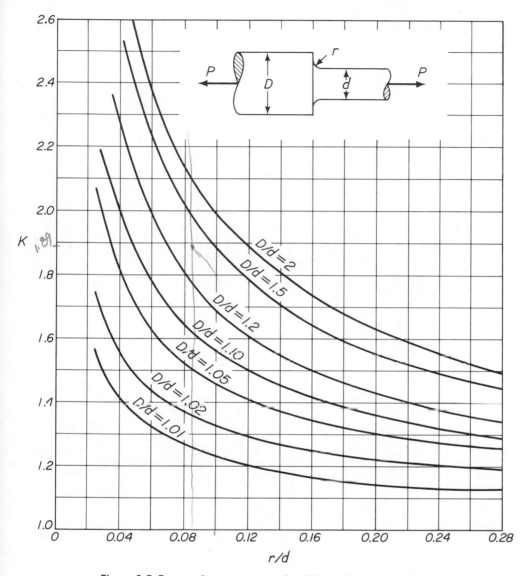

Figure 2-5 Factors of stress concentration K for various sizes of fillets for round bar in tension or compression to be applied to the stress in the section diameter d.

Solution. (a) $D = 2.25$ in., $d = 1.25$ in., $r = 0.25$ in., $D/d = 1.80$, $r/d = 0.2$.

From Fig. 2-4: $K = 1.78$.

The maximum tension stress in the bar is found by taking 1.78 times the P/A value for the narrower width.

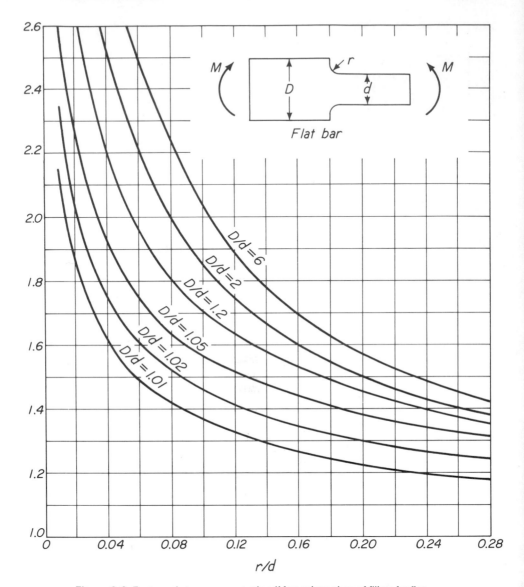

Figure 2-6 Factors of stress concentration K for various sizes of fillets for flat bar in bending to be applied to the stress in the section of width d.

(b) From Fig. 2-6: $K = 1.48$.

The maximum bending stress in the bar is found by taking 1.48 times the Mc/I value for the narrower width.

The stress concentration factor for this bar would be increased if the width

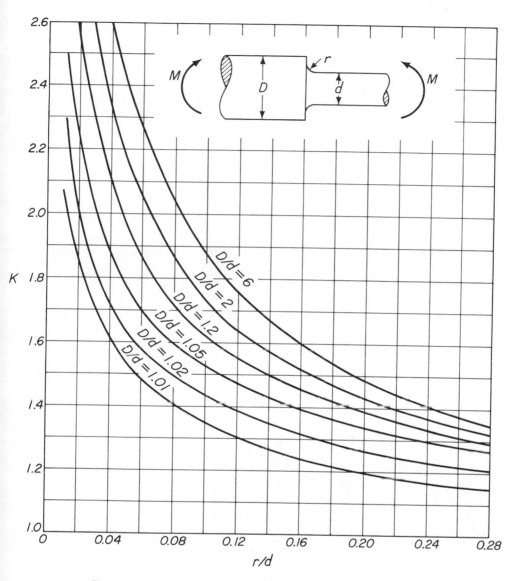

Figure 2-7 Factors of stress concentration K for various sizes of fillets for round bar in bending to be applied to the stress in the section of diameter d.

of the left portion were increased. A decrease in fillet radius also causes an increase in the factor.

In general, a stress concentration factor is applied to the stress computed for the net or smallest cross section.

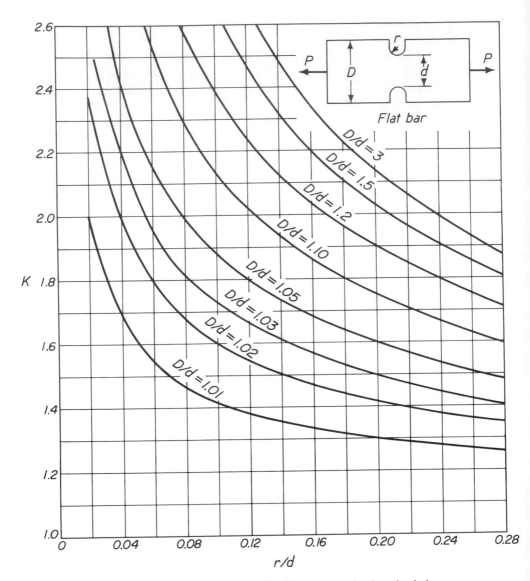

Figure 2-8 Factors of stress concentration K for grooves of various depths in tension or compression to be applied to the stress in the section of the flat bar of width d.

The effect of notches on the stresses for tension can be found in Figs. 2-8 and 2-9, respectively. The stress concentration factors for a bar containing a circular hole[2] are shown in Fig. 2-10. If a circular hole in a plate contains

[2]See Frocht, M. M., "Factors of Stress Concentration Photoelastically Determined," *Trans. ASME,* **57,** 1935, p. A-67.

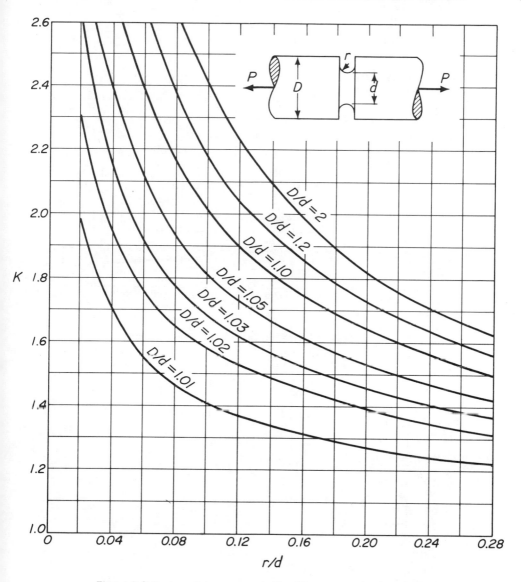

Figure 2-9 Factors of stress concentration K for grooves of various depths for round bar in tension or compression to be applied to the stress in the section of diameter d.

a pin through which the load is applied, the stress concentration factors will be as shown in Fig. 2-11.

Inspection of the curves in Figs. 2-4 to 2-10 indicates that stress concentration factors are reduced by the use of larger fillets or by more gradual transitions from cross sections of one size to those of another. Sometimes

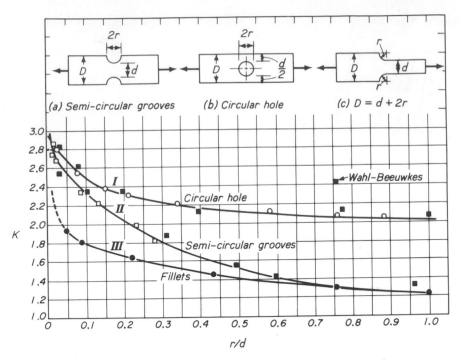

Figure 2-10 Invariant cases in tension and compression.

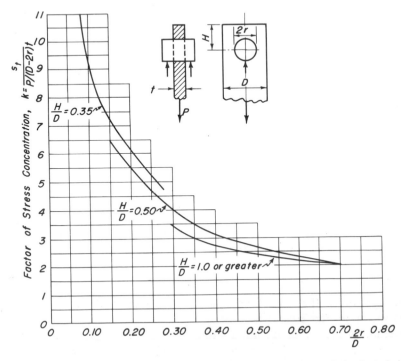

Figure 2-11 Stress concentration factors around a central circular hole in a plate loaded through a pin in the hole.

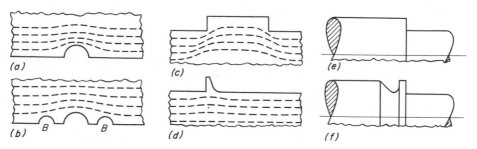

Figure 2-12 Reduction of stress concentration by removal of material.

the designer can specify the removal of material in such a way as to secure a more gradual transition in size. Thus in Fig. 2-12(b) it is easy to visualize that the stress concentration would be less when the notches B are present than when the main notch stands alone. For the same reason, a bolt with a continuous thread shows fewer concentration effects than a bar with a single circumferential groove. The narrow projection of Fig. 2-12(d), into which the force cannot spread, has less increase in stress than the wide projection of Fig. 2-12(c). It may be beneficial to use a stress-relieving groove, as shown in Fig. 2-12(f), on a shaft with a sudden change in diameter if it is not possible to use a fillet of suitable size at the junction. A reduction of stress concentration can be had by using fillets of elliptical shape as shown in Fig. 2-13. Fillets are needed only in regions of high stress. At points of low stress, undercuts may

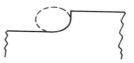

Figure 2-13 Fillet of elliptical form.

simplify machining and grinding operations. The designer can frequently reduce the harmful effects of a stress concentration by carefully studying the details and by making minor changes in the outline of the parts.

4. Endurance Limit of Materials

Working stresses that have been determined from the ultimate or yield point values of the material by a factor of safety give safe and reliable results only for static loading. Many machine parts, however, are subjected to a loading cycle in which the stress is not steady but continuously varying. Failures in machine parts are generally caused by such repeated loadings and at stresses that are considerably below the yield point. For many materials, long experience has proved that when the stress is below a certain value called the *fatigue* or *endurance limit*, the part will last indefinitely so far as ill effects from the stress are concerned. However, for slightly greater values of the stress, failure can be expected after a certain number of repetitions of the stress cycle. Fracture occurs without perceptible stretching and resembles the failure of a brittle material. Although such breaks are called fatigue failures, no change has taken place in the material except in the immediate neighborhood of the fracture itself. Failure has been brought about by a tiny crack that started at a stress concentration or at a flaw in the material at a highly stressed point. The crack itself serves as a stress concentration, and grows continually larger until the failure of the part occurs.

Methods of testing have been developed that aid in evaluating the ability of a material to resist failure by fatigue. The rotating-beam test is in widest use. It applies a pure bending moment to the specimen, shown schematically in Fig. 2-14(a). As the specimen rotates, the bending stress varies continuously

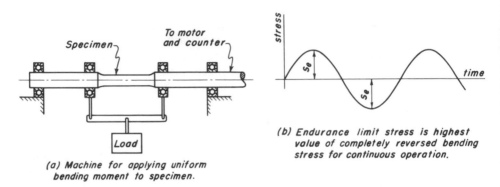

(a) Machine for applying uniform bending moment to specimen.

(b) Endurance limit stress is highest value of completely reversed bending stress for continuous operation.

Figure 2-14 Rotating beam type of fatigue testing machine (schematic drawing).

from a maximum tension to a maximum compression, which can be represented on the time-stress axes by the curve of Fig. 2-14(b). The *endurance-limit stress* s_e is frequently defined as the maximum value of the completely reversed bending stress, which a plain specimen can sustain for 10 million or more load cycles without failure. If a specimen of ferrous material lasts for this number of cycles, in general it can be assumed that it will last indefinitely.

Fatigue testing is also done for types of loading other than the completely reversed bending stress described above. Many kinds of stresses can be obtained by superposing a variable or fluctuating component on a static or steady stress. Testing machines are built that will give fluctuating axial stresses of various magnitudes to the specimen. It is customary to test spring materials with a pulsating shearing stress that varies continuously from zero to some maximum value.

Fatigue failures due to bending are the most common. Torsion failures are next, and failure due to axial loading are rare. Once started, a fatigue crack follows a general direction normal to the tension stress. A fatigue failure usually takes place across the crystals.

5. Interpretation of Service Fractures

The appearance of the fractured section gives information about the magnitude of the stress that caused the failure. For example, the fracture of a shaft in bending is usually composed of smooth and coarse areas as shown in Fig. 2-15. The smooth area was caused by the opening and closing of the crack during its development. The coarse area was caused by the final rupture. A fatigue crack starts under cyclic loading if the stress exceeds the strength of the weakest grain on a cross section. Continued operation causes the crack to grow as the strength of adjacent grains is exceeded by the high stress at the end of the crack.

If the stress was originally at a high value, the strength at a number of grains around the edge of the cross section was probably exceeded. Cracks started at these, and spread until they united with one another. The circumferential crack then progressed toward the center until final rupture occurred. It can therefore be concluded, when the ruptured area was near the center of the cross section, that the stress was considerably greater than the endurance limit for the material. Failure may have been caused by only a few hundred thousands of stress cycles.

Figure 2-15 Fatigue failure of shaft.

When the final ruptured area was far to one side, the stress was probably only slightly above a safe value, so that the endurance limit was exceeded at but a single point on the boundary. Several million cycles may have been required to produce failure, and a small improvement in the design may make the part safe for indefinite operation.

A fatigue crack at approximately 45° to the axis indicates that the stress was mainly alternating torsion.

6. Factors Affecting Fatigue Strength

The value of the endurance limit is dependent on the condition of the surface of the specimen. The *endurance stress* s_e for ground and polished specimens, when no stress concentrations are present, is frequently found to be approximately equal to one-half the ultimate strength for wrought steels. For a somewhat rougher surface, as produced by machining, the endurance limit may be only 35 or 40 per cent of the ultimate strength. The endurance limit is further reduced if the surface is covered with scale from hot rolling or forging. Corrosion from water or acids may reduce the

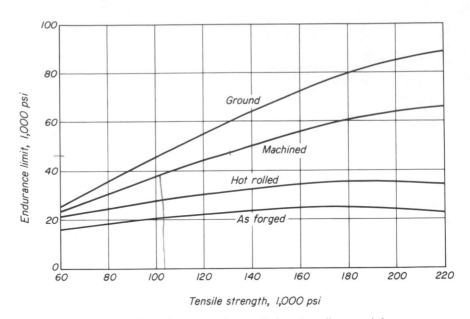

Figure 2-16 Relation between endurance limit and tensile strength for un-notched specimens in reversed bending.

endurance limit to a very low value. The small pits that form on the surface act as stress raisers. The effect of the condition of the surface[3] on the endurance limit for steels of various tensile strengths is shown in Fig. 2-16. The endurance limit is reduced for temperatures above room temperature. Carbon and alloy wrought steels give the most consistent results with respect to fatigue strength. For steel castings and cast iron, the endurance limit is about 40 per cent of the ultimate strength. Apparently no relationship exists between the endurance limit and the yield point, impact strength, or ductility. Experiments have shown that the endurance limit for reversed torsion is about 0.56 of that for reversed bending.

Fatigue cracks can start not only at easily recognized changes of form, but also at frequently overlooked stress raisers, such as file and tool marks, accidental and grinding scratches, quenching cracks, or part number and inspection stamps, which produce a high value for the stress and serve as the starting point for the progressive failure. The attention of the designer must therefore be focused on such "sore spots" whenever they are located in a region of high tension stress.

Since fatigue cracks are due to tensile stress, a residual stress of tension on the surface of the part constitutes an additional fatigue hazard. Such a tensile stress, for example, may arise from a cold-working operation on the

[3]See Noll, G. C., and C. Lipson, "Allowable Working Stresses," *Proc. Soc. Exp. Stress Anal.*, **3**, No. 2, 1946, p. 89.

part without stress relieving. Parts that are finished by grinding frequently have an extremely thin surface layer, which is highly stressed in tension. Such residual stresses, combined with the tensile stress from the loading, may give a resultant stress sufficiently great to cause a fatigue crack to start.

Any residual surface tension should be removed, or better still, converted into a layer of compression. A prestressed surface layer of compression can be secured by such shop operations as shot blasting, peening, tumbling, or cold working by rolling between hardened steel rolls. When the surface layer is in compression, the resultant working stress in tension may have a low value. Sometimes the part can be subjected to an excessive load and the yield point stress exceeded in such a way that a residual stress of compression can be obtained at a point where the maximum working stress is tension. Any cold working of the part must, of course, be done under controlled conditions. Sand blasting must be avoided since the scratches serve as stress raisers. Carburized and nitrided parts have a compressive surface layer, which may account for the effectiveness of such parts in resisting fatigue. A finish grinding operation, however, may leave the surface in tension, as previously mentioned. Additional discussion of residual stress is given in Chapter 14.

A weak decarburized layer on the surface of a heat-treated part has a low endurance limit. This condition is especially harmful in springs. Reductions in strength are also brought about by residual stresses, such as those produced by press and shrink fits. A press-fitted, antifriction bearing race also causes a reduction in the fatigue strength of the part to which it is fitted. The rough surface of a weld or an internal void also serves as a stress concentration. A layer of hot-dipped galvanizing causes a considerable reduction in the fatigue strength. The same is true for chromium platings. Electroplated zinc coatings have been found to be harmless. Many nonferrous materials do not have a definitely defined endurance limit.

The par value for the material is the endurance limit for a plain polished specimen. Fatigue tests with various kinds of notches show how much of the potential strength is being sacrificed by a particular type of notch. It is usually difficult if not impossible to devise a notched specimen that has the same fatigue strength as a machine part of a particular shape. Apparently such fatigue values can be obtained only from tests on the actual part. The results of tests on notched specimens may, however, serve as a better guide for the selection of a suitable steel than endurance tests made with plain specimens.

If proper attention has been given to the effect of stress concentration, failure in service is due mainly to accidental overload or abuse of the part, which could not be anticipated by the designer. Thus the ability to withstand overload is a very desirable quality in engineering materials. Sometimes a material of low endurance limit exhibits better resistance to overloads in fatigue than do high-strength, heat-treated materials. The ability to resist crack propagation after a crack has started is another desirable quality. Experiments have shown that killed steels are superior to rimmed steels with respect to crack propagation. Killed steels are deoxidized in the furnace or ladle before being poured into the ingot mold. Rimmed steels are not so degassified.

The life of a part can be reduced by factors other than fatigue, such as wear, corrosion, and high temperatures. If, for example, the useful life of a machine will be limited by wear of some of the parts, it would be uneconomical to design the other parts for infinite life in fatigue. The use of higher working stresses in fatigue can sometimes be justified in order to make the life of all the parts approximately equal. Precise information on the *S-N* curve for the material is required for such exact designing, and extensive testing must be conducted on the finished product. Ball bearings, and sometimes automotive parts, are designed on the basis of finite life.

7. Types of Failure. Ductile Materials and Brittle Materials

Two types of mechanical failures occur in materials: yielding and fracture. Yielding or permanent deformation is a pronounced sliding along certain angular planes in the material. It takes place without rupture. The engineering usefulness for most machine parts is ended after a sufficient amount of yielding has taken place. Therefore, yielding can properly be termed failure. Fracture is a separation failure that occurs on a cross section normal to the tension stress.

A ductile material can be defined as one whose resistance to sliding is smaller than its resistance to separation. Failure takes place by yielding. Many ductile materials have the same yield point value for compression as for tension.

A brittle material is one whose resistance to separation is less than its resistance to sliding. Failure takes place by fracture. A limit of about 5 per cent elongation is usually taken as the dividing line between ductile materials and brittle materials. Most brittle materials have a considerably higher value for the ultimate strength in compression than for tension.

Under certain conditions, a material ordinarily said to be ductile will undergo a fracture or separation failure similar to that of a brittle material. Some of these conditions are (a) cyclic loading at normal temperatures (fatigue); (b) longtime static loading at elevated temperatures (creep); (c) impact or very rapidly applied loading, especially at low temperatures; (d) work hardening by a sufficient amount of yielding; (e) severe quenching in heat treatment if not followed by tempering; and (f) a three-dimensional state of stress in which sliding is prevented, as at the bottom of the narrow groove in the bar shown in Fig. 2-17. Internal cavities or voids in castings or forgings may have a similar effect.

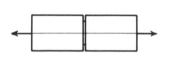

Figure 2-17 Bar in tension with deep groove.

8. Ductile Materials with Steady Stress

Under steady or static loading, a machine part made of a ductile material fails by yielding. The working stress is therefore based on the yield point stress.

It is possible for the yield point to be exceeded by the stress concentration as a result of a sudden change of form even though the elementary equation indicates that the average stress at the cross section has a safe value. In general, no damage occurs provided that the load is steady and the material is ductile. The material merely yields locally in the overstressed regions, and the stress is thereby relieved. Hence, it is customary for designers to neglect the effects of stress concentration when the loads are steady and the material is ductile.[4]

(a) *Simple Tension or Compression.* When the material is subjected to simple tension or compression the working stress s is given by the following equation.

Working stress:
$$s = \frac{S_{yp}}{FS} \tag{2}$$

where FS is the factor of safety.

(b) *Pure Shear.* For pure shear loading, the equation for working stress in shear is

Working stress:
$$S_{smax} = \frac{S_{syp}}{FS} \tag{3}$$

The expressions maximum stress or working stress are usually used interchangeably.

9. Maximum Shear Theory of Failure

Because ductile materials fail by shearing, the *maximum shear theory of failure* is in wide use by designers. The theory is applied by first finding the maximum shearing stress for the given loading and then dividing it into the yield point stress in shear to find the factor of safety. It is thus an application of Eq. (3).

The Mohr circle in Fig. 2-18 indicates that a body with simple tension stress s has shear stresses equal to one-half this value at directions 45° to the direction of s.

$$S_{smax} = \tfrac{1}{2}s \tag{4}$$

If stress s in this figure is increased to the yield point value, the maximum shear theory postulates that the material will then be at the yield point value in shear. Thus

[4]See p. 10 of Reference 7, p. 447 of Reference 3, and p. 298 of Reference 4, end of chapter.

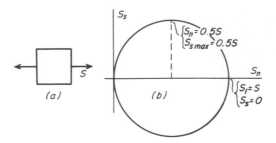

Figure 2-18 Element in simple tension and corresponding Mohr circle.

$$S_{syp} = \tfrac{1}{2}S_{yp} \tag{5}$$

This can be substituted into Eq. (3) to give

$$S_{smax} = \frac{0.5s_{yp}}{FS} \tag{6}$$

Failure in shear is assumed to occur along the 45° directions of Fig. 2-19(a).

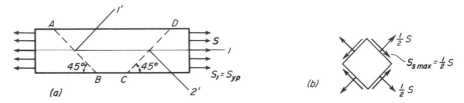

Figure 2-19 Planes of maximum shear stress for bar with tensile load.

Equations (5) and (6) are valid only when it is understood that the maximum shear theory of failure is being employed.

In addition to the shear stresses, an element oriented at 45° to the direction of s in Fig. 2-18(a) has normal stresses on all sides of 0.5s. The complete loading for this element is then given by Fig. 2-19(b).

Example 2. Suppose the part of Example 1 is 0.5 in. thick and is loaded by a steady tensile force of 18,750 lb. The material is soft steel with a yield point value 45,000 psi. Find the value of the factor of safety based on the yield point.

Solution. By maximum shear theory: $s_{syp} = 0.5 \times 45{,}000 = 22{,}500$ psi

$$A = 1.25 \times 0.5 = 0.625 \text{ in.}^2$$

In right portion: $s = \dfrac{P}{A} = \dfrac{18{,}750}{0.625} = 30{,}000$ psi

By Eq. (4): $s_{smax} = 0.5 \times 30{,}000 = 15{,}000$ psi

By Eq. (3): $FS = \dfrac{s_{syp}}{s_{smax}} = \dfrac{22{,}500}{15{,}000} = 1.5$

As found in Example 1, the stress concentration factor for the fillets is equal to 1.78. The stress thus reaches the yield point value on the fillets, and local yielding occurs in these regions. The stress for the cross section as a whole, however, remains at a safe value. It is thus justified to neglect the effect of stress concentration for ductile materials and steady loads.

Consider the element of Fig. 2-20(a) loaded in pure shear. The corresponding Mohr circle is shown in sketch (b). For such loading, an element

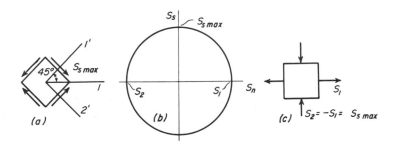

Figure 2-20 Elements and Mohr circle for pure shear stress.

oriented at 45° to the pure shearing stresses is loaded only by the normal stresses of Fig. 2-20(c). The stresses of sketches (a) and (c) can thus be considered as being interchangeable.

10. Normal Stresses in Two Directions

For two- and three-dimensional states of stress, the failure of an engineering material is a complicated phenomenon. In addition, test data for combined loading is nearly always lacking, and the design must therefore be based solely on the yield point or ultimate strength values as found by the simple tension test. It is under such conditions that a theory of failure is most useful.

Let the theories for combined stress of Chapter 1 be applied to the case of general loading for stresses s_x, s_y, and s_{xy}, and thus obtain the principal stresses s_1 and s_2 of Fig. 2-21. For convenience, these are shown in the horizontal and vertical directions. The algebraically larger of the two stresses is designated s_1. If shear stress s_{xy} is equal to zero, s_x and s_y are principal stresses.

To arrive at suitable values for the working stresses, it is necessary to know how the element in Fig. 2-21 will fail. The presence of two stresses makes the situation more complicated than the case of simple tension of Fig. 2-18. Figure 2-22 shows a perspective of the element in Fig. 2-21. Since all bodies are three-dimensional, three planes of failure must be investigated.

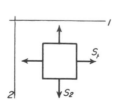

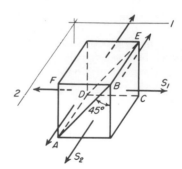

Figure 2-21 Principal stresses in two dimensions.

Figure 2-22 Plane of failure for two dimensional stress. All stresses tension.

(a) Both stresses are tension as shown in Fig. 2-22. The weakest plane is $BADE$ because s_2 has no effect on this plane. Failure is determined solely by stress s_1. Equations (2)–(6) apply.

(b) Both stresses are compression as in Fig. 2-23. The weakest plane is $BCDF$ because stress s_1 has no effect on this plane. Since s_2 is numerically the larger, failure is determined solely by stress s_2. Equations (2)–(6) apply.

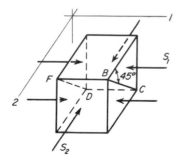

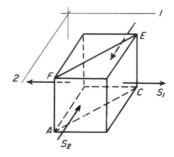

Figure 2-23 Plane of failure for two dimensional stress. All stresses compression.

Figure 2-24 Plane of failure for two dimensional stress, s_1 tension, s_2 compression.

(c) Stress s_1 is tension, and stress s_2 is compression as in Fig. 2-24. The weakest plane is $ACEF$. Both stresses contribute to the shear stress on this plane. The value of the maximum shear stress is

$$s_{smax} = \tfrac{1}{2}(s_1 - s_2) = \sqrt{\left(\frac{s_x - s_y}{2}\right)^2 + s_{xy}^2} \qquad (7)$$

Equation (6) applies.

The maximum shear theory of failure is thus very easy to apply to two-dimensional stress problems.

11. Mises–Hencky or Distortion Energy Theory

Another criterion, and one that fits experimental results closely, is the *Mises–Hencky* or *distortion energy theory*. For two-dimensional stress the equation is

$$S^2 = s_1^2 + s_2^2 - s_1 s_2 \tag{8}$$

Substitution of Eqs. (41) and (42) of Chapter 1 for s_1 and s_2 gives

$$S^2 = s_x^2 - s_x s_y + s_y^2 + 3s_{xy}^2 \tag{9}$$

Equivalent stress S can be considered a simple normal stress, and can be used as stress s in Eq. (2).

Example 3. The stresses at a point in a body are $s_x = 13{,}000$ psi, $s_y = 3{,}000$ psi, and $s_{xy} = 12{,}000$ psi. The material tests $s_{yp} = 40{,}000$ psi.
(a) Find the factor of safety by the maximum shear theory of failure.
(b) Find the factor of safety by the Mises–Hencky theory.

Solution. (a) By the Mohr circle or the combined stress equations the following values for the principal stresses s_1 and s_2 are obtained.

$$s_1 = 21{,}000 \text{ psi}; \qquad s_2 = -5{,}000 \text{ psi}$$

The weakest shearing plane is the one to which both s_1 and s_2 contribute to the shearing stress as in Fig. 2-24.

By Eq. (7): $s_{smax} = \frac{1}{2}[21{,}000 - (-5{,}000)] - 13{,}000$ psi

By Eq. (5): $s_{syp} = 0.5 \times 40{,}000 = 20{,}000$ psi

By Eq. (3): $FS - \dfrac{20{,}000}{13{,}000} = 1.54$

(b) By Eq. (8): $S = \sqrt{21{,}000^2 + (-5{,}000)^2 - 21{,}000(-5{,}000)}$

$$= 23{,}900 \text{ psi}$$

Hence $FS = \dfrac{40{,}000}{23{,}900} = 1.67$

Example 4. The same body as that used in Example 3 has stresses $s_x = 20{,}000$ psi, $s_y = 4{,}000$ psi, and $s_{xy} = 6{,}000$ psi.
(a) Find the factor of safety by the maximum shear theory of failure.
(b) Find the factor of safety by the Mises–Hencky theory.

Solution. (a) The principal stresses are found to be

$$s_1 = 22{,}000 \text{ psi}; \qquad s_2 = 2{,}000 \text{ psi}$$

The weakest shearing plane is the one affected by s_1 alone as in Fig. 2-22. For this plane

By Eq. (4): $s_{smax} = 0.5 \times 22{,}000 = 11{,}000$ psi

By Eq. (3): $FS = \dfrac{20{,}000}{11{,}000} = 1.82$

(b) By Eq. (8): $S = \sqrt{22{,}000^2 + 2{,}000^2 - 22{,}000 \times 2{,}000}$

$$= 21{,}070 \text{ psi}$$

Hence $$FS = \frac{40,000}{21,070} = 1.90$$

Example 5. A 2-in.-diam shaft is loaded statically in pure torque at a shearing stress of 10,000 psi. Find the *FS* if the material is 4140 hot-rolled steel. Use the Mises–Hencky theory.

Solution. By Fig. 2-20(c): $s_{smax} = s_1 = -s_2 = 10,000$ psi

By Eq. (8): $S^2 = 10,000^2 + 10,000^2 - 10,000(-10,000)$

$$S = 10,000\sqrt{3\cdot} = 17,320 \text{ psi}$$

By Table 14-5: $s_{yp} = 62,000$ psi

By Eq. (2): $FS = \dfrac{s_{yp}}{S} = \dfrac{62,000}{17,320} = 3.58$

12. Ductile Materials with Completely Reversing Stress

As mentioned at the beginning of Secton 8, local yielding under steady load takes place if the yield point is exceeded at certain points of stress concentration. However, when the load is fluctuating, such local relief cannot be obtained, and a suitable stress concentration factor K must be applied. Fatigue failure from such loading will be by fracture.

When the load is alternating or completely reversing, the endurance-limit stress s_e, as determined by testing, is the criterion used for determining the factor of safety. For such loading, the working stress can be called the range stress s_r. Hence

$$FS = \frac{s_e}{Ks_r} \tag{10}$$

A similar equation could be written for completely reversed shearing stress.

Example 6. A Class E freight car axle, Fig. 2-25, carries a load of 52,500 lb. The material is medium-carbon forged steel. The Brinell hardness number is 180, and the ultimate strength is 90,000 psi. The material surface is in the as-forged condition.

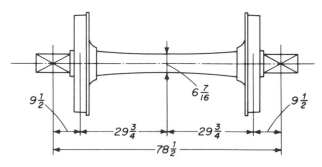

Figure 2-25 Example 6.

Find the value of the FS at the center of the axle as the car is proceeding along a smooth straight-level track.

Solution. $$I = \frac{\pi d^4}{64} = \frac{\pi 6.4375^4}{64} = 84.302 \text{ in.}^4$$

$$s = \frac{Mc}{I} = \frac{26{,}250 \times 9.5 \times 3.219}{84.302} = 9{,}520 \text{ psi}$$

Since the bending stress is completely reversing, Eq. (10) will apply.

By Fig. 2-16: $s_e = 19{,}500 \text{ psi}$

By Eq. (10): $$FS = \frac{s_e}{s_r} = \frac{19{,}500}{9{,}520} = 2.05$$

13. Ductile Materials with Combined Steady and Alternating Stress

In many strength problems, the major components of stress are static, with less accurately known alternating stresses superposed. Most failures originate with stresses of this type. The problem presents great difficulties because of the fundamentally different mechanisms of failure in the two sources of stress.

Suppose the tensile load P on the bar of Fig. 2-26(a) is continuously varying in magnitude as shown by the graph of Fig. 2-26(b). This load can be

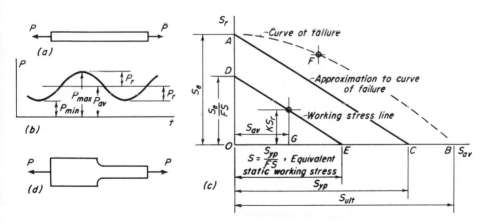

Figure 2-26 Working stress diagram for non-steady loading.

considered as being made up of two parts, the steady or average load P_{av}, and the variable or range load P_r. As illustrated by the figure, the maximum load is equal to the average load plus the range load; the minimum load is equal to the average minus the range load. Normal stresses s_{av} and s_r are found by dividing loads P_{av} and P_r by the cross-sectional area A. When the average stress is high, the material will safely carry only a small additional range component. However, if the average stress is small, a larger range component can be permitted.

To take care of the unlimited number of combinations of range and average stress, the line of failure of the material must be used. Specimens are tested with fluctuating loads that are low enough to permit continuous operation but high enough so that any increase in either the average or range load will eventually cause failure. The s_{av} stress for the test is plotted as the abscissa, and the s_r stress as the ordinate. Thus, a typical point F is located, as in Fig. 2-26(c). After other combinations of s_{av} and s_r have been determined, the points are plotted to form a curve of failure. Point A, where the average stress is zero, represents the endurance limit for completely reversed stress as given by the s_e value of Fig. 2-16. Point B, where the range stress is zero, represents the static ultimate stress for the material.

Since experimental data for the line of failure are usually not at hand, it is customary to make the conservative approximation that it is a straight line.[5] To be still further on the safe side, the line is drawn from the endurance limit value at A to point C, representing the yield point, rather than to point B for the ultimate stress. The line representing actual working stresses in the material is drawn after both s_e and s_{yp} have been divided by the factor of safety FS as shown in Fig. 2-26(c).

If a stress concentration exists, as illustrated by Fig. 2-26(d), at the cross section for which the stresses are computed, it is commonly neglected so far as the average stress s_{av} is concerned. However, since stress concentration must be taken into account for alternating stresses, the range stress s_r must be multiplied by the stress concentration factor K before plotting. If a point determined by s_{av} and Ks_r as coordinates falls on or below the working-stress line, the part is assumed to be safe for continuous operation.

The situation can be handled conveniently by equations as follows. All points along line DE can be assumed to be equally safe. This includes point E. Stress OE, or $s = s_{yp}/FS$, can then be considered as the static stress equivalent to the fluctuating stress $s_{av} \pm Ks_r$. By similar triangles, it is easy to show that GE is equal to $s_{yp}Ks_r/s_e$. Then

$$\frac{s_{yp}}{FS} = s = s_{av} + \frac{Ks_{yp}}{s_e}s_r \tag{11}$$

This is sometimes called Soderberg's equation. When stress s is obtained, it can be used in Eq. (2) to determine the factor of safety.

Example 7. (a) Find the area required for the safe continuous operation of a uniform bar in tension if $P_{max} = 50,000$ lb and $P_{min} = 20,000$ lb. Material tests

---

[5]See Soderberg, C. R., "Factors of Safety and Working Stress." *Trans. ASME*, **52**(1), p. APM-13, 1930; also **55**, p. APM-131, 1933; and **57**, p. A-106, 1935.

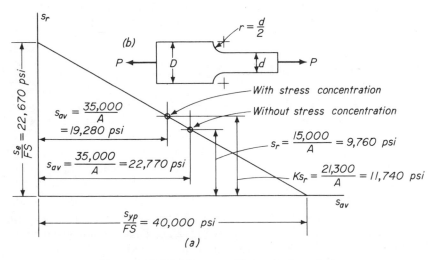

Figure 2-27 Working stress diagram for Example 7.

$s_{ult} = 90,000$ psi and $s_{yp} = 60,000$ psi. Take the factor of safety FS equal to 1.5 based on the yield point. The bar has a machined surface.
(b) Repeat (a) using the bar of two widths shown in Fig. 2-27(b) for $r = d/2$.

Solution. (a) By Fig. 2-16, $s_e = 34,000$ psi. Divide s_{yp} and s_e each by FS 1.5 and plot as shown in Fig. 2-27. $P_{av} = 35,000$ lb and $P_r = 15,000$ lb. $s_{av} = 35,000/A$ and $s_r = 15,000/A$, where A is the required area.

By Eq. (11): $s = \dfrac{60,000}{1.5} = \dfrac{35,000}{A} + \dfrac{60,000}{34,000} \times \dfrac{15,000}{A}$

From which: $A = 1.537$ in.²
(b) Sketch (b) agrees with Fig. 2-10(c). Then for $r/d = 0.5$, $K = 1.42$:

By Eq. (11): $s = \dfrac{60,000}{1.5} = \dfrac{35,000}{A} + \dfrac{60,000}{34,000} \times \dfrac{1.42 \times 15,000}{A}$

From which: $A = 1.815$ in.²
The stress concentration caused by widening the left part has increased the possibility of failure so that the right portion must be increased also.

Although Eq. (11) refers to normal stress, the development could have been made equally well for shear stress. The equation for static shear stress s_s equivalent to the variable shear loading $s_{sav} \pm s_{sr}$ is

$$\frac{s_{syp}}{FS} = s_s = s_{sav} + \frac{K s_{yp}}{s_e} s_{sr} \tag{12}$$

Stress s_s can be used in Eq. (3) for determination of the factor of safety. In the equation above, ratio s_{yp}/s_e is taken as approximately equal to s_{syp}/s_{se}, data for which are usually not available.

The case of fluctuating loading for combined normal and shear stress is discussed in Section 4 of the following chapter on shafting.

14. Cumulative Damage in Fatigue. Miner's Equation

By definition, the endurance limit stress s_e is the maximum completely reversing stress that the material can sustain for an unlimited number of cycles without fatigue failure. If the completely reversed stress is higher than s_e, then failure can be expected to take place after some finite number of cycles. The higher the stress the fewer will be the cycles before failure can be expected.

The relationship between stress s and the number of cycles N usually plots as a straight line on log-log paper. A typical set of such curves, for different types of surfaces, is shown[6] in Fig. 2-28. Thus, for any given stress, the number of expected cycles to failure can be found.

When more than one stress level above the endurance limit is present

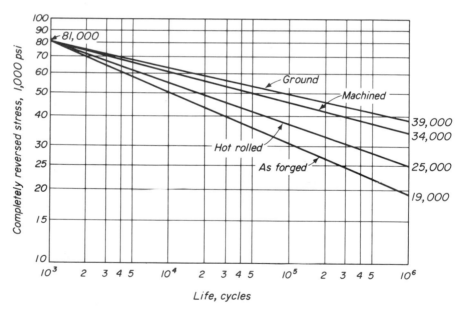

Figure 2-28 Allowable stresses for limited life for steel parts with Brinell hardness in the range 187 to 207.

[6] For additional curves covering the entire hardness range, see Noll, G. C., and M. A. Erickson, "Allowable Stresses for Steel Members of Finite Life," *Proc. Soc. Exp. Stress Anal.*, **5**, No. 2, 1948, p. 132. See also Lipson, C., G. C. Noll, and L. S. Clock, *Stress and Strength of Manufactured Parts*, New York: McGraw-Hill Book Company, 1950, p. 216, and Lipson, C., and R. C. Juvinall, *Handbook of Stress and Strength*, New York: The Macmillan Company, 1963, p. 285.

within a work cycle, it becomes necessary to combine the effects of the various stresses to obtain an overall estimate of the expected life of the parts.

Suppose at stress s_1 the fatigue life is N_1 and that there will be N_1' cycles at this stress. Each cycle of stress consumes $1/N_1$ of the life of the part, and the N_1' cycles will consume N_1'/N_1 of the life.

Similar reasoning will apply to stress s_2, which has a life of N_2 cycles. If N_2' cycles are actually used, the proportion of life consumed at this stress is N_2'/N_2. The process can be continued for all the different stress levels in the work cycle.

It is obvious that the sum of the different proportions must be equal to unity. Thus

$$\frac{N_1'}{N_1} + \frac{N_2'}{N_2} + \frac{N_3'}{N_3} + \cdots = 1 \tag{13}$$

Usually the values for $N_1', N_2', \ldots$, are unknown. However the proportion α_1 of the total life at stress s_1 will be known, or can be estimated. Then $\alpha_1 N = N_1'$. Similarly, $\alpha_2 N = N_2'$, and so on. Substitution into Eq. (13) gives

$$\frac{\alpha_1}{N_1} + \frac{\alpha_2}{N_2} + \cdots = \frac{1}{N} \tag{14}$$

This is known as Miner's equation.[7] It should be noted that

$$\alpha_1 N + \alpha_2 N + \cdots = N, \qquad \text{so that} \qquad \alpha_1 + \alpha_2 + \cdots = 1$$

Example 8. A part with a machined surface is subjected to a normal bending stress of 60,000 psi for 10 per cent of the time, a reversed bending stress of 50,000 psi for 40 per cent of the time, and a reversed bending stress of 40,000 psi for 50 per cent of the time. The Brinell Hardness Number (BHN) of the material is 190.

Find the expected life of the part in cycles.

Solution. By Fig. 2-28:   for $s_1 = 60,000$ psi, $N_1 = 10,000$ cycles

for $s_2 = 50,000$ psi, $N_2 = 45,000$ cycles

for $s_3 = 40,000$ psi, $N_3 = 260,000$ cycles

In Eq. (14): $\dfrac{0.10}{10,000} + \dfrac{0.40}{45,000} + \dfrac{0.50}{260,000} = \dfrac{1}{N}$

From which: $N = 48,000$ cycles

The Miner equation assumes that the damage to the material is directly proportional to the number of cycles at a given stress. This has not been completely verified in practice. In addition, the order in which the stresses are applied is important. Should the higher stresses be applied first, the material may be damaged early in its life and the total number of stress cycles it can sustain will be reduced. On the other hand, if the lower stresses are

[7]See Miner, M. A., "Cumulative Damage in Fatigue," *Trans. ASME,* **67**, 1945 p. A-159, and p. 278 of Reference 9, end of chapter. For cumulative damage design when the loading consists of an average stress plus and minus a range component, see the author's *Mechanical Design Analysis*, Englewood Cliffs, N.J.: Prentice-Hall, Inc., 1964, p. 33.

first applied, the material may be actually strengthened by the strain-hardening effects. Sometimes specifications are used in which the right side of Eq. (13) is taken at somewhat less than unity, say, about 0.8.

15. The Modified Goodman Diagram

Other types of diagrams have been devised for determining the values of the working stresses for parts subjected to fluctuating loads. The modified

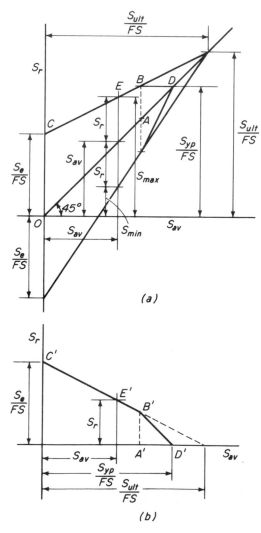

Figure 2-29 Modified Goodman diagram for fluctuating stress.

Goodman diagram is one of these. When the line for average stress is inclined at 45° as in Fig. 2-29(a), values for s_{max} and s_{min} can be scaled directly from the figure. Instead of drawing the complete figure, a diagram consisting only of line *CBD* can be drawn, as shown in Fig. 2-29(b). Such a diagram will give the same values for the average and range stress for a point such as *E* providing *A'B'* is made equal to *AB*. This diagram permits somewhat higher stress values than Fig. 2-26 because line *C'B'* is directed toward the ultimate stress rather than toward the yield point stress.

16. Brittle Materials with Steady Stress

Failure in brittle materials takes place by fracture. The ultimate strength is used as the basis for determining the working stress. It is necessary to have separate equations for the factor of safety in tension and for the factor of safety in compression. Brittle materials are unable to yield locally at points of high stress caused by changes of form, and stress concentration factors are customarily applied even when the loading is steady.

(a) *Simple Tension or Compression.* The following equations can be written.

For tension, s_t:
$$FS = \frac{S_{ut}}{Ks_t} \qquad (15)$$

where s_{ut} represents the ultimate stress in tension.

For compression, s_c:
$$FS = \frac{S_{uc}}{Ks_c} \qquad (16)$$

where s_{uc} represents the ultimate stress in compression.

(b) *Normal Stress in Two Directions.* Two cases must be considered.

(1) Both principal stresses of same sign. In accord with Figs. 2-22 and 2-23, failure is assumed to be due only to the principal stress of larger magnitude without regard to the stress at right angles thereto. Equations (15) and (16) apply.

(2) Principal stresses have opposite signs as shown in Fig. 2-30. The problem is as yet poorly understood. The best-known rational method is due to Mohr and is based on maximum shear stress theory, which yields the following equation.[8]

$$\frac{s_1}{S_{ut}} + \frac{s_2}{S_{uc}} = \frac{1}{FS} \qquad (17)$$

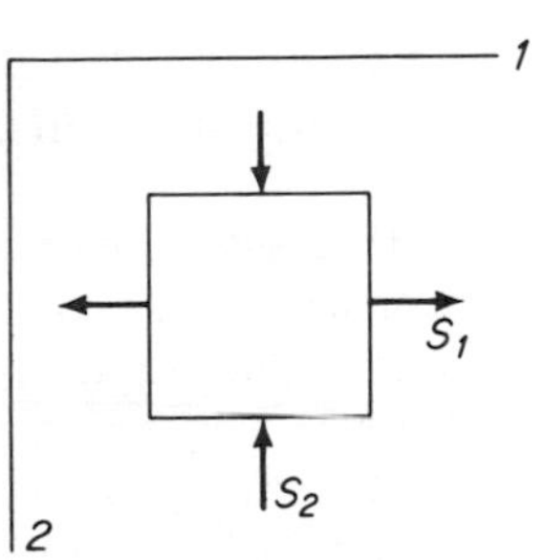

Figure 2-30 Brittle material with tension and compression.

[8]For derivation, see earlier editions of this book. See also p. 438 of Reference 3, end of chapter.

In using this equation, s_{uc} must be substituted as a negative number. Stress concentration factors can be applied as shown in the following example.

Example 9. Assume that the computed stresses at a point in a cast-iron body are as follows: $s_x = 2,000$ psi, $s_y = -6,000$ psi, and $s_{xy} = 3,000$ psi. The stress concentration factor is equal to 2 for all stresses. Material tests $s_{ut} = 20,000$ psi and $s_{uc} = -80,000$ psi. Find the value of the factor of safety.

Solution. In Eq. (41), Chapter 1: $Ks_1 = 2(-2,000 + \sqrt{4,000^2 + 3,000^2})$

$$= 6,000 \text{ psi}$$

In Eq. (42), Chapter 1: $Ks_2 = 2(-2,000 - \sqrt{4,000^2 + 3,000^2})$

$$= -14,000 \text{ psi}$$

In Eq. (17): $$\frac{6,000}{20,000} + \frac{-14,000}{-80,000} = \frac{1}{FS}$$

$$FS = \frac{80}{38} = 2.1$$

The mechanism of failure for brittle materials is very complex. Unfortunately, Eq. (17) does not fit test results very well, and it should therefore be considered only as a rough approximation.

(c) *Pure Shear.* Equation (17) is applicable to pure shear loading by taking $s_1 = -s_2 = s_s$ as indicated by Fig. 2-20.

17. Brittle Materials with Fluctuating Loads

Although successful applications can be cited, brittle materials are usually considered unsatisfactory where the load is fluctuating. Large values for the factor of safety must be used.

18. Sensitivity to Stress Concentration

The actual reduction in fatigue strength, as indicated by the foregoing theoretical stress concentration factors, is approached only by large parts made of fine-grained heat-treated steel. The effect of stress concentration in coarse-grained annealed steels may be considerably less. Small specimens are affected less by stress concentration than larger parts made of the same material. The size effect in steel is attributed mainly to the grain size of the material. When the crystal size is taken into account, it is seen that there is not complete geometric similarity between large and small specimens of the same material. Although a heat-treated part of expensive alloy steel may have a higher endurance limit than one made of a softer nonheat-treated material, the advantage may be largely lost in the presence of a stress concentration.

A wide variation exists in the *notch sensitivity* of different materials. For some quenched and tempered steels the effect of a sharp notch may be so great that a high-strength material may be no better in fatigue than one of lower strength. Materials that work harden rapidly, such as the 18-8 stainless steels, may show great resistance to loss of fatigue strength due to notches. Notches have but little effect on the fatigue strength of cast iron, but may have a large effect as far as impact loads are concerned. However, the impact strength of some steels is but little affected by notches.

Methods are available for making a quantitative estimate of the sensitivity of a steel to stress concentration,[9] but the methods are beyond the scope of this book. However, when the full theoretical stress concentration factor is applied to the fluctuating component, the result will usually be on the safe side.

19. Factor of Safety

It is sometimes difficult to evaluate accurately the various factors that are involved in an engineering design problem. It is particularly difficult, in some cases, to determine the magnitude of the various forces to which a machine part is subjected. Sometimes the shape of the part is such that no design equations are available for accurate computation of the stresses. Variations and nonuniformity in the strength of the material must be kept in mind, as well as the consequences that might result from failure of the parts.

Engineers employ a so-called "factor of safety" to insure against uncertain or unknown conditions as mentioned above. The working stress is determined from the yield or ultimate strength of the material by means of a factor of safety, as demonstrated by the equations of this chapter. For columns and other elements where load and stress are not proportional, the factor of safety should be applied to the load on the member rather than to the stress.

Static loads can sometimes be determined accurately, but the values of fluctuating loads are more uncertain. The effects of impact loads and residual stress are particularly difficult to evaluate. Consideration must also be given to the long-term effects of corrosion and high temperatures. Sometimes the shape of the parts, or the method of support, can be modified in such a way that the design equations may fit the conditions more accurately. Perhaps a more uniform and reliable construction material can be used.

The strength of materials is usually obtained by laboratory tests. It must be kept in mind, however, that conventional laboratory tests rarely reproduce the conditions that the material must meet in service. Surface conditions in

[9]See Peterson, R. E., "Relation Between Life Testing and Conventional Tests of Materials," *ASTM Bull.*, **133**, Mar. 9, 1945; also p. 8 of Reference 7, Chapter 4 of Reference 5, and Chapter 13 of Reference 9, end of chapter.

particular may be different for the test specimen than for an actual part. A much greater amount of knowledge of engineering materials than is now available must be gained before the designer will be able to apply test data to conditions that differ appreciably from the conditions of the test. The best method, whenever it is possible, is to make final adjustment in the proportioning of the components by tests on the completed product.

In general it is not economical to use safety factors large enough to eliminate all possibility of failure resulting from the worst possible combination of circumstances. The designer attempts to reduce the probability of failure to a suitable level, which necessarily depends on each particular application. A failure that involves only a little inconvenience or loss of time might be allowed more frequently than one involving large financial loss or human life. Provision should be made for easy and rapid replacement of failed parts. If the product operates under conditions of frequent service inspection, smaller values for the factor of safety can sometimes be used. A more thorough and detailed analysis of the problem may show that smaller factors of safety can be used, and may justify the additional engineering expense involved.

When building one general product, the engineer usually does not think in terms of the factor of safety. He has learned from experience that certain materials under certain conditions, with working stresses of given values, will lead to satisfactory results. Although the determination of suitable values for the factor of safety is a matter of great importance, the subject has been much neglected and is in an unsatisfactory state. Experience, which can only be accumulated as a result of a long period of trial and error, is the ultimate basis for the prediction of failure in engineering designs. Many failures are due to circumstances that the designer failed to consider.

REFERENCES

1. Battelle Memorial Institute, *Prevention of Fatigue of Metals*, New York: John Wiley & Sons, Inc., 1941.

2. Findley, W. N., "Fatigue of Metals under Combinations of Stresses," *Trans. ASME*, **79**, 1957, p. 1337.

3. Hetényi, M., Ed., *Handbook of Experimental Stress Analysis*, New York: John Wiley & Sons, Inc., 1950.

4. Horger, O. J., Ed., *Metals Engineering Design, ASME Handbook*, New York: McGraw-Hill Book Company, 1953.

5. Murray, W. M., Ed., *Fatigue and Fracture of Metals*, New York: John Wiley & Sons, Inc., 1952.

6. Marin, Joseph, *Mechanical Properties of Engineering Materials*, Englewood Cliffs, N.J.: Prentice-Hall, Inc., 1962.

7. Peterson, R. E., *Stress Concentration Design Factors*, New York: John Wiley & Sons, Inc., 1953.

8. *Proc. Intern. Conf. Fatigue Metals*, London: the Institution of Mechanical Engineers, 1956.

9. Sines, George, and J. L. Waisman, Eds., *Metal Fatigue*, New York: McGraw-Hill Book Company, 1959.

10. Starkey, W. L., and S. M., Marco, "Effects of Stress-Time Cycles on the Fatigue Properties of Metals," *Trans, ASME*, **79**, 1957, p. 1329.

PROBLEMS

1. Find the value of the stress at each hole in Fig. 2–31.

Ans.   *A, Ks* = 29,710 psi; *B, Ks* = 31,330 psi; *C, Ks* = 35,680 psi.

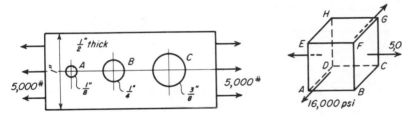

Figure 2-31   Problem 1.   **Figure 2-32**   Problem 2.

2. The loading on an element of 1045 steel in the hot rolled condition is shown in Fig. 2-32. Loads are steady and no stress concentrations are present. Assume maximum shear theory to be valid.

(a) Find the value of the *FS* for the *CDEF* plane.

(b) Find the value of the *FS* for the *ACGE* plane.

(c) Find the value of the *FS* for the *AFGD* plane.

Ans.   (a) *FS* = 3.69; (b) *FS* = 5.36; (c) *FS* = 11.8.

3. A plate of 1045 steel in the hot-rolled condition is subjected to the following stresses: s_x = 3,300 psi; s_y = −29,000 psi; s_{xy} = 0.

(a) Find the value of the *FS* by the maximum shear theory.

(b) Find the value of the *FS* by the Mises–Hencky theory.

(c) Find *FS* if the plate is made of Class 25 cast iron.

Ans.   (a) *FS* = 1.83; (b) *FS* = 1.92; (c) *FS* = 2.37.

4. A machine part has an *FS* of 3 by the Mises–Hencky theory. The material tests s_{yp} = 63,000 psi. If s_x is equal to 24,000 psi, find the value of s_y. s_{xy} = 0.

Ans. s_y = 15,000 or 9,000 psi.

5. A shaft is loaded by a torque of 40,000 in. lb. The material has a yield point of 50,000 psi. *FS* is equal to 2.

(a) Find the required diameter by the maximum shear theory.

(b) Find the required diameter by the Mises–Hencky theory.

Ans.   (a) *d* = 2.535 in.; (b) *d* = 2.417 in.

6. A 15 × 15-in. plate of material with $s_{yp} = 60,000$ psi has normal stresses only acting on all edges. Stress s_x is tension and s_y is compression. The length in the y-direction is reduced by 0.008 in. FS is equal to 2 by the maximum shear theory. Find the values of the stresses. *Ans.* $s_x = 20,000$ psi; $s_y = -10,000$ psi.

7. A 12 × 12-in. steel plate has normal stresses only acting on all edges. Stress s_x equals 12,000 psi tension. s_y is compression. Increase in length in the x-direction is 0.006 in. If FS is equal to 2.5 by the maximum shear theory, what is the yield pointvalue for the material? *Ans.* $s_{yp} = 55,000$ psi.

8. A 15 × 15-in. plate of 1045 steel in the as-rolled condition has normal stresses only acting on all edges. Stress s_x equals 6,000 psi tension. s_y is compression. For an FS equal to 3 by the maximum shear theory, find the change in length in the x-direction. *Ans.* $\delta_x = 0.0051$ in.

9. The link shown in Fig. 2-33 is subjected to a completely reversing load of 20,000 lb. Find the maximum value of stress at each hole.
Ans. Top, $Ks_r = 17,880$ psi; bottom, $Ks_r = 17,540$ psi.

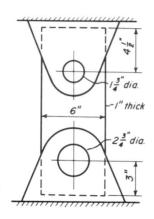

Figure 2-33 Problem 9.

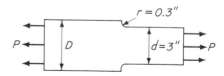

Figure 2-34 Problem 10.

10. The plate shown in Fig. 2-34 is $\frac{1}{2}$ in. thick. Load P varies from 20,000 to 10,000 lb. Material tests $s_{yp} = 42,000$ psi; $s_e = 24,000$ psi. Factor of safety based on yield point is 2. Find the maximum permissible value of width D.
Ans. $D = 3.81$ in.

11. The plate of Fig. 2-35 is 1 in. thick. The load varies from 50,000 to 30,000 lb. Factor of safety = 2; $s_{yp} = 40,300$ psi; $s_e = 28,000$ psi. (a) Find the value of d. (b) Find d if the minimum load is 20,000 lb, other data remaining unchanged.
Ans. $d = 3.13$ and 3.45 in.

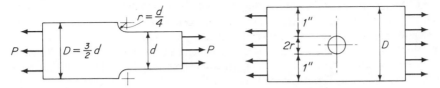

Figure 2-35 Problem 11.   **Figure 2-36** Problem 12.

12. Find the diameter of the hole and the total width of the plate of Fig. 2-36 if the part is to be safe for continuous operation. The load varies from 36,000 to 20,000 lb. Use stress values of $s_{yp}/FS = 30,000$ psi and $s_e/FS = 18,000$ psi. The plate is 1 in. thick.   *Ans.* $D = 2.55$ in.

13. A part is designed as shown in Fig. 2-37. Check the design by plotting the line of failure and points representing the stress values for the material at the hole and fillet. Is the part safe for continuous operation? The load varies from 12,000 to 2,000 lb. $s_{yp} = 41,000$ psi; $s_e = 28,500$ psi. Plate is $\frac{7}{16}$ in. thick.

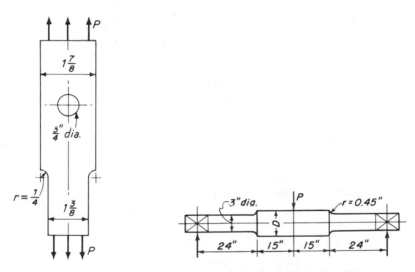

Figure 2-37 Problem 13.   **Figure 2-38** Problem 14.

14. Figure 2-38 shows a nonrotating shaft with load P varying from 1,000 to 3,000 lb. The material tests $s_{yp} = 42,000$ psi, and $s_e = 24,000$ psi. Factor of safety is equal to 2. Find the permissible value for D if stress conditions at the fillets are to be satisfactory for continuous operation.   *Ans.* $D = 4.50$ in.

15. Material for the eyebolt of Fig. 2-39 tests $s_{yp} = 39{,}000$ psi and $s_e = 26{,}000$ psi. The factor of safety equals 2. Threads are American National. Let stress concentration factor for the threads be equal to 2.5. Upon assembly, the spring is given an initial stretch. During operation, the lower end of the spring moves 1 in. each way from its initial position. Find the permissible amount of initial stretch that may be given to the spring if the eyebolt is to be safe for continuous operation.

Ans. 5.45 in.

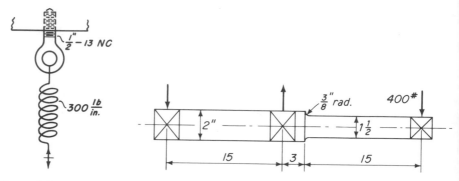

Figure 2-39 Problem 15.   Figure 2-40 Problem 16.

16. The shaft of Fig. 2-40 rotates but carries no torque. For the material, $s_e = 35{,}000$ psi. Determine the value of the bending stress at the fillet. (a) Is the shaft safe for continuous operation? What is the value of the factor of safety? (b) Suppose in the turning operation the radius of the fillet is made $\frac{1}{8}$ in. and that the inspectors do not discover the mistake. Is the part now safe for continuous operation?

Ans. (a) $FS = 1.45$.

17. The shaft in Fig. 2-41 rotates. Find the length l if the bending stress at the fillet is equal to the stress at the center.

Ans. $l = 91.4$ in.

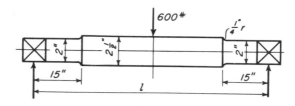

Figure 2-41 Problem 17.

18. The part in Fig. 2-42 is made of 1045 steel, quenched and drawn at 1,000°F. Bending moment varies from 10,000 to 50,000 in. lb machined surfaces. Find the factor of safety.

Ans. $FS = 1.53$.

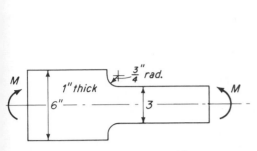

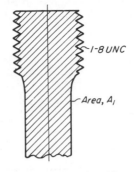

Figure 2-42 Problem 18.

Figure 2-43 Problem 19.

19. The tensile load on the turned bolt of Fig. 2-43 varies from 12,000 to 20,000 lb. The stress concentration factor for the threads is equal to 3 and for the fillet, 1.2. Material is 8742 steel, oil quenched and drawn at 1,200°F. If threads are safe for continuous operation, find the value of area A_1 necessary to make the fillet safe for continuous operation.   *Ans.* $A_1 = 0.374$ in.²

20. The beam shown in Fig. 2-44 is made of a brittle material. Find the value of the maximum bending stress at cross section A.   *Ans.* $s = 5,625$ psi.

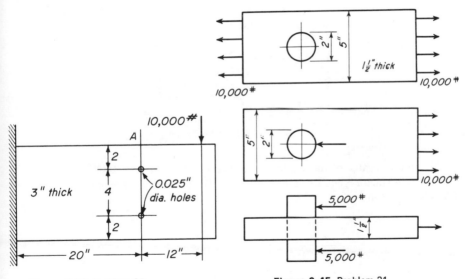

Figure 2-44 Problem 20.

Figure 2-45 Problem 21.

21. The plates in Fig. 2-45 are made of a brittle material and are identical. The loading for (a) consists of a uniform tension applied at each end. In (b) the hole contains a pin that carries the load and a uniform tension is applied to the other end of the plate. Find the maximum tensile stress for each type of loading. Note that loading through a pin as for (b) gives a higher stress. $H > D$ in Fig. 2-11.   *Ans.*   (a) $Ks = 4,890$ psi; (b) $Ks = 6,110$ psi.

22. The plates shown in Fig. 2-46 are made of a brittle material. Assume each rivet transfers one-half the load. Find the value of the maximum tensile stress in the the thicker plate.   *Ans.* $s = 8,740$ psi.

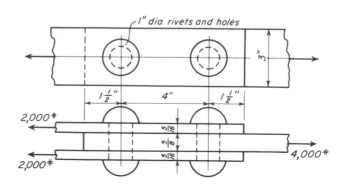

Figure 2-46 Problem 22.

23. Find the value of the maximum stress on the fillet in Fig. 2-47 if the stress concentration factor is equal to 1.6. What is the *FS* if the part is made of Class 25 cast iron?   *Ans.* $FS = 3.12$.

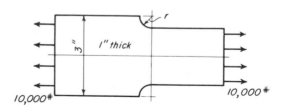

Figure 2-47 Problem 23.

24. Find the value of the maximum stress on the fillet in Fig. 2-48 if the stress concentration factor is equal to 1.75. $D/d = 1.5$. What is the *FS* if the part is made of Class 30 cast iron?   *Ans.* $FS = 3.57$.

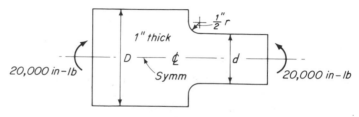

Figure 2-48 Problem 24.

25. (a) Find the value of the *FS* if the part in Fig. 2-49 is made of 1045 steel in the as-rolled condition.

(b) Find the value of the *FS* if the same part in Fig. 2-49 is made of Class 20 cast iron.

Figure 2-50 Problem 26.

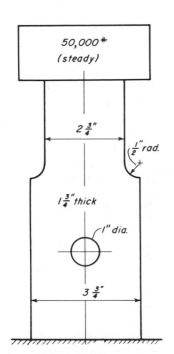

Figure 2-49 Problem 25.

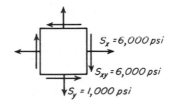

Figure 2-51 Problem 27.

26. The element in Fig. 2-50 is located at the inner edge of a hub that has been press fitted on a shaft. Stress s_r or p is equal to $-6{,}000$ psi, and stress s_t is equal to 10,000 psi.

(a) Find the *FS* by maximum shear theory if the hub is made of steel with a yield point value of 60,000 psi.

(b) Find the *FS* if the hub is made of Class 25 cast iron.

Ans.   (a) $FS = 3.75$; (b) $FS = 2.17$.

27. The state of stress for a material is shown in Fig. 2-51.

(a) Find the *FS* by the maximum shear theory if the material is 1045 steel in the hot-rolled condition.

(b) Find the *FS* by the Mises–Hencky theory.

(c) Find the *FS* if the material is Class 30 cast iron.

Ans.   (a) $FS = 4.54$; (b) $FS = 5.00$; (c) $FS = 2.77$.

28. Same as Problem 27 except for the element of Fig. 2-52.

Ans.   (a) $FS = 5.90$; (b) $FS = 6.64$; (c) $FS = 3.84$.

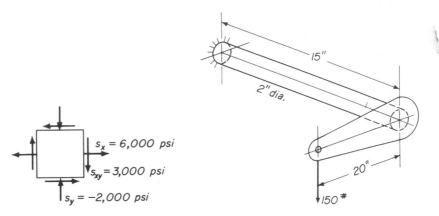

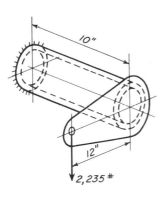

Figure 2-52 Problem 28.

Figure 2-53 Problem 29.

29. The shaft in Fig. 2-53 is rigidly attached to the wall and is made of Class 25 cast iron. Stress concentration at the wall for bending and torsion is equal to 2. Find the value of the *FS*.   *Ans.* $FS = 3.08$.

30. The hollow cylinder in Fig. 2-54 is fixed at the wall and has an OD of 6 in. and an ID of 4.5 in. Stress concentration factor at wall is equal to 3.

(a) Find value of *FS* if material is Class 25 cast iron.

(b) Find value of *FS* if material is 0.33 carbon cast steel normalized.

Ans.   (a) $FS = 4$; (b) $FS = 20.8$.

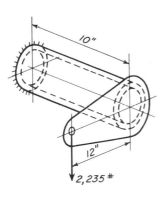

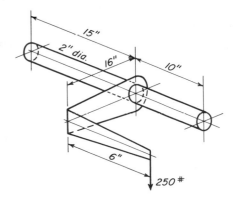

Figure 2-54 Problem 30.

Figure 2-55 Problem 31.

31. The shaft in Fig. 2-55 is made of Class 20 cast iron. The ends are simply supported, but are keyed against rotation. The stress concentration factor at bracket is equal to 2. Find the *FS* for the shaft on either side of the bracket.

Ans.   Left, *FS* = 6.07; right, *FS* = 2.82.

32. Find the diameter of the hole in Fig. 2-56 if the stress concentration factor there is to be the same as at the fillet.

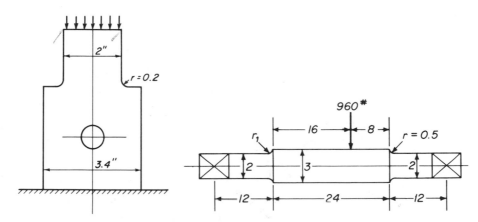

Figure 2-56 Problem 32.   **Figure 2-57** Problem 33.

33. Find the value of the fillet radius r_1 on the left in Fig. 2-57 if the stress is to be the same as on the fillet on the right.

34. In Fig. 2-53, let the 15-in. dimension become 12 in., and the 20-in. dimension become 10 in. Shaft material tests s_{yp} = 60,000 psi. Find the permissible load at end of arm if *FS* by the Mises–Hencky theory is to be 2.5.

35. A steel plate 20 × 10 in. has normal stresses only on the four edges. After the loads are applied, the dimensions become 20.0066 and 9.9981 in. Material is 1060 hot-rolled steel. Find the *FS* by the maximum shear theory.

36. A 2-in.-diam nonrotating shaft of 1045 hot-rolled steel has a steady bending moment of 10,000 in. lb. Find the permissible steady torque that can be superposed on the bending moment if the *FS* is to be 3 by the Mises–Hencky theory.

37. An element of 1045 hot-rolled steel has a steady axial stress of 10,000 psi. Find the permissible shearing stress that can be superposed if the *FS* is to be 2.5 by the Mises–Hencky theory.

38. Work Problem 7 by the Mises–Hencky theory.

39. Work Problem 7, but with 0.006 in. as 0.0048 in. and s_x as 9,600 psi.

40. A 15 × 20 × 0.5-in. plate of 1045 hot-rolled steel carries a uniformly distributed tension of 90,000 lb on the 15-in. edge. Find the compressive stress on the 20-in. edge if the *FS* is 2 by the maximum shear theory. Find the dimensions of the plate after the loads are acting.

41. In Fig. 2-55, the 16-in. dimension becomes 8 in. with the load at the end of the bracket equal to 2,500 lb. The shaft material tests 54,000 psi yield and 90,000 psi ultimate. Find the *FS* by the maximum shear theory for an element on the top surface 4 in. to the left of the bracket. Do the same by the Mises–Hencky theory.

42. The part shown in Fig. 2-58 is made from 8742 steel quenched and tempered at 1,200°F. The load varies continuously from 60,000 to 100,000 lb machined surfaces. Find the *FS*.

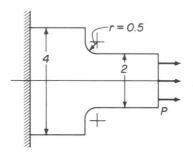

Figure 2-58 Problem 42.

43. A 2-in.-diam shaft carries a torque that fluctuates 25 per cent each way from the average value of 20,000 in. lb. The stress concentration factor is 2.5. The *FS* based on the yield point is 2. If the endurance limit in shear is equal to 0.6 of the yield point in shear, determine the minimum value of the yield point in shear that the material must have if the shaft is to be safe for continuous operation.

44. A body loaded by normal stresses s_1 and s_2 has an *FS* by the Mises–Hencky theory that is 10 per cent greater than the *FS* when computed by the maximum shear theory. If s_1 is 10,000 psi, find the value of s_2.

45. In Fig. 2-59, the leaf spring is straight and unstressed when the cam and shaft are removed. The material is cold-drawn 3140 steel. The stress concentration is zero. Spring surfaces are ground. If the cam rotates continuously, find *FS* for the spring.

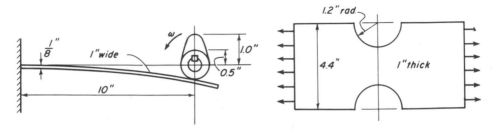

Figure 2-59 Problem 45.   **Figure 2-60** Problem 47.

46. Work Problem 27(c) with $s_x = 17,000$ psi.

47. For an *FS* of 4, find the total force *P* in Fig. 2-60 if the material is Class 30 cast iron.

48. Work Problem 29, but find the *FS* for an element at the wall at midheight on the surface of the shaft on the side toward the observer.

49. A 10 × 20-in. plate of Class 50 cast iron has normal stresses only acting on the edges.

(a) Find the *FS* if the dimensions become 9.9981 × 20.0066 in. after the loads are applied.

(b) Do the same, but with dimensions of 10.0011 × 20.0048 in.

50. Find the relationship between s_{syp} and s_{yp} by the Mises–Hencky theory.

51. When the motor is turning in Fig. 2-61, it is observed that the end of the beam has a total up and down vibration of 0.25 in. The beam is of 1035 cold-rolled steel, but assume that the surface is equivalent to a hot-rolled surface. Find the factor of safety at the wall where the stress concentration factor is 1.5.

Ans. *FS* = 4.16.

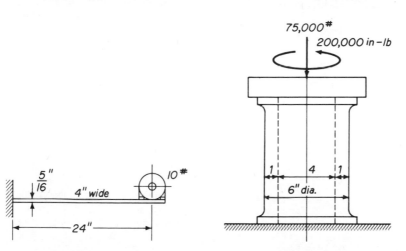

Figure 2-61 Problem 51.   **Figure 2-62** Problem 52.

52. The hollow cylinder in Fig. 2-62 is of Class 50 cast iron. If the stress concentration factor is 1.5, find the value of the factor of safety.   *Ans.* *FS* = 5.00.

53. A part with a machined surface is subjected to a reversed bending stress of 70,000 psi for 5 per cent of the time, a reversed bending stress of 55,000 psi for 15 per cent of the time, and a reversed bending stress of 40,000 psi for 80 per cent of the time. BHN for the material is 190.

Find the expected life of the part in cycles.   *Ans.* 36,700 cycles.

54. A part with an as-forged surface was found to have a fatigue life of 200,000 cycles. For 30 per cent of the time the reversed bending stress had a value of 30,000 psi, and for 60 per cent of the time the stress had a value of 20,000 psi. BHN is 190.

Find the probable value of the reversed bending stress for the remainder of the time.   *Ans.* $s = \pm 34,000$ psi.

55. Consider the freight car axle of Example 6, Fig. 2-25. The center of gravity of the given load is located 72 in. above the top of the rails. The wheels have a diameter of 33 in. Suppose that the car is going around a sharp curve to the right and that the horizontal centrifugal force is sufficient to reduce the reaction at the right rail to zero. Find the value of the bending stress at the center of the axle.

Ans. $s = 16,350$ psi.

56. The following procedure can effect a large reduction in the time required for testing.

(a) A device, tested under heavy loading, has a life N_2 of 7.5 hr.

(b) The device is now tested for a period under the heavy load, and for the remainder of its life under the normal operating load. The life N is found to be 12 hr, of which 4.8 hr was at the heavy load, and 7.2 hr was at the normal operating load. If Miner's equation is assumed to be applicable, find the life at the normal operating load.

Ans. $N_1 = 120$ hr.

57. Same as above except $N_2 = 50$ hr, $N = 99$ hr, and $\alpha_1 = \alpha_2 = 0.50$. If Miner's equation is applicable, find the life at the normal operating load.

Ans. $N_1 = 4,950$ hr.

3   Shafting

Shafts are used in all kinds of machinery and mechanical equipment. Although the elementary theory for a circular shaft with static torsional loads is useful, most shafts are subjected to fluctuating loads of combined bending and torsion with various degrees of stress concentration. For such shafts the problem is fundamentally one of fatigue loading. In addition to the shaft itself, the design usually must include the calculation of the necessary keys and couplings. The normal operating speed of a shaft should not be close to a critical speed, or large vibrations are likely to develop. Equations are given for finding the deflections of shafts of nonuniform diameters. Machine parts with noncircular cross sections are sometimes loaded in torsion; the designer must therefore be able to determine the stresses and deformations sustained by such bodies.

d, diameter
E, modulus of elasticity, normal stress
fpm, feet per minute
FS, factor of safety
G, modulus of elasticity in shear
hp, horsepower

I, moment of inertia
J, polar moment of inertia
K, stress concentration factor, normal stress
K_t, stress concentration factor, shear stress
l, length

M, bending moment

n, revolutions per minute

r, radius

rpm, revolutions per minute

s_e, endurance limit stress, reversed bending

s_s, shearing stress

s_{yp}, yield point stress, tension

SAE, Society of Automotive Engineers

T, torque

V, velocity, feet per minute

y, deflection

φ, (phi) angular deformation

ω, (omega) angular velocity, radians per second

1. Torsion of Circular Shaft

Figure 3-1 shows a circular shaft of uniform cross section loaded at the ends by the torques T, which twist it about the longitudinal axis. The shaft is assumed to be much longer with respect to the diameter than is indicated by the figure. It can be shown experimentally that cross sections perpendicular to the axis before loading remain plane and perpendicular to the axis after the loads T have been applied. The diameter of the bar is unchanged and radial lines remain straight and radial after twisting.

The only deformation in the bar is the rotation of the cross sections with respect to each other. As shown in Fig. 3-1, the bottom cross section has been rotated with respect to the top through the angle φ.

The sides of an element on the cylindrical surface of radius r_1 are unchanged in length, but the angles in the corners are changed by angle γ from their original 90° values. The element is thus stressed in pure shear. As shown by Fig. 3-1, $r_1\varphi = l\gamma$. Substitution of Hooke's law, $\gamma = s_s/G$, where G is the modulus of elasticity in shear, gives

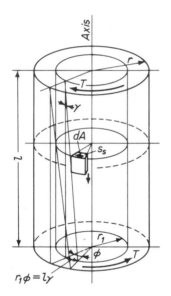

Figure 3-1 Circular shaft twisted by torques at ends.

$$s_s = \frac{\varphi G r_1}{l} \tag{1}$$

Since φ, G, and l are constants in Fig. 3-1, the value of the shearing stress s_s varies directly with the radius r_1.

If the portion of the bar above the element dA in Fig. 3-1 were removed, the torque of the shearing stress s_s, if summed or integrated over the whole cross section, would be equal to the applied torque T. Hence,

$$T = \int_0^r s_s r_1 \, dA \tag{a}$$

The right side is multiplied and divided by r_1; by Eq. (1), ratio s_s/r_1 is a constant and can be removed from the integral. Thus

$$T = \int_0^r \frac{S_s}{r_1} r_1^2\, dA = \frac{S_s}{r_1} \int_0^r r_1^2\, dA = \frac{S_s}{r_1} J \tag{b}$$

In the last form of Eq. (b), the symbol J, called *polar moment of inertia*, has been substituted for the integral $\int r_1^2\, dA$.

The maximum value of the shearing stress occurs at the outer surface, where $r_1 = r$. Hence, from Eq. (b),

$$S_s = \frac{Tr}{J} = \frac{16T}{\pi d^3} \tag{2}$$

The similarity of Eq. (2) to Eq. (11) of Chapter 1 for bending stress, $s = Mc/I$, should be noted. The ratio J/r is called the *section modulus* of the shaft.

For a solid circular cross section,

$$J = \frac{\pi d^4}{32} = \frac{\pi r^4}{2} \tag{3}$$

It should be noted that the value of J for a circle is twice as great as the corresponding value of I. For a hollow shaft with outside diameter d_o and inside diameter d_i, the net value for the polar moment of inertia is equal to the value of J for the outer circle minus the J for the inner circle. Hence, for a hollow shaft,

$$J = \frac{\pi}{32}(d_o^4 - d_i^4) = \frac{\pi}{2}(r_o^4 - r_i^4) \tag{4}$$

Elimination of s_s between Eqs. (b) and (1) gives

$$\varphi = \frac{Tl}{JG} \tag{5}$$

This equation can be easily committed to memory when its analogy to Eq. (4), Chapter 1, $\delta = Pl/AE$, for axial deformation is noticed.

Angle φ is in radian measure. It should be recalled that $1°$ is equal to $\pi/180$ rad, or 1 rad $= 57.296°$.

In order that the foregoing equations may be valid in the neighborhood of the ends, moments T must be applied by means of stresses that vary in intensity with the distance from the axis. Since this condition rarely occurs in practice, Eq. (2) gives correct results only at cross sections somewhat removed from the points where loads T are applied.

Example 1. The shaft in Fig. 3-2 does not rotate; the loads are steady. Assume simple supports.

(a) Make a sketch for the element at A on the bottom surface of the shaft and show the values of the stresses.

(b) Do the same for the element at B at the elevation of the shaft axis.

Solution. (a) *Element at A.*

By Eq. (15), Chapter 1: $s = \dfrac{32M}{\pi d^3} = \dfrac{32 \times 3{,}500}{\pi 8} = 4{,}460$ psi, tension

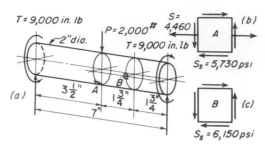

Figure 3-2 Example 1.

By Eq. (2):
$$s_s = \frac{16T}{\pi d^3} = \frac{16 \times 9,000}{\pi 8} = 5,730 \text{ psi}$$

If it is assumed that the shaft is cut at the right edge of the element, the given torque causes the shear stress s_s to have the direction shown in Fig. 3-2(b). The transverse shear stress for element A is equal to zero.

(b) *Element at B.* The torsional shear stress has the same value as in part (a). The transverse shear stress is found by Eq. (29) of Chapter 1.

$$s_s = \frac{4V}{3A} = \frac{4 \times 1,000}{3\pi} = 420 \text{ psi}$$

If it is assumed that the shaft is cut at the right edge of the element, the transverse shear stress acting upon the element is directed upward. The total shear stress has the value shown in Fig. 3-2(c). The bending stress is zero for the element at B.

Example 2. A hollow shaft must carry a torque of 30,000 in. lb at a shearing stress of 8,000 psi. The inside diameter is to be 0.65 of the outside diameter. Find the value of the outside diameter.

Solution. $d_i = 0.65d_o$

In Eq. (4): $J = \frac{\pi}{32}(d_o^4 - 0.65^4 d_o^4) = 0.08065 d_o^4$

In Eq. (2): $J = \frac{30,000 \times 0.5d_o}{8,000} = 0.08065 d_o^4$

$d_o^3 = 23.2486$

$d_o = 2.854 \text{ in.}$

Example 3. Suppose it is specified that the angular deformation in a shaft should not exceed $1°$ in a length of 6 ft. The permissible shearing stress is 12,000 psi. Find the diameter of the shaft. The material is steel.

Solution. $\varphi = 1° = \frac{\pi}{180} \text{rad}$

By Eqs. (2) and (5): $T = \frac{s_s J}{r} = \frac{\varphi J G}{l}$

or
$$r = \frac{s_s l}{\varphi G} = \frac{12{,}000 \times 72 \times 180}{11{,}500{,}000\pi} = 4.305 \text{ in.}$$

$$d = 8.609 \text{ in.}$$

Example 4. The shaft in Fig. 3-3 carries the torque of 10,000 in. lb at the location shown. If the ends of the shaft are fixed against rotation, find the values of the torque reactions T_1 and T_2.

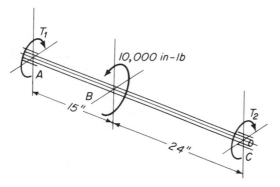

Figure 3-3 Example 4.

Solution. By Eq. (5), Ang. def. for AB: $\varphi_1 = \dfrac{15T_1}{JG}$

By Eq. (5), Ang. def. for BC: $\varphi_2 = \dfrac{24T_2}{JG}$

The angular deformation in AB is equal to the angular deformation in BC. The values for φ_1 and φ_2 should be equated.

$$15T_1 - 24T_2 = 0 \tag{c}$$

By statics: $T_1 + T_2 = 10{,}000$

The equations above should now be solved simultaneously to give

$$T_1 = 6{,}154 \text{ in. lb}; \qquad T_2 = 3{,}846 \text{ in. lb}$$

2. Horsepower

Power is defined as the rate at which work is performed. The unit is the horsepower, which is equal to 33,000 ft lb/min. If a force of F lb acts at a velocity of V ft/min, the work done per minute is FV, and the equation for horsepower is

$$hp = \frac{FV}{33{,}000} \tag{6}$$

In machinery where power is transmitted by shafting, it is necessary to transform Eq. (6) into angular dimensions. If force F is acting at radius r in.

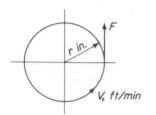

Figure 3-4 Relationship for deriving horsepower equation.

as shown in Fig. 3-4, the angular velocity, ω rad/sec, is equal to

$$\omega = \frac{V}{60} \times \frac{12}{r} = \frac{2\pi n}{60} \tag{7}$$

where n is the speed in revolutions per minute.

The value of V from Eq. (7) should now be substituted in Eq. (6) to give

$$hp = \frac{T\omega}{12 \times 550} \tag{8}$$

In Eq. (8) the product Fr has been replaced by torque T, which has dimensions in inch-pounds.

Velocity V, in feet per minute, in Fig. 3-4 is equal to

$$V = \frac{\pi dn}{12} \tag{9}$$

When this substitution is made in Eq. (6) the following equation results.

$$hp = \frac{Tn}{63,025} \tag{10}$$

Torque T in this equation has dimensions in inch-pounds.

3. Maximum Static Shearing Stress

Many shafts carry combined loads of bending and torque. The bending moment M causes a normal stress in the axial direction of the shaft as shown by s in Fig. 3-5(a), and the torque T produces the shearing stress s_s. The normal stress in the y-direction, or at right angles to the shaft axis, is in general equal to zero. From the Mohr circle for this element, shown in Fig. 3-5(b), the value of the maximum shearing stress for static loading is given by the equation

$$s_{smax} = \frac{0.5 s_{yp}}{FS} = \sqrt{\left(\frac{s}{2}\right)^2 + s_s^2} \tag{11}$$

The equations for the stresses for a solid circular shaft,

$$s = \frac{32M}{\pi d^3} \quad \text{and} \quad s_s = \frac{16T}{\pi d^3} \tag{12}$$

should now be substituted into Eq. (11) to give the following equation for the maximum shearing stress for *static* loads.

$$s_{smax} = \frac{0.5 s_{yp}}{FS} = \frac{16}{\pi d^3}\sqrt{M^2 + T^2} \tag{13}$$

Substitution of $s_{smax} = 0.5 s_{yp}/FS$ implies that the maximum shear

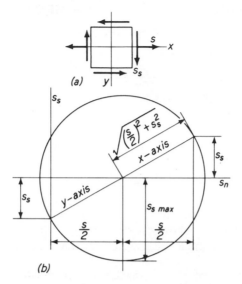

Figure 3-5 Stresses on element of shaft surface.

theory of failure has been assumed to the applicable. This is in accord with Eq. (6) of Chapter 2.

4. ASME Code for Design of Transmission Shafting

Since the loads on most shafts in machinery are not constant, it is necessary to make proper allowance for the harmful effects of the fluctuation. The ASME Code for the Design of Transmission Shafting, B17c—1927, does this by inserting constants C_m and C_t into the equation for stress as follows:

$$s_{smax} = \frac{0.5 s_{yp}}{FS} = \sqrt{\left(\frac{C_m s}{2}\right)^2 + (C_t s_s)^2} \tag{14}$$

or

$$s_{smax} = \frac{0.5 s_{yp}}{FS} = \frac{16}{\pi d^3}\sqrt{(C_m M)^2 + (C_t T)^2} \tag{15}$$

where C_m = numerical combined shock and fatigue factor to be applied in every case to the computed bending moment

and C_t = the corresponding factor to be applied to the computed torque.

The recommended values for the shock and fatigue factors are given in Table 3-1. For rotating shafts, the bending stress s is not constant; it varies continuously from maximum tension to maximum compression as the shaft rotates. For steady loads, the table indicates that suitable compensation can

Table 3-1   CONSTANTS FOR ASME CODE

Nature of Loading	Values for	
	C_m	C_t
Stationary shafts		
Gradually applied load	1.0	1.0
Suddenly applied load	1.5–2.0	1.5–2.0
Rotating shafts		
Gradually applied or steady load	1.5	1.0
Suddenly applied loads, minor shocks only	1.5–2.0	1.0–1.5
Suddenly applied loads, heavy shocks	2.0–3.0	1.5–3.0

be made for the alternating nature of the bending stress by using a value of 1.5 for C_m. If the shaft is hollow, factor $16/\pi d^3$ can be replaced by r/J, where r is the outside radius, and J is computed for the net area.

Example 5. Find the diameter by the ASME Code for a rotating shaft subjected to a maximum steady torque of 16,200 in. lb, and a steady bending moment of 27,000 in. lb. The shaft has a keyway.

Solution. $s_{smax} = 0.75 \times 8,000 = 6,000$ psi. See Section 21, this chapter. From Table 3-1:

$$C_m = 1.5, \quad \text{and} \quad C_t = 1.0$$

In Eq. (15): $d^3 = \dfrac{16}{6,000\pi}\sqrt{(1.5 \times 27,000)^2 + 16,200^2} = 37.026$

$$d = 3.333 \text{ in.}$$

Example 6. A2 in.-diam rotating shaft carries a torque of 12,000 in. lb, which may be applied suddenly, and a bending moment of 8,000 in. lb, which also may be applied suddenly. Material tests $s_{yp} = 70,000$ psi. Find the value of the *FS* by the ASME Code.

Solution. Assume $C_m = 2$ and $C_t = 1.5$.

By Eq. (15): $\dfrac{0.5 \times 70,000}{FS} = \dfrac{16}{\pi 2^3}\sqrt{(2 \times 8,000)^2 + (1.5 \times 12,000)^2}$

$$\dfrac{35,000}{FS} = \dfrac{2}{\pi} \times 24.083$$

or $FS = 2.28$

5. Maximum Shear Theory when Loads Are Fluctuating

It will now be shown that the maximum shear theory of failure can be applied when the normal and shear stresses in a shaft are fluctuating. The loading for an element on the shaft surface is shown in Fig. 3-6. Stresses normal to the shaft axis are zero. It is assumed that the normal and shear stresses reach their maximum and minimum values simultaneously.

Soderberg's Eq. (11), presented in Chapter 2, applied to the fluctuating normal stress of Fig. 3-6 gives the equivalent static normal stress s as

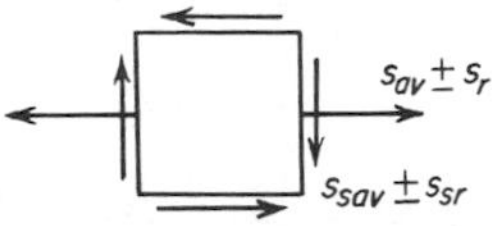

Figure 3-6 Element loaded by fluctuating stresses.

$$s = s_{av} + \frac{Ks_{yp}}{s_e} s_r \qquad \text{(a)}$$

The equivalent static shear stress s_s for Fig. 3-6 has the value

$$s_s = s_{sav} + \frac{K_t s_{yp}}{s_e} s_{sr} \qquad \text{(b)}$$

Equations (a) and (b) are now substituted into Eq. (11) to give the resultant static shear stress s_{smax}.

$$s_{smax} = \frac{0.5 s_{yp}}{FS} = \sqrt{\frac{1}{4}\left(s_{av} + \frac{Ks_{yp}}{s_e} s_r\right)^2 + \left(s_{sav} + \frac{K_t s_{yp}}{s_e} s_{sr}\right)^2} \qquad (16)$$

or

$$s_{smax} = \frac{0.5 s_{yp}}{FS} = \frac{16}{\pi d^3}\sqrt{\left(M_{av} + \frac{Ks_{yp}}{s_e} M_r\right)^2 + \left(T_{av} + \frac{K_t s_{yp}}{s_e} T_r\right)^2} \qquad (17)$$

Stress concentration factor K can be had from Fig. 2-7 and factor K_t from Fig. 3-9.

Stress concentration factors for the appropriate change of form and type of loading can be determined from the curves of Chapter 2 and from Section 8 of this chapter.

The equations above apply to solid circular shafts. When a keyway is present at the section for which the calculations are made, the strength is reduced not only because of stress concentration but because of loss of cross section as well. A theoretical determination of this latter quantity would be very involved, but its magnitude may be estimated by the principles explained in Section 16. Perhaps the best way of taking care of the situation in a design is to use a lower value for the working stress in shear. Fortunately, the form of the equation is such that this correction could be considerably in error without causing very much difference in the resulting diameter of the shaft.

Example 7. A revolving shaft carries a bending moment of 27,000 in. lb and a steady torque of 80,000 in. lb. Assume the torque fluctuates 20 per cent each way from the mean value. Stress concentration factor for bending and torsion is equal to 1.35. Factor of safety is to be equal to 2. Material has an ultimate strength of 120,000 psi and a yield strength of 90,000 psi.

Find the required diameter by the maximum shear theory.

Solution. $M_{av} = 0$ (revolving shaft), $M_r = 27,000$ in. lb

$T_{av} = 80,000$ in. lb, $T_r = 16,000$ in. lb

By Fig. 2-16: $s_e = 44,000$ psi

Working stress, $s_{smax} = \dfrac{0.5 s_{yp}}{FS} = \dfrac{0.5 \times 90,000}{2} = 22,500$ psi

By Eq. (17):

$$d^3 = \frac{16}{22,500\pi}\sqrt{\left(0 + \frac{1.35 \times 90}{44} \times 27,000\right)^2 + \left(80,000 + \frac{1.35 \times 90}{44} \times 16,000\right)^2}$$

$$= \frac{16}{22,500\pi}\sqrt{74,560^2 + 124,180^2} = 32.786$$

From which: $d = 3.201$ in.

6. Mises–Hencky Theory for Shafting

As before, the element is loaded as in Fig. 3-6. The Mises–Hencky equation, Eq. (9) of Chapter 2, for the equivalent static normal stress S for this element is

$$S^2 = s^2 + 3s_s^2$$

Substitution of Eqs. (a) and (b) of the preceding section gives

$$S = \frac{S_{yp}}{FS} = \sqrt{\left(s_{av} + \frac{K s_{yp}}{S_e} s_r\right)^2 + 3\left(s_{sav} + \frac{K_t s_{yp}}{S_e} s_{sr}\right)^2} \tag{18}$$

$$S = \frac{S_{yp}}{FS} = \frac{32}{\pi d^3}\sqrt{\left(M_{av} + \frac{K s_{yp}}{S_e} M_r\right)^2 + 0.75\left(T_{av} + \frac{K_t s_{yp}}{S_e} T_r\right)^2} \tag{19}$$

Example 8. Use the data of Example 7 and calculate the required diameter by the Mises–Hencky theory.

Solution. By Eq. (19): $d^3 = \dfrac{32FS}{\pi s_{yp}}\sqrt{74,560^2 + 0.75 \times 124,180^2}$

$$= \frac{32 \times 2}{90,000\pi} \times 130,860 = 29.621$$

From which: $d = 3.094$ in.

A slightly smaller shaft is required when the Mises–Hencky criterion is used. However, unless the torsional loading on the shaft predominates, the two methods give practically the same results.

7. Keys

Shafts and hubs are usually fastened together by means of keys. Several different kinds of keys are shown in Fig. 3-7. The square and flat type of keys are in wide use for general machine construction. Dimensions for square keys are given in Table 3-2. Kennedy keys are usually made tapered and are driven tightly into place upon assembly. They are adapted for rough, heavy service. The Woodruff key is much used in the automotive and machine tool

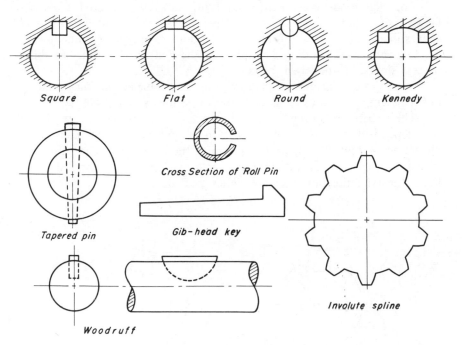

Figure 3-7 Types of keys.

Table 3-2 DIMENSIONS OF SQUARE KEYS. ASA B17.1–1943

Diameter Shaft	Size Key	Diameter Shaft	Size Key	Diameter Shaft	Size Key
1/2–9/16	1/8	1-7/16–1-3/4	3/8	3-3/8–3-3/4	7/8
5/8–7/8	3/16	1-13/16–2-1/4	1/2	3-7/8–4-1/2	1
15/16–1-1/4	1/4	2-5/16–2-3/4	5/8	4-3/4–5-1/2	1-1/4
1-5/16–1-3/8	5/16	2-7/8–3-1/4	3/4	5-3/4–6	1-1/2

industries. The gib-head key facilitates removal, although the projecting head for some applications constitutes a hazard for workmen.

In addition to a key, setscrews are usually employed to keep the hub from shifting axially on the shaft. Generally, two screws are placed in the hub: one screw bears on the key and the other bears on the shaft. For light service, rotation between shaft and hub may be prevented by setscrews alone.

Movement between shaft and hub can be prevented by a taper pin driven tightly into place. The so-called "roll pin" is not solid. It has the cross section shown in Fig. 3-7. It has sufficient flexbility to accommodate itself to small amounts of misalignment and variation in hole diameters, and will not come loose under vibrating loads.

For high-grade construction, and for cases where axial movement between shaft and hub is required, relative rotation is prevented by means of splines machined on the shaft and into the bore. One type of spline uses the involute curve as the outline. The spline on the shaft can be cut by a hobbing process similar to that used for cutting gears.

Tables of dimensions for the foregoing machine elements may be found in engineering handbooks and cataglogs of various supply houses.

The distribution of the force on the surfaces of a key is very complicated. It is dependent upon the fit of the key in the grooves of shaft and hub, as illustrated by Figs. 3-8(a) and (b), in which the distributed loads are represented by single arrows. In addition, the stresses are not uniform along the key in the axial direction; they are highest near the ends.

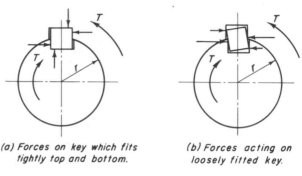

(a) Forces on key which fits   (b) Forces acting on
tightly top and bottom.   loosely fitted key.

Figure 3-8 Forces on key.

Because of many uncertainties, an exact analysis of the stresses usually cannot be made. Engineers commonly assume that the entire torque is carried by a tangential force F located at the shaft surface. That is,

$$T = Fr \tag{20}$$

Shearing and compressive stresses are computed for the key from force F, and a sufficiently large factor of safety is employed.

Example 9. A $3\frac{7}{16}$-in.-diam shaft is made from material with a yield point value of 58,000 psi. A $\frac{7}{8} \times \frac{7}{8}$-in. key is to be used of material with a yield point value of 48,000 psi. Let $s_{syp} = 0.5 s_{yp}$. The factor of safety is equal to 2.

Find the required length of key based on the torque value of the gross shaft.

Solution.

Shaft: $s_{yp} = 58,000$ psi, working stress, $s = \dfrac{58,000}{2} = 29,000$ psi

$s_{syp} = 29,000$ psi, working stress, $s_s = \dfrac{29,000}{2} = 14,500$ psi

Key: $s_{yp} = 48,000$ psi, working stress, $s = \dfrac{48,000}{2} = 24,000$ psi

$s_{syp} = 24,000$ psi, working stress, $s_s = \dfrac{24,000}{2} = 12,000$ psi

$J = \dfrac{\pi d^4}{32} = 13.708$ in.^{4}

In Eq. (2), torque in shaft: $T = \dfrac{s_s J}{r} = \dfrac{14,500 \times 13.708}{1.719} = 115,650$ in. lb

Force at shaft surface: $F = \dfrac{T}{r} = \dfrac{115,650}{1.719} = 67,280$ lb

For length of key:

Based on bearing on shaft: $l = \dfrac{67,280}{29,000 \times 0.438} = 5.30$ in.

Based on bearing on key: $l = \dfrac{67,280}{24,000 \times 0.438} = 6.41$ in.

Based on shear in key: $l = \dfrac{67,280}{12,000 \times 0.875} = 6.41$ in.

8. Stress Concentration

Stress concentration factors[1] for a shaft with two diameters joined by fillets and loaded in torsion are given in Fig. 3-9. When a shaft has a transverse hole with a bending load, the stress concentration factors[2] are as shown in Fig. 3-10. The stress concentration factors[3] for a transverse hole and torsional loading are given in Fig. 3-11.

The results of some tests to determine the fatigue strength reduction factors[4] for alternating bending stresses for shafts with keyways are given in Table 3-3. Two kinds of steel were used: a medium-carbon steel and a heat-treated, chrome–nickel steel. Specimens tested were 1 in. in diameter with two types of keyways: sled-runner and profile, which are shown in Fig. 3-12. To simulate conditions at an oil hole, tests were also made on carbon steel specimens with a $\frac{1}{4}$-in. transverse hole; the results are shown in Table 3-3.

For fluctuating loads, the fatigue stress concentration factor K is defined as follows:

$$K = \frac{\text{endurance limit for plain specimen}}{\text{endurance limit with keyway or hole}}$$

[1] See Peterson, R. E., *Machine Design*, **23**, June 1951, p. 177; also p. 76, Reference 4, end of chapter.

[2] See Peterson, R. E., *Machine Design*, **23**, July 1951, p. 156; also p. 104, Reference 4, end of chapter.

[3] See Seely, F. B., and T. J. Dolan, *Univ. Illinois Eng. Exp. Sta. Bull. 276*, 1935.

[4] See Peterson, R. E., "Fatigue of Shafts Having Keyways," *Proc. ASTM*, **32**, Part II, 1932, p. 413.

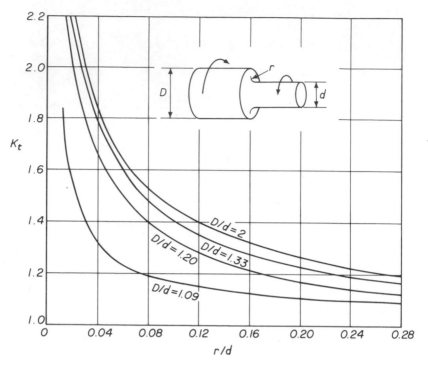

Figure 3-9 Factors of torsional stress concentration K for circular shafts of two diameters to be applied to the shear stress in section of diameter d.

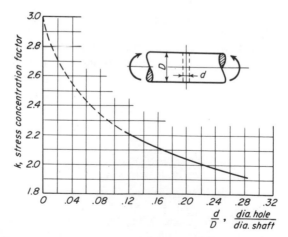

Figure 3-10 Stress concentration factors for shaft with transverse hole loaded in bending. Based on section modulus of the net area.

Table 3-3 FATIGUE STRESS CONCENTRATION FACTORS IN BENDING FOR SHAFTS WITH KEYWAYS BASED ON SECTION MODULUS OF FULL AREA

Steel	Tensile Strength, psi	Yield Strength (Plastic Def. 0.2 per cent)	For Reversed Bending Stress	Chrome-nickel, Heat-treated		Medium-carbon, Normalized	
				Endurance Limit, psi	Stress Concentration Factor, K	Endurance Limit, psi	Stress Concentration Factor, K
Chrome-nickel (About SAE 3140)	103,500	70,000	No keyway, or-dinary tapered specimen	58,000		37,000	
			Sled-runner keyway	36,000	1.61	28,000	1.32
Medium-carbon	80,000	45,000	Profile keyway	28,000	2.07	23,000	1.61
(About SAE 1045)			½-in. transverse hole			12,100	3.06

In these tests a fatigue crack started on the outer surface of the shaft near the end of the keyway. The results of the tests indicate that the sled-runner keyway is preferable to the profile. It should also be noted that the heat-treated specimens had larger stress concentration factors than the plain carbon. As was mentioned in Section 18, Chapter 2, a heat-treated steel exhibits greater sensitivity to notch effects than does plain carbon steel. This fact unfortunately causes a reduction in the advantages that would otherwise be obtained by the use of a high-strength alloy steel. The tests also show that a designer should avoid locating an oil hole at a highly stressed point on a shaft.

Because of the lack of available data, Table 3-3

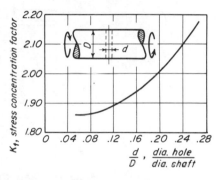

Figure 3-11 Stress concentration factors for shaft with transverse hole. Torsional loading. Based on full cross sectional area. No reduction for hole.

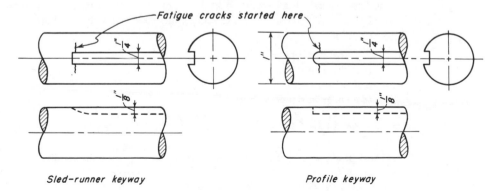

Sled-runner keyway   Profile keyway

Figure 3-12 Types of keyways tested for stress concentration effects.

may also be used for the stress concentration factors for torsion in the equations of Sections 5 and 6.

All stress concentration factors so far presented in this book are the geometric or full theoretical values. Section 18 of Chapter 2 discussed the fact that under fatigue loading the actual effect of stress concentration was usually less severe than indicated by the theoretical values.

9. Couplings

A wide variety of devices is available for connecting the ends of two shafts together. The solid coupling shown in Fig. 3-13 is a typical example.

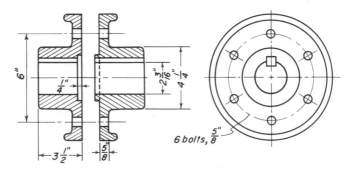

Figure 3-13 Solid coupling.

It is inexpensive and will withstand rough usage. Good alignment between the ends of the shafts is necessary, however, to avoid inducing bending stresses in the shafts or loads in the bearings.

Example 10. For the coupling shown in Fig. 3-13, the key is $\frac{1}{2} \times \frac{1}{2}$ in. The shaft carries a steady load of 50 hp at 150 rpm. For all parts, $s_{yp} = 60,000$ psi, and $s_{syp} = 30,000$ psi. Find the following stresses and the *FS* based on the yield point:

    (a) Shear and bearing in key.
    (b) Shear in bolts.
    (c) Bearing on bolts in flange.
    (d) Shear in flange at hub.

Solution. (a) From Eq. (10): $T = \dfrac{63,000\ hp}{n} = \dfrac{63,000 \times 50}{150} = 21,000$ in. lb

Tangential force at shaft surface: $F = \dfrac{T}{r} = \dfrac{21,000}{1.094} = 19,200$ lb

Area in bearing for key $= 0.25 \times 3.25 = 0.8125$ in.^{2}

Compressive stress: $s_c = \dfrac{19,200}{0.8125} = 23,630$ psi

$$FS = \frac{60,000}{23,630} = 2.54 \text{ in bearing}$$

$$\text{Area in shear for key} = 0.5 \times 3.25 = 1.625 \text{ in.}^2$$

Shearing stress in key: $\qquad s_s = \frac{19,200}{1.625} = 11,820 \text{ psi}$

$$FS = \frac{30,000}{11,820} = 2.54 \text{ in shear}$$

(b) $\qquad$ Area in shear for bolts $= 6 \times \frac{\pi}{4} \times 0.625^2 = 1.841 \text{ in.}^2$

Force at bolt circle: $\qquad F = \frac{21,000}{3} = 7,000 \text{ lb}$

Shear stress in bolts: $\qquad s_s = \frac{7,000}{1.841} = 3,800 \text{ psi}$

$$FS = \frac{30,000}{3,800} = 7.89$$

(c) $\qquad$ Area in bearing for bolts $= 6 \times 0.625 \times 0.625 = 2.344 \text{ in.}^2$

Compressive stress on bolts: $\qquad s_c = \frac{7,000}{2.344} = 2,990 \text{ psi}$

$$FS = \frac{60,000}{2,990} = 20.1$$

(d) $\quad$ Area in shear at edge of hub $= 4.25\pi \times 0.625 = 8.345 \text{ in.}^2$

Force at edge of hub $= \frac{21,000}{2.125} = 9,880 \text{ lb}$

Shear stress in web: $\qquad s_s = \frac{9,880}{8.345} = 1,180 \text{ psi}$

$$FS = \frac{30,000}{1,180} = 25.4$$

Many types of flexible couplings are available that provide for some misalignment. Such couplings are often provided with springs or rubber inserts to cushion the shock of suddenly applied loads. Details, dimensions, and load ratings based on long experience may be found in the catalogs of the various manufacturers.[5] Information is also given in the mechanical engineering handbooks.

10. Bending Loads in Two Planes

Shafts are sometimes subjected to loads applied at different angles. To find the resulting bending moment at any cross section, it is necessary to have the components of the loads in two perpendicular axial planes. The following example illustrates a typical method for solving such problems.

[5] See, for example, Reference 1, end of chapter.

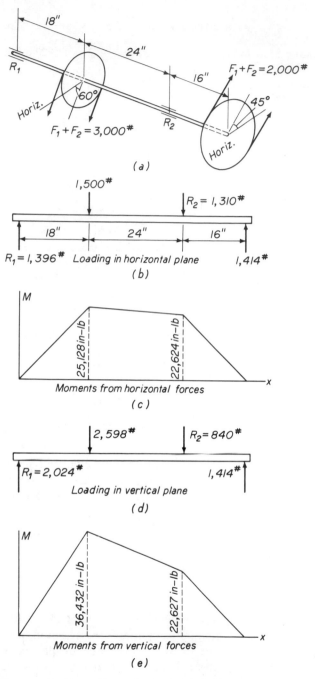

Figure 3-14 Shaft with bending loads in two planes. Example 11.

Example 11. Figure 3-14 shows a shaft with the belts making angles of 45° and 60° with the horizontal. Find the value of the maximum bending moment for the shaft.

Solution. The components of the belt forces in the horizontal plane and the corresponding bearing reactions are shown in Fig. 3-14(b). These loads give the bending moment diagram of sketch (c).

     The loads and reactions for the vertical plane are given in Fig. 3-14(d), and the bending moment diagram is shown in sketch (e).

     The maximum bending moment occurs at the left pulley and has the following value.

$$M_{max} = \sqrt{25{,}130^2 + 36{,}430^2} = 44{,}290 \text{ in. lb}$$

11. Shaft on Three Supports

     Shafts are sometimes supported on three bearings as illustrated in Fig. 3-15. Such a problem is statically indeterminate with three unknown reactions

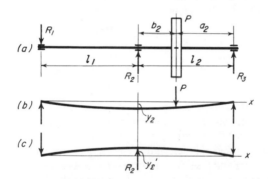

Figure 3-15 Shaft on three supports.

R_1, R_2, and R_3. It is possible to write only two independent equations from statics, one for the summation of the vertical forces and one for the summation of the moments. The additional equation required for a solution to the problem can be obtained by taking into account the deformation of the body. This can be done in a variety of ways. The following example illustrates a typical method of solution.

Example 12. Derive the equation for reaction R_2 of Fig. 3-15.

Solution. Assume that support R_2 has been temporarily removed as shown by Fig. 3-15(b). Deflection y_2 can be found by the equation for y_1 in No. 7 of Fig. 1-14. When substitutions from Fig. 3-15(a) are made in this equation, the result is as follows.

$$y_2 = \frac{Pa_2 l_1}{6(l_1 + l_2)EI}(l_2^2 + 2l_1 l_2 - a_2^2) \tag{a}$$

Now assume that reaction R_2 of sketch (a) is acting, but that load P has been temporarily removed as shown in sketch (c). Deflection y'_2 then is

$$y'_2 = \frac{2R_2 l_1^2 l_2^2}{6(l_1 + l_2)EI} \tag{b}$$

If sketches (b) and (c) are combined, the original shaft of Fig. 3-15(a) is obtained in which the deflection at R_2 is zero. In other words, the downward deflection y_2 of Eq. (a) must be equal to the upward deflection y'_2 of Eq. (b). When these are equated the following expression for R_2 results.

$$R_2 = \frac{Pa_2}{2l_1 l_2^2}(l_2^2 + 2l_1 l_2 - a_2^2) \tag{c}$$

Reactions R_1 and R_3, as well as other information concerning the shaft, can now be easily obtained by statics.

Example 13. In Fig. 3-15 let $a_2 = 20$ in., $l_1 = l_2 = 50$ in., $P = 300$ lb, and $EI = 26,000,000$ lb in.[2] Find the value of R_2 if the elevation of the center bearing is made 0.05 in. lower than the others.

Solution. From the given conditions y'_2 in Fig. 3-15(c) will be 0.05 in. smaller than y_2 in Fig. 3-15(b). Hence

$$\frac{2R_2 l_1^4}{6 \times 2l_1 EI} + 0.05 = \frac{Pa_2 l_1}{6 \times 2l_1 EI}(l_1^2 + 2l_1^2 - a_2^2)$$

The given numerical data should now be substituted in this equation to give

$$R_2 = 108 \text{ lb}$$

When all bearings are at the same elevation, Eq. (c) gives a value for R_2 of 170.4 lb, a larger value than the 108 lb obtained above. Thus a small change in the geometry of the system made a relatively large change in the forces. This is characteristic of indeterminate structures, and care must be exercised in the manufacture and assembly of equipment where such a condition is present. This feature has been used to good advantage in the example above to effect a reduction in the force at the center bearing.

12. Crankshafts

To determine the stresses in a crankshaft, the loading on the separate parts of the mechanism must be secured. A typical example is shown in Fig. 3-16(a), which illustrates a single-cylinder, belt-driven air compressor. Suppose the dimensions of the machine are known, and it is desired to find the stresses in the cheek CD of the crankshaft. From the cylinder bore and the air pressure, the force on the piston can be found. By making a force triangle, the force in the connecting rod can be determined. This force also acts on the crankpin at A. As shown in Fig. 3-16(b), it is divided into components tangential and normal to the plane of the crank. The normal component gives the torque that the belt must exert at the given crank position. The forces in the tight and slack sides of the belt are now determined from the

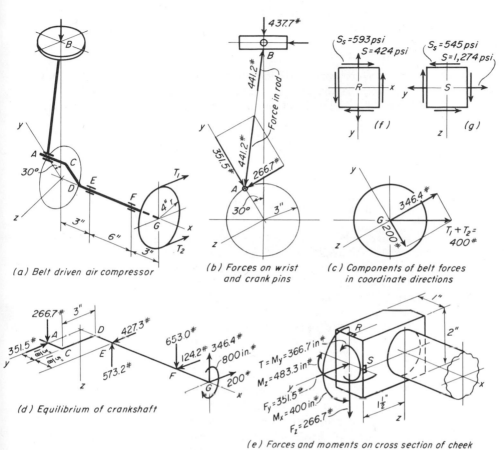

(a) Belt driven air compressor

(b) Forces on wrist and crank pins

(c) Components of belt forces in coordinate directions

(d) Equilibrium of crankshaft

(e) Forces and moments on cross section of cheek

Figure 3-16 Forces and moments for crankshaft. Example 14.

torque, and the sum is divided into components in the coordinate directions, as shown in Fig. 3-16(c).

The free-body diagram for the crank should now be made as shown in Fig. 3-16(d), utilizing the rod and belt forces, and with the bearing reactions determined by statics. It is customary to assume that the entire bearing load acts at the center of the bearing. The crank can now be cut apart, and the forces and moments for each portion can be determined in the usual way. Thus Fig. 3-16(e) shows the cheek after cutting at the midpoint between C and D with the various forces and moments that act on the cut surface.

Example 14. Let the bore of the compressor of Fig. 3-16 be 4.75 in., the stroke 6 in., and the length of rod 12 in. For a crank angle of 30°, the air pressure is 24.7 psi gage. The belt is horizontal with tension T_1 in the tight side 3 times as great as tension T_2 in the slack side. The coordinate system is as shown.

Find the forces and moments for the cross section of the cheek midway between C and D.

Solution. Force on piston $= \dfrac{\pi}{4} \times 4.75^2 \times 24.7 = 437.7 \, \text{lb}$

By taking components, the force in the rod is found to be 441.2 lb, as shown in Fig. 3-16(b). At the crankpin, the rod force divides into components 351.5 lb and 266.7 lb, tangential and normal, respectively, to the plane of the crank.

Torque on crank: $T = 266.7 \times 3 = 800 \, \text{in. lb}$

$$T = 4(T_1 - T_2) = 4(3T_2 - T_2) = 800 \, \text{in. lb}$$

Hence $T_2 = 100 \, \text{lb}$

$$T_1 = 300 \, \text{lb}$$

The total belt load of 400 lb is now divided into components, as shown in Fig. 3-16(c).

The free-body diagram for the crank is shown in Fig. 3-16(d), in which the previously determined crankpin and belt loads are acting. The bearing reactions are determined in the usual way by taking moments.

The forces and moments for the desired cross section of the cheek are shown in Fig. 3-16(e). The reader should check the values shown thereon. Moment arms are the distances from the line of action of the force to the center of gravity of the cross section affected.

The arrows in Fig. 3-16(e) represent the resultant values for the forces and moments on the entire cross section. Such loads actually are distributed as stresses in some fashion over the surface of the cheek where it was cut from the balance of the shaft. Because of the irregular shape of the crank, the stress equations previously derived may or may not be applicable. If the crank of Fig. 3-16 is of lightweight construction with relatively small-diameter pins, the stress equations might be valid for the cross section at the midpoint of the cheek. However, the stresses for large, massive crankshafts, such as those used in internal combustion engines, are usually determined experimentally by direct strain gage measurements.[6]

13. Critical Speed of Rotating Shaft

Rotating shafts become dynamically unstable at certain speeds, and large vibrations are likely to develop. The speed at which this phenomenon occurs is called a *critical speed*. It is shown in books on vibration theory that the frequency for free lateral vibration when the shaft is not rotating will be the same as its critical speed.

Vibration difficulties frequently arise at the lowest or fundamental

[6]See Timoshenko, S., "Torsion of Crankshafts," *Trans. ASME*, **44**, 1922, p. 653; also **45**, 1923, p. 449.

critical speed. The equation for finding this speed for a shaft on two supports is as follows.

$$f = \frac{1}{2\pi}\sqrt{\frac{g(W_1y_1 + W_2y_2 + W_3y_3 + \cdots)}{W_1y_1^2 + W_2y_2^2 + W_3y_3^2 + \cdots}}\text{cycles/sec} \qquad (21)$$

where W_1, W_2, etc., represent the weights of the rotating bodies, and y_1, y_2, etc., represent the respective static deflections of the weights. The gravitational constant of 386 in./sec² is represented by g.

Example 15. Find the value of the fundamental critical speed for the shaft shown in Fig. 3-17. $E = 30,000,000$ psi.

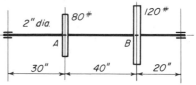

Solution. The static deflections at the weights can be found by the equations for No. 7 of Fig. 1-14.

Figure 3-17 Example 14.

$$I = \frac{\pi d^4}{64} = \frac{\pi}{4} = 0.7854 \text{ in.}^4$$

At A, due to 80 lb: $y = \dfrac{80 \times 60 \times 30}{6 \times 90 \times 30,000,000 \times 0.7854}(90^2 - 60^2$

$$- 30^2) = 0.04074 \text{ in.}$$

At A, due to 120 lb: $y = \dfrac{120 \times 20 \times 30}{6 \times 90 \times 30,000,000 \times 0.7854}(90^2 - 20^2$

$$- 30^2) = 0.03848 \text{ in.}$$

Total deflection at A: $y = 0.04074 + 0.03848 = 0.07922 \text{ in.}$

At B, due to 80 lb: $y = \dfrac{80 \times 30 \times 20}{6 \times 90 \times 30,000,000 \times 0.7854}(90^2 - 30^2$

$$- 20^2) = 0.02565 \text{ in.}$$

At B, due to 120 lb: $y = \dfrac{120 \times 20 \times 70}{6 \times 90 \times 30,000,000 \times 0.7854}(90^2 - 70^2$

$$- 20^2) = 0.03697 \text{ in.}$$

Total deflection at B: $y = 0.02565 + 0.03697 = 0.06262 \text{ in.}$

In Eq. (21): $f = \dfrac{1}{2\pi}\sqrt{\dfrac{386(80 \times 0.07922 + 120 \times 0.06262)}{80 \times 0.07922^2 + 120 \times 0.06262^2}}$

$$= 11.80 \text{ cycles/sec}$$

$$n_{cr} = 11.80 \times 60 = 708 \text{ rpm}, \qquad \text{critical speed}$$

The normal operating speed for a shaft should be considerably above or below the value of a critical speed.

It should be noted that Eq. (21) ignores the effect of the weight of the shaft and also assumes that all weights are concentrated. The equation does not take into account any effect of the flexibility of the bearings or supports. This additional flexibility may in some cases lower the value of the critical speed below that indicated by the equation.

A shaft will have as many critical speeds as there are rotating masses. The determination of the higher critical speeds is beyond the scope of this book.

14. Deflection of Shaft of Nonuniform Diameter

Another way to determine the deflection y of a shaft is by elastic energy. The equation is as follows.

$$EFy = \int \frac{M_p M_f \, dx}{I} \tag{22}$$

As shown in Fig. 3-18, force F represents an auxiliary load placed on the

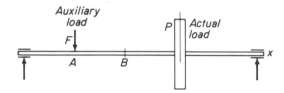

Figure 3-18 Deflection of shaft by elastic energy.

shaft at point A where the deflection y is desired; M_p is the bending moment in terms of distance x for any general point B as caused by the actual loads P; M_f is the bending moment at B caused by the auxiliary load F. The modulus of elasticity of the material is represented by E; I is the moment of inertia of the cross section. The integration must extend throughout the entire volume of the shaft.

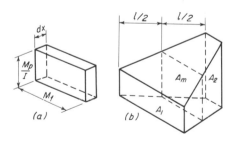

Figure 3-19 Deflection of shaft by elastic energy.

Equation (22) can be given a geometric interpretation[7] that eliminates the necessity of performing mathematical integrations. The integrand can be considered a solid with the three dimensions M_p/I, M_f, and dx, as shown in Fig. 3-19(a). If the moment diagrams for M_p and M_f consist of segments of straight lines, it is not necessary to consider a solid of thickness dx, but solids of finite lengths along the shaft can be taken and the volumes found by solid geometry. When this is done for the full length of the shaft, the integration of Eq. (22) will have been performed.

Odd-shaped solids will, in general, be present, and for these the prismoidal equation is useful.

$$\text{Vol.} = \frac{l}{6}(A_1 + 4A_m + A_2) \tag{23}$$

[7]See Spotts, M. F., "Critical Speed of Shafts," *Prod. Eng.*, **12**, 1941, p. 20.

As shown in Fig. 3-19(b), A_1 and A_2 refer to the end surfaces of the solid that are perpendicular to the centerline of the shaft. Length l refers to the distance between A_1 and A_2. Area A_m is the area of the cross section midway between A_1 and A_2. Note that in general it is not $(A_1 + A_2)/2$.

Equations (22) and (23) are especially useful when the shaft is non-uniform in diameter as illustrated by the following example.

Example 16. Find the deflection for point A where the diameter of the shaft of Fig. 3-20 changes. $E = 30{,}000{,}000$ psi.

Solution. The bending moment diagram for the P load is shown in Fig. 3-20(b). The values of the moments are divided by the corresponding values for I to give the M_p/I diagram of sketch (c). Let load F be taken as 1 lb acting

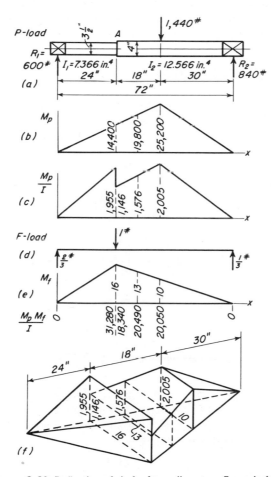

Figure 3-20 Deflection of shaft of two diameters. Example 16.

at point A, as shown in sketch (d). This load has the bending moment diagram of sketch (e). The various solids formed by the M_p/I and M_f diagrams are given by sketch (f). The solid on each end is a pyramid with a volume equal to one-third of the product of the base by the altitude. The prismoidal equation, however, must be applied to the solid in the middle. The calculations are as follows.

$$\tfrac{1}{3} \times 24 \times 31,280 = 250,250$$

$$\tfrac{18}{6}(18,340 + 4 \times 20,490 + 20,050) = 361,050$$

$$\tfrac{1}{3} \times 30 \times 20,050 = \underline{200,500}$$

$$Ey = 811,800$$

$$y = \frac{811,800}{30,000,000} = 0.02706 \text{ in.}$$

In solving problems, it is not necessary to make a sketch such as (f). The $M_p M_f/I$ values can be entered on sketch (e) and used directly. It should be noted that each solid extends between concentrated loads or points where the diameter changes.

15. Slope of Shaft by Elastic Energy

The slope of the shaft can also be found by elastic energy, but the auxiliary load must be a moment applied at the point where the slope is desired. Thus a moment M' is applied to the shaft at point A in Fig. 3-21.

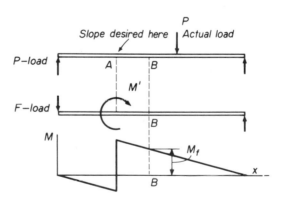

Figure 3-21 Slope of shaft by elastic energy.

The M_f diagram is given by Fig. 3-21(c). Let θ be the rotation or change of slope at A caused by load P. The external work from load M' due to rotation θ is then $M'\theta$. The expression for internal work is the same as that derived for Eq. (22). The energy equation thus becomes

$$EM'\theta = \int \frac{M_p M_f \, dx}{I} \qquad (24)$$

Moment M' is usually taken as 1 in. lb. Positive and negative signs must be observed as indicated by Fig. 3-21(c).

16. Torsion of Noncircular Shafts

It is sometimes necessary to make a design for a shaft of noncircular cross section. For example, the designer might have to know the torsional stress in the rectangular cheek of a crankshaft. Also, miscellaneous machine parts, such as brackets and supports, although not shafts, are sometimes loaded in torsion.

The theory of torsion for shafts of noncircular cross section is complicated because the assumptions that are valid for circular shafts do not apply. Cross sections are no longer plane and perpendicular to the shaft axis after twisting; rather, they are warped, as shown in Fig. 3-22, and the equations for the stresses are therefore more involved.

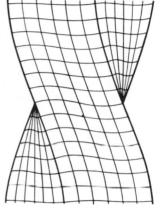

Complicated problems may be solved experimentally by a method known as the *membrane analogy*.[8] A thin homogeneous membrane, such as a soap film, is stretched over a hole in a closed box. The hole is geometrically similar to the cross section of the shaft being studied. The film is slightly bowed by air pressure and the elevations of the resulting surface are measured. The measurements obtained will permit the contours, or lines of uniform elevations, to be plotted. For example, see Fig. 3-24(c). By mathematical analysis it can be shown that

Figure 3-22 Rectangular bar in torsion.

(1) The maximum shearing stress in the shaft has the same direction as the contour line at the corresponding point in the film.

(2) The maximum shearing stress at any point is proportional to the slope of the membrane at right angles to the contour at the corresponding point in the film.

(3) The torque carried by the shaft is proportional to twice the volume enclosed between the membrane and the plane of its base.

Although complicated problems can be solved by taking measurements on the membrane, experimental apparatus, which is rarely available to the designer, is required. Nevertheless, the idea of the membrane analogy is very useful since it gives a mental picture of the state of stress. The designer can visualize the points of greatest slope, hence greatest stress, on the bowed film, and often, by making small changes in shape, can cause a reduction in stress.

[8]See p. 266 of Reference 7 and p. 268 of Reference 8, end of chapter.

For example, the membrane may have a steep slope, indicating a high stress at the internal corners of a keyway as shown in Fig. 3-23(a). The membrane analogy indicates that this stress concentration can be reduced by rounding off the bottom corners. However, as was mentioned in Section 8, unless the keyway runs the full length of the shaft, a still greater stress occurs on the shaft surface at the end of the keyway.

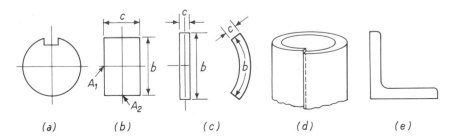

(a)   (b)   (c)   (d)   (e)

Figure 3-23 Typical cross sections of bars loaded in torsion.

The membrane for a rectangular shaft has its greatest slope at point A_1 in Fig. 3-23(b); hence the stress is at a maximum at this point, and not at the corners, as is sometimes supposed. In the corner, the membrane has a zero slope along both edges, indicating a zero stress.

As shown by Fig. 3-23(c), the volume enclosed by the membrane for a thin, narrow, rectangular cross section is practically the same whether the rectangle is straight or formed into a curve. The torques carried by either section are the same. However, a slight concentration of stress will exist on the inner side of the curved section. If the flat bar is bent into a full circle to make a slit tube, the deformations at the ends, when torques are applied, will be as indicated in sketch (d).

The volume enclosed by the membrane for a composite section, as in Fig. 3-23(e), is approximately equal to the sum of the volumes for the separate parts. Hence the membrane analogy permits the torque that such sections will carry to be found easily. However, a stress concentration will exist on the fillet in the reentrant corner.

17. Torsion of Wide Rectangular Bar

As was previously mentioned, a mathematical solution for the stresses in a rectangular shaft is difficult to obtain. If the bar is very wide, however, the stress situation is much simplified. This is illustrated by Fig. 3-24(a), where it is assumed that side b of the cross section is much greater than width c. The membrane for this shaft has the contour lines shown in Fig. 3-24(c).

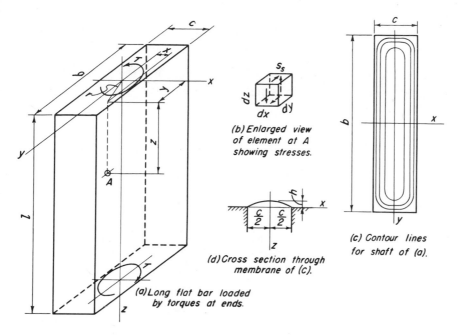

(b) Enlarged view
of element at A
showing stresses.

(c) Contour lines
for shaft of (a).

(d) Cross section through
membrane of (c).

(a) Long flat bar loaded
by torques at ends.

Figure 3-24 Torsion of thin, wide, rectangular shaft.

These lines are practically straight and parallel to side b for almost the entire surface. Hence, except for small regions in the ends, the shearing stress is directed parallel to side b.

In accordance with the foregoing, the only stresses that act on an element at an interior point A of the bar of Fig. 3-24(a) are those shown in Fig. 3-24(b). The maximum value of the stress occurs on the long side b of the cross section. It can be shown that the value of this stress is given by the following equation.

$$s_{smax} = \frac{T}{0.333bc^2} \tag{25}$$

The angular rotation between two cross sections 1 unit distance apart is given by the following equation.

$$\theta_1 = \frac{T}{0.333Gbc^3} \tag{26}$$

One of the equations above should now be divided by the other, to give

$$s_{smax} = Gc\theta_1 \tag{27}$$

Reference to Eq. (1) shows that the stress in the bar shown in Fig. 3-24(a) is twice as great as for a round bar of diameter c with equal angular deformation θ_1.

18. Torsion of Rectangular Bars, General Case

When side b of the cross section is not relatively great as compared with width c, the foregoing equations cannot be used. For the bar in Fig. 3-22, the general equations for stress and deformation may be written in the following forms:

$$s_s = \frac{T}{\alpha_1 bc^2} \qquad \text{for point } A_1, \text{ Fig. 3-23(b)} \tag{28}$$

$$s_s = \frac{T}{\alpha_2 bc^2} \qquad \text{for point } A_2, \text{ Fig. 3-23(b)} \tag{29}$$

$$\theta_1 = \frac{T}{\beta Gbc^3} \qquad \text{ang. def., radians per inch of length} \tag{30}$$

Values of the constants α_1, α_2, and β have been computed for various ratios of b/c and are given in Table 3-4.

Table 3-4   CONSTANTS FOR TORSION OF RECTANGULAR BARS

b/c	1.00	1.20	1.50	1.75	2.00	2.50	3.00	4.00	5.00	6.00	8.00	10.00	∞
α_1	0.208	0.219	0.231	0.239	0.246	0.258	0.267	0.282	0.291	0.299	0.307	0.312	0.333
α_2	0.208	0.235	0.269	0.291	0.309	0.336	0.355	0.378	0.392	0.402	0.414	0.421	$\cdots$
β	0.1406	0.166	0.196	0.214	0.229	0.249	0.263	0.281	0.291	0.299	0.307	0.312	0.333

With the values from the table, computations can be made for the shear stress at the midpoint of both the long and the short sides, as well as the angular deformation, for rectangular cross sections starting from the square $b/c = 1$ to $b/c = \infty$. The maximum shear stress on the cross section occurs at the center A_1 of the long side, and is found by using α_1.

Example 17. Find the stresses at points R and S at the center of the sides of the cheek for the cross section midway between points C and D in Fig. 3-16.

Solution. Element at R:

Direct stress: $s = \dfrac{P}{A} = \dfrac{351.5}{2} = 176 \text{ psi}$,   compression

Bending: $s = \dfrac{6M}{bh^2} = \dfrac{6 \times 400}{1 \times 2^2} = 600 \text{ psi}$,   tension

The net tension of 424 psi is shown acting on the element in Fig. 3-16(f).

$$\frac{b}{c} = 2$$

From Table 3-4: $\alpha_2 = 0.309$

Shear stress, Eq. (29): $s_s = \dfrac{T}{\alpha_2 bc^2} = \dfrac{366.7}{0.309 \times 2 \times 1^2} = 593 \text{ psi}$

This stress, properly directed, is also shown on the element in Fig. 3-16(f).
Element at S:

As before, direct stress: $s = 176$ psi, compression

Bending: $$s = \frac{6M}{bh^2} = \frac{6 \times 483.3}{2 \times 1^2} = 1,450 \text{ psi, tension}$$

The net tension of 1,274 psi is shown on the element in Fig. 3-16(g).

$$\frac{b}{c} = 2$$

From Table 3-4: $\alpha_1 = 0.246$

Because of torque: $$s_s = \frac{T}{\alpha_1 bc^2} = \frac{366.7}{0.246 \times 2 \times 1^2} = 745 \text{ psi}$$

Because of transverse shear: $$s_s = \frac{3V}{2A} = \frac{3 \times 266.7}{2 \times 2} = 200 \text{ psi}$$

These shear stresses have opposite signs. The resultant, 545 psi, properly
directed, is shown in Fig. 3-16(g).

19. Composite Sections

The membrane analogy indicates that the torsional moment carried by a
cross section consisting of a number of areas joined together is equal to the
sum of the torques of the separate parts. The angle θ_1 applies to each of the
parts as well as to the whole section. Therefore, the total torque T for the
cross section of Fig. 3-25 is equal to the sum of the torques T_1, T_2, and T_3
for the separate parts 1, 2, and 3, respectively. Hence by Eq. (30):

Torque carried by part 1: $T_1 = \theta_1 G \beta' b_1 c_1^3$

Torque carried by part 2: $T_2 = \theta_1 G \beta'' b_2 c_2^3$ (a)

Torque carried by part 3: $T_3 = \theta_1 G \beta''' b_3 c_3^3$

Adding: $T = \theta_1 G(\beta' b_1 c_1^3 + \beta'' b_2 c_2^3 + \beta''' b_3 c_3^3)$ (b)

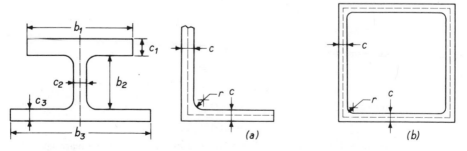

Figure 3-25 Composite section loaded in torsion.

Figure 3-26 Structural angle and square tube of Table 3-5.

Here β', β'', and β''' are the β values for parts 1, 2, and 3, respectively. It should be noted that the right-hand side of Eq. (b) will contain as many terms as there are rectangles in the cross section under consideration.

The maximum value of the shearing stress occurs in the bar of greatest width. Let this bar be No. 1 in Fig. 3-25. Hence, from Eq. (28),

$$s_{s1} = \frac{T_1}{\alpha_1 b_1 c_1^2} = \frac{\theta_1 G \beta' c_1}{\alpha_1} \tag{c}$$

Elimination of $\theta_1 G$ between Eqs. (b) and (c) gives

$$s_{s1} = \frac{T \beta' c_1}{\alpha_1 (\beta' b_1 c_1^3 + \beta'' b_2 c_2^3 + \beta''' b_3 c_3^3 + \cdots)} \tag{31}$$

The angular deformation per inch of length is found from Eq. (b).

$$\theta_1 = \frac{T}{G(\beta' b_1 c_1^3 + \beta'' b_2 c_2^3 + \beta''' b_3 c_3^3 + \cdots)} \tag{32}$$

As was previously mentioned, a concentration of stress exists at the reentrant corners of a composite section. This factor depends upon the ratio between the radius of the fillet and the thickness of the member. Values[9] are shown in Table 3-5, for the structural angle, Fig. 3-26(a), and for the thin-walled square tube in sketch (b).

Table 3-5 STRESS CONCENTRATION FACTOR K_t FOR STRUCTURAL ANGLE AND THIN-WALLED SQUARE TUBE

r/c	0.125	0.25	0.50	0.75	1.00	1.25	1.50
Angle	2.72	2.00	1.63	1.57	1.56	1.57	1.60
Square tube	2.46	1.70	1.40	1.25	1.14	1.25	1.07

Example 18. Find the torque that a 30-in. piece of $2 \times 2 \times \frac{3}{8}$-in. angle iron can carry if the maximum shearing stress on the fillet is to be 12,000 psi. Radius of the fillet is $\frac{1}{4}$ in. Find the angular deformation sustained with such loading.

Solution. $\dfrac{r}{c} = \dfrac{0.25}{0.375} = 0.67$

By Table 3-5: $K_t = 1.59$

Developed center line: $b = 3\frac{5}{8}$ in.

$$\frac{b}{c} = \frac{3.625}{0.375} = 9.7$$

By Table 3-4: $\alpha_1 = 0.312$

$$\beta = 0.312$$

[9] See Huth, J. H., "Torsional Stress Concentration in Angle and Square Tube Fillets," *Trans. ASME*, **72**, 1950, *Jour. Appl. Mech.*, p. 388. See also p. 131, Reference 4, end of chapter.

By Eq. 28, at center of 2-in. leg:

$$s_s = \frac{T}{\alpha_1 bc^2} = \frac{T}{0.312 \times 3.625 \times 0.375^2} = \frac{T}{0.159}$$

On fillet: $\qquad\qquad s_s = 1.59\frac{T}{0.159} = 12{,}000$

From which: $\qquad\quad T = \frac{0.159 \times 12{,}000}{1.59} = 1{,}200$ in. lb

By Eq. (30): $\quad \theta_1 = \frac{T}{\beta Gbc^3} = \frac{1{,}200}{0.312 \times 11{,}500{,}000 \times 3.625 \times 0.375^3}$

$$= 0.0017495 \text{ rad/axial in.}$$

Total deformation: $\quad \theta = 30 \times 0.0017495 = 0.052485 \text{ rad} = 3.007°$

20. Thin-Walled Tube

Thin-walled tubes are frequently loaded in torsion. It can be shown[10] that the shear stress in the wall is given by the following equation

$$s_s = \frac{T}{2Ac} \qquad (33)$$

Here A is the area enclosed by the line running through the center of the wall section, and c is the width of the narrowest wall. The length of the wall, center to center, is designated by a.

The deformation per axial inch of the tube is given by

$$\theta_1 = \frac{T}{4A^2G} \Sigma \frac{a}{c} \qquad (34)$$

The summation is to extend around the entire circumference of the tube.

Example 19. Find the value of the maximum stress in the tube with the cross section of Fig. 3-27. Applied torque is 100,000 in. lb. Also find the angular deformation per axial inch if the tube is steel.

Solution. $\qquad A = 4.5 \times 9 = 40.50 \text{ in.}^2$

By Eq. (33): $\quad s = \frac{100{,}000}{2 \times 40.50 \times 0.5} = 2{,}470 \text{ psi}$

By Eq. (34): $\quad \theta_1 = \frac{100{,}000}{4 \times 40.5^2 \times 11{,}500{,}000}\left[2 \times \frac{9}{0.5} + 2 \times \frac{4.5}{1}\right]$

$$= 0.0000596° \text{ rad per in. or } 0.0034° \text{ per in.}$$

A stress concentration occurs at the reentrant corners of tubes. For square tubes of uniform wall thickness, values of the concentration factor may be had from Table 3-5. These results are valid only when the length of the sides of the tube is at least 15 or 20 times as great as the thickness.

[10] See p. 278, Reference 7, end of chapter.

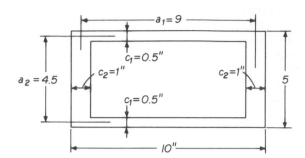

Figure 3-27 Rectangular tube in torsion. Example 19.

If a tube is slit longitudinally, calculations for stress and angular deformation can be made by use of the equations for a wide rectangular bar as described in Section 17. Slitting a tube greatly reduces the torque capacity and increases the angular deformation.

21. Materials Used for Shafting

When service requirements are not too severe, the least expensive shaft material is hot-rolled, plain-carbon steel. For maximum machinability, a normalizing or annealing treatment may be necessary to improve the grain structure and to secure uniformity. Since hot-rolled bars as received from the mill are usually covered with scale, the shaft must be machined all over if a smooth surface is desired.

Cold-drawn bars, in contrast, have a smooth, bright finish and have diameters held to tolerances of a few thousandths of an inch. This material is sometimes erroneously called cold-rolled shafting. It is available in both plain-carbon and alloy compositions, and is in wide use in the field of general power transmission, since the amount of machining required is minimal. Cold drawing improves the physical properties; it raises the values for tensile strength and the yield point. When greater accuracy is required, shafting that has been turned and ground can be secured from the steel warehouses.

If greater strength is needed than can be secured by the use of a low-carbon steel in the as-rolled condition, a steel of somewhat higher-carbon content can be used. After the machining has been completed, the tensile and yield strengths and hardness can be increased by a quenching and tempering heat treatment. To respond to quenching, the carbon content must be about 0.30 per cent or more. For forged shafts, such as are used in internal combustion engines and railroad cars, the carbon content is usually 0.45 or 0.50 per cent. A widely used steel for such service is plain-carbon steel 1045.

When service conditions are more severe, or when certain desirable physical properties are to be obtained, an alloy steel can be used. As a rule, such steels are not used unless the part is to be heat treated, since full advantage of the expensive alloying elements can be secured only in this way. When heat treated to high strength and hardness, alloy steels are tougher, more ductile, and better adapted to shock and impact loads than are plain-carbon steels. The effect of the quenching penetrates deeper in alloy steels than in carbon steels, and a greater volume of the part is strengthened than if the hardening were confined to a shallow zone over the surface. Alloy steels warp and distort less in heat treatment, have less tendency to crack, and have smaller residual stresses than have carbon steels. Although practically all the alloy steels find application in the field of shafting, chromium-molybdenum steel 4140 and chromium-nickel-molybdenum steels 4340 and A8640 are in wide use as general-purpose alloy steels.

For equal hardness, alloy steels are superior in machining qualities. Where considerable machining is required, shop costs can sometimes be reduced by use of a free-cutting steel, such as 1137. This material is high in manganese and has a relatively high machinability rating for a heat-treating alloy steel.

If the service requirements demand resistance to wear rather than extreme strength, it is customary to harden only the surface of a shaft. The case-hardening or carburizing process is in wide use. Carbon is absorbed by a relatively thin layer while the part is held at a red heat in the furnace. Low-carbon alloy steels, such as 4320, 4820, and A8620, are frequently used for carburizing. The cyaniding and nitriding processes are also used to produce a hard surface. Sometimes it is necessary to localize the wear to a relatively small area. The hardening treatment is applied to those surfaces requiring it; the remainder of the shaft is left in its original condition.

ASME Code B17c-1927 recommends that the working stress s_{smax} in shear be taken at 8,000 psi for "Commercial Shafting" but without any definite specifications for physical and chemical properties of the material. The ultimate strength of such steels may range from 45,000 to 70,000 psi. The corresponding elastic limits would be from 22,500 to 55,000 psi. When there is a keyway at the section for which the stress calculations are made, the working stress is to be reduced to 75 per cent of the value for a solid circular shaft. This reduction can be considered as making allowance for both loss of section and stress concentration.

REFERENCES

1. *ASME Annual Catalog and Directory.*

2. Faupel, J. H., *Engineering Design*, New York: John Wiley & Sons, Inc., 1964.

3. Lessels, J. M., *Strength and Resistance of Metals*, New York: John Wiley & Sons, Inc., 1954.

4. Peterson, R. E., *Stress Concentration Factors in Design*, New York: John Wiley & Sons, Inc., 1953.

5. Roark, R. J., *Formulas for Stress and Strain*, 3d ed., New York: McGraw-Hill Book Company, 1954.

6. Seely, F. B., and J. O. Smith, *Advanced Mechanics of Materials*, 2d ed., New York: John Wiley & Sons, Inc., 1952.

7. Timoshenko, S., *Strength of Materials*, Part II, 2nd ed., New York: Reinhold Van Nostrand, 1941.

8. Timoshenko, S., and J. N. Goodier, *Theory of Elasticity*, 2d ed., New York: McGraw-Hill Book Company, 1957.

PROBLEMS

1. A shaft carries a steady torque of 30,000 in. lb at a shearing stress of 8,000 psi. What is the diameter of the shaft?   *Ans.* $d = 2.673$ in.

2. The torsional deformation of a steel shaft is to be 1° in a length of 2 ft when the shearing stress is equal to 10,000 psi. Find the diameter of the shaft.
Ans. $d = 2.392$ in.

3. Suppose it is specified that the deflection at the center of a simply supported shaft under its own weight should not exceed 0.010 in. per foot of span.

(a) Find the maximum permissible span for a 3-in.-diam steel shaft.

(b) Find the stress caused by its own weight when the span length is determined as in part (a).   *Ans.* $l = 13$ ft; $s = 2,300$ psi.

4. The bent bar *ABCDE* in Fig. 3-28 lies in the horizontal plane and is simply supported at *A* and *E*. Angles at *B* and *D* are 90° each. Draw a view of the element lying on the top surface midway between *C* and *D* and show the values of the stresses. Mark the *CD* direction on your sketch. Diam = 2 in.
Ans. $s = 1,590$ psi; $s_s = 3,820$ psi.

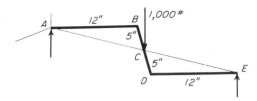

Figure 3-28 Problem 4.

5. Make a free-body diagram for the unit consisting of shaft *AB* and disk *C* (see Fig. 3-29), and show all forces and torques necessary for equilibrium. Coefficient of friction $\mu = 0.3$. Sufficient force is exerted between the disks to develop the full coefficient of friction. The entire torque is carried by the torque reaction at *A*. Force reactions are applied at the bearings as shown.

6. Find the value of dimension *b* in Fig. 3-30 that causes the slope of the shaft to be zero at the bearings.   *Ans.* $b = 40$ in.

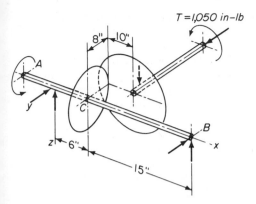

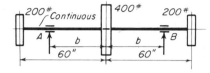

Figure 3-29 Problem 5.   **Figure 3-30** Problem 6.

7. (a) Apply the equation for combined stress and find the value of the maximum shearing stress for the element at A in Fig. 3-2.

(b) If this shaft rotates, find the value of the maximum shearing stress by the ASME Code.   *Ans.*   (a) 6,150 psi; (b) 6,630 psi.

8. The element at A in Fig. 3-31 is located on the bottom of the shaft; the element at B is at the elevation of the shaft axis.

(a) Compute the value of the bending and shear stresses in directions parallel and perpendicular to the shaft axis for the element at A. Draw a view of the element showing stress arrows and mark the value of each. Compute the value of the maximum shearing stress for this location on the shaft. Loads are steady and the shaft does not rotate.

(b) Repeat (a) for the element at B. Take into account the transverse shear for this location.

(c) Find the value of the maximum shear stress for the element at A for steady loads and a rotating shaft by the ASME Code.   *Ans.*   (c) 6,890 psi.

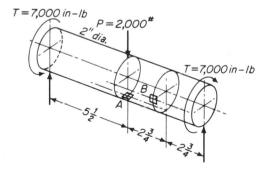

Figure 3-31 Problems 8 and 9.

9. Work Problem 8 using Fig. 3-31 but with the length of the shaft not given. Find the value of length l that will make the maximum value of the shear stress at both elements equal. Take into account the transverse shear. The loads are steady and the shaft is nonrotating.   *Ans.*   6.25 in.

10. The shaft of Fig. 3-32 does not rotate, and is simply supported at A and B. The element at C is on the top surface; the element at D is at the elevation of the shaft axis.

(a) Draw a view of the element at C with sides parallel and perpendicular to the shaft axis; show arrows representing stresses, together with numerical values.

(b) Draw a view for the element at C, properly oriented with respect to the shaft axis, which has the maximum shearing stress. Show arrows and numerical values for all stresses acting.

(c) Work (b) for the element at D.   *Ans.*   (b) $s_{smax} = 9,180$ psi.

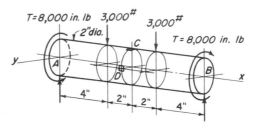

Figure 3-32 Problem 10.

11. The shaft of Fig. 3-33 is simply supported at A and B and is keyed against rotation at A.

(a) Draw a view of the element on the top surface of the shaft at B with sides parallel to the x- and y-axes. Show arrows and numerical values for all stresses acting.

(b) Draw the element at B properly oriented to give the maximum shearing stresses. Show arrows and numerical values for all stresses.

Ans.   (b) $s_{smax} = 12,010$ psi.

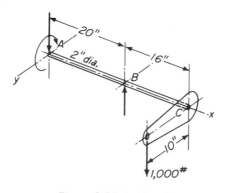

Figure 3-33 Problem 11.

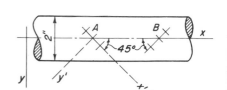

Figure 3-34 Problem 12.

12. Resistance-wire strain gages giving deformations at 45° with the shaft axis are attached at A and B in Fig. 3-34. Lead wires are carried from the gages through slip rings and brushes to electric instruments, which permit the deformations to be determined. The shaft carries torque only. Find its value if the elongation at A in x'-direction is 0.0006 in./in. positive, and if elongation at B in the y'-direction is the same amount but negative. Draw a view of the element with sides parallel to the x- and y-axes and show stresses acting. Also draw a view for the element with sides parallel to x'- and y'-axes. $\mu = \frac{1}{3}$; $E = 30,000,000$ psi.

Ans. $T = 21,200$ in. lb.

13. Find the permissible weight of the flywheel in Fig. 3-35 if the value of the maximum shearing stress in the shaft is to be 9,000 psi. The shaft rotates and carries a steady torque of 40,000 in. lb. Use the ASME Code.   *Ans.* $W = 1,600$ lb.

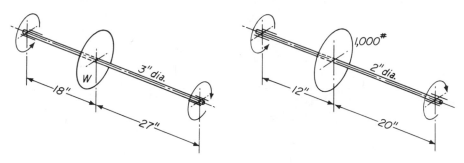

Figure 3-35 Problem 13.   Figure 3-36 Problem 14.

14. How much torque will the shaft of Fig. 3-36 carry if the maximum shearing stress is not to exceed 8,000 psi? Loads are steady and the shaft rotates. Use the ASME Code.   *Ans.* $T = 5,600$ in. lb.

15. The shaft of Fig. 3-37 carries a torque of 20,000 in. lb. What must be the distance x in order to make the maximum shearing stress 8,000 psi? The shaft rotates and loads are steady. Use the ASME Code.   *Ans.* $x = 7.84$ in.

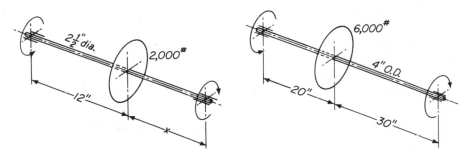

Figure 3-37 Problem 15.   Figure 3-38 Problem 16.

16. The shaft of Fig. 3-38 carries a torque of 50,000 in. lb. How large a hole may be drilled through the shaft so that the maximum shearing stress does not exceed 10,000 psi? Loads are steady and the shaft rotates. Use the ASME Code.

Ans.   1.92 in.

17. A hollow shaft has a hole of diameter λd, where d is the outside diameter of the shaft, and λ is the appropriate constant. Show that the equation for the shearing stress by the ASME Code will be the same as Eq. (14) except that factor $(1 - \lambda^4)$ will appear in the denominator on the right-hand side.

18. What diameter hollow shaft is required to carry a bending moment of 16,200 in. lb together with a torque of 40,000 in. lb if the diameter of the hole is equal to 0.6 of the outside diameter of the shaft? Loads are steady and the shaft rotates. Maximum shearing stress equals 10,000 psi.   *Ans.* $d = 3.014$ in.

19. Determine the required diameter for the hollow shaft of Fig. 3-39 having a hole diameter 0.6 as great as the shaft diameter. The maximum shear stress is to be 12,000 psi. The shaft rotates and loads are steady. Use the ASME Code.

Ans. $d = 1.982$ in.

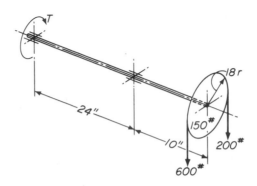

Figure 3-39 Problem 19.

20. A hollow shaft has a hole of diameter λd_o, where d_o is the outside diameter and λ is a constant. Find the value of the ratio of d_o to the diameter d of a solid shaft for equal shearing stresses caused by the same torques for values of λ of 0.5, 0.55, and 0.6.   *Ans.* $d_o/d = 1.022$, 1.033, and 1.047.

21. What is the ratio of the weight of the hollow shaft per unit length to the weight of the solid for the foregoing values of λ?   *Ans.*   0.783, 0.744, and 0.702.

22. Find the required shaft diameter (a) by the maximum shear theory, and (b) by the Mises–Hencky theory for the following conditions. The torque varies from 0 to 12,000 in. lb. The bending moment varies from 6,000 to 10,000 in. lb. Stress concentration due to fillet for both bending and torque is equal to 2.5. The shaft does not rotate. The material tests $s_{ult} = 60,000$ psi and $s_{yp} = 40,000$ psi. Get s_e from Fig. 2-16 for a machined surface. The factor of safety equals 2 based on yield point.   *Ans.*   (a) 2.63 in.; (b) 2.53 in.

23. The belt tensions for the pulleys of Fig. 3-40 fluctuate from the values shown in (a) to those given in (b). Details of shaft and hubs are shown in (c). Consider the stress concentration as being due to fillets only. Material tests $s_{yp} = 69,000$ psi. and $s_{ult} = 104,000$ psi. Get s_e from Fig. 2-16 for a machined surface. Let $D/d = 1.20$ and $r/d = 0.125$. The factor of safety is 1.9 based on yield point.

Draw and dimension the bending moment diagram. Based on the moment just left of the right hub, find the value of d (a) by the maximum shear theory, and (b) by the Mises–Hencky theory. Include the dead weight of the pulleys.

Ans.   (a) 2.271 in.; (b) 2.251 in.

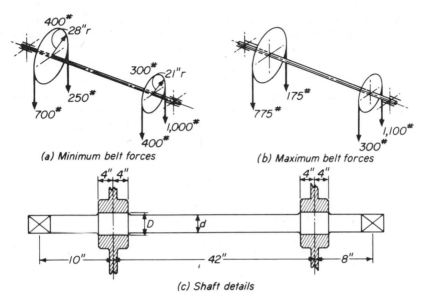

(a) Minimum belt forces   (b) Maximum belt forces

(c) Shaft details

Figure 3-40 Problem 23.

24. Solve Problem 13, Fig. 3-35, by the maximum shear theory. Let the shaft be of uniform diameter, but let it have a keyway at the flywheel of the usual proportions with a stress concentration factor of 1.35. Let $s_e = 0.5s_{yp}$ for the shaft material. Take working stress equal to 90 per cent of the value for the shaft without a keyway.

Ans. $W = 536$ lb.

25. A $2\frac{1}{2}$-in.-diam shaft is made of steel with $s_{ult} = 120,000$ psi and $s_{yp} = 90,000$ psi, machined surface. The shaft rotates and carries a steady 2,500-lb load at the center of a 40-in. simply supported span. Average torque is 20,000 in. lb. Assume $T_r = 0.1T_{av}$. Let $K = K_t = 1.7$. On the basis of (a) the maximum shear theory and (b) the Mises–Hencky theory, find the *FS* for this shaft.

Ans.   (a) $FS = 1.517$; (b) $FS = 1.534$.

26. A $2\frac{1}{2}$-in.-diam shaft has a key 0.625×0.625 in. The shaft material tests 60,000 psi at yield point. Let $s_{syp} = 0.5s_{yp}$. The factor of safety equals 2. The shaft fits into a cast-iron hub for which the working stress in compression is 18,000 psi. What length of key in the hub material will be required to carry the torque of the solid shaft? The key material is assumed to be amply strong.   *Ans.*   6.55 in.

27. A 3-in.-diam shaft of material with a yield point value of 50,000 psi has a $0.75 \times 0.75 \times 5$-in. key. What must the minimum yield point value be for the material in the key in order to transmit the torque of the shaft? The factor of safety equals 2. $s_{syp} = 0.5s_{yp}$.   *Ans.*   47,100 psi.

28. A square key has a breadth equal to one-fourth of the shaft diameter. The shaft and key are of materials that are equally strong with a yield point value in shear equal to one-half the yield point value in tension. Find the required length of the key in terms of shaft diameter necessary to transmit the shaft torque.

Ans. $l = 1.57d.$

29. A 3-in.-diam shaft is transmitting 400 hp at 600 rpm. A solid coupling similar to that shown in Fig. 3-13 has 6 bolts, each $\frac{3}{4}$ in. in diameter. Find the required diameter of the bolt circle based on an average shearing stress of 4,000 psi in the bolts.

Ans. Diam = 7.93 in.

30. Make the horizontal and vertical load and moment diagrams for the shaft shown in Fig. 3-41. Find the location and value of the minimum bending moment in the shaft for the portion lying between the left pulley and the right bearing.

Ans. Min $M = 17,050$ in. lb for $x = 7.88$ in.

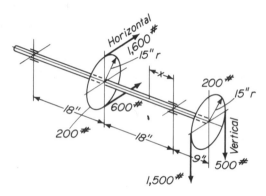

Figure 3-41 Problem 30.

31. Repeat Problem 30 for the shaft in Fig. 3-42. The minimum moment should be for the portion between the two pulleys.

Ans. Min $M = 10,540$ in. lb for $x = 19.96$ in.

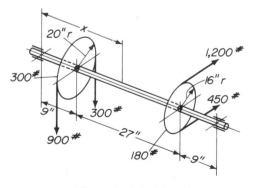

Figure 3-42 Problem 31.

32. The shaft shown in Fig. 3-43 is fixed at the ends. Find the reactions at the ends and the stresses in each portion of the shaft. Draw views of the elements for each portion and show the stresses acting.   *Ans.* $s_s = 17,190$ and 9,550 psi.

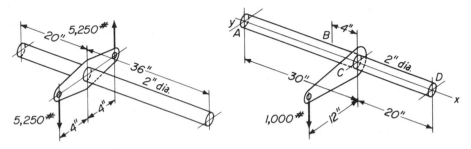

Figure 3-43 Problem 32.
Figure 3-44 Problem 33.

33. The shaft in Fig. 3-44 is simply supported at A and D but is keyed against rotation at both points.

(a) Find reactions at the ends, and draw a view of the element at B on the top surface of the shaft. Show all stresses acting and their numerical values.

(b) Compute the value of the maximum shearing stress at B and the angle at which it acts. Make a view of the element properly oriented showing maximum shearing stresses acting as well as the normal stresses on all faces.

Ans.   (a) $s_x = 13,240$ psi; $s_{xy} = 3,060$ psi.

34. The shaft in Fig. 3-45 is simply supported at A and C, but is keyed against rotation. Draw and dimension the bending moment diagram, and find all reactions at the ends.

Find the resultant stress from all causes for the elements on the top surface of the shaft at D and F. Do the same for the elements at E and G at the elevation of the center line. Draw sketches for the elements with arrows properly directed for the stresses and show numerical values.   *Ans.*   At D, $s_s = 3,060$ psi; $s = 9,170$ psi.

At E, $s_s = 3,230$ psi; $s = 0$.

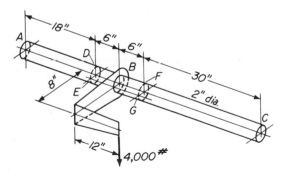

Figure 3-45 Problem 34.

35. The shaft of Fig. 3-46 is built in at A and D. Find the value of the torque reactions at the ends.   *Ans.*   39,230 in. lb; 60,770 in. lb.

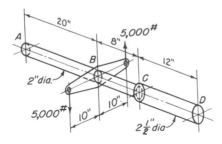

Figure 3-46 Problem 35.

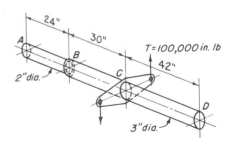

Figure 3-47 Problem 36.

36. The shaft of Fig. 3-47 is built in at A and D. Find the value of the torque reactions at A and D.   *Ans.*   21,706 in. lb; 78,294 in. lb.

37. (a) Find the reactions, and draw and dimension the bending moment diagram for the shaft of Fig. 3-48. All bearings are on immovable supports at the same elevation. Include the effect of the dead load of the shaft. For the shaft, $E = 30,000,000$ psi: $\gamma = 0.283$ lb/in.^{3}

(b) Find the values of the reactions if the bearing at C is $\frac{1}{8}$ in. lower than the others. Include the effect of the dead load of the shaft.

(c) Let the center support consist of a 6-in. I-beam, 12.5 lb per ft $(I = 21.8$ in.$^4)$, 12 ft long, simply supported, with bearing C at its center. Find the reactions for the shaft, and draw and dimension the bending moment diagram. Neglect effects of the dead loads.

(d) What must the moment of inertia be for a beam supporting the bearing at C if the value of the bending moments, as caused by the 1,000-lb loads, at points B, C, and D, are to be numerically equal? What will be the deflection of point C? Neglect effects of the dead loads.   *Ans.*   (a) $R_1 = 253$ lb; $R_2 = 1,734$ lb.
(b) $R_1 = 460$ lb; $R_2 = 1,320$ lb.
(c) $R_1 = 398$ lb; $R_2 = 1,204$ lb.
(d) $I = 123$ in.^{4}

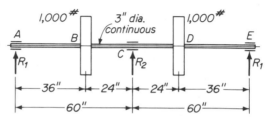

Figure 3-48 Problem 37.

38. (a) What moment of inertia would be required for the center beam of Fig. 3-49 if the bending moments for the shaft, as caused by the 1,000-lb loads, at points B, C, and D, are to be numerically equal? Neglect effects of the dead loads.

(b) If all three beams are 8-in. I-beams, 18.4 lb per ft, find the three bearing reactions, and draw the bending moment diagram for the shaft. Neglect effects of the dead loads.

$$\textit{Ans.} \quad \text{(a)} \quad I = 90.2 \text{ in.}^4$$
$$\text{(b)} \ R_1 = 278 \text{ lb}; \ R_2 = 1{,}443 \text{ lb}.$$

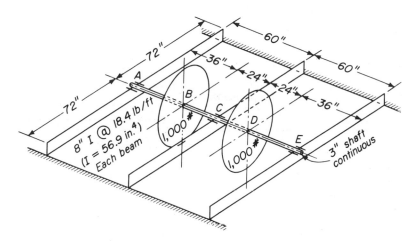

Figure 3-49 Problem 38.

39. Bearings A and B in Fig. 3-50 rest on unyielding supports. The bearing at C is located at the center of a simply supported 6-in. I-beam, 12 ft long. Ignore the effects of the dead loads.

(a) Find reactions at A, B, and C.

(b) If the bearing at C is resting on an unyielding support, find the three reactions.

(c) What change in elevation of bearing C of part (b) must be made if bending moments in the shaft at load and at C are to be numerically equal?

$$\textit{Ans.} \quad \text{(a)} \ R_a = 322 \text{ lb}; \ R_c = 533 \text{ lb}; \ R_b = 105 \text{ lb downward}.$$
$$\text{(b)} \ R_a = 281 \text{ lb}; \ R_c = 656 \text{ lb}; \ R_b = 187 \text{ lb downward}.$$
$$\text{(c)} \ 0.039 \text{ in. higher}.$$

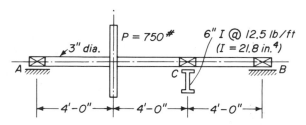

Figure 3-50 Problem 39.

40. A rotating shaft has the 1,000-lb centrifugal forces acting as shown in Fig. 3-51.

(a) Draw the bending moment diagram for the shaft. Note that large stresses are possible, even though bearing reactions are equal to zero.

(b) Suppose an additional bearing is placed in the center of the shaft. Find the reactions, and draw and dimension the bending moment diagram. Ignore the effect of the dead-load deflection. Note how the presence of the central bearing can affect the bearing loads of an engine crankshaft.

    Ans.   (b) Reactions: 427 lb downward; 854 lb upward; 427 lb downward.

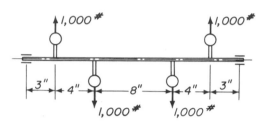

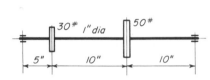

Figure 3-51 Problem 40.   Figure 3-52 Problem 41.

41. Find the lowest critical speed for the steel shaft shown in Fig. 3-52.
                            Ans. $n_{cr} = 1,700$ rpm.

42. The static deflection at the center of the steel shaft in Fig. 3-53 is equal to 0.0125 in. Find the value of the critical speed.   *Ans.* $n_{cr} = 1,900$ rpm.

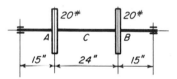

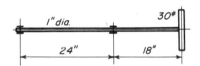

Figure 3-53 Problem 42.   Figure 3-54 Problem 43.

43. Find the value of the critical speed for the steel shaft of Fig. 3-54.
                            Ans. $n_{cr} = 617$ rpm.

44. Find the value of the critical speed for the steel shaft in Fig. 3-55.

Ans. $n_{cr} = 426$ rpm.

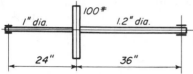

Figure 3-55 Problem 44.   **Figure 3-56** Problem 45.

45. Find the deflection at the load and the slope at the end for the steel shaft shown in Fig. 3-56.   *Ans.* $y_c = 0.0593$ in.; $\theta = 0.132°$.

46. Find the deflection at the load and the slope at the left end of the steel shaft in Fig. 3-57.   *Ans.* $y = 0.0136$ in.; $\theta = 0.050°$.

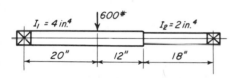

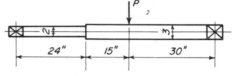

Figure 3-57 Problem 46.   **Figure 3-58** Problem 47.

47. In Fig. 3-58, find the value of load P if the slope at the left bearing is $0.25°$. Steel shaft.   *Ans.* $P = 776$ lb.

48. Find the torque that a long piece of $3 \times 3 \times \frac{1}{2}$-in. angle iron can carry if the maximum shearing stress at the fillet is 12,000 psi. The radius of the fillet is 0.5 in.   *Ans.* $T = 3,320$ in. lb.

49. (a) What percentage more torque will a square shaft carry than a round shaft of the same diameter if both have the same shear stress?

(b) If the cost per pound is the same, what will be the percentage increase in cost of the square over the round shaft?   *Ans.* (a) 5.9%; (b) 27.3%.

50. Two pieces of shafting have the cross sections shown in Fig. 3-59. If the shear stresses in the two shafts are equal, find the ratio between the torques that the shafts are carrying. Approximate the stress concentration factor for the shaft of (a) from Table 3-5.

What will be the ratio between the angular displacements if the applied torques are equal?   *Ans.*   6.27; 8.04.

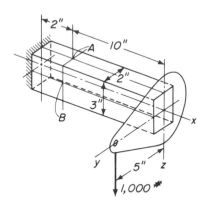

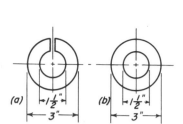

(a)   (b)

Figure 3-59 Problem 50.   **Figure 3-60** Problem 51.

51. The rectangular shaft in Fig. 3-60 is fixed at the wall.

(a) Cut the shaft at a point 2 in. from the wall, remove the portion to the right, and show all forces and moments on the end surface of the part that remains.

(b) Find the stresses from all causes on an element at *A* at the center of the top surface of the shaft at the cut. Take sides of the element parallel to the coordinate axes.

(c) Repeat (b) for an element at *B* at the center of the vertical side of the cut on the near side.   *Ans.*   At *A*, s_s = 1,550 psi; s = 3,330 psi.

52. Work Problem 51, using Fig. 3-60, with the same data and dimensions except that the shaft is now oriented as shown in Fig. 3-61.

Ans.   At *B*, s_s = 2,020 psi; s = 2,500 psi.

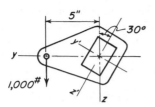

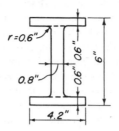

Figure 3-61 Problem 52.   **Figure 3-62** Problem 53.

53. If the torque applied to a long bar having the cross section shown in Fig. 3-62 is equal to 5,000 in. lb, find the value of the maximum shearing stress.

Ans. $s_s = 4,890$ psi.

54. The beam shown in Fig. 3-63 is simply supported but is keyed at the ends to prevent rotation.

   (a) Find all reactions at the ends.

   (b) Find the value of the maximum shearing stress due to torque.

   (c) Find the angular rotation of the bracket in degrees.

Ans. $s_s = 8,520$ psi; $\varphi = 0.416°$.

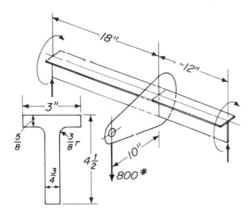

Figure 3-63 Problem 54.

55. (a) Make an isometric drawing of the crankshaft shown in Fig. 3-64, and place all reactions thereon that are necessary for equilibrium.

(b) Cheek CD is rectangular in cross section, 2 in. wide in the x-direction, and 3 in. deep in the z-direction. Cut the cheek midway between C and D, and show all forces and moments acting on the cut surface.

(c) Draw the element at R lying at the midpoint of the top surface of the cheek with sides parallel to the coordinate axes, and show all stresses acting. Mark directions of the axes on the sketch.

(d) Repeat (c) for the element at S at the center of the near-side vertical surface of the cheek.

(e) Suppose the loading consists of a force of 3,000 lb in the xy-plane acting to the left at A parallel to the y-axis. Find the bearing reactions for the shaft. Cut the cheek midway between C and D, and show all forces and moments acting on the cut surface.

(f) Draw the element at R showing all stresses acting.

(g) Draw the element at S showing all stresses acting.

$Ans.$ (c) $s = 4,667$ psi; $s_s = 3,717$ psi.
(d) $s = 0$; $s_s = 5,329$ psi.

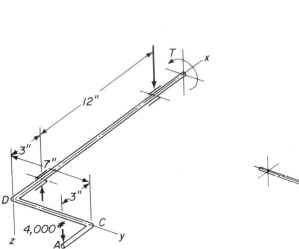

Figure 3-64 Problem 55.

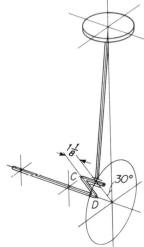

Figure 3-65 Problem 56.

56. The stroke of the air compressor shown in Fig. 3-65 is 3 in., and the length of the connecting rod is 5 in. Crank cheek CD is circular in cross section and is 1 in. in diameter. If the torsional shearing stress in the cheek is equal to 12,000 psi, find the value of the gas force on the piston. $Ans.$ Gas force $= 3,320$ lb.

57. (a) The bore of the air compressor of Fig. 3-66 is 4 in., and the stroke is 6 in. The value of the air pressure is 300 psi gage for a crank angle of 45°. The length of the rod is 11 in. Axes x, y, and z are mutually perpendicular. The shaft is turned

by a pure torque applied at the right bearing as shown. Make sketches similar to
Figs. 3-16(b), (d), and (e) for this problem, and place the value of all necessary forces
and moments thereon.

(b) The cheek is $\frac{7}{8}$ in. wide in the x-direction and $1\frac{3}{4}$ in. deep in the z-direction.
Make an enlarged sketch for the element with sides parallel to the x- and y-axes
lying at the center of the top surface of the cheek. Show all stresses acting.

(c) Repeat (b) for the element lying at the center of the right vertical side of the
cheek.   *Ans.* (b) $s = 15,370$ psi; $s_s = 5,060$ psi.
                    (c) $s = 6,990$ psi; $s_s = 7,910$ psi.

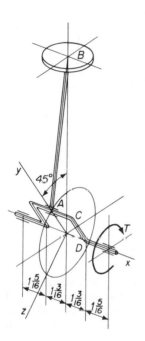

Figure 3-66 Problem 57.

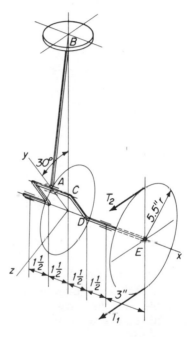

Figure 3-67 Problem 58.

58. A belt-driven air compressor has the crank mechanism shown in Fig. 3-67.
The axes x, y, and z are mutually perpendicular. The bore is 4 in.; stroke is 5 in.;
length of the rod is 9 in. For a crank angle of 30° the air pressure is 100 psi gage.
The belts are perpendicular to the axis OB of the cylinder. The tight-side tension
is three times that of the slack side.

Make sketches similar to Figs. 3-16(b), (c), (d), and (e) for this problem, and
place all necessary forces and moments thereon.

Ans.   Axial force, $F_y = 322$ lb.
        Transverse shear, $F_z = 698$ lb.
        Moment, $M_x = 1,080$ in. lb.
        Torque, $M_y = 797$ in. lb.
        Moment, $M_z = 1,549$ in. lb.

59. An air compressor has a crank arrangement similar to that shown in Fig. 3-16. The stroke is 4 in. and length of the rod is 7 in. The crank cheek is circular in cross section, and is 1 in. in diameter. Distance AC is equal to $1\frac{1}{8}$ in. If the torsional shearing stress at the midlength of the cheek is equal to 10,000 psi for a crank angle of 30°, find the value of the gas force on the piston.

Ans.   Gas force = 2,792 lb.

60. Find the deflection of the weight in Fig. 3-68. The bending moments are negligible since bearings are located close to the crank arms. Shafts and cranks lie in a horizontal plane. Ignore the effect of bending in cranks. The material is steel.

Ans. $\delta = 0.146$ in.

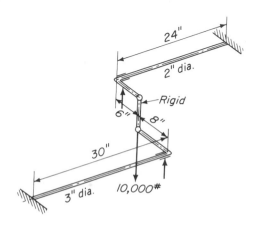

Figure 3-68 Problem 60.

61. Find the deflection of the weight in Fig. 3-69. The shaft and beam lie in horizontal planes, and are built in at the walls. Bending in crank and shaft is negligible, as is the extension in link connecting the members together. All joints are frictionless. The material is steel.   *Ans.* $\delta = 0.244$ in.

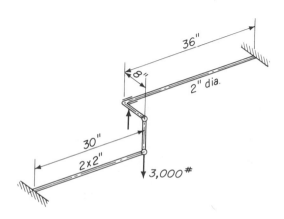

Figure 3-69 Problem 61.

62. Work Problem 61 but with the connection between the end of the crank and the beam replaced by a flexible member with a spring rate equal to 16,000 lb/in.
Ans. $\delta = 0.309$ in.

63. Find the three reactions for the shaft in Fig. 3-70. Assume the effect of the dead-load deflection of the beam to be negligible. *Ans.* $13wl/16$; $33wl/16$; $wl/8$.

64. If reactions R_1, R_2, and R_3 of Fig. 3-70 have values of $15wl/16$, $27wl/16$, and $3wl/8$, respectively, find the amount that support R_2 is lower than R_1 and R_3. Ignore the effect of dead-load deflection of the shaft. *Ans.* $\delta = wl^4/6EI$.

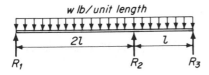

Figure 3-70 Problems 63 and 64.

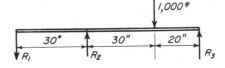

Figure 3-71 Problem 65.

65. The three supports in Fig. 3-71 are all at the same elevation. Ignore the dead load of the shaft. Find the values of the three reactions.
Ans. $R_1 = 175$ lb, down; $R_2 = 680$ lb, up; $R_3 = 495$ lb, up.

66. The three supports in Fig. 3-72 are all at the same elevation. Ignore the dead load of the shaft. Find the values of the three reactions.
Ans. $R_1 = 422$ lb, down; $R_2 = 1,063$ lb, up; $R_3 = 359$ lb, up.

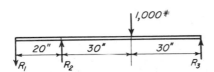

Figure 3-72 Problem 66.

67. The center bearing in Fig. 3-73 rests on an immovable support. End bearings rest on structural beams with the spring rates shown. For the shaft, $EI = 7,200,000$ lb in.² Find the load carried by an end bearing. *Ans.* 391 lb

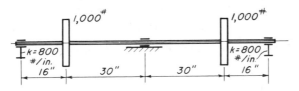

Figure 3-73 Problem 67.

68. Bearings A and B in Fig. 3-74 rest on immovable supports. The bearing at C rests on an I-beam with a spring rate of 5,000 lb/in. The shaft is steel. Find the values of the bearing reactions.   *Ans.* $R_3 = 295$ lb.

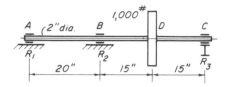

Figure 3-74 Problem 68.

69. Work Problem 19 except with the 10-in. dimension as 12 in. and the 18-in. radius as 15 in.

70. Make a figure like that in Fig. 3-45 except that the 12-in. length becomes 16 in., and the 8-in. length becomes 7 in. Mark numerical values for all the reactions on the drawing. Draw and dimension the bending moment diagram for the shaft.

71. Work Problem 37(c) but with only one pulley, on the right, present. Find the values of the bearing loads.

72. Show that the deflection at the center of the shaft of Fig. 3-75 is given by the following equation.

$$y = \frac{Pl^3}{162E}\left(\frac{1}{I_1} + \frac{19}{8I_2}\right)$$

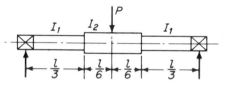

Figure 3-75 Problem 72.

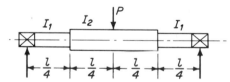

Figure 3-76 Problem 73.

73. Show that the deflection at the center of the shaft of Fig. 3-76 is given by the following equation.

$$y = \frac{Pl^3}{384E}\left(\frac{1}{I_1} + \frac{7}{I_2}\right)$$

74. Find the deflection at point A in Fig. 3-77. Consider the bearings as simple supports.

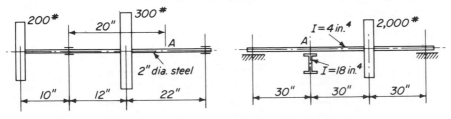

Figure 3-77 Problem 74.

Figure 3-78 Problem 75.

75. In Fig. 3-78 the end bearings rest on immovable supports. The steel shaft $I = 4$ in.[4] The beam is 9 ft long, simply supported, with the bearing at the center, $I = 18$ in.[4] Find the bearing load at A.

76. The numerical values for the bending moments at B, C, and D in Fig. 3-79 are equal. Find the difference in elevation between the bearing at C and those at A and E.

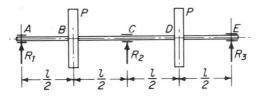

Figure 3-79 Problem 76.

77. The shaft of Fig. 3-80 has the bearing reactions shown. Find the change in elevation of the center bearing with respect to those at the ends.

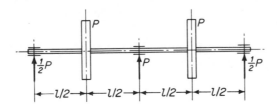

Figure 3-80 Problem 77.

78. Show that the deflection at the center of the shaft of Fig. 3-81 is given by the following equation.

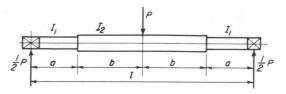

Figure 3-81 Problem 78.

$$y = \frac{P}{24E}\left[\frac{4a^3}{I_1} + \frac{b}{I_2}(8a^2 + 4ab + l^2)\right]$$

79. A square steel tube is 8 in. on the side with wall thickness 0.2 in. Find the shearing stress in the wall if a torque of 100,000 in. lb is applied to the tube. Find the angular deformation per axial inch.

4 Springs

When flexibility or deflection in a mechanical system is specifically desired, some form of spring can be used. Otherwise, the elastic deformation of an engineering body is usually a disadvantage. Springs are employed to exert forces or torques in a mechanism or to absorb the energy of suddenly applied loads. Springs frequently operate with high values for the working stresses, and with loads that are continuously varying.

Helical and leaf springs are in widest use. A number of other types, such as Belleville, disk, spiral, ring, and volute springs, are shown in Fig. 4-1.

A, area
c_1, spring index
d, diameter of wire
E, modulus of elasticity, normal stress
FS, factor of safety
G, modulus of elasticity, shear stress
I, moment of inertia
J, polar moment of inertia
k, spring rate
K_c, stress concentration factor due to curvature

N, number of active coils
P, load
R, mean radius of helix
s_s, shearing stress
s_{syp}, yield point stress in shear
s_{ult}, ultimate tensile stress
s_{yp}, yield point stress in tension
s'_{se}, endurance limit in shear, zero to maximum stress
T, torque
δ, (delta) deflection of helical spring

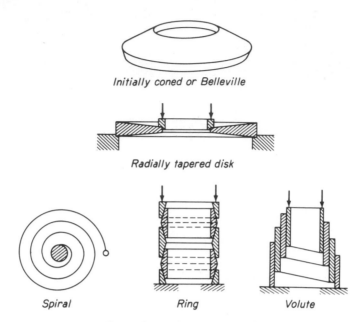

Initially coned or Belleville

Radially tapered disk

Spiral   Ring   Volute

Figure 4-1 Various types of springs.

1. Helical Springs

The equations for the stress and deformation of a closely coiled helical spring are derived directly from the corresponding equations for the torsion of a round bar, as shown in Fig. 4-2(a). The bar of length l and diameter d is

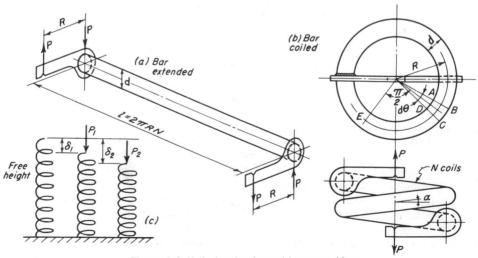

Figure 4-2 Helical spring formed from round bar.

fitted with brackets at each end of length R, and is in equilibrium under the action of the loads P. Assume that the straight bar is bent into the helix of N coils of radius R as shown in Fig. 4-2(b). The coiled bar is in equilibrium under the action of the two equal and opposite forces P.

The stress in the straight bar is shear caused by a torque equal to PR. The major stress in the helix is also torsional shearing stress. From Eq. (2) of the preceding chapter, the stress in Fig. 4-2 is

$$\text{torsional shearing stress} = \frac{Tr}{J} = \frac{16PR}{\pi d^3} \tag{a}$$

After being coiled into the helix, the cross sections have an additional stress from the transverse shear. An exact analysis shows that this stress at the midheight has the value $1.23P/A$. Then

$$\text{transverse shearing stress} = 1.23\frac{P}{A} = \frac{16PR}{\pi d^3} \times \frac{0.615}{c_1} \tag{b}$$

where

$$c_1 = \frac{2R}{d} \tag{1}$$

The total shearing stress s_s on the inside of the coil at the midheight from static load P is given by the sum of Eqs. (a) and (b).

$$s_s = \frac{16PR}{\pi d^3}\left(1 + \frac{0.615}{c_1}\right) \tag{2}$$

By substitution of Eq. (1), this equation assumes the forms

$$s_s = \frac{8Pc_1}{\pi d^2}\left(1 + \frac{0.615}{c_1}\right) \tag{3}$$

$$s_s = \frac{2Pc_1^3}{\pi R^2}\left(1 + \frac{0.615}{c_1}\right) \tag{4}$$

The deflection of the spring can be found by considering the rotation of the cross sections with respect to each other caused by the torque PR. Assume temporarily that element $ABCD$ in Fig. 4-2(b) is flexible, but that the remainder of the spring is rigid. Thus, from Eq. (5), Chapter 3, the rotation $d\varphi$ of section CD with respect to the adjacent cross section AB is equal to

$$d\varphi = \frac{PR\,dl}{JG}$$

This rotation of the differential length of the spring causes point E, located 90° away, to be moved an amount equal to

$$d\delta = R\,d\varphi = \frac{PR^2\,dl}{JG}$$

The total deflection caused by the torque when the entire spring is elastic is found by integration of this equation over the entire length of the spring.

$$\delta = \frac{PR^2 l}{JG} = \frac{64PR^3 N}{d^4 G} \tag{5}$$

The last form of Eq. (5) is obtained by substituting the equivalent values for l and J.

Substitution of Eq. (1) causes the equation above to assume the additional forms:

$$\delta = \frac{8Pc_1^3 N}{Gd} = \frac{4Pc_1^4 N}{GR} \tag{6}$$

An equation for the spring rate k, or the force required for a deflection of 1 in., can be had by considering Fig. 4-2(c).

$$k = \frac{P_1}{\delta_1} = \frac{P_2}{\delta_2} = \frac{P_2 - P_1}{\delta_2 - \delta_1} \tag{7}$$

Another equation for k can be had from Eqs. (5) and (6) by replacing P by k and δ by unity.

$$k = \frac{Gd^4}{64R^3 N} = \frac{Gd}{8c_1^3 N} = \frac{GR}{4c_1^4 N} \tag{8}$$

Example 1. A helical compression spring is made from 0.225-in.-diam steel wire, and has an outside coil diameter of 2 in. There are 8.6 active coils. Find the static load that will cause a shearing stress of 50,000 psi. Find the deflection of the spring.

Solution. $R = \frac{1}{2}(2.0 - 0.225) = 0.8875$ in.,   mean radius

By Eq. (1): $c_1 = \dfrac{2R}{d} = \dfrac{2 \times 0.8875}{0.225} = 7.89$

By Eq. (2): $50,000 = \dfrac{16P0.8875}{\pi 0.225^3}\left(1 + \dfrac{0.615}{7.89}\right) = 428P$

$P = 116.9$ lb

By Eq. (5): $\delta = \dfrac{64 \times 116.9 \times 0.8875^3 \times 8.6}{0.225^4 \times 11,500,000} = 1.526$ in.

No allowance is made in the design for stress concentration because of curvature caused by loads that are static or by loads that may fluctuate only a small number of times during the expected life of the spring.

The average value of G for steels used for springs is equal to 11,500,000 psi.

Torsion bar springs similar to Fig. 4-2(a) are sometimes used. For example, the springs of motor buses can be arranged to run longitudinally along the sides of the bus beneath the floor.

If a helical compression spring is relatively long as compared with its diameter, danger of column action or lateral buckling may exist at loads smaller than the desired working load. Sometimes it is practical to prevent the buckling of a long compression spring by placing it over a loosely fitting bar or in a tube, which serves as a guide. Calculations for buckling loads can be made, but the theory is beyond the scope of this book.[1]

2. Properties of Spring Materials

Helical springs are either cold formed or hot formed depending on the size of the wire. Small sizes are wound cold, but when the bar has a diameter larger than about $\frac{3}{8}$ in., the spring is wound from a heated bar.

Numerous materials are used for cold-formed springs. Sizes and strength properties for several materials are shown in Table 4-1. These are cold-drawn

Table 4-1 STEEL SPRING WIRE. MINIMUM TENSILE STRENGTHS, s_{ult}, AND ASTM DESIGNATIONS

W & M Gage* No.	Diameter d, in.	Hard-Drawn, Class 1, A227-64	Music Wire† A228-63T	Oil-Tempered Class 1, A229-64	302 Stainless Steel,† 18-8 A313-67
25	0.0204	283,000	349,000	293,000	299,000
24	0.0230	279,000	343,000	289,000	294,000
23	0.0258	275,000	337,000	286,000	289,000
22	0.0286	271,000	332,000	283,000	285,000
21	0.0317	266,000	327,000	280,000	278,000
20	0.0348	261,000	323,000	274,000	274,000
19	0.0410	255,000	314,000	266,000	269,000
18	0.0475	248,000	306,000	259,000	262,000
17	0.0540	243,000	301,000	253,000	260,000
16	0.0625	237,000	293,000	247,000	255,000
15	0.0720	232,000	287,000	241,000	250,000
14	0.0800	227,000	282,000	235,000	245,000
13	0.0915	220,000	275,000	230,000	240,000
12	0.1055	216,000	269,000	225,000	232,000
11	0.1205	210,000	263,000	220,000	225,000
10	0.1350	206,000	258,000	215,000	217,000
9	0.1483	203,000	253,000	210,000	210,000
8	0.1620	200,000	249,000	205,000	205,000
7	0.1770	195,000	245,000	200,000	195,000
6	0.1920	192,000	241,000	195,000	190,000
5	0.2070	190,000	238,000	190,000	185,000
4	0.2253	186,000	235,000	188,000	180,000
1/4 in.	0.2500	182,000	230,000	185,000	175,000
5/16 in.	0.3125	174,000		183,000	160,000
3/8 in.	0.3750	167,000		180,000	140,000

*Washburn and Moen.

†Values interpolated to nearest gage size.

[1]See p. 43, Reference 3, end of chapter. See also Burdick, W. E., F. S. Chaplin, and W. L. Sheppard, "Deflection of Helical Springs Under Transverse Loadings," *Trans. ASME*, **61**, 1939, p. 623.

wires and the smaller sizes are stronger because of the greater penetration of the hardening from the drawing. Hard-drawn wire is the cheapest and is commonly used for general-purpose springs where cost is an important factor. It is used where the stresses are low and where only static loading is present. Springs of this material should not be used at temperatures above 250°F nor for subzero applications.

Music wire is a high-quality high-carbon steel wire that is widely used in the smaller diameters. The temperature restrictions are the same as for hard-drawn wire.

Oil-tempered wire is cold drawn to size and then quenched and tempered. It can be used in many places where the cost of music wire is prohibitive, but is ordinarily not considered suitable where long fatigue life is required.

Stainless steel 302 (18 per cent chromium, 8 per cent nickel) has high tensile strength and corrosion resistance. It can be used up to about 550°F and at subzero temperatures. It has superior creep resistant properties at the higher temperatures.

The properties for alloy spring steel wires are given by Table 4-2. These materials are used where the service conditions are severe, and where shock and impact loads may be present. For severe forming, the wire should be in the annealed condition. When wound from annealed stock, the spring must be heat treated after coiling. These materials are also available in the oil-tempered condition.

Valve spring wire, Table 4-3, has been developed especially for valve springs and other applications requiring high fatigue properties.

Table 4-2 ALLOY STEEL SPRING WIRE, MINIMUM TENSILE STRENGTHS, s_{ult}

Table 4-3 CARBON STEEL VALVE SPRING WIRE. MINIMUM TENSILE STRENGTHS, s_{ult}, ASTM SPEC. A230-63T

Diameter, in.	Cr-V A231-63T	Cr-Si A401-63T	Diameter, in.	Cr-V A231-63T	Cr-Si A401-63T	Diameter, in.	Class 1
0.020	300,000	300,000	0.177		260,000	0.093–0.128	210,000
0.032	290,000		0.192	220,000	260,000	0.129–0.162	205,000
0.041	280,000	298,000	0.218		255,000	0.163–0.192	200,000
0.054	270,000	292,000	0.244	210,000		0.193–0.250	195,000
0.062	265,000	290,000	0.250		250,000		Class 2
0.080	255,000	285,000	0.283	205,000			
0.092		280,000	0.312	203,000	245,000	0.093–0.128	230,000
0.105	245,000		0.375	200,000	240,000	0.129–0.162	225,000
0.120		275,000	0.437	195,000		0.163–0.192	220,000
0.135	235,000	270,000	0.500	190,000		0.193–0.225	215,000
0.162	225,000	265,000				0.226–0.250	210,000

Phosphor bronze, Table 4-4, has good corrosion resistance and electrical conductivity, and is frequently used for contact fingers in switches. Spring brass, Table 4-5, has similar properties and generally costs less than phosphor bronze. It should not be used above 150°F, but is suitable for subzero applications.

Table 4-4 PHOSPHOR BRONZE WIRE. COPPER ALLOY NO. 510, SPRING TEMPER, ASTM SPEC. B159-66

Diameter, in., or Distance Between Parallel Surfaces	Minimum Tensile Strength
0.025 and under	145,000
Over 0.025 to 0.0625	135,000
Over 0.0625 to 0.125	130,000
Over 0.125 to 0.250	125,000
Over 0.250 to 0.375	120,000
Over 0.375 to 0.500	105,000

Sn, 4.2–5.8%; Ph 0.03–0.35%; Cu + Sn + Ph, 99.5% min

Table 4-5 BRASS WIRE. MINIMUM TENSILE STRENGTH, s_{ult}, FOR 0.20-IN. DIAM AND LARGER (OR DISTANCE BETWEEN PARALLEL SURFACES), ASTM SPEC. B134-66

Copper Alloy No.	Copper per cent	Total Other Elements	Zinc	Minimum Tensile Strength
210	94.0–96.0	0.20	Remainder	72,000
220	89.0–91.0	0.20	Remainder	84,000
230	84.0–86.0	0.25	Remainder	100,000
240	78.5–81.5	0.25	Remainder	116,000
260	68.5–71.5	0.27	Remainder	120,000
268	63.0–68.5	0.30	Remainder	120,000
274	61.0–64.0	0.35	Remainder	120,000

The maximum strength values for these materials are somewhat higher than the minimum values shown in the tables.

Tables 4-1 to 4-5 pertain only to the tensile strength of spring materials, while designs are usually based on the yield strength in shear. Experimental results indicate that the relationship between the tensile strength s_{ult} and the yield strength in shear, s_{syp}, is as given[2] in Table 4-6.

Table 4-6 RATIO OF YIELD STRENGTH IN SHEAR, s_{syp}, AND ENDURANCE LIMIT IN SHEAR (ZERO TO MAXIMUM), s'_{se}, TO ULTIMATE STRENGTH, s_{ult}

Type	$\dfrac{s_{syp}}{s_{ult}}$	$\dfrac{s'_{se}}{s_{ult}}$
Hard-drawn wire	0.42	0.21
Music wire	0.40	0.23
Oil-tempered wire	0.45	0.22
302 stainless steel wire, 18-8	0.46	0.20
Cr-V and Cr-Si alloy wire	0.51	0.20

Example 2. A helical compression spring of oil-tempered wire is to carry a maximum load of 40 lb. The mean radius of the helix is 0.5 in. The factor of safety is 1.5. Find a suitable standard-size diameter of wire. Assume $s_{yp} = 0.75s_{ult}$ and $s_{syp} = 0.6s_{yp}$.

Solution. Assume tentatively a No. 11 wire.

By Table 4-1: $d = 0.1205$ in., $s_{ult} = 220,000$ psi

By Table 4-6: $s_{syp} = 0.45 \times 220,000 = 99,000$ psi

Permissible stress: $s_s = \dfrac{s_{syp}}{FS} = \dfrac{99,000}{1.5} = 66,000$ psi

[2]See Elmendorf, H. J., "Ratio of Spring Torsional Elastic Limit to Wire Tensile Strength," *Metal Progr.* **73**, April 1958, p. 84. See also p. 165, Reference 3, end of chapter.

By Eq. (1):
$$c_1 = \frac{2R}{d} = \frac{1}{0.1205} = 8.3$$

By Eq. (2), actual stress:
$$s_s = \frac{16 \times 40 \times 0.5}{\pi 0.1205^3}\left(1 + \frac{0.615}{8.3}\right) = 62{,}530 \text{ psi}$$

The assumed wire size is satisfactory.

Example 3. A helical compression spring of music wire has a maximum load that is is 4 lb greater than the minimum load. The deflection under the maximum load is 0.25 in. greater than the deflection under the minimum load. Assume tentatively that the number of active coils is 10. The factor of safety is 1.5. $R = 0.25$ in.

Determine a suitable standard-size wire, and find the exact number of active coils. Find the initial deflection of the spring.

Solution.

By Eq. (7):
$$k = \frac{P_2 - P_1}{\delta_2 - \delta_1} = \frac{4}{0.25} = 16 \text{ lb per in.}$$

By Eq. (8):
$$d^4 = \frac{64 \times 0.25^3 \times 10 \times 16}{11{,}500{,}000}$$
$$= 0.0000139, \qquad d = 0.0611 \text{ in.}$$

By Table 4-1, use No. 16 wire: $d = 0.0625$ in., $s_{ult} = 293{,}000$ psi

By Table 4-6: $s_{syp} = 0.4 \times 293{,}000 = 117{,}200$ psi

Working stress:
$$s_s = \frac{117{,}200}{1.5} = 78{,}130 \text{ psi}$$

By Eq. (1):
$$c_1 = \frac{2R}{d} = \frac{0.50}{0.0625} = 8$$

By Eq. (2):
$$78{,}130 = \frac{16P_2 0.25}{\pi 0.0625^3}\left(1 + \frac{0.615}{8}\right) = 5{,}616 P_2$$
$$P_2 = 13.9 \text{ lb}, \; P_1 = 9.9 \text{ lb}$$

By Eq. (8):
$$N = \frac{d^4 G}{64 R^3 k} = \frac{0.0625^4 \times 11{,}500{,}000}{64 \times 0.25^3 \times 16}$$
$$= 11.0 \text{ active coils}$$

By Eq. (7):
$$\delta_1 = \frac{P_1}{k} = \frac{9.9}{16} = 0.620 \text{ in.}$$

Cold-formed springs must be wound to a smaller diameter than the desired size because of the spring back[3] or expansion that occurs after coiling. Cold-formed springs must be given a stress-relieving heat treatment before being put in service.

Cold-formed springs can also be wound from plain-carbon or alloy steel wire in the annealed condition. Afterwards the spring is heat treated to develop suitable strength values.

[3]Gardiner, F. J., and H. C. R. Carlson, "The Spring-back of Coil Springs," *Mech. Eng.*, **80**, 1958, p. 74.

Compression springs are sometimes wound with considerably greater free height and pitch of coil than is desired. They are then compressed solid several times, which permanently sets the final height at the desired position. This operation is known as presetting or cold setting, and reduces the tendency of the spring to take a permanent set in service. The yield point is exceeded, and residual stresses are retained in the material with a sign opposite to those produced by normal operation. Such a residual stress permits a spring to carry larger loads than one in which the material was originally stress free.

3. Hot-Formed Springs

Helical springs made of $\frac{3}{8}$-in. bar and larger are usually hot wound to avoid the high residual stresses that would be induced by cold forming.

Table 4-7 TENSILE STRENGTH OF HEAT-TREATED STEELS FOR HOT-FORMED SPRINGS. $1\frac{5}{8}$-IN.-DIAM SPECIMENS QUENCHED AND DRAWN AT TEMPERATURES SHOWN

Draw Temp. F	1095 Plain Carbon		6150 Chromium-Vanadium		8660 Chromium-Nickel-Molybdenum		9262 Silicon-Manganese	
	Ultimate	Yield	Ultimate	Yield	Ultimate	Yield	Ultimate	Yield
850	192,000	128,000	220,000	203,000	206,000	193,000	243,000	212,000
950	188,000	120,000	198,000	185,000	190,000	170,000	214,000	182,000
1050	172,000	107,000	180,000	168,000	171,000	150,000	188,000	156,000
1150	151,000	92,000	162,000	152,000	150,000	129,000	167,000	137,000

For torsion, use 0.60 of corresponding tensile value.

After forming, the spring is heat treated by quenching and tempering to produce the desired physical properties. Both plain-carbon and alloy steels are used for hot-formed springs. The materials in widest use[4] are shown in Table 4-7. Compositions for these steels are given in Chapter 14.

Plain-carbon steel 1095 is in wide use because of its availability and relatively low cost. This material, however, is shallow hardening, and sections larger than $\frac{3}{8}$ in. will not harden completely through. To obtain depth of hardening and a material with a higher yield point, an alloy steel must be used, even though the cost is greater. Silicon-manganese steel 9262 has been widely used as a low-priced alloy spring material. It has the disadvantages of being subject to decarburization in heat treatment and of being inclined to have excessive quantities of nonmetallic inclusions as well as a poor surface.

[4]See Keysor, H. C., "Carbon and Alloy Materials for Hot-Formed Springs," *Prod. Eng.*, **17**, Nov. 1946, p. 86. See also Bittner, E. T., "Alloy Spring Steels," *Trans. ASM*, **40**, 1948, p. 263.

Because of its higher cost, chromium-vanadium steel 6150 is being supplanted by other steels. Use of chromium-nickel-molybdenum steel 8660 is increasing because of its many desirable qualities such as good impact resistance, freedom from decarburization and surface defects, and relatively low cost. Alloy steels in general undergo less permanent set in service than plain-carbon steels, and are better suited to impact conditions, especially for service conditions where the surface of the spring becomes scratched and pitted.

4. Optimum Design of Helical Spring

The multiplicity of symbols in the equations for the stress and deflection of helical springs makes the general problem of spring design largely one of trial and error, with the consequent possibility of the use of more material than necessary. For many applications, a spring must be designed to fulfill the following conditions.

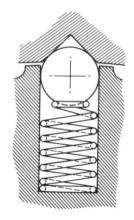

Figure 4-3 Detent spring.

(1) At the most extended condition, the spring must be capable of exerting a given force P_1.

(2) At its most compressed condition, the torsional shear stress must not exceed a specified value s_{s_2}.

The detent spring of Fig. 4-3 is an example that fulfills these requirements. When the ball is in the notch, the spring must exert sufficient force to hold the parts in position with respect to one another. When the lower member is rotated, and the ball is out of the notch, the stress in the spring must not exceed a safe value.

If only the torsional shear stress is considered, it is easy to show[5] that if the spring is designed so that the minimum load, stress, and deflection are exactly one-half the maximum load, stress, and deflection, the spring will contain the least possible amount of material. This will be demonstrated in the following example.

Example 4. Suppose a steel spring must be capable of exerting a force P_1 of 30 lb in its most extended condition. After being compressed an additional amount ($\delta_2 - \delta_1$ of 0.375 in.) the torsional shearing stress s_{s_2} must not exceed the value 75,000 psi. Space limitations indicate that the mean helix radius should be about 0.40 in.

(a) Find the wire diameter, number of active coils, and the volume of material for the optimum spring.

(b) Suppose the designer arbitrarily decides to make the spring of No. 9

[5]See *Trans. ASME*, **75**, 1955, p. 435. See also the author's *Mechanical Design Analysis*, Englewood Cliffs, N.J.: Prentice-Hall, Inc., 1964, p. 72.

wire. Find the volume of material contained in the active coils and compare with the volume of part (a). Retain the value for R of 0.40 in.

Solution. Neglect the transverse shear stress.

(a) For the optimum spring: $s_{s_1} = 0.5 \times 75,000 = 37,500$ psi

By Eq. (2):
$$d^3 = \frac{16P_1R}{\pi s_{s_1}} = \frac{16 \times 30 \times 0.4}{37,500\pi}$$

$$= 0.001630$$

$$d = 0.1177 \text{ in.}$$

Use No. 11 wire: $\qquad d = 0.1205$ in.

By Eq. (2):
$$R = \frac{\pi s_{s_1} d^3}{16P_1} = \frac{\pi 37,500 \times 0.1205^3}{16 \times 30}$$

$$= 0.4294 \text{ in. adjusted value}$$

By Eq. (5):
$$N = \frac{d^4 G \delta_1}{64 P_1 R^3} = \frac{0.1205^4 \times 11,500,000 \times 0.375}{64 \times 30 \times 0.4294^3}$$

$$= 5.98 \text{ coils}$$

Volume:
$$V_{min} = \tfrac{1}{4}\pi d^2 2\pi RN = \tfrac{1}{2}\pi^2 d^2 RN \qquad\qquad \text{(a)}$$

$$= \tfrac{1}{2}\pi^2 \times 0.1205^2 \times 0.4294 \times 5.98$$

$$= 0.1840 \text{ in.}^3$$

(b) No. 9 wire: $\qquad d = 0.1483$ in.

By Eq. (2):
$$P_2 = \frac{\pi d^3 s_{s_2}}{16R} = \frac{\pi 0.1483^3 \times 75,000}{16 \times 0.4}$$

$$= 120.1 \text{ lb}$$

Because of the change in $P_2 - P_1$, the number of active coils becomes

By Eq. (5): $\quad N = \dfrac{d^4 G(\delta_2 - \delta_1)}{64(P_2 - P_1)R^3} = \dfrac{0.1483^4 \times 11,500,000 \times 0.375}{64(120.1 - 30) \times 0.4^3}$

$$= 5.65 \text{ coils}$$

Volume:
$$V = \tfrac{1}{2}\pi^2 d^2 RN$$

$$= \tfrac{1}{2}\pi^2 0.1483^2 \times 0.4 \times 5.65$$

$$= 0.2455 \text{ in.}^3$$

Volume ratio:
$$\frac{V}{V_{min}} = \frac{0.2455}{0.1840} = 1.33$$

Hence the arbitrarily selected wire diameter turned out to be a poor choice as the resulting spring contains 33% more material than the spring of minimum volume.

When Eqs. (2) and (5) are substituted into Eq. (a), a convenient equation for the minimum volume results: $V_{min} = 8P_1(\delta_2 - \delta_1)G/s_{s_2}^2$. It should be noted that this equation is independent of both d and R. Either quantity can be selected arbitrarily and the other adjusted by means of Eq. (2) without affecting the volume. The volume is a minimum because s_{s_1} is taken equal to one-half of s_{s_2}. When space is limited and R is made small, the number of coils N will increase until there is danger of buckling of a compression spring.

5. Fatigue of Springs

Since most failures are caused by fatigue, a poor surface is the worst handicap of hot-formed springs. Figure 4-4 shows a typical fatigue failure

of a helical spring. A fatigue crack usually starts at a surface imperfection in a region of stress concentration. The endurance-limit stress for steel bars in the as-rolled condition may be from 30,000 to 45,000 psi for both plaincarbon and low-alloy steels. If the surface is badly pitted, the endurance limit may be as low as 18,000–20,000 psi. These values for actual springs are thus seen to be much lower than the endurance limit for the same material when polished specimens are tested in the laboratory. A layer of decarburized material on the surface, resulting from the heat treatment, is also a source of weakness since the endurance limit for the surface may then be less than the working stress for the spring. Decarburization can be avoided if the heat treatment is conducted in a controlled atmosphere. Corrosion, even in a mild form, greatly reduces the fatigue strength. Cadmium plating offers some degree of relief against corrosion.

Figure 4-4 Fatigue failure of helical spring. (From Wahl, "Mechanical Springs," *Penton Publishing Co.* p. 31.)

Shot peening, which leaves the surface in compression, has proved to be very successful in raising the fatigue life of springs. A good surface can also be secured by the use of ground stock and controlled atmospheres, but the costs are relatively high. Such a surface, however, will be spoiled if the spring is subjected to rough usage.

If the spring operates under conditions of elevated temperature, there is a danger of creep or permanent deformation unless very low stress values are used. Such effects become noticeable above 350°F, and the ordinary spring steels cannot be used at temperatures above 400°F. Stainless steel of the 18-8 type resists high temperature better than other spring steels. For temperatures of 500–800°F, high-speed steel (18W, 4Cr, 1V) must be used.

For low-temperature service, the alloy steels, listed in Table 4-7, are in use. Failure under impact loads at low temperatures can frequently be guarded against by providing a stop to limit the deflection to a safe value.

Many special-type springs for appliances and other products are stamped from flat stock. For high stresses and severe service conditions, the sheared edges of flat springs must be polished to prevent the formation of fatigue cracks.

6. Design for Fluctuating Loads

If the loading on the spring is continuously fluctuating, due allowance must be made in the design for fatigue and stress concentration. The working-stress triangle, as explained in Chapter 2, should be modified, as shown in Fig. 4-5, when the spring material is tested in pulsating shear zero to maximum

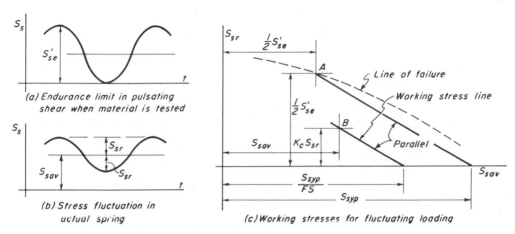

(a) Endurance limit in pulsating shear when material is tested

(b) Stress fluctuation in actual spring

(c) Working stresses for fluctuating loading

Figure 4-5 Working stress diagram for springs.

stress s'_{se}. The relationship between s'_{se} and the ultimate tensile strength, s_{ult}, is given[6] in Table 4-6. For such test loading, the range stress is equal to the average stress or $\frac{1}{2}s'_{se}$. Thus the straight line approximating the line of failure can be drawn from point A in Fig. 4-5(c). The line giving actual working stresses is parallel to this line and can be drawn after dividing the yield point stress in shear s_{syp} by the factor of safety FS.

The stress concentration factor K_c for curvature can be calculated by the following equation.

$$K_c = \frac{4c_1 - 1}{4c_1 - 4} \tag{9}$$

The range stress s_{sr} as determined by Eqs. (2), (3), or (4) for range load P_r is multiplied by K_c before plotting in Fig. 4-5(c).

As is customary when designing for fluctuating loading, stress concentration is ignored when the average stress s_{sav}, as determined by the average load P_{av}, is plotted. It is also assumed that the residual stresses from coiling and from eccentric application of the load are small enough to be neglected.

By making a proportion of corresponding sides of similar triangles in Fig. 4-5, the following equation can be written.

[6]See p. 164, Reference 3, end of chapter.

$$\frac{K_c S_{sr}}{(S_{syp}/FS) - S_{sav}} = \frac{\frac{1}{2}s'_{se}}{S_{syp} - \frac{1}{2}s'_{se}} \qquad (10)$$

This is the basic equation for the design of springs with continuously fluctuating loading.

Example 5. A helical compression spring, made of No. 4 music wire, carries a fluctuating load. The spring index is 6, and the factor of safety is 1.5. If the average load on the spring is 120 lb, find the permissible values for the maximum and minimum loads.

Solution.

From Eq. (1):
$$R = \frac{c_1 d}{2} = \frac{6 \times 0.2253}{2} = 0.676$$

By Eq. (9):
$$K_c = \frac{4 \times 6 - 1}{4 \times 6 - 4} = 1.15$$

By Eq. (2):
$$S_{sav} = \frac{16 \times 120 \times 0.676}{\pi 0.2253^3}\left(1 + \frac{0.615}{6}\right)$$
$$= 39,820 \text{ psi}$$

By Table 4-1:
$$S_{ult} = 235,000 \text{ psi}$$

By Table 4-6:
$$S_{syp} = 0.4 \times 235,000 = 94,000 \text{ psi}$$
$$s'_{se} = 0.23 \times 235,000 = 54,050 \text{ psi}$$

By Eq. (10):
$$\frac{1.15 S_{sr}}{(94,000/1.5) - 39,820} = \frac{27,025}{94,000 - 27,025}$$
$$S_{sr} = 8,020 \text{ psi}$$

Then:
$$P_r = \frac{S_{sr}}{S_{sav}} P_{av} = \frac{8,020}{39,820} \times 120 = 24.2 \text{ lb}$$
$$P_2 = 120 + 24.2 = 144.2 \text{ lb}$$
$$P_1 = 120 - 24.2 = 95.8 \text{ lb}$$

The reader should plot a figure similar to Fig. 4-5(c) for this problem and check the stress values above by measuring lengths on the figure.[7]

7. Vibration or Surging of Helical Springs

A sudden compression of the end of a helical spring may form a compression wave that travels along the spring and is reflected at the far end. The material in the compressed wave is subjected to higher stresses, which may cause early fatigue failure. The natural frequency for a round trip of the wave is given by the following equation.

$$f = \frac{d}{2\pi R^2 N}\sqrt{\frac{Gg}{32\gamma}} \text{ cycles/sec} \qquad (11)$$

[7]Other fatigue diagrams are in use for the design of springs. In some of these the range stress is denoted as the difference between the maximum and minimum stresses. See p. 102, Reference 5, and p. 164, Reference 3, end of chapter.

Here g represents the acceleration constant of gravity, 386 in./sec^2, and γ represents the weight in pounds per cubic inch for the material of the spring.

For a steel spring, $G = 11,500,000$ psi and $\gamma = 0.285$ lbs/in.^{3} The equation above then becomes

$$f = \frac{3,510d}{R^2 N} \text{ cycles/sec} \tag{12}$$

Example 6. Find the lowest natural frequency for a valve spring of No. 4 steel wire with 10 active coils and mean diameter of helix of 2 in.

Solution. $R = 1$ in.

By Table 4-1: $d = 0.2253$ in.

In Eq. (12): $f = \dfrac{3,510 \times 0.2253}{1^2 \times 10} = 79.1$ cycles/sec

$= 4,745$ cycles/min

A spring can exhibit higher modes of vibration whose frequencies are 2, 3, 4, . . ., times the value given by Eqs. (11) and (12).

8. Commercial Tolerances

Sizes of helical springs are not standardized. They must therefore be made to order. Production costs can be kept low if liberal tolerances for

Table 4-8 TOLERANCES ON SPRING WIRE DIAMETERS, PLUS AND MINUS, INCHES

Diameter, in.	Music Wire	Diameter, in.	Hard-Drawn, Oil-Tempered	Diameter, in.	Alloy Steel, Valve Spring
Up to 0.026	0.0003	Up to 0.075	0.001	Up to 0.148	0.001
0.027 to 0.063	0.0005	0.076 to 0.375	0.002	0.149 to 0.177	0.0015
0.064 to 0.250	0.0010	0.376 and up	0.003	0.178 to 0.375	0.002
				0.376 and up	0.003

Table 4-9 TOLERANCES ON COIL DIAMETER, FREE LENGTH, LOAD AND LOAD RATE. COLD-WOUND COMPRESSION OR EXTENSION SPRINGS

Coil Diameter Tolerance, in., ±			Free-Length Tolerance, in., ±			Load and Load Rate Tolerance		
Mean Coil Diameter, in.	D/d = 3–7.9	D/d = 8–15	Free Length, in.	D/d = 3–7.9	D/d = 8–15	No. Active Coils	Load Tol. %, ±	Rate Tol. %, ±
1/8 or less	0.003	0.004	1/2 or less	0.025	0.040	3 or less	15	10
Over 1/8 to 1/4	0.004	0.006	Over 1/2 to 1	0.035	0.060	Over 3 to 9	10	8
Over 1/4 to 1/2	0.006	0.010	Over 1 to 2	0.050	0.080	Over 9 to 15	8	6
Over 1/2 to 1	0.010	0.016	Over 2 to 4	0.080	0.12	Over 15	7	5
Over 1 to 2	0.016	0.025	Over 4 to 8	0.12	0.19			
Over 2 to 4	0.025	0.042	Over 8 to 16	0.22	0.30			
Over 4 to 8	0.042	0.063	Over 16 to 32	0.35	0.45			

dimensions and loading are allowed. Tolerances for commercial grade springs are given[8] in Tables 4-8 and 4-9.

9. Effect of End Turns for Compression Springs

Several different types of end turns used for helical compression springs are shown in Fig. 4-6. The equations are derived for springs assuming that the loading is axial—a condition difficult to secure in practice. The end coils produce an eccentric application of the load, increasing the stress on one

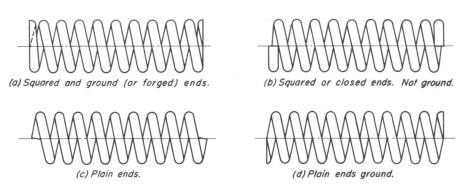

(a) Squared and ground (or forged) ends.   (b) Squared or closed ends. Not ground.

(c) Plain ends.   (d) Plain ends ground.

Figure 4-6 Types of end turns used for compression springs.

side of the spring. Under certain conditions, especially where the number of coils is small, this effect must be taken into account. The nearest approach to an axial load is secured by the spring shown in Fig. 4-6(a), where the end turns are squared and then ground perpendicular to the helix axis.

Equation (5) for the deflection requires the use of the proper number of active coils N. A deduction must be made from the total number of coils to take care of the turns at the ends, which do not affect the deflection. It is impossible to say definitely just how much this deduction should be. However, an average value for the number of active coils, based on experimental results, is found by deducting 1.75 turns from the total number, tip to tip, when both ends are squared and ground as shown in Fig. 4-6(a). For plain ends, Fig. 4-6(c), the deduction from the total turns should be approximately one-half turn; and for plain ends ground, Fig. 4-6(d), the deduction should be one turn.

10. Helical Extension Springs

In helical tension springs, the shape of the hooks or end turns for applying the load must be designed so that the stress concentration effects caused by the presence of sharp bends are decreased as much as possible. In Fig. 4-7

[8]See Reference 2, end of chapter.

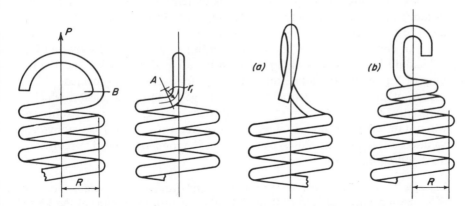

Figure 4-7 End of tension spring made by turning up half loop.

Figure 4-8 Types of end turns which reduce stress concentration.

the end of the tension spring has been formed by merely bending up a half-loop. If the radius of the bend is small, the stress concentration at cross section A will be large.

The most obvious method for avoiding these severe stress concentrations is to make the radius r_1 of the bend larger. Figure 4-8(a) shows one method of doing this. Here the end has been formed by turning up a complete loop in a gradual sweeping curve; stress concentration has been greatly reduced. Another expedient is to reduce the radius of the end turns gradually from the maximum value R, as shown in Fig. 4-8(b). Although the curvature of the end turns has been increased, the moment arm for the force has been correspondingly reduced.

Table 4-10   SHEAR STRESS INDUCED
BY INITIAL TENSION IN HELICAL
EXTENSION SPRING*

$c_1 = 2R/d$	Shear Stress, s_s, psi
4	20,000
5	18,550
6	17,300
7	16,100
8	14,950
9	14,000
10	13,100
11	12,450
12	12,000

*Data by Spring Mfrs.' Assn., Inc.

Special hooks or loops on the ends of tension springs, as well as the grinding of the ends of compression springs, add to the cost of manufacture and should be avoided whenever possible.

For computing the deflection of a tension spring, the end shown in Fig. 4-7 should be counted as about 0.1 turn for each of the hooks, and the full loop of Fig. 4-8(a) should count as 0.5 turn for each end so formed.

Another feature of helical tension springs is the initial tension which is induced at the time the spring is coiled. Such springs are wound solid and will not deflect until this initial tension is overcome. Table 4-10 gives average values of shear stress as caused by initial tension for springs produced on standard coiling machines.

Example 7. A helical extension spring is wound from 0.080-in.-diam wire with mean diameter of helix of 0.50 in. Find the approximate value of the load P that the spring can sustain before noticeable deflection occurs.

Solution.

By Eq. (1):
$$c_1 = \frac{2R}{d} = \frac{0.50}{0.08} = 6.25$$

By Table 4-10:
$$s_s = 17{,}000 \text{ psi}$$

By Eq. (2):
$$17{,}000 = \frac{16P0.25}{\pi(0.08)^3}\left(1 + \frac{0.615}{6.25}\right)$$

$$P = 6.22 \text{ lb}$$

Working stresses for helical extension springs are usually about 75 per cent as great as for corresponding compression springs.

11. Helical Springs of Rectangular Wire

When rectangular wire is used for helical springs, the value of the shearing stress can be found by use of the equations for rectangular shafts. For the springs of Figs. 4-9(a) and (b) the stresses at points A_1 and A_2 are found by Eqs. (28) and (29) of Chapter 3, which are as follows:

$$s_s = \frac{PR}{\alpha_1 bc^2}, \qquad \text{for point } A_1 \tag{13}$$

$$s_s = \frac{PR}{\alpha_2 bc^2}, \qquad \text{for point } A_2 \tag{14}$$

Values of α_1 and α_2 for various b/c ratios are found in Table 3-4.

To these stresses must be added the transverse shearing stress of $1.5P/A$ to point A_1 in Fig. 4-9(a) and point A_2 in Fig. 4-9(b).

An equation for deflection may be derived from Eq. (30) of Chapter 3 by substituting $\theta_1 = \delta/Rl$, and $l = 2\pi RN$. Hence,

$$\delta = \frac{2\pi PR^3 N}{\beta Gbc^3} \tag{15}$$

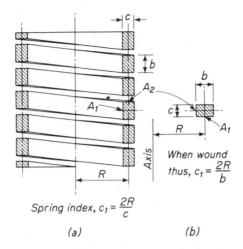

When wound

thus, $c_1 = \dfrac{2R}{b}$

Spring index, $c_1 = \dfrac{2R}{c}$

(a)   (b)

Figure 4-9 Helical spring of rectangular wire.

Since a rectangular bar tends to become trapezoidal in cross section after it is wound into a spring, equations for the stresses and deflections for the true shape are more complicated. The equations above are only approximate since it is assumed that the cross section remains rectangular.

Example 8. Find the stresses at points A_1 and A_2 of the spring of Fig. 4-9(b) for $b = \frac{3}{4}$ in., $c = \frac{1}{2}$ in., and $R = 1.2$ in. The static load on the spring is 2,500 lb. Find the deflection if the number of active coils is 6.

Solution. $\qquad\qquad c_1 = \dfrac{2R}{b} = \dfrac{2 \times 1.2}{\frac{3}{4}} = 3.2$

From Table 3-4: $b/c = 1.50$; $\alpha_1 = 0.231$; $\alpha_2 = 0.269$; $\beta = 0.196$

By Eq. (13): $\qquad s_s = \dfrac{PR}{\alpha_1 bc^2} = \dfrac{2,500 \times 1.2}{0.231 \times 0.75 \times 0.5^2}$

$\qquad\qquad\qquad = 69,270$ psi $\qquad$ for point A_1

By Eq. (14): $\qquad s_s = \dfrac{PR}{\alpha_2 bc^2} = \dfrac{2,500 \times 1.2}{0.269 \times 0.75 \times 0.5^2}$

$\qquad\qquad\qquad = 59,480$ psi, $\qquad$ for point A_2

Transverse shear stress at A_2:

$$s_s = 1.5\frac{P}{A} = 1.5\frac{2,500}{0.75 \times 0.5} = 10,000 \text{ psi}$$

Total shear stress at A_2:

$$s_s = 59,480 + 10,000 = 69,480 \text{ psi}$$

In Eq. (15): $\qquad \delta = \dfrac{2\pi \times 2,500 \times 1.2^3 \times 6}{0.196 \times 11,500,000 \times 0.75 \times 0.5^3} = 0.771$ in.

For fluctuating loads, the design can be made as in Section 6, with the stress concentration factor K_c estimated by Eq. (9).

12. Helical Springs with Torsional Loading

A helical spring can be loaded by a torque about the axis of the helix. Such loading, as shown in Fig. 4-10(a), is similar to the torsional loading of a shaft. The torque about the axis of the helix acts as a bending moment on each section of the wire as shown in Fig. 4-10(b). The material is thus stressed in flexure, and the usual equation for bending stress can be used.

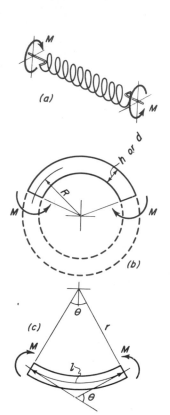

Because of the curved shape, the stress is larger on the inner edge of the coil and smaller on the outer edge. The adjustment can be made by multiplying the stress, as given by $s = Mc/I$, by an appropriate stress concentration factor. For wire of rectangular cross section, these factors are

$$K_1 = \frac{3c_1^2 - c_1 - 0.8}{3c_1(c_1 - 1)}, \qquad \text{for inner edge} \qquad (16)$$

$$K_2 = \frac{3c_1^2 + c_1 - 0.8}{3c_1(c_1 + 1)}, \qquad \text{for outer edge} \qquad (17)$$

where $c_1 = 2R/h$; h is the depth of section perpendicular to the axis.

For circular cross sections, the stress concentration factors are

$$K_3 = \frac{4c_1^2 - c_1 - 1}{4c_1(c_1 - 1)}, \qquad \text{for inner edge} \qquad (18)$$

$$K_4 = \frac{4c_1^2 + c_1 - 1}{4c_1(c_1 + 1)}, \qquad \text{for outer edge} \qquad (19)$$

where $c_1 = 2R/d$.

Figure 4-10 Loading and angular deformation of torsional spring.

Cold forming leaves a residual stress of compression in the outer region of the cross section. The normal loading of the spring should therefore be such as to cause it to wind up and thus produce tensile stress in the outer portion of the wire. If the spring is mounted over an arbor, sufficient clearance must be provided to allow for the decrease in radius of the helix as the spring is wound. For fluctuating loads, care should be exercised in the design of the end turns or hooks to avoid stress concentration caused by sharp bends.

For commonly used values of the spring index, the curvature has no effect on the angular deformation. The deformation of the wire in the spring is the same as for a straight bar of the same length l. By No. 1 of Fig. 1-13 the total angular deformation θ between tangents drawn at the ends of the bar, as shown in Fig. 4-10(c), is

$$\theta = \frac{Ml}{EI} \tag{20}$$

Angle θ in some cases may amount to a number of revolutions.

Example 9. A torsional window-shade spring is made from No. 17 music wire. The mean diameter of helix is 0.875 in. and the number of coils is 400. Assume $s_{yp} = 0.6s_{ult}$ and $FS = 2$ based on the yield point. Compute stresses on the inside of the helix, taking into account the stress concentration due to curvature. Find the torque that the roller can exert after unwinding 12 revolutions from the most highly stressed condition.

Solution.

By Table 4-1: $d = 0.0540,$ $s_{ult} = 301{,}000$ psi

$$c_1 = \frac{2R}{d} = \frac{0.875}{0.054} = 16.2$$

By Eq. (18): $K_3 = \dfrac{4 \times 16.2^2 - 16.2 - 1}{4 \times 16.2(16.2 - 1)} = 1.048$

$$s_{yp} = 0.6 \times 301{,}000 = 180{,}600 \text{ psi}$$

$$K_3 s = \frac{s_{yp}}{FS} = \frac{180{,}600}{2} = 90{,}300 \text{ psi}$$

$$s = \frac{90{,}300}{1.048} = 86{,}140 \text{ psi}$$

$$I = \frac{\pi d^4}{64} = \frac{\pi \times 0.054^4}{64} = 0.0000004174 \text{ in.}^4$$

At maximum stress:

$$M = \frac{sI}{c} = \frac{86{,}140 \times 0.0000004174}{0.027} = 1.3317 \text{ in. lb}$$

$$l = 2\pi RN = \pi \times 0.875 \times 400 = 1{,}100 \text{ in.}$$

At maximum stress by Eq. (20):

$$\theta = \frac{1.3317 \times 1{,}100}{30{,}000{,}000 \times 0.0000004174} = 116.94 \text{ rad}$$

Deflection, after unwinding 12 revolutions:

$$\theta_2 = 116.94 - 12 \times 2\pi = 41.54 \text{ rad}$$

Remaining torque, by Eq. (20):

$$M = \frac{30{,}000{,}000 \times 0.0000004174 \times 41.54}{1{,}100} = 0.473 \text{ in. lb}$$

13. Leaf Springs

The stresses and deformations of rectangular leaf springs for small deflections can be found by the appropriate equations for beams as given in Chapter 1. For cases where the width of the cross section is large compared with the thickness, it is necessary to multiply the deflection as given by the equation for a narrow beam by $(1 - \mu^2)$, where μ is Poisson's ratio, as explained in Section 10 of Chapter 1.

Example 10. Find the sidewise deflection of the spring-mounted table shown in Fig. 4-11. The steel springs are fixed at top and bottom so that tangents at the ends remain vertical at all times.

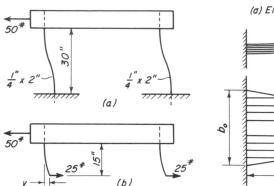

Figure 4-11 Example 10.

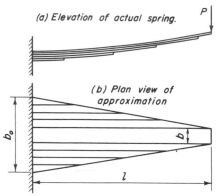

Figure 4-12 Multiple-leaf spring approximated by trapezoid.

Solution. The springs are deformed into symmetrical curves with points of inflection at the midheight. A point of inflection has a zero value for the bending moment because the moment changes sign at such a point. If the springs are cut at the midheights, shear forces only are present as shown in the free-body diagram of Fig. 4-11(b). Hence,

$$I = \frac{bh^3}{12} = \frac{2}{64 \times 12} = \frac{1}{384} \text{ in.}^4$$

$$y = (1 - \mu^2)\frac{Pl^3}{3EI} = \frac{(1 - 0.3^2) \times 25 \times 15^3 \times 384}{3 \times 30,000,000} = 0.3276 \text{ in.}$$

Total deflection of table $= 2 \times 0.3276 = 0.6552$ in.

Multiple-leaf springs are in wide use, especially in motorcar and railway service. An exact analysis of this type of spring is complicated. An approximate solution can be obtained if the shorter leaves are tapered to a sharp point, and if it is assumed that the leaves remain in contact with one another throughout their length when the spring deflects. If these conditions are fulfilled, the curvature, and therefore the stress, may be obtained by replacing the actual spring shown in Fig. 4-12(a) by the trapezoid in Fig. 4-12(b). To form this trapezoid, it is assumed that each of the shorter leaves has been split along its center, that each half has been placed on either side of the longest leaf, and that all edges have then been welded together. The width b_0 at the support is equal to the width of each leaf multiplied by the number of leaves. When the trapezoid deflects under the load, the leaves on either side of a welded edge have the same deflection and curvature, and hence the same stress.

For small deflections, the equation for the cantilever trapezoid with a load on the end is as follows.

$$\delta = K_1 \frac{Pl^3}{3EI_o} \tag{21}$$

Factor K_1 depends on the ratio of the widths b/b_o, and is given by the curve of Fig. 4-13. The moment of inertia I_o refers to the section at the wall and is

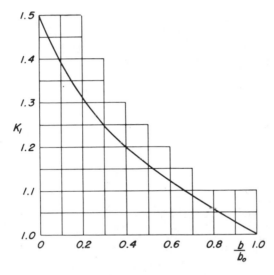

Figure 4-13 K_1 factor for deflection of trapezoidal spring.

equal to $b_o h^3/12$, where h is the thickness of the leaf. If b is large in comparison with h, the previous remark relative to the deflection of wide beams should be observed.

When the deflection is large compared to the thickness, the usual equations for small deflections of beams can no longer be used. The theory is complicated and beyond the scope of this book.[9]

Leaf springs frequently have a hole through the leaves for the tie bolt. This reduction of area, together with the stress concentration, usually occurs at the point of maximum bending stress and is frequently the cause of fatigue failures. The stress concentration factor for a hole, and also for semicircular

[9]See Bisshopp, K. E., and D. C. Drucker, "Large Deflection of Cantilever Beam," *Quart. Appl. Math.*, **3**, No. 3, Oct. 1945, p. 272. See also the author's *Mechanical Design Analysis*, Englewood Cliffs, N.J.: Prentice-Hall, Inc., 1964, p. 90.

notches, may be taken from Fig. 2-10 for a bar in tension. Another harmful effect is due to the clamping pressure used in attaching the spring. This compression also usually occurs at the location of maximum bending stress and causes a reduction in the fatigue strength.

Materials suitable for leaf springs are listed in Table 4-7. The previous remarks concerning condition of surface and decarburized surface layer apply equally well to leaf springs. If sharp bends occur in the region of high stress, the concentration must be taken into account, particularly for non-steady loading.

14. Energy Storage by Springs

Springs are frequently used for storing energy in spring motors, or for absorbing the energy of shock or impact loads. When the force and deformation of a spring are proportional, the energy stored is equal to $\frac{1}{2}P\delta$, where P is the load and δ is the deformation. This equation assumes that the spring was carrying no load at the time P was applied.

From Eq. (5) for the deflection of a helical spring of round wire, and the value for s_s from Eq. (a), Section 1, the expression for the energy for static loads can be written, after rearrangement of terms, as

$$\text{energy} = \tfrac{1}{2}P\delta = \frac{s_s^2}{4G} \times \text{volume} \tag{22}$$

For a given stress, some types of springs can store more energy, per unit volume of material, than can other types. For the purpose of comparing the helical spring with other kinds, Eq. (22) will be transformed into units of axial stress. Within the elastic range $s_s = 0.5s$ and $G = E/2(1 + \mu)$, where $\mu = 0.3$. These substitutions are made in Eq. (22) to give

$$\text{energy} = \frac{s^2}{6.15E} \times \text{volume} \tag{23}$$

For a torsional spring, the energy is equal to $\frac{1}{2}\theta M$. Equation (20) is now substituted for θ; for M, its value $M = sI/c$ is substituted. When the wire is rectangular, the further substitution, $I = bh^3/12$, is made. Hence, after the terms have been rearranged,

$$\text{energy} = \frac{s^2}{6E} \times \text{volume} \tag{24}$$

If the torsional spring is made of round wire, $I = \pi d^4/64$, and the equation for energy becomes

$$\text{energy} = \frac{s^2}{8E} \times \text{volume} \tag{25}$$

The energy storage capacity is thus smaller when round wire is used.

For a rectangular cantilever with load P on the end, the deflection is

$Pl^3/3EI$, and the expression for the energy in terms of the maximum stress, which occurs at the support, is

$$\text{energy} = \frac{s^2}{18E} \times \text{volume} \tag{26}$$

If the cantilever is triangular ($b/b_o = 0$) the deflection, from Eq. (21) and Fig. 4-13, is equal to $1.5Pl^3/3EI_o$, and the equation for the energy becomes

$$\text{energy} = \frac{s^2}{6E} \times \text{volume} \tag{27}$$

Since the multiple-leaf spring may be approximated at the triangular cantilever, Eq. (27) indicates that such a spring is efficient from the standpoint of energy storage.

15. Belleville Spring

The initially coned or Belleville spring, Fig. 4-14, is characterized by the nonlinear relationship between load and deflection. The dimensions can be arranged so that a large deflection can be obtained with practically no change in load. This is sometimes a useful characteristic in machine construction.

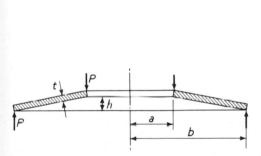

Figure 4-14 Cross section through initially coned or Belleville spring.

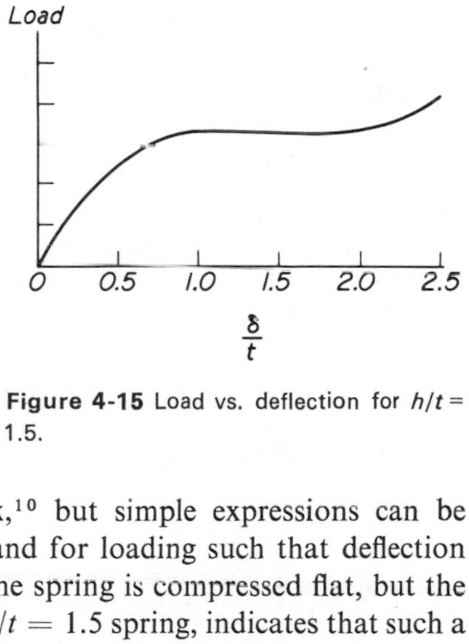

Figure 4-15 Load vs. deflection for $h/t = 1.5$.

The theory is exceedingly complex,[10] but simple expressions can be obtained for the spring with $h/t = 1.5$ and for loading such that deflection δ is equal to the height h. At this time the spring is compressed flat, but the load-deflection curve, Fig. 4-15, for the $h/t = 1.5$ spring, indicates that such a load will practically correspond to δ/t values between 1.0 and 2.0. The spring can thus undergo a large deflection range with practically no change in loading. Table 4-11 gives values by which calculations can be made. The following equation is also used.

[10] See Chapter 13, Reference 5, end of Chapter, Also, p. 80 of the author's book mentioned in footnote 9.

Table 4-11 CONSTANTS FOR STEEL BELLEVILLE SPRING FOR $h/t = \delta/t = 1.5$

$\dfrac{b}{a}$	K_1	$\dfrac{bs_t}{\sqrt{P}}$
1.25	−8.83	−22,090
1.50	−6.29	−19,430
1.75	−5.63	−19,050
2.00	−5.44	−19,350
2.50	−5.54	−20,630

$$s_t = K_1 \frac{Et^2}{b^2} \tag{28}$$

High stresses are used in the design of Belleville springs. The maximum stress is compression on the upper surface and inner boundary in Fig. 4-14. For static loads, or loads that may be repeated but a few times, experience has shown that a maximum stress of 200,000 psi can be used even though the yield strength of the material may be considerably less. If the yield point is exceeded, a redistribution of the stresses occurs because of localized yielding. Residual stresses of the opposite sign are produced, which give lower values for the load stresses than the calculated values.

Example 11. Find the outside radius b and thickness t for a steel Belleville spring, $h/t = 1.5$, that carries a load of 1,000 lb at a maximum compressive stress of 200,000 psi (compressed flat). Let $b/a = 1.75$.
Solution.

From Table 4-11: $bs_t/\sqrt{P} = -19,050$ and $K_1 = -5.63$

Then:
$$b = \frac{-19,050\sqrt{P}}{s_t} = \frac{-19,050\sqrt{1,000}}{-200,000} = 3,012 \text{ in.}$$

$$a = \frac{b}{1.75} = \frac{3,012}{1.75} = 1.721$$

By Eq. (28):
$$t^2 = \frac{b^2 s_t}{K_1 E} = \frac{3.012^2(-200,000)}{-5.63 \times 30,000,000} = 0.01074$$

or
$$t = 0.1036 \text{ in.}$$

and
$$h = 1.5t = 0.1554 \text{ in.}$$

The spring dimensions are thus completely determined.

16. Rubber Springs

Rubber springs and cushioning devices are finding an increasing range of application in industry. Knowledge of the behavior of rubber under stress is not clearly understood, and the results of calculations must be considered only approximate. The modulus of elasticity for rubber depends on the durometer hardness number, as shown by Fig. 4-16. Rubber does not follow Hooke's law, but becomes increasingly stiff as the deformation is increased. The relationship between load and deflection for a 1-in. cube of 55 durometer rubber in compression is shown by Fig. 4-17.

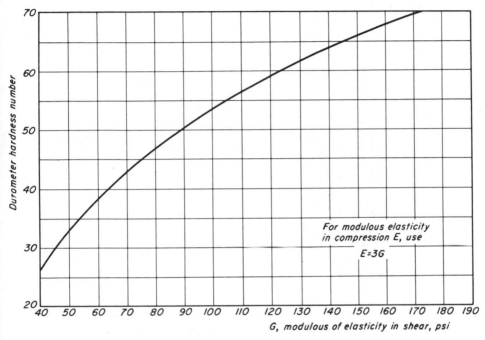

Figure 4-16 Relationship between durometer hardness number and modulus of elasticity for rubber.

A rectangular block of rubber as shown in Fig. 4-18 is frequently used as a compression spring. Based on a considerable number of tests, an empirical equation[11] has been derived for finding the percentage of deflection α for such a spring.

$$\alpha = \frac{\alpha_{55}E_{55}}{E} \cdot \frac{(h\beta)^{2/3}}{\sqrt{A}} \tag{29}$$

Here α_{55} represents the percentage of deformation of a 1-in. cube of 55 durometer rubber for the given compressive stress as shown by Fig. 4-17; E_{55} represents the modulus of elasticity, 313 psi, for 55 durometer rubber; modulus E refers to the durometer hardness of the slab under consideration,

[11]See Smith, J. F. D., "Rubber Mountings," *Trans. ASME*, **60**, 1938, p. A-13 and **61**, 1939, p. A-159.

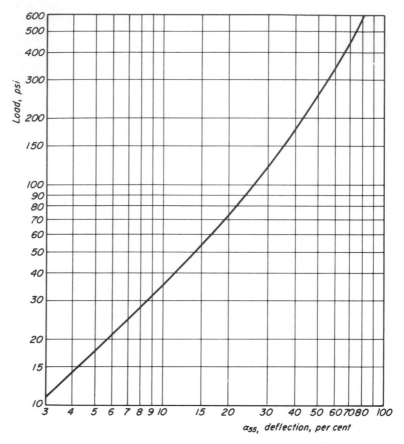

Figure 4-17 Load deflection curve for 1 in. cube of 55 durometer rubber between steel plates bonded to the rubber.

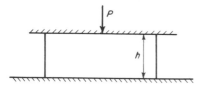

Figure 4-18 Rubber block spring with compression loading.

as given by Fig. 4-16. Thickness of the slab is h in., and β is the ratio of the length to the breadth of the slab. The area normal to the load is given by A.

Example 12. Find the percentage of deflection of a 3 × 6-in. slab of 65 durometer rubber 1 in. thick that carries a load of 5,000 lb.

Solution.

Unit load: $p = \dfrac{5,000}{3 \times 6} = 278$ psi

By Fig. 4-17: $\alpha_{55} = 54\%$

$\beta = \tfrac{6}{3} = 2$

By Fig. 4-16: $E = 3 \times 145 = 435$ psi for 65 durometer

In Eq. (29): $\alpha = \dfrac{54 \times 313(1 \times 2)^{2/3}}{435\sqrt{3 \times 6}} = 14.54\%$

A load-deflection diagram can be plotted for a rubber spring by finding the value of α for other loads. The foregoing discussion assumes that the rubber in Figs. 4-17 and 4-18 is bonded to metal plates top and bottom.

Figure 4-19 shows a simple shear spring of two blocks of rubber. Hooke's law holds reasonably well for rubber in shear, and the following equation can be used for the shearing deformation γ

$$\gamma = \frac{s_s}{G} = \frac{P}{2AG} \tag{a}$$

where A represents the area of one of the surfaces in shear.

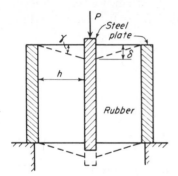

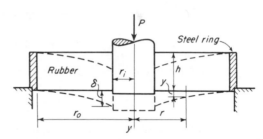

Figure 4-19 Simple rubber shear spring.

Figure 4-20 Cylindrical rubber spring with shear loading.

Deformation δ is equal to $h \tan \gamma$, and for moderate values of γ the angle can be substituted for the tangent, which gives $\delta = h\gamma$. Substitution should now be made for γ in Eq. (a) to obtain an approximate equation for the deflection δ.

$$\delta = \frac{Ph}{2AG} \tag{30}$$

A shear spring of cylindrical form is shown in Fig. 4-20. The rubber is bonded to a steel ring on the outside and a steel shaft in the center. Shear stress in the rubber at radius r is equal to

$$s_s = \gamma G = \frac{P}{2\pi rh} \tag{b}$$

Deformation angle γ is then equal to

$$\gamma = \frac{P}{2\pi rhG} = \frac{b}{r} \tag{c}$$

where
$$b = \frac{P}{2\pi hG} \qquad (d)$$

The slope of the deformed rubber at radius r is given by

$$\frac{dy}{dr} = \tan \gamma = -\tan \frac{b}{r} \qquad (e)$$

The minus sign is due to the fact that the slope is negative for the chosen coordinate system. The tangent term should now be expanded into a series and integrated term by term.

$$\frac{dy}{dr} = \frac{-b}{r} - \frac{b^3}{3r^3} - \cdots \qquad (f)$$

$$y = --b \log_e r + \frac{b^3}{6r^2} + \cdots + C \qquad (g)$$

When $\tan \gamma$ is less than about 0.4, only the $b \log_e r$ term need be retained. The constant of integration C can be evaluated from the condition that $y = 0$ when $r = r_o$. Then

$$C = b \log_e r_o \qquad (h)$$

and
$$y = b \log_e \frac{r_o}{r} \qquad (i)$$

The maximum value of y or δ occurs for $r = r_i$. Hence,

$$\delta = b \log_e \frac{r_o}{r_i} = \frac{P}{2\pi hG} \log_e \frac{r_o}{r_i} \qquad (31)$$

A cylindrical spring loaded in torsional shear is shown in Fig. 4-21. At radius r the area in shear is $2\pi rh$. The shear force is this area multiplied by shearing stress s_s, and the torque or moment is the force multiplied by the radius r.

$$M = 2\pi r^2 h s_s \qquad \text{or} \qquad s_s = \frac{M}{2\pi r^2 h} \qquad (j)$$

The shear stress has a maximum value for the minimum value of r. Hence,

$$s_{smax} = \frac{M}{2\pi r_i^2 h} \qquad (32)$$

Figure 4-21(b) shows, to an enlarged scale, the deformation angle γ for the element at radius r from the axis. Since $r\, d\theta = dr \tan \gamma$, the contribution $d\theta$ to the total angular rotation of the shaft, given by the element at radius r, is

$$d\theta = \frac{dr \tan \gamma}{r} \qquad (k)$$

Angle γ can be found by the following equation.

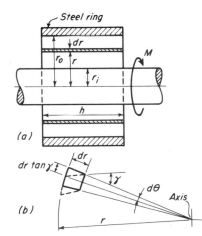

Figure 4-21 Cylindrical rubber spring with torsion loading.

$$\gamma = \frac{S_s}{G} = \frac{M}{2\pi r^2 hG} = \frac{c}{r^2} \tag{l}$$

where
$$c = \frac{M}{2\pi hG} \tag{m}$$

This value of γ should now be substituted in Eq. (k), and the tangent term expanded into a series. The resulting equation should be integrated term by term.

$$d\theta = \frac{1}{r}\left(\tan\frac{c}{r^2}\right) dr = \frac{1}{r}\left(\frac{c}{r^2} + \frac{c^3}{3r^6} + \cdots\right) dr \tag{n}$$

$$\theta = \left(-\frac{c}{2r^2} - \frac{c^3}{18r^6}\right)_{r_i}^{r_o} \tag{o}$$

This series converges rapidly, and the first term will usually be sufficient. When the limits for r are substituted, the following equation for θ is obtained.

$$\theta = \frac{M}{4\pi hG}\left(\frac{1}{r_i^2} - \frac{1}{r_o^2}\right) \tag{33}$$

The equations for many other types of rubber springs can be found in the literature.

Working stresses are generally limited to 25–50 psi in shear. For rubber compression springs, the deformation should not be greater than 10–20 per cent of the free height to avoid excessive creep. Under fatigue loading even smaller values should be used. For satisfactory curing during vulcanization the maximum thickness should be limited to 2 in., with a preferred thickness of not over 1 in.

REFERENCES

1. *Handbook of Mechanical Spring Design*, Bristol, Conn.: Associated Spring Corp.

2. *Manual on the Design and Application of Helical and Spiral Springs—SAE J 795, Handbook Supplement* **9**, New York: Society of Automotive Engineers, Inc., July 1962.

3. *Metals Handbook*, 8th ed., Vol. 1, Metals Park, Ohio: American Society for Metals, 1961.

4. Rothbart, H. A., Ed., *Mechanical Design and Systems Handbook*, New York: McGraw-Hill Book Company, 1964, Sec. 33.

5. Wahl, A. M., *Mechanical Springs*, 2d ed., New York: McGraw-Hill Book Company, 1963.

PROBLEMS

Spring material for the problems is steel.

1. A helical spring must sustain a static load of 100 lb at a deflection of 1 in. The spring index is equal to 6. Find the value of the maximum stress if the spring is made of No. 8 wire. Find the required number of active coils.   *Ans.* $N = 10.8$.

2. A helical spring is made of No. 4 wire and has 10 active coils. The spring index is equal to 6. Find the stress when the deflection is equal to 1 in.

Ans.   49,760 psi.

3. A helical spring is made of No. 8 wire with the spring index equal to 6. The load is steady and the spring rate is 150 lb/in. Maximum stress is 60,000 psi. Find the number of active coils.   *Ans.* $N = 7.2$ coils.

4. A helical spring of rate 60 lb/in. is mounted on top of another spring of rate 42 lb/in. Find the force required to give a total deflection of 2 in.

Ans. $F = 49.4$ lb.

5. The outer of two concentric helical springs has a rate of 2,400 lb/in. The inner spring has a rate of 1,750 lb/in. The outer spring is $\frac{1}{2}$ in. longer than the inner. If the total load is 8,000 lb, find the weight carried by each spring.

Ans.   Outer $= 5,132.5$ lb; inner $= 2,867.5$ lb.

6. The larger of two concentric compression helical springs is made of $1\frac{1}{2}$-in.-diam round bar stock, is 9-in. OD of helix, and has 6 active coils. The inner spring is of 1-in. bar, is $5\frac{1}{2}$-in. OD helix, and has 9 active coils. The free height of the outer spring is $\frac{3}{4}$ in. more than the inner. Find the deflection of each spring for a load of 20,000 lb. Find the load carried by each spring.

Ans. $P_o = 13,242$ lb; $\delta_o = 4.606$ in.

7. A compression spring is to be made of 0.120-in.-diam wire with 10 active coils, and 1-in. mean diameter of helix. Commercial limits for the wire diameter are 0.1185 and 0.1215 in. Helix diameter may vary from 0.980 to 1.020 in. The number of active coils may vary from $9\frac{3}{4}$ to $10\frac{1}{4}$. The modulus may vary from 11,400,000 to 11,800,000 psi. Find the spring rate when all variables tend to give the weakest spring. Find the spring rate when all variables tend to give the stiffest spring.

Ans. $k = 25.8$ lb/in.; $k = 35.0$ lb/in.

8. A helical spring must be capable of exerting a force of 150 lb after being released 0.4 in. from its most highly compressed position. $\frac{1}{4}$ in. wire is used; the spring index is equal to 6. The loading is static. The maximum stress is 60,000 psi. Find the required number of active coils.   *Ans.* $N = 9.17$ coils.

9. A helical compression spring is made from $\frac{1}{4}$-in.-diam wire and carries a static load of 250 lb. The maximum shearing stress is to be 60,000 psi. Find the radius of the helix and the number of active coils for a deflection of $\frac{3}{4}$ in.

Ans. $R = 0.660$ in.; $N = 7.34$ coils.

10. A helical spring of music wire is subjected to a continuously varying load. The length varies between $2\frac{1}{2}$ and $2\frac{3}{4}$ in. with corresponding loads of 114 and 78 lb.

The spring is made of No. 7 wire and has a mean helix diameter of 1 in. Find the value of the factor of safety and the number of active coils.

Ans. $FS = 1.30$; $N = 9.8$ coils.

11. A helical spring is subjected to a continuously varying load and is made from $\frac{1}{4}$-in.-diam hard-drawn wire. The mean radius of the helix is $\frac{3}{4}$ in. The average load is 100 lb. The factor of safety is 1.5. Find the values of the maximum and minimum loads.   Ans. $P_{max} = 125.8$ lb; $P_{min} = 74.2$ lb.

12. A helical spring is subjected to a continuously varying load. No. 7 oil-tempered wire is used with a mean radius of helix of $\frac{1}{2}$ in. In the most compressed condition the force is 82 lb. After 0.32 in. of release the minimum force is 58 lb. Find the factor of safety and the number of active coils.

Ans. $FS = 1.56$; $N = 18.5$ coils.

13. A helical spring is made from rectangular wire $\frac{1}{2} \times \frac{1}{4}$ in. with the narrow side parallel to the helix axis. Mean helix radius is 1.5 in. The static load is 150 lb.

(a) Find the value of the maximum stress, and the deflection, if the number of active coils is eight.

(b) Repeat (a) with the spring wound with the long side of the cross section parallel to the axis.   Ans. (a) $s_s = 29,270$ psi; $\delta = 1.24$ in.

14. A helical spring carries a static load of 250 lb: There are 10 active coils, and the radius of the helix to the center of the wire is 1 in.

(a) What size square wire must be used and what will be the stress if the deflection is to be 2 in.?

(b) If this spring is made of round wire, find the required diameter and the stress. What percentage of weight of material is saved by the use of round wire?   Ans. (a) $c = 0.264$ in.; $s_s = 70,700$ psi.

15. A torsional helical spring for a 30-in. door (see Fig. 4-22) is to have a pull at the handle of 1 lb when the door is closed, and a pull of 3 lb after the door has rotated through 180°. No. 6 wire is to be used, and maximum stress is 150,000 psi.

Find the required diameter of the helix, the required initial angular deformation, and the number of active coils in the spring. Take into account the stress concention.

Ans. Diam helix $= 1.047$ in.; $\theta_0 = 90°$; $N = 31.8$.

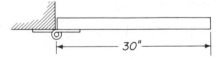

Figure 4-22 Problem 15.

16. How many coils must there be in a window-shade spring that is to exert a pull on the shade of 3 lb after being wound up 15 revolutions? The spring is made of 0.0475-in. square wire. The helix diameter is 0.75 in. center to center of coils, and the roller diameter is 1.25 in. What is the value of the maximum flexural stress?

Ans. $N = 272$; $s = 109,600$ psi.

17. A flexible coupling has an active element in the form of a torsional helical spring. Twenty-five horsepower are transmitted at 900 rpm. To what radius of helix must the spring be wound if made of $\frac{3}{4}$-in. square wire, and if the permissible stress is 30,000 psi? How many active coils must there be if the torsional deflection is not to exceed 5°?   *Ans.* $R = 1.47$ in.; $N = 4.27$.

18. A torsional spring is to be located within the hub of the reel for raising the weight shown in Fig. 4-23. The spring is wound from $\frac{5}{16}$-in. square wire. The OD of helix is $6\frac{1}{2}$ in. Ignore the effect of curvature on stress. The number of active coils is 20.

(a) Find the initial angular deformation required by the spring in order to support the load in the upper position.

(b) Find the additional force that must be applied to the weight in order to hold it in the lower position.

(c) Using Eq. (25), compute the strain energy stored in the spring because of lowering the weight; check the result by totaling the work done by the external forces.   *Ans.* $\theta_0 = 3.47$ rad; force $= 81.5$ lb.

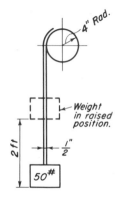

Figure 4-23 Problem 18.

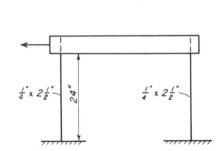

Figure 4-24 Problem 19.

19. (a) Find the force F necessary to move the top of the shaking table shown in Fig. 4-24 a distance of 1.75 in. to one side of the midposition. Tangents to the springs at the top and bottom remain vertical at all times. Find the value of the bending stress in the springs.

(b) Repeat (a) except that the right spring is now $\frac{5}{16}$ in. thick. The left spring remains $\frac{1}{4}$ in. thick.   *Ans.*   (a) $F = 326$ lb; (b) $F = 481$ lb.

20. A cantilever, multiple-leaf spring similar to that in Fig. 4-12 is composed of five leaves $2\frac{1}{2}$ in. wide and $\frac{1}{2}$ in. thick, and has a clear span to the point of application of the load of 20 in. Assume that the shorter leaves are of the proper length and end taper to permit the spring to be approximated as a trapezoid. What static load will the spring support at a stress in the material of 60,000 psi? What will be the deflection of the load?   *Ans.* $P = 1,560$ lb; $\delta = 1.40$ in.

21. A spring mount for a machine is built up of steel plates and rollers as shown in Fig. 4-25. Find the deflection caused by the load and the bending stress in the plates.   *Ans.* $y = 1.084$ in.; $s = 11,850$ psi.

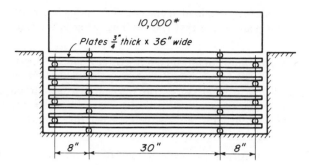

Figure 4-25 Problem 21.

22. Find the value of force P if the force between the contact points in Fig. 4-26 is to be 3 lb. Find the deflection of point B. Leaves are steel, 0.03 in. thick, and 0.40 in. wide. Neglect effects of any friction.   *Ans.* $P = 5$ lb; $y = 0.369$ in.

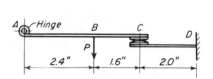

Figure 4-26 Problem 22.

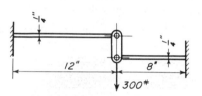

Figure 4-27 Problem 23.

23. Find the load carried by each spring in Fig. 4-27. The left spring is trapezoidal in plan, 2 in. wide at the hinge, and 10 in. wide at the wall. The right spring has a uniform width of 3 in. Assume that the hinges are frictionless and the connecting link is inextensible. Find the deflection of the load.

Ans.   Left $= 129$ lb; right 171 lb; $\delta = 0.2267$ in.

24. Forces R in Fig. 4-28 are just sufficient to bring the ends of the upper spring into contact with the lower spring. The material is steel. Find the value of R.

Ans. $R = 941$ lb.

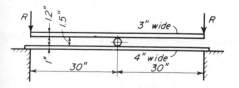

Figure 4-28 Problem 24.

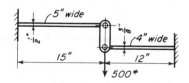

Figure 4-29 Problem 25.

25. Find the deflection of the load in Fig. 4-29. The springs are of steel. Assume that the hinges are frictionless and the connecting link is inextensible.

Ans. $y = 0.1975$ in.

26. A cantilever spring has a load at the end. The spring is of uniform thickness, but is triangular in plan. Prove that the bending stress is the same for all cross sections.

27. Let the expression for the strain energy U for a simply supported leaf spring with load P in the center be given by $\frac{1}{2}Py$, where y is the deflection. Prove that stress $s = c\sqrt{6EU/Il}$, where c is the distance from the neutral axis to the edge of the cross section. Use this relationship to prove that for a rectangular cross section the energy absorbed for a constant stress s is the same regardless of whether the long or short dimension is vertical.

28. A helical spring is to carry a static load of 100 lb at a stress of 70,000 psi. The spring index equals 6. Find the theoretical diameter of the wire.

29. A steel helical spring of nine active coils is to have a deflection of $\frac{5}{8}$ in. under a static load of 50 lb. The spring index equals 5. Find the theoretical diameter of the wire.

30. A steel helical spring of index 4 has 10 active coils. The deflection is 1.5 in. when the stress is 70,000 psi. Find the static load the spring is carrying and the mean radius of the helix.

31. A steel helical spring must exert a force of 2.5 lb after being released 0.10 in. from its most highly compressed condition. No. 19 wire is used with a spring index of 6. Find the static loading for a maximum stress of 60,000 psi. Find the number of active coils.

32. The maximum and minimum static loads on a steel helical spring are 86 and 71 lb, respectively. These loads cause a change in the deflection of 0.25 in. The spring index is 8, and the maximum stress is 70,000 psi. Find the theoretical value of d. Also find R and N.

33. Find the FS for a spring with the same data as Problem 12 except that R equals 0.6 in.

34. An engine valve spring must exert a force of 60 lb when the value is closed and 100 lb when the valve is open. The lift is $\frac{5}{16}$ in. Material tests $s_{syp} = 90,000$ psi and $s'_{se} = 48,000$ psi. $FS = 1.5$. If the spring index is 6, find the theoretical wire diameter. Also find the number of active coils and the initial compression of the spring.

35. A helical spring is made from No. 10 hard-drawn wire. The load varies continuously from 25 to 50 lb. The radius of the helix is $\frac{5}{16}$ in. Find the value of the factor of safety.

36. Work Problem 22 but with the force between the contacts 2 lb. The 1.6 dimension becomes 2.4 in.

37. In Fig. 4-30 each steel spring has 10 active coils of No. 12 wire. The mean radius of the helix is 0.3 in. The right spring has a free height 0.5 in. greater than the left spring. When the load is acting, the bar is level. Find the distance x.

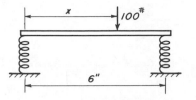

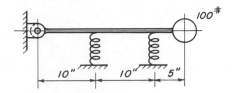

Figure 4-30 Problem 37.   **Figure 4-31** Problem 38.

38. In Fig. 4-31 assume the bar to be rigid and weightless, and the joint at the left to be frictionless. Each steel spring has 10 active coils of No. 10 wire wound with a spring index of 5. Springs are stress free when the bar is level. Find the force carried by each spring and the deflection of the weight.

39. The bar in Fig. 4-32 is made of steel. Find the force in the spring.

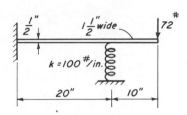

Figure 4-32 Problem 39.

40. The difference between the maximum and minimum deflections of a helical spring is 0.625 in. The minimum load is 60 lb and the maximum shearing stress is is 62,500 psi. The mean radius of the helix is approximately $\frac{1}{2}$ in.

(a) Find the values of d, N, and the volume for the spring of minimum volume.

(b) Suppose the spring had been wound arbitrarily from No. 5 wire. What percentage more of material would this spring contain than the spring of part (a)?
                              Ans.   (a) V_{min} = 0.883 in.^{3}; (b) 25 per cent.

41. Find the outside diameter b and thickness t for a Belleville spring, $h/t = 1.5$, that carries a load of 500 lb when compressed flat. Maximum compressive stress is 150,000 psi. Let $b/a = 2$.   *Ans.* b = 2.88 in.; t = 0.087 in.

5 Screws

Bolts and screws are used to fasten the various parts of an assembly together. The designer should be familiar with the different kinds of threads in commercial use and with the method of specifying the desired tolerances for the fit between screw and nut. He should understand the reasons for the increase of fatigue strength obtained by the application of initial tension in the bolt. Power screws are employed in machines for obtaining motion of translation and also for exerting forces.

1. Kinds of Threads

The Unified thread is in general service for studs and bolts. This thread represents the agreement of standardization committees of Canada, Great Britain, and the United States. It is a replacement for the American National form of thread, which was formerly the standard in our country. Bolts and nuts of the two systems are interchangeable. The systems differ only in minor details, mainly in the arrangement of the tolerances. The included angle of the thread is 60°. The basic form of the thread is shown in Fig. 5-1(a), although actual threads are made with a rounded root. The

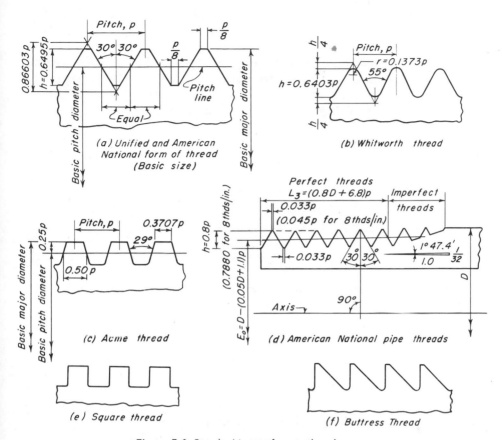

Figure 5-1 Standard types of screw threads.

crest is flat in the American National and may be either flat or rounded in the Unified thread. The Whitworth thread with 55° included angle, shown in Fig. 5-1(b), was formerly the standard in Great Britain.

The remaining countries of the world have adopted the International Metric thread. Although the included angle is 60°, this thread, unfortunately, is not interchangeable with the Unified. A single worldwide system for screw threads apparently lies in the distant future.

For lead screws and power transmission the Acme screw shown in Fig. 5-1(c) is in wide use. It has an included angle of thread of 29°. The standard proportions of the American National pipe thread are given in Fig. 5-1(d). The taper, together with the smaller flat at crest and root, assists in producing a fluid-tight joint. Square and buttress threads shown in Fig. 5-1(e) and (f) are used to a limited extent for power transmission.

If an imaginary cylinder, coaxial with the screw, intersects the thread at the height that makes the width of thread equal to the width of space, the diameter of this cylinder is called the *pitch diameter* of the screw. See Fig. 5-1(a). The distance measured parallel to the axis from a point on one thread to the corresponding point on the adjacent thread is called the *pitch*. A screw made by cutting a single helical groove on the cylinder is called a *single-thread* screw. If the helix angle is somewhat steeper, and a second thread is cut in the space between the grooves of the first, a *double-thread* screw is formed, Fig. 5-2. For certain applications, triple and quadruple threads are in use.

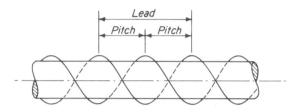

Figure 5-2 Double-threaded screw.

For multiple-thread screws the *lead* is the distance the nut advances in one revolution. The pitch is defined as for a single-thread screw. The helix may be cut either right hand or left hand. The screw shown in Fig. 5-11 is a right-hand screw.

2. Standardized Threads

Table 5-1 gives a summary of the various sizes and pitches for Unified and American National forms of threads. The size of a screw refers to the major diameter, or the size of the stock upon which the helix is cut. It should be noticed that sizes less than $\frac{1}{4}$ in. are designated by number. In general, there are two pitches, coarse and fine, that may be cut on each diameter. For certain sizes, there is also an extra-fine series of pitches. Complete information on sizes, pitches, tolerances, and allowances is given in *Unified and American Screw Threads, ASA B1.1-1960*, published by the American Society of Mechanical Engineers. There are three additional series of threads of 8, 12, and 16 threads per in., which may be cut on a wide variety of diameters.

Example 1. Compute the basic pitch diameter for a 1-in. coarse-thread series screw. Find the value of the helix angle α.

Solution. By Table 5-1, the 1-in. coarse series has 8 threads per in. The basic pitch diameter is equal to the nominal outside diameter minus the height of thread h.

By Fig. 5-1(a): $\qquad\qquad\qquad h = 0.6495p = 0.6495 \times \frac{1}{8} = 0.0812$ in.

Basic pitch diameter: $\qquad\qquad d = 1.0000 - 0.0812 = 0.9188$ in.

Circumference of pitch circle: $\quad \pi d = 0.9188\pi = 2.8865$ in.

Table 5-1 DIMENSION OF UNIFIED AND AMERICAN NATIONAL SCREW THREADS

Size	Outside or Major Diameter, in.	Coarse-Thread Series			Fine-Thread Series			Extra-Fine-Thread Series			Hex. Nut Width Across Flats
		Threads per in.	Basic Pitch Diameter, in.	Stress Area, in.^{2}	Threads per in.	Basic Pitch Diameter, in.	Stress Area, in.^{2}	Threads per in.	Basic Pitch Diameter, in.	Stress Area, in.^{2}	
0	0.0600				80	0.0519	0.0018				
1	0.0730	64	0.0629	0.0026	72	0.0640	0.0028				
2	0.0860	56	0.0744	0.0037	64	0.0759	0.0039				
3	0.0990	48	0.0855	0.0049	56	0.0874	0.0052				
4	0.1120	40	0.0958	0.0060	48	0.0985	0.0066				
5	0.1250	40	0.1088	0.0080	44	0.1102	0.0083				
6	0.1380	32	0.1177	0.0091	40	0.1218	0.0102				
8	0.1640	32	0.1437	0.0140	36	0.1460	0.0147				
10	0.1900	24	0.1629	0.0175	32	0.1697	0.0200				
12	0.2160	24	0.1889	0.0242	28	0.1928	0.0258	32	0.1957	0.0270	
1/4	0.2500	20	0.2175	0.0318	28	0.2268	0.0364	32	0.2297	0.0379	7/16
5/16	0.3125	18	0.2764	0.0524	24	0.2854	0.0580	32	0.2922	0.0625	1/2
3/8	0.3750	16	0.3344	0.0775	24	0.3479	0.0878	32	0.3547	0.0932	9/16
7/16	0.4375	14	0.3911	0.1063	20	0.4050	0.1187	28	0.4143	0.1274	5/8
1/2	0.5000	13	0.4500	0.1419	20	0.4675	0.1599	28	0.4768	0.170	3/4
9/16	0.5625	12	0.5084	0.182	18	0.5264	0.203	24	0.5354	0.214	13/16
5/8	0.6250	11	0.5660	0.226	18	0.5889	0.256	24	0.5979	0.268	15/16
3/4	0.7500	10	0.6850	0.334	16	0.7094	0.373	20	0.7175	0.386	1-1/8
7/8	0.8750	9	0.8028	0.462	14	0.8286	0.509	20	0.8425	0.536	1-5/16
1	1.0000	8	0.9188	0.606	12	0.9459	0.663	20	0.9675	0.711	1-1/2
1-1/8	1.1250	7	1.0322	0.763	12	1.0709	0.856	18	1.0889	0.901	1-11/16
1-1/4	1.2500	7	1.1572	0.969	12	1.1959	1.073	18	1.2139	1.123	1-7/8
1-3/8	1.3750	6	1.2667	1.155	12	1.3209	1.315	18	1.3389	1.370	2-1/16
1-1/2	1.5000	6	1.3917	1.405	12	1.4459	1.581	18	1.4639	1.64	2-1/4
1-3/4	1.7500	5	1.6201	1.90				16	1.7094	2.24	2-5/8
2	2.0000	4½	1.8557	2.50				16	1.9594	2.95	3
2-1/4	2.2500	4½	2.1057	3.25							3-3/8
2-1/2	2.5000	4	2.3376	4.00							3-3/4
2-3/4	2.7500	4	2.5876	4.93							4-1/8
3	3.0000	4	2.8376	5.97							4-1/2
3-1/4	3.2500	4	3.0876	7.10							4-7/8
3-1/2	3.5000	4	3.3376	8.33							5-1/2
3-3/4	3.7500	4	3.5876	9.66							5-5/8
4	4.0000	4	3.8376	11.08							6

See References, end of chapter.

For helix angle:

$$\tan \alpha = \frac{0.125}{2.8865} = 0.04331$$

$$\alpha = 2°29'$$

Since it is impossible to manufacture parts exactly to a specified size, tolerances have been placed on the dimensions of the screw and nut. Such tolerance zones are illustrated in Figs. 5-3 and 5-4. When the measured

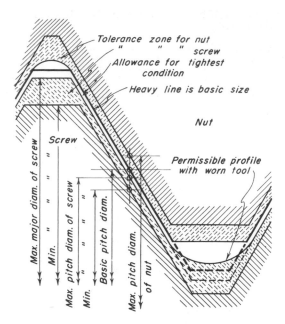

Figure 5-3 Tolerance zones and allowances for classes 1A, 1B, 2A, and 2B Unified Threads.

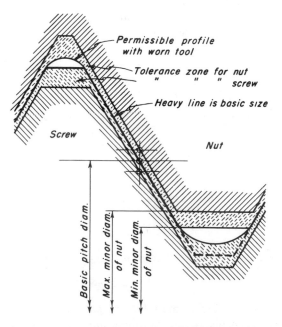

Figure 5-4 Tolerance zones for classes 3A and 3B Unified threads and classes 2 and 3 American National threads.

dimension of a part falls within the tolerance zone, it is accepted as meeting the dimensional specifications.

The term *maximum metal* refers to the screw of largest pitch diameter and the nut of smallest pitch diameter permitted by the tolerance zones. Similarly, *minimum metal* refers to the screw of smallest pitch diameter and the nut of largest pitch diameter. *Allowance* is the clearance between the pitch diameters of screw and nut when both parts of the assembly are at the maximum metal condition.

In general, the quality of a product is highest when the tolerance zones are smallest. Manufacturing costs, however, rise as tolerance zones are made smaller.

Unified and American National threads are divided into various classes, which depend on the size of the tolerance zones. Figure 5-1(a) illustrates the manner in which the basic size is obtained for both screw and nut.

3. Unified Threads

Unified threads are divided into classes 1A, 2A, and 3A for external threads and classes 1B, 2B, and 3B for internal threads.

Classes 1A and 1B are provided when bolts and nuts of the largest tolerances and the largest clearance after assembly are required. An assembly of such parts is illustrated in Fig. 5-3. The pitch diameter of the maximum metal nut is at the basic size, and the pitch diameter of the maximum metal screw is smaller than basic. Bar diagrams representing the tolerances and allowances for a typical screw are shown in Fig. 5-5.

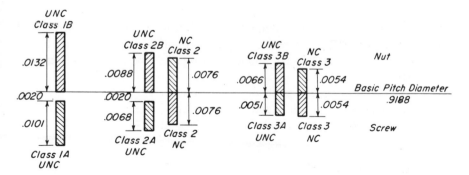

Figure 5-5 Pitch diameter tolerances and allowances for different classes of fits for a 1 in. screw and nut of 8 threads per inch.

Classes 2A and 2B are similar to classes 1A and 1B but have smaller tolerances and a smaller clearance after assembly. These classes are well suited for the vast majority of bolts.

Classes 3A and 3B have the smallest tolerances and zero allowance for an assembly at the maximum metal condition at which both screw and nut have the basic pitch diameter. Such an assembly is illustrated in Fig. 5-4. These classes are intended for applications when closeness of fit and high accuracy are essential.

If desired, the screw can be of one class and the nut can be of another.

4. American National Threads

These threads are divided into Classes 2 and 3 for screws and nuts.

Class 2 has the larger tolerances for screw and nut but no allowance for a maximum metal assembly as shown in Fig. 5-4.

Class 3 has smaller tolerances and produces tighter fits upon assembly. At the maximum metal condition there is no allowance between screw and nut as shown in Fig. 5-4.

It should be noted that Classes 2A and 2B Unified threads have an allowance but that Class 2 American National does not.

As illustrated in Fig. 5-6, a variation in the thread angle of a screw

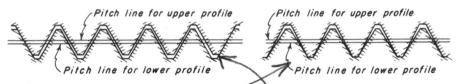

Figure 5-6 Variation in pitch diameter caused by changes in thread angle.   **Figure 5-7** Variation in pitch diameter caused by difference in lead.

causes a change in its pitch diameter. As shown in Fig. 5-7, a screw and nut, if made with slight differences in lead, may still be assembled if the extent of engagement is not too great, and if the two parts have a difference in pitch diameter.

The tolerance on the pitch diameter permitted by the standards must provide for errors from all causes. The effects of angle and lead variation must be included, as well as discrepancies in the pitch diameter itself.

5. Identification Symbols

Following are examples of approved identification symbols for use on drawings, tools, and specifications.

(a) Unified coarse-thread series, $\frac{1}{2}$-in.-diam external thread, 13 threads per in., Class 2A fit:

$$\tfrac{1}{2}\ ''-13UNC-2A$$

(b) American National coarse-thread series, $\frac{1}{2}$-in.-diam, 13 threads per in., Class 2 fit:

$$\tfrac{1}{2}\text{''--13NC--2}$$

(c) Unified fine-thread series, 1 in.-diam internal thread, 12 threads per in., Class 1B fit:

$$1\text{''--12UNF--1B}$$

For a left-hand thread the designations above should be followed by the letters LH.

6. Effect of Initial Stress

Figure 5-8 shows a load P carried by the part that is attached to the support by means of a bolt. The bolt center is located at the center of gravity of the cross section of the part. Let it be assumed that the nut is merely made snug against the part but with no initial tension in the bolt. Also let it be temporarily assumed in Fig. 5-8 that the part is welded to the support, making it possible for both bolt and part to carry tensile loads.

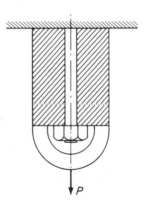

Figure 5-8 Bolted part carrying load.

When load P is placed on the assembly, the part immediately stretches, and since it is in contact with the nut, the bolt is stretched also. Both part and bolt have tension stresses, because load P divides and is carried partly by the bolt and partly by the part. In finding the portion of load carried by each, it is convenient to work with the so-called "spring constants" of these members. The spring constant k is the value of the force required to give a deformation of 1 in. It may be found by using the equation for deformation, $\delta = Pl/AE$, and letting P be equal to k when δ becomes equal to unity. When these substitutions are made, the equations for k_b for the bolt and k_p for the part become

$$k_b = \frac{A_b E_b}{l_b}; \quad \text{and} \quad k_p = \frac{A_p E_p}{l_p} \tag{1}$$

where subscripts b and p refer to bolt and part, respectively, and where A is the cross-sectional area, E is the modulus of elasticity, and l is the length in the direction of the force. If the thread stops immediately above the nut, the gross area of the bolt must be used in computating k_b, since it is the un-threaded portion that is stretched by the load.[1]

---

[1]When a greater length of bolt is threaded, suitable allowance must be made for the increased flexibility that thereby results.

Let P_b be the portion of load P that is carried by the bolt, and P_p be the remainder that is taken by the part. The deformation of the bolt is equal to P_b/k_b, and of the part, P_p/k_p. Because of the arrangement of the members in Fig. 5-8, these deformations are equal, and the following equation can be written.

$$\frac{P_b}{k_b} = \frac{P_p}{k_p} \tag{a}$$

Substitution of $P - P_b$ for P_p gives

$$P_b = \frac{k_b}{k_b + k_p} P \tag{b}$$

Similarly,

$$P_p = \frac{k_p}{k_b + k_p} P \tag{c}$$

Suppose now that the nut is tightened to the extent that an additional tensile load equal to F_o is placed in the bolt. The total tension in the bolt is now

$$F_b = \frac{k_b}{k_b + k_p} P + F_o \tag{2}$$

At the same time that the bolt was receiving the additional tension F_o from the tightening, the part was receiving an equal and opposite force of compression F_o. The resultant force in the part now is

$$F_p = \frac{k_p}{k_b + k_p} P - F_o \tag{3}$$

It is usual practice to induce force F_o initially upon assembly of the parts before load P is applied. Equations (2) and (3) are, of course, unaffected. Initial force F_o is substituted as a positive number in Eq. (3), and is always made sufficiently large to keep force F_p for the part negative or compression, and the assumed welding is thus not required. Should P be so great as to remove all the initial compression from the part, then neither Eq. (2) nor (3) can be used, for the bolt will then be carrying the total load P.

The maximum bolt force results when load P is a maximum, and the minimum bolt force results when the load is a minimum. The average bolt force $F_{b\,av}$ is equal to half the sum of $F_{b\,max}$ and $F_{b\,min}$. Thus

$$F_{b\,av} = \frac{k_b}{k_b + k_p} P_{av} + F_o \tag{4}$$

The range of the bolt force is equal to one-half the difference between $F_{b\,max}$ and $F_{b\,min}$. Then

$$F_{br} = \frac{k_b}{k_b + k_p} P_r \tag{5}$$

The minimum compressive force in the part occurs when load P is a maximum.

$$F_{p\,min} = \frac{k_p}{k_b + k_p} P_{max} - F_o \tag{6}$$

The use of sufficient initial tension is very advantageous in reducing the fatigue effects in the bolt if load P is not steady but fluctuating. For varying load, the maximum bolt force is found by substituting the maximum value of P in Eq. (2), and the minimum bolt force is obtained by use of the minimum value of P. Whereas the maximum value of the bolt force is somewhat larger than if no initial force were used, the presence of F_o gives a considerably larger value of the minimum bolt force than if no tightening-up tension were used. The resultant effect is to give a much smaller range of variation[2] in the bolt force F_b than takes place in the applied load P, and the fatigue effects, which depend on the extent of the variation of the stress, are correspondingly reduced.

Example 2. Let the bolt in Fig. 5-8 be $\frac{1}{2}$ in. $\times$ 13UNC. Bolt and part are of the same length; the threads stop immediately above the nut. The bolt material has ground threads and a yield strength of 70,000 psi. Take the stress concentration factor for the threads as 3.85. The steel part has a net area of 0.5 in.[2] The load fluctuates continuously between 0 and 2,400 lb. Take s_e as 30,000 psi.

(a) Find the FS for the bolt when no initial force is present.

(b) Find the minimum required value of F_o to prevent loss of compression in the part.

(c) Find the FS for the bolt when F_o is taken as 2,500 lb.

(d) Find the minimum force in the part for the given loading and F_o equal to 2,500 lb.

Solution.

(a) By Table 5-1: Stress area $= 0.1419$ in.[2]

$$P_{av} = 1,200 \text{ lb} \qquad S_{av} = \frac{1,200}{0.1419} = 8,460 \text{ psi}$$

$$P_r = 1,200 \text{ lb} \qquad Ks_r = 3.85 \times \frac{1,200}{0.1419} = 32,560 \text{ psi}$$

When s_{av} and Ks_r are plotted as in Fig. 5-9, it is seen that the bolt loading is unsafe.

(b) For bodies of equal lengths and moduli, the spring constants are proportional to the cross-sectional areas.

$$\text{gross area of bolt} = \frac{\pi}{4}\left(\frac{1}{2}\right)^2 = 0.1964 \text{ in.}^2$$

In Eq. (6), when $F_{p\ min}$ is equal to zero,

$$F_o = \frac{k_p}{k_b + k_p}P_{max} = \frac{0.5}{0.6964} \times 2,400 = 1,720 \text{ lb}$$

This is the theoretical minimum value for F_o.

(c) By Eq. (4): $\qquad F_{b\ av} = \frac{k_b}{k_b + k_p}P_{av} + F_o$

$$= \frac{0.1964}{0.6964} \times 1,200 + 2,500 = 2,840 \text{ lb}$$

[2]See Almen, J. O., *Machine Design*, **15**, Aug. 1943, p. 133, and **16**, Feb. 1944. p. 158.

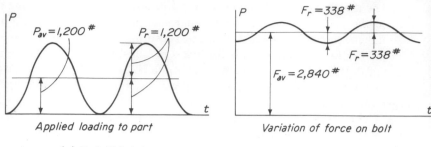

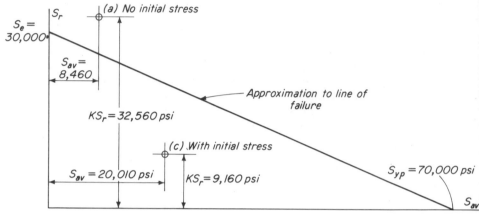

Figure 5-9 Loads and stresses on bolt. Example 2.

$$s_{av} = \frac{2,840}{0.1419} = 20,010 \text{ psi}$$

By Eq. (5):
$$F_{br} = \frac{k_b}{k_b + k_p} P_r = \frac{0.1964}{0.6964} \times 1,200 = 338 \text{ lb}$$

$$s_r = \frac{338}{0.1419} = 2,380 \text{ psi}$$

By Fig. 2-16:
$$Ks_r = 3.85 \times 2,380 = 9,160 \text{ psi}$$

By Eq. (11), Chapter 2:
$$\frac{S_{yp}}{FS} = s_{av} + \frac{Ks_{yp}}{s_e} s_r$$

$$\frac{70,000}{FS} = 20,010 + \frac{3.85 \times 70 \times 2,380}{30}$$

$$= 20,010 + 21,380 = 41,390 \text{ psi}$$

$$FS = \frac{70,000}{41,390} = 1.69$$

(d) By Eq. (6):
$$F_{pmin} = \frac{k_p}{k_b + k_p} P_{max} - F_o$$

$$= \frac{0.5}{0.6964} \times 2,400 - 2,500 = -780 \text{ lb.}$$

It should be noted that the use of initial stress has changed the situation to the extent that the bolt now has a reasonable value for the factor of safety.

Equations can also be derived for the case where the bolt is located eccentrically to the center of gravity of the part.[3]

7. Effect of Spring Washers and Gaskets

Spring washers and gaskets are frequently incorporated in bolted assemblies. Their effect must be taken into account in the design, since the resulting assembly may be made either weaker or stronger.

Consider first the bolt in Fig. 5-10(a). If the area of the part is very large

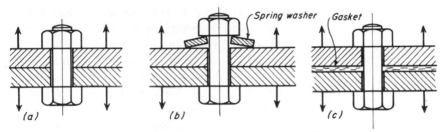

Figure 5-10 Various types of bolted assemblies.

as compared with the cross section of the bolt, the value of k_p in Eq. (2) will be much larger than k_b, and variations in load P will have but small effect on the value of F_b. The force in the bolt remains substantially constant at the value F_o as long as initial compression is retained by the part. Although this is a desirable feature, it must be remembered, when the part is rigid, that the lengthening of the bolt required to induce force F_o is a very small quantity. Should the loading cause any creep or recession at the high spots of the contact surfaces, the initial stretch of the bolt may be lost and force F_o will disappear. Shortening of the part may also be caused by corrosion, wear, or displacement of platings and coatings.

The situation can be improved by the use of a spring washer under the nut, as shown in Fig. 5-10(b). The deformation sustained by the washer from force F_o may be many times as great as the stretch of the bolt. Any small shortening of the part during service then has but small effect in reducing the value of F_o, and the initial force can be expected to be retained. It is assumed that the washer is acting in its elastic range and is not compressed solid. Ordinary lock washers, which are compressed solid upon assembly, are of no help in reducing the fluctuations of the load in the bolt. In fact,

[3] See the author's *Mechanical Design Analysis*, Englewood Cliffs, N. J.: Prentice-Hall, Inc., 1964, p. 96.

such devices may constitute an additional hazard from loss of initial stress, caused by plastic flow at high spots or burrs on the surfaces.

A gasket is shown between the parts in Fig. 5-10(c). The stiffness of the gasket, hard or soft, influences the overall value of k_p. A soft gasket reduces the value of k_p and causes a larger proportion of the load P to be taken by the bolt. It should be kept in mind that any set or permanent deformation of the gasket may cause a loss of the initial force F_o.

8. Power Screws

A power screw can be used for raising weights or exerting forces in machines. The weight W shown in Fig. 5-11, into which the supporting screw is threaded, can be raised or lowered by the rotation of the screw. It is, of course, assumed that W is prevented from turning when the screw rotates. An expression will now be derived for finding the value of the torque needed to raise the load.

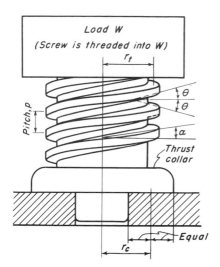

Figure 5-11 Power screw.

The total force on the threads can be represented by a single force F_n, Fig 5-12(a), which is normal to the thread surface. Force F_n is the diagonal of the parallelopiped. Side $ABEO$ is an axial section through the bolt. The projection of F_n into this plane is inclined at θ or half the included thread angle. Side $ACHO$ lies in the plane tangent to the pitch cylinder. The projection of F_n into this plane is inclined at the helix angle α calculated at the pitch radius of the screw. The length of this component is $F_n \cos \theta_n$ as indicated in sketch (c). It has the vertical and horizontal components shown.

The developed helix of a screw becomes merely an inclined plane with angle α equal to the helix angle of the screw. The lower triangular block thus represents the thread, which raises the weight when it is pushed to the left by force F.

The total friction force along the thread is $\mu_1 F_n$, where μ_1 is the coefficient of thread friction. Force $\mu_1 F_n$ has the vertical and horizontal components shown.

The upward reaction at the base is equal to weight W. During motion there is a resisting friction force $\mu_2 W$, where μ_2 is the coefficient of friction for the base or collar.

Let force F be just sufficient to cause motion to the left. Summation of the vertical forces in sketch (c) gives

$$F_n \cos \theta_n \cos \alpha = \mu_1 F_n \sin \alpha + W$$

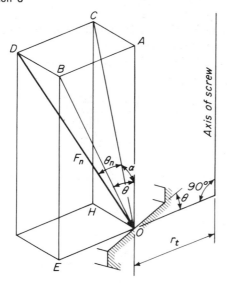

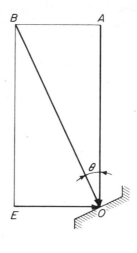

(a) Force parallelepiped   (b) Axial section

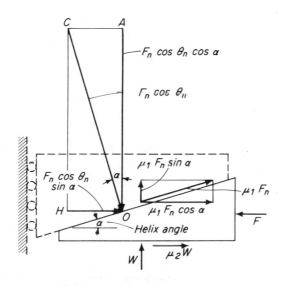

(c) Tangential plane

Figure 5-12 Force acting on screw when raising weight W.

or
$$F_n = \frac{W}{\cos\theta_n \cos\alpha - \mu_1 \sin\alpha} \tag{7}$$

All thread forces, as well as force F, act at the pitch radius of the thread r_t. The force of the collar friction acts at radius r_c to the midpoint of the collar surface. The torque required to raise the load is found by multiplying the horizontal forces by the appropriate radii. Hence,

$$T = Fr_t = r_t(F_n \cos \theta_n \sin \alpha + \mu_1 F_n \cos \alpha) + r_c \mu_2 W$$

Substitution of the above value for F_n gives

$$T = r_t W \left(\frac{\cos \theta_n \sin \alpha + \mu_1 \cos \alpha}{\cos \theta_n \cos \alpha - \mu_1 \sin \alpha} + \frac{r_c}{r_t} \mu_2 \right) \tag{8}$$

or

$$T = r_t W \left(\frac{\cos \theta_n \tan \alpha + \mu_1}{\cos \theta_n - \mu_1 \tan \alpha} + \frac{r_c}{r_t} \mu_2 \right) \tag{9}$$

The relationship between θ_n and thread angle θ is obtained as follows. In sketch (a),

$$\tan \theta_n = \frac{CD}{OC} = \frac{AB}{OC} \tag{a}$$

$$AB = AO \tan \theta \quad \text{and} \quad OC = \frac{OA}{\cos \alpha}$$

These should now be substituted into Eq. (a) to give

$$\tan \theta_n = \tan \theta \cos \alpha \tag{10}$$

Equations (8) and (9) give the value of the torque required to raise the load when friction for both thread and collar are included. Sometimes the collar consists of an antifriction bearing, in which case μ_2 may be sufficiently small to be neglected. The equations then contain the μ_1 terms only.

Example 3. A quadruple Acme thread, 1-in. OD, has a pitch of 0.200 in. Collar diameters are 1.5 and 0.5 in. Find the required torque when the screw is exerting a force of 1,000 lb. Let μ_1 and $\mu_2 = 0.12$.

Solution. $\theta = \frac{1}{2} \times 29° = 14°30'$

By Fig. 5-1:   Pitch diameter $= 1.000 - 0.5p = 0.900$ in.

Helix angle: $\tan \alpha = \dfrac{4 \times 0.200}{0.900\pi} = 0.28294$

$$\alpha = 15°48'$$

By Eq. (10): $\tan \theta_n = 0.25862 \times 0.96222 = 0.24885$

$$\theta_n = 13°58.5'$$

By Eq. (9): $T = 0.45 \times 1,000 \left(\dfrac{0.97040 \times 0.28294 + 0.12}{0.97040 - 0.12 \times 0.28294} + \dfrac{0.50}{0.45} \times 0.12 \right)$

$$= 450(0.42133 + 0.13333) = 249.6 \text{ in. lb}$$

For standard screws that have small values for the helix angle α, Eq. (10) indicates that θ_n has almost the same value as θ. When this is so, θ_n can be replaced by one-half the included thread angle in the foregoing equations.

Computations indicate that the torque required to induce load W in the bolt, for standard threads of 60° angle and coefficient of friction 0.15, can be found approximately by the following equation.

$$T = 0.2dW \tag{11}$$

where d is the nominal or outside diameter of the screw. In arriving at Eq. (11), the collar radius r_c was taken at the midpoint for the bearing surface of the nut.

If the weight is being lowered by the application of force F acting to the right in Fig. 5-12, the sign of F, together with the signs of all the friction terms, are reversed. The torque required to lower the load then becomes

$$T = r_t W \left(-\frac{\cos \theta_n \tan \alpha - \mu_1}{\cos \theta_n + \mu_1 \tan \alpha} + \frac{r_c}{r_t} \mu_2 \right) \tag{12}$$

If the helix angle is sufficiently great, the screw will overhaul, or the weight will revolve the screw. The inclined plane in Fig. 5-12(a) will then move to the right and force F must act to the left to preserve uniform motion. The torque equation for the overhauling screw is found to be

$$T = r_t W \left(\frac{\cos \theta_n \tan \alpha - \mu_1}{\cos \theta_n + \mu_1 \tan \alpha} - \frac{r_c}{r_t} \mu_2 \right) \tag{13}$$

If all friction could be eliminated in both screw and collar, Eq. (9) shows that the torque required to raise the load would be

$$T' = r_t W \tan \alpha \tag{14}$$

The efficiency of a power screw with collar friction when it is raising the load is equal to the ratio of the torques of Eqs. (9) and (14). Hence for both screw and collar,

$$\text{efficiency} = \frac{T'}{T} \tag{15}$$

If the collar friction is negligible, the following equation for the efficiency of the screw alone results.

$$\text{efficiency} = \frac{\cos \theta_n - \mu_1 \tan \alpha}{\cos \theta_n + \mu_1 \cot \alpha} \tag{16}$$

Efficiencies, as given by Eq. (16), have been plotted in Fig. 5-13 for several values of μ_1. It should be noted that the power screw has very low mechanical efficiency when the helix angle is in the neighborhood of either 0° or 90°.

The friction between the threads and between the nut and the abutment depends on a variety of factors, such as surface finish, degree of lubrication, alignment, materials, plating, burrs, and the like. Calculations for the induced force in the bolt will therefore vary as these factors affect the coefficient of friction. The torque wrench, although widely used, is not considered a very reliable means for obtaining an accurate value of the induced force.

A better way is to measure the bolt elongation when this is possible. The angle of twist of the nut can also be used. The nut is first firmly tightened to seat the parts on each other. It is then loosened and made finger tight. The nut is then turned through a specified angle, which has been calculated to give the desired axial force in the bolt.

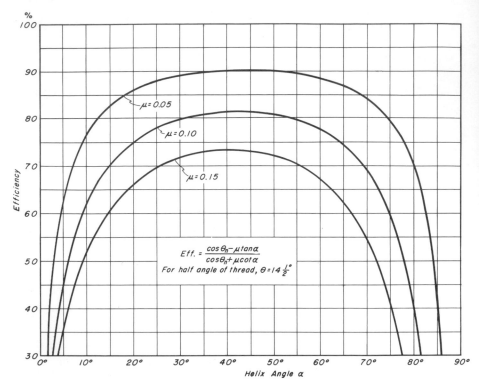

Figure 5-13 Efficiency of power screw when collar friction is negligible.

9. Friction of Screws

Published articles on experiments for screw-thread friction indicate that the following values would be a good estimate for the coefficient of friction.[4]

$$\text{Mean value of } \mu = 0.15$$

$$\text{Range of variation} = \pm 33\%$$

The above is to apply to threads lubricated with mineral oil but with no special control over surface finish other than that normally found under conditions of large quantity production.

10. Stress Concentration

A stress concentration is present in the screw where the load is transferred through the nut to the adjoining member. This arises because the force in

---

[4]See Lambert, T. H., "Effects of Variations in the Screw-Thread Coefficient of Friction on the Clamping Force of Bolted Connections," *J. Mech. Eng. Sci.*, **4**, 1962, p. 401.

the screw must shift outwardly to the region near the boundary as it is transferred from the screw to the nut. Under ideal conditions, the tension in the screw and the compression in the nut should be reduced uniformly starting from full load at the first contact between screw and nut. However, the tension increases the pitch in the screw, and the compression decreases the pitch in the nut so that correct compliance between the loaded parts is not maintained. The major portion of the load is transferred at the first pair of contacting threads and a large stress concentration is present here. Although the stress concentration is somewhat relieved by bending of the threads and expansion of the nut, most bolt failures occur at this point.

Stress concentration factors for threads with static loads are usually determined by photoelastic analysis. Three-dimensional tests[5] have indicated a stress concentration factor of 3.85 at the root of the first engaged thread. Other investigators,[6] however, have found higher as well as lower values. Various methods are used for increasing the flexibility of the nut and thus increasing the area over which this transfer of force takes place. A tension nut, or a nut with a tapered lip, as shown in Fig. 5-14(a), has been successfully

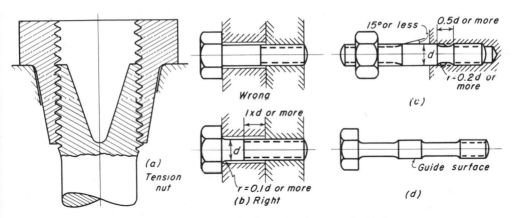

Figure 5-14 Typical methods for increasing the strength of bolts.

used in fatigue service. The reduced cross-sectional area of the lip deforms in tension with the bolt, and the load is accordingly spread over a greater number of threads. Another method is to cut the thread of the nut on a very small taper, thus reducing the contact area for the first few threads. Since

[5]See Hetenyi, M., "A Photoelastic Study of Bolt and Nut Fastenings," *Trans. ASME,* **65,** 1943, p. A-93.
[6]See Brown, A. F. C., and V. M. Hickson, "A Photo-Elastic Study of Stresses in Screw Threads," *Inst. Mech. Engr., Proc. (B),* **1B,** 1952–53, p. 605.

these threads will bend and carry less load, additional threads must come into service.[7] This method is expensive because of the close dimensional tolerance required. Nuts made of a material with a modulus smaller than the bolt have also been successful in spreading the load over a larger area. The material in the nut, however, must have a sufficient reserve of ductility to deform without breaking.

An increase in the flexibility of the bolt is also beneficial under fatigue conditions. Accordingly, there should be a considerable length of free thread beyond that required for engagement of the nut, as shown in Fig. 5-14(b). A radius between shank and head is helpful in reducing the stress concentration at this point. The run-out angle for the thread should have a small value, as shown in Fig. 5-14(c), since the stress concentration is highest at the first threads. For the same reason, a stress-relieving groove is shown on the right end of this stud. It should have a diameter equal to that of the root of the thread or slightly less, and should be merged to the shank by radii of considerable size. A large radius causes less stress concentration in transferring the load from the larger to the smaller cross section. In addition, if the stud is bent because of misfits in assembly of the parts, the smaller diameter will produce a smaller bending stress in the material. Flexibility of the bolt is increased by reducing the entire length between the head and the threads to the root diameter. If the bolt is holding two or more members together, short lengths of the original diameter must be left at each junction to serve as guide surfaces for the bolted parts, as shown in Fig. 5-14(d).

Heat-treated alloy steel is usually more sensitive to changes of form than plain-carbon steel. When an actual thread is tested in fatigue loading, it is more correct to speak of a stress reduction factor. This factor is defined as the ratio of the endurance limit of the material for a plain specimen to the endurance limit for a specimen with threads.

It has been estimated that bolt failures are distributed about as follows: 15 per cent under the head, 20 per cent in the bolt at the end of the threads, and 65 per cent in the bolt at the nut face.

11. Locknuts

Many different types of locknuts have been devised to prevent a nut from loosening in service due to vibration. Several different forms of locknuts are illustrated in Fig. 5-15. The speed nut, shown in Fig. 5-16, acts as a single-thread nut of spring steel strip. This type of nut can be rapidly assembled. When tightened, the prongs press against the stud, and the strip acts as an arched spring lock.

Many kinds of lock and spring washers, placed beneath an ordinary nut,

[7]See Stoeckly, E. E., and H. J. Macks, "The Effect of Taper on Screw Thread Load Distribution," *Trans. ASME*, **74**, 1952, p. 103.

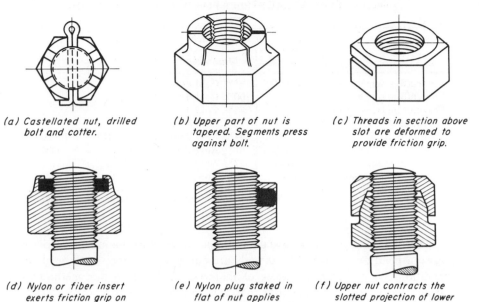

(a) *Castellated nut, drilled bolt and cotter.*

(b) *Upper part of nut is tapered. Segments press against bolt.*

(c) *Threads in section above slot are deformed to provide friction grip.*

(d) *Nylon or fiber insert exerts friction grip on threads.*

(e) *Nylon plug staked in flat of nut applies friction to bolt threads.*

(f) *Upper nut contracts the slotted projection of lower nut against bolt.*

Figure 5-15 Several forms of locknuts.

are also in use. Success depends on the type of washer and the conditions of the particular application. Various kinds of adhesives are also in use for this purpose.

Self-tapping and drive screws are available for fastener service in sheet metal, plastics, nonferrous metals, and other materials.

Figure 5-16 Standard flat-type speed nut.

12. Materials and Methods of Manufacture

Bolts and screws may be turned on automatic screw machines, using bar stock of the same dimensions as the head, or they may be cold- or hot-headed from stock of the same diameter as the shank. A number of typical steels for bolts and screws are shown in Table 5-2. Free-cutting screw stock is preferred for use on automatic screw machines because the chips break short and, as a result, there is less danger of fouling the working parts of the machine. These advantages are indicated by the high machinability ratings for these materials. Steels 1112 and 1113 are Bessemer steels of high sulphur and phosphorous content; 1117 and 1137 are high-manganese, open-hearth steels. Alloy steel bolts are used for severe service conditions where the properties of high strength and ductility are required. Bolts of alloy steels are usually heat treated to secure the benefits of the alloy content of the material.

Table 5-2   TYPICAL SAE STEELS FOR BOLTS AND SCREWS

Steel	Automatic Screw Machine		Cold Heading	Hot Heading
	Material	*Machinability*		
Free-cutting screw stock	1113	135% Cold drawn		
Free-cutting screw stock	1112	100% Cold drawn		
Free-cutting screw stock	1117	85% Cold drawn		
Free-cutting screw stock	1137	73% Cold drawn		
Plain-carbon	1015–1035	57% Cold drawn	1010–1045	1010–1045
Nickel	2330	61% Annealed	2330	2330
Nickel-chromium	3140	57% Annealed	3140	3140
Chromium-molybdenum	4140	61% Annealed	4140	4140
Chromium-nickel-molybdenum	A8640	55% Annealed	A8640	A8640

Turned bolts generally have good dimensional control and concentricity between head and shank. The threads are cut as one of the screw machine operations. A considerable waste of material results from this method in turning the shank from material that was originally the size of the head. Because of this loss, the automatic screw machine is in widest use for bolts of about $\frac{1}{4}$-in. diam and less.

The cold-heading process uses slightly oversize stock, which is first drawn through a die to secure dimensional uniformity and to cold work the surface. The low-carbon steels, because of better workability, give longer die life than alloy steels. The threads on cold-headed bolts are usually formed by rolling the shank between dies which depress part of the steel to form the root and which force the remainder up to make the top of the thread. The outside diameter of the thread thus is larger than the stock on which it was rolled. If the threaded portion of the bolt is to have the same diameter as the unthreaded part, the rolling must be done on a reduced diameter. Forging machines form this portion of smaller cross section at the same time that the head is upset. Rolled threads, because of favorable grain structure at the root of the thread, are stronger than cut threads in fatigue and impact.

For larger-size heads and more complicated shapes, hot heading is generally employed. Hot heading is especially suitable for the tougher and less ductile alloy steels. The process is more expensive than cold heading because of the operations involved in heating the end of the bar to the forging temperature. Bolts and screws of nonferrous materials are produced in large quantities by the same methods used for steel.

13. Stress Due to Impact Load

Bolts are sometimes subjected to suddenly applied or impact loads. The stress caused by such loads can be found from the energy of impact U that the bolt must absorb.

The force-deformation diagram for a bolt under a tensile load is a triangle, as shown in Fig. 5-17. The area represents the strain energy U stored in the bolt. Thus

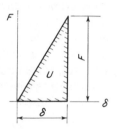

$$U = \tfrac{1}{2}F\delta \qquad (17)$$

where F is the force caused by impact, and δ is the corresponding deformation. Deformation δ is equal to F/k, where k is the spring constant for the bolt, AE/l. Substitution for δ gives

Figure 5-17 Strain energy diagram for bolt

$$U = \frac{F^2}{2k} \qquad (18)$$

The stress in the bolt is equal to force F divided by the smallest cross-sectional area. For a threaded bolt, this area is the stress area of the threaded portion.

Example 4. A $\frac{1}{2}$ in.-13UNC $\times$ 12 in. long steel bolt must carry an impact load of 40 in. lb.

    (a) Find the stress in the stress area for a standard bolt.

    (b) Find the stress if the entire body of the bolt between head and nut is reduced to that of the stress area.

Solution. (a) If the threads stop adjacent to the nut, the full $\frac{1}{2}$-in. diameter is stretched by the impact force. Hence,

$$A = \frac{\pi}{4}\left(\frac{1}{2}\right)^2 = 0.1964 \text{ in.}^2$$

$$k = \frac{AE}{l} = \frac{0.1964E}{12} = 0.01637E$$

In Eq. (18):
$$F = \sqrt{2kU} = \sqrt{2 \times 0.01637 \times 30{,}000{,}000 \times 40}$$
$$= 6{,}267 \text{ lb}$$

By Table 5-1: stress area $= 0.1419$ in.^{2}

In stress area:
$$s = \frac{6{,}267}{0.1419} = 44{,}160 \text{ psi}$$

It is possible for stress concentration to cause a large increase in the stress above.

    (b) The value of k now depends on the stress area; hence,

$$k = \frac{AE}{l} = \frac{0.1419E}{12} = 0.01182E$$

In Eq. (18):
$$F = \sqrt{2 \times 0.01182 \times 30{,}000{,}000 \times 40} = 5{,}320$$

$$s = \frac{5{,}320}{0.1419} = 37{,}530 \text{ psi}$$

By careful attention to details, this reduced diameter can be gradually blended into the threaded portion and the stress concentration factor reduced practically to unity.

This example illustrates how a smaller impact stress results from increasing the flexibility by making the diameter equal to the root diameter of the thread. Flexibility can also be increased by using a longer bolt. If the bolt is made of a ductile material, and the yield point is exceeded by the impact force, some permanent stretching results.

Experiments have indicated that the yield strength of a material under impact conditions is somewhat higher than when the loading is static.[8]

14. Relaxation

Bolts that connect heavy parts operate under conditions of constant elongation. In high-temperature service the initial tension will diminish until, after a sufficiently long time, the joint will no longer be tight. This phenomenon is known as *relaxation*. The initial tension must be sufficiently great to maintain a tight joint at the end of the expected life of the assembly.[9] Similar conditions prevail in shrink or press-fitted joints in high-temperature service.

REFERENCES

1. *Guide to World Screw Thread Standards*, Bedford, England: W. H. A. Robertson & Co. Ltd, Lyton Works.

2. Rothbart, H. A., Ed., *Mechanical Design and Systems Handbook*, New York: McGraw-Hill Book Company, 1964, Sections 20, 21, and 26.

3. *Screw-Threads for Federal Services, Handbook H-28*, Washington, D.C.: U.S. Government Printing Office, 1957.

4. *Unified and American Screw Threads, ASA B1.1-1960*, New York: American Society of Mechanical Engineers.

PROBLEMS

In Problems 1–5, let it be assumed that the threads stop immediately above the nut.

1. In an assembly with loading arrangement equivalent to Fig. 5-8, the bolt is $\frac{3}{4}$ in.–16UNF, and the total load varies from 2,000 to 10,000 lb. The material in

[8]See Forkois, H. M., R. W. Conrad, and I. Vigness, "Properties of Bolts under Shock Loading," *Proc. Soc. Exp. Stress Anal.*, **10**, No. 1, 1952, p. 165. See also Davidenkoff, N. N., "Allowable Working Stresses under Impact," *Trans. ASME*, **56**, 1934, APM-56, p. 97.

[9]For design equations see Mechanical Design Analysis. loc. cit. footnote 3, p. 296.

the bolt tests $s_{ult} = 130,000$ psi and $s_{yp} = 110,000$ psi, with stress concentration factor for the threads equal to 3.85. Let the factor of safety based on the yield point be 2, and let E for both part and bolt be equal to 30,000,000 psi. The threads are machined. The cross-sectional area of the part is equal to 1.25 in.²

(a) Draw the working-stress triangle for the bolt material, and plot the stresses when there is no initial stress in the bolt.

(b) If the initial force in the bolt is 8,000 lb, plot the stress values for the bolt on the working-stress triangle.

It will be necessary to determine whether or not the part will have a compressive force when the maximum load is acting. Should all the initial compression be removed, the force in the bolt will have the same value as the load.

(c) Repeat (b) with an initial bolt force of 13,000 lb. Note the effect of excessive initial stress.

(d) Suppose the bolt is made of steel, but the part is made of aluminum, $E = 10,000,000$ psi. Plot the stress values for the bolt if other data are the same as in (b).

(e) Suppose the bolt were turned to the stress area of the thread over its entire length except for the nut. Plot the stress values for the bolt; other data are the same as in (b). *Ans.* (b) $s_{av} = 25,650$ psi; $Ks_r = 10,790$ psi.

2. The connecting rod bolt of Fig. 5-18 is $\frac{3}{8}$ in.–24UNF and is drawn up to an initial force of 3,500 lb. The bolt material has ground threads, $s_{ult} = 110,000$ psi, and $s_{yp} = 90,000$ psi. The average cross-sectional area of the bolted parts is equal to 0.50 in.² Stress concentration factor for the threads is equal to 3.85. The load for the part varies continuously from 0 to 2,500 lb. Find the value of the factor of safety for the bolt. Find the value of the minimum force in the part. The material of the part has the same modulus as the bolt. *Ans.* $FS = 1.49.$

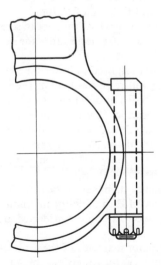

Figure 5-18 Problem 2.

3. Let the bolt in Fig. 5-8 be a 1 in.–12UNF made from cold-drawn 2317 steel. The load on the part varies continuously from 10,000 to 20,000 lb. The threads are machined. $K = 3.85$. Area of the steel part is 1.2 in.^{2} Find the maximum permissible value of F_o if bolt is to be safe for continuous operation with an FS equal to 2.

Ans. $F_o = 3,060$ lb.

4. The bolt in Fig. 5-8 is $\frac{3}{4}$ in.–10UNC, and is made from 1137 cold-drawn steel. Threads are machined. $FS = 2.5$; $K = 3.85$. Area of the steel part is 1.0 in.^{2} Initial force F_o equals 7,000 lb. If the average stress in the stress area is 25,000 psi, find the maximum and minimum values of the varying load P on the part.

Ans. $P_{max} = 7,040$ lb; $P_{min} = 3,740$ lb.

5. (a) In Fig. 5-8 let the part undergo a permanent shortening in length equal to ϵ after the bolt has been tightened to an initial force of F_o. Derive an equation for the new value F'_o for the initial force in the bolt.

(b) Find the value of F'_p for the bolt of Problem 1 (b) if the part for some reason should shorten 0.0003 in. Bolt and part are 5.0 in. long.

Ans. (a) $F'_o = F_o - \dfrac{\epsilon k_b k_p}{k_b + k_p}$; (b) $F'_p = -23$ lb.

6. Find the force of tension in a $\frac{5}{16}$ in.–24NF bolt if it is tightened by a wrench with a torque equal to 240 in. lb. Assume the coefficient of friction for the screw and the collar to be the same, and make computations for coefficients of 0.10, 0.15, and 0.20. Take the outer collar diameter as being equal to 0.50 in.

Ans. For $\mu = 0.10$, $F_o = 5,510$ lb.

7. A load of 10,000 lb is carried by a 2.5-in. single-thread Acme screw of standard proportions. The pitch is $\frac{1}{3}$ in. and the pitch diameter is $2\frac{1}{3}$ in. The OD of the collar is equal to 4 in., and the ID is equal to 1.25 in.

(a) For $\mu_1 = \mu_2 = 0.15$, find the horsepower required to rotate the screw if the weight is to be raised at the rate of 10 fpm.

(b) What is the efficiency when friction of both screw and collar is considered? What is the efficiency if the collar friction were made negligible by use of an antifriction bearing?

(c) Find the horsepower required to lower the load at the same rate.

(d) What horsepower will raise the load at the given rate when the collar is supported on a ball thrust bearing for which $\mu_2 = 0.003$? Let the collar radius be the same as for the plain bearing. What will now be the efficiency?

(e) Find the pitch of the thread at which overhauling will take place, using a ball thrust bearing. The pitch diameter is the same.

(f) Suppose the screw is made with pitch just sufficient to overhaul. What will be the efficiency for the screw alone?

(g) If the minimum major diameter of the screw is 2.483 in. and the maximum minor diameter of the nut is 2.183 in., find the minimum length of the nut that must be engaged if the compressive stress on the projected area of the threads is equal to 600 psi.

(h) If the minimum minor diameter of the screw is 2.130 in., find the average value of the compressive stress at the root of the thread. What is the bearing pressure for the collar?   *Ans.* (a) $hp = 24.7$; (b) eff $= 12.3$ and 22.5%; (e) $p = 1.16$ in.; (g) $l = 5.05$ in.

8. (a) Write the expression for tan α for a square thread with negligible collar friction at which the maximum efficiency will occur.

(b) What is the value of the maximum efficiency, and the angle at which it occurs, for $\mu_1 = 0.1$?   *Ans.* $\tan \alpha = -\mu_1 + (1 + \mu_1^2)^{1/2}$; eff $= 81.9\%$.

9. What pitch must be provided on a square thread power screw to raise a 2,000-lb weight at 40 fpm with power consumption of 4 hp? The pitch diameter is 1.375 in., μ_1 is 0.15, and collar friction is negligible.   Ans. $p = 1.106$ in.

10. A square thread screw has an efficiency of 65 per cent when raising a load. The coefficient of friction for the threads is 0.15 with collar friction negligible. Pitch diameter is 2.75 in. When lowering a load, a uniform velocity is maintained by a brake mounted on the screw. If the load is equal to 10 tons, what torque must be exerted by the brake?   *Ans.* $T = 4,550$ in. lb.

11. A square thread screw has an efficiency of 70 per cent when raising a weight. The coefficient of friction $\mu_1 = 0.12$ with collar friction negligible. The load is 8,000 lb and the pitch diameter is 1.15 in. Find the torque that a brake mounted on the screw must exert when lowering the load at a uniform rate.

Ans. $T = 890$ in. lb.

12. A square thread screw, 0.9-in. pitch diam, has an efficiency of 70 per cent when raising a 4,000-lb weight. The coefficient of friction for the threads is 0.10, with collar friction negligible. Find the torque that a brake mounted on the screw must exert when the load is being lowered at a constant rate.

Ans. $T = 270$ in. lb.

13. A square thread screw has a pitch diameter of 1.5 in. and a lead of 1 in. The screw consumes 4 hp when raising a 2,800-lb weight at the rate of 30 fpm. Collar friction is negligible. Find the coefficient of friction for the threads.

Ans. $\mu_1 = 0.113$.

14. A square thread screw is at the point of overhauling when at rest. When raising a 1,000-lb weight at 15 fpm the input is 0.93 hp. Find the pitch of the screw if the pitch diameter is 0.8 in. Collar friction is negligible.   *Ans.* $p = 0.377$ in.

15. A screw of 9 threads per in. is to be cut on a $1\frac{3}{16}$-in.-diam bar. Proportions are similar to the Unified thread. Find the value of the basic pitch diameter. If the screw is made double thread, find the helix angle.

16. The minimum major diameter of a 1 in.-8UNC screw is 0.9755 in. The maximum minor diameter of the nut is 0.8797 in. Find the required length of nut that must be engaged with the screw if the tensile load is 8,000 lb and the permissible bearing stress on the projected area of the threads is 10,000 psi.

17. The OD of the aluminium part in Fig. 5-8 is 1.00 in. The bolt is steel, $\frac{1}{2}$ in. in diameter. The nut is torqued to 600 in. lb by the approximate equation. Find the maximum load P that can be placed on the part without losing all the initial compression in the part. Threads stop immediately above the nut. Part and bolt are of equal length.

18. A square thread power screw has μ_1 equal to 0.14 and negligible collar friction. When holding a load, it is just at the point of overhauling. What is the efficiency for this screw when raising a load?

19. Work Problem 11, but with the efficiency at 75 per cent.

20. Work Problem 13, but with the horsepower of 4.5.

21. Work Problem 14, but with the input horsepower of 0.95.

22. A square thread screw is at the point of overhauling. Its efficiency when raising a load is 0.49. What is the coefficient of friction for the threads if the collar friction is negligible?

23. A square thread screw has an efficiency of 60 per cent when mounted with an antifriction collar. The helix angle is 12°. Find the torque required to exert a 2,000-lb force if the pitch diameter of the screw is 2 in.

24. The bolts in Fig. 5-19 are $\frac{3}{4}$ in.-16UNF made from cold-drawn 8740 steel. Machined threads; $K = 3.85$. The steel cylinder is 4-in. ID. Fluid pressure varies continuously from 0 to 2,000 psi. Assume the end plates are rigid. Find the value of initial force F_o if the bolts are to have an *FS* equal to 4.   *Ans.* $F_o = 8,290$ lb.

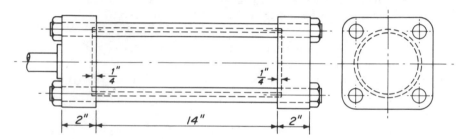

Figure 5-19 Problem 24.

6

Belts, Clutches, Brakes, and Chains

Belts, brakes, clutches, and chains are examples of machine elements that employ friction as a useful agent. A belt provides a convenient means for transferring power from one shaft to another. Belts are frequently necessary to reduce the high rotative speeds of electric motors to the lower values required by mechanical equipment. The function of the brake is to turn mechanical energy into heat. Clutches are required when shafts must be frequently connected and disconnected. The design of frictional devices is subject to uncertainties in the value of the coefficient of friction that must necessarily be used. Chains provide a convenient and effective means for transferring power between parallel shafts.

A, area
b, width of lining, inches
c, center distance, inches
CCW, counterclockwise
CW, clockwise
d, diameter, inches
$e = 2.718$ base of natural logarithms

fpm, feet per minute
fps, feet per second
F, force (pounds)
F_n, force normal to surface
$g = 32.2$ ft/sec^2 = 386 in./sec^2 acceleration due to gravity
hp, horsepower
l, length of belt, inches

237

M_f, moment due to friction force

M_n, moment due to normal force

n, revolutions per minute, rpm

N, number

p, pressure psi; pitch, inches

p_n, pressure normal to surface psi

P, force, pounds

r, radius, inches

R, reactive force, pounds

T, torque, inch-pounds

T_1, force in tight-side, pounds

T_2, force in slack side, pounds

V, velocity, feet per minute

α, (alpha) angle of contact

δ, (delta) wear, inches

δ_n, wear normal to surface

μ, (mu) coefficient of friction

ψ, (psi) half-angle of contact for belt

1. V-belts

The rayon and rubber V-belt is widely used for power transmission. Such belts are made in two series: the standard V-belt shown in Fig. 6-1 and the high-capacity V-belt of Fig. 6-2. The belts can be used with short

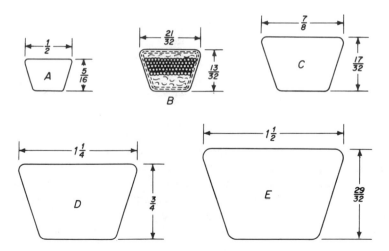

Figure 6-1 Cross sections of multiple V-belts.

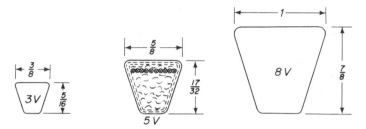

Figure 6-2 Cross sections of high-capacity V-belts (full size).

Table 6-1   PITCH LENGTHS OF MULTIPLE V-BELTS

Cross Section							
A	*B*	*C*	*A*	*B*	*C*	*D*	*E*
27.3			97.3		98.9		
32.3				98.8			
36.3	36.8		106.3	106.8	107.9		
39.3	39.8		113.3	113.8	114.9		
43.3	43.8		121.3	121.8	122.9	123.3	
47.3	47.8		129.3	129.8	130.9	131.3	
52.3	52.8	53.9		145.8	146.9	147.3	
56.3	56.8			159.8	160.9	161.3	
61.3	61.8	62.9		174.8	175.9	176.3	
69.3	69.8	70.9		181.8	182.9	183.3	184.5
76.3	76.8	77.9		196.8	197.9	198.3	199.5
81.3				211.8	212.9	213.3	214.5
	82.8	83.9		240.3	240.9	240.8	241.0
86.3	86.8	87.9		270.3	270.9	270.8	271.0
91.3	91.8	92.9		300.3	300.9	300.8	301.0

Longer belts are also available in *C*, *D*, and *E* Sections. Above lengths are subject to nominal manufacturing tolerances.

Table 6-2   OUTSIDE CIRCUMFERENCES OF HIGH-CAPACITY V-BELTS

3V	3V, 5V	3V, 5V, 8V	5V, 8V
25	50	100	150
$26\frac{1}{2}$	53	106	160
28	56	112	170
30	60	118	180
$31\frac{1}{2}$	63	125	190
$33\frac{1}{2}$	67	132	200
$35\frac{1}{2}$	71	140	212
$37\frac{1}{2}$	75		224
40	80		236
$42\frac{1}{2}$	85		250
45	90		265
$47\frac{1}{2}$	95		280

center distances and are made endless so that difficulty with splicing devices is avoided. The lengths are standardized and are given in Table 6-1 for multiple V-belts, and in Table 6-2 for high-capacity V-belts.

First cost is low, and power output may be increased by operating several belts side by side. All belts in the drive should stretch at the same rate in order to keep the load equally divided among them. When one of the belts breaks, the entire group must usually be replaced. The drive may be inclined at any angle with tight side either top or bottom. Since belts can operate on relatively small pulleys, large reductions of speed in a single drive are possible.

The included angle for the belt groove is usually from 34 to 38°. The wedging action of the belt in the groove gives a large increase in the tractive force developed by the belt.

Pulleys may be made of cast iron, sheet steel, or die cast metal. Sufficient clearance must be provided at the bottom of the groove to prevent the belt from bottoming as it becomes narrower from wear. Sometimes the larger pulley is not grooved when it is possible to develop the required tractive force by running on the inner surface of the belt. The cost of cutting the grooves is thereby eliminated. Pulleys are on the market that permit an adjustment in the width of the groove. The effective pitch diameter of the pulley is thus varied, and moderate changes in the speed ratio can be secured.

2. Design of V-Belt Drive

During a circuit around the pulleys, the force on the belt varies over a considerable range, as is illustrated by Fig. 6-3. Here T_c is the centrifugal force, and T_1 and T_2 are the tensions in the tight and slack sides, respectively. Forces T_{b1} and T_{b2} are additional forces arising from bending around the pulleys.

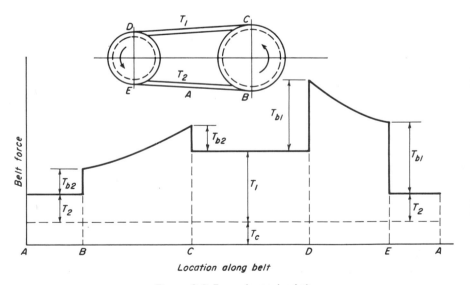

Figure 6-3 Forces in moving belt.

In general, failure of a V-belt occurs from fatigue in some portion of the structure as affected by the force peaks at C and D in Fig. 6-3. The peak or total force F in the belt is the sum of the forces T_1, T_b, and T_c.

$$F = T_1 + T_b + T_c \qquad (1)$$

Here T_1 is found from the horsepower equation

$$\text{hp} = \frac{(T_1 - T_2)V}{33,000} \qquad (2)$$

Velocity V of the belt is in units of feet per minite (fpm). Bending force T_b is given by

$$T_b = \frac{K_b}{d} \qquad (3)$$

where K_b is a constant from Table 6-3, and d is the pitch diameter of the pulley. The centrifugal force T_c is given by

Table 6-3 DESIGN CONSTANTS FOR V-BELTS

Section	K_b	K_c	Peak Force F		
			At 10^8	At 10^9	At 10^{10}
A	220	0.561	128	104	
B	576	0.965	221	179	
C	1,600	1.716	392	319	
D	5,680	3.498	801	651	
E	10,850	5.041	1,153	937	
3V	230	0.425	166	138	110
5V	1,098	1.217	383	319	255
8V	4,830	3.288	846	705	564

Data courtesy Gates Rubber Co., Denver, Colorado.

$$T_c = K_c \left(\frac{V}{1,000} \right)^2 \qquad (4)$$

where K_c is given by Table 6-3. Belt velocity V, fpm, can be found from

$$V = \frac{\pi d n}{12} \qquad (5)$$

where diameter d is in inches and n is the speed of the pulley, in revolutions per minute.

Permissible values for the peak force F for fatigue lives of 10^8, 10^9, and 10^{10} force peaks are given in Table 6-3.

The fatigue curve for F vs. the number of force peaks N to failure can be conveniently represented by the following equation.

$$N = \left(\frac{Q}{F}\right)^x \tag{6}$$

Values for constants Q and x are given in Table 6-4.

Table 6-4   DESIGN CONSTANTS FOR V-BELTS

Section	10^8–10^9 Force Peaks		10^9–10^{10} Force Peaks		Minimum Sheave Diameter
	Q	x	Q	x	
A	674	11.089			3.0
B	1,193	10.924			5.0
C	2,038	11.173			8.5
D	4,208	11.105			13.0
E	6,061	11.100			21.6
3V	728	12.464	1,062	10.153	2.65
5V	1,654	12.593	2,394	10.283	7.1
8V	3,638	12.629	5,253	10.319	12.5

A service factor must usually be applied to the nominal horsepower requirements to take care of fluctuations in the loading and the prime mover. A recommended set of such factors[1] is given in Table 6-5.

When the pulley diameters are equal, it is customary to assume that the tight-side tension T_1 is 5 times as great as the slack-side tension T_2.

Example 1. A C-section V-belt has a pitch length of 70.9 in. and carries a load horsepower of 9, but a service factor of 1.6 must be used. Pulley diameters are 10 in. and operate at a speed of 1,160 rpm. Find the expected life of the belt in hours.

Solution.   Design hp $= 9 \times 1.6 = 14.4$

By Eq. (5): $V = \dfrac{\pi d n}{12} = \dfrac{\pi 10 \times 1,160}{12} = 3,037$ fpm

By Eq. (2), since $T_2 = 0.2T_1$:

$$T_1 - T_2 = 0.8T_1 = \frac{33,000 \text{ hp}}{V} = \frac{33,000 \times 14.4}{3,037}$$

Then: $T_1 = 195.6$ lb

By Table 6-3 and Eq. (3): $T_b = \dfrac{K_b}{d} = \dfrac{1,600}{10} = 160.0$ lb

By Table 6-3 and Eq. (4):

$$T_c = K_c \left(\frac{V}{1,000}\right)^2 = 1.716 \times 3.037^2 = 15.8 \text{ lb}$$

Peak force: $F = 371.4$ lb

[1] See *Engineering Standard—Specifications for Drives Using Classical Multiple V-belts*, New York: Rubber Manufacturers Assn., 1968.

Table 6-5 SUGGESTED SERVICE FACTORS

	AC Motors: Normal Torque, Squirrel Cage, Synchronous, Split Phase DC Motors: Shunt Wound Engines: Multicylinder Internal Combustion			AC Motors: High-Torque, High-Slip, Repulsion-Induction, Single-phase, Series Wound, Slipring DC Motors: Series Wound, Compound Wound Engines: Single-cylinder Internal Combustion Lineshafts: Clutches		
Hours in daily service	3–5	8–10	16–24	3–5	8–10	16–24
Agitators for liquids Blowers and exhausters Centrifugal pumps and compressors Fans up to 10 hp Light-duty conveyors	1.0	1.1	1.2	1.1	1.2	1.3
Belt conveyors for sand, grain, etc. Dough mixers Fans over 10 hp Generators Lineshafts Laundry machinery Machine tools Punches, presses, shears Printing machinery Positive displacement rotary pumps Revolving and vibrating screens	1.1	1.2	1.3	1.2	1.3	1.4
Brick machinery Bucket elevators Exciters Piston compressors Conveyors (drag, pan, screw) Hammermills Papermill beaters Piston pumps Positive displacement blowers Pulverizers Sawmill and woodworking machinery Textile machinery	1.2	1.3	1.4	1.4	1.5	1.6
Crushers (gyratory, jaw, roll) Mills (ball, rod, tube) Hoists Rubber calenders, extruders, mills	1.3	1.4	1.5	1.5	1.6	1.8

By Eq. (6):
$$N = \left(\frac{Q}{F}\right)^x = \left(\frac{2,038}{371.4}\right)^x = 5.487^{11.173}$$

$$\log N = 11.173 \log 5.487 = 11.173 \times 0.73933 = 8.26053$$

$$N = 182{,}200{,}000 \text{ force peaks}$$

$$\frac{\text{belt passes}}{\text{min}} = \frac{12V}{l} = \frac{3,037 \times 12}{70.9} = 514$$

For equal pulleys there are two equal force peaks per belt pass.

$$\frac{\text{force peaks}}{\text{min}} = 514 \times 2 = 1{,}028$$

$$\text{expected life} = \frac{182{,}200{,}000}{1{,}028 \times 60} = 2{,}950 \text{ hr}$$

Example 2. A 5V high-capacity V-belt operates on pulleys 10.3 in. in diameter. Length is 125 in. and speed is 1,750 rpm. For an expected life of 20,000 hr, find the number of belts required to transmit 60 hp.

Solution. By Eq. (5): $\qquad V = \dfrac{\pi dn}{12} = \dfrac{\pi 10.3 \times 1{,}750}{12} = 4{,}720 \text{ fpm}$

$$\frac{\text{Force peaks}}{\text{min}} = \frac{12V}{l} \times 2 = \frac{4{,}720 \times 12}{125} \times 2 = 906$$

Number of force peaks: $\quad N = \left(\dfrac{\text{force peaks}}{\text{min}}\right) \times 60 \times 20{,}000$

$$= 1{,}087{,}600{,}000$$

Note that N lies between 10^9 and 10^{10}.

By Eq. (6): $\qquad N = \left(\dfrac{Q}{F}\right)^x \qquad \text{or} \qquad F = \dfrac{Q}{N^{1/x}}$

By Table 6-4: $\quad \log N^{1/x} = \dfrac{\log N}{x} = \dfrac{9.03645}{10.283} = 0.87878$

$$N^{1/x} = 7.5645$$

Peak force: $\qquad F = \dfrac{Q}{N^{1/x}} = \dfrac{2{,}394}{7.5645} = 316.5 \text{ lb}$

By Eq. (3): $\qquad T_b = \dfrac{K_b}{d} = \dfrac{1{,}098}{10.3} = 106.6 \text{ lb}$

By Eq. (4): $\qquad T_c = K_c\left(\dfrac{V}{1{,}000}\right)^2 = 1.217 \times 4.720^2 = \underline{27.1 \text{ lb}}$

$$T_b + T_c = 133.7 \text{ lb}$$

By Eq. (1): $\qquad T_1 = F - (T_b + T_c) = 316.5 - 133.7 = 182.8 \text{ lb}$

By Eq. (2): $\qquad \text{hp} = \dfrac{(T_1 - T_2)V}{33{,}000} = \dfrac{0.8(T_1)V}{33{,}000} = \dfrac{0.8 \times 182.8 \times 4{,}720}{33{,}000}$

$$= 20.92$$

Number of belts required:

$$= \frac{60}{20.92} = 2.87 \qquad \text{(Use 3 belts.)}$$

3. Center Distance for Pulleys of Unequal Diameters

When the pulley diameters and center distances are given, the length of belt can be easily calculated. For most applications, however, a belt of standard length will be used, and it becomes necessary to calculate the center

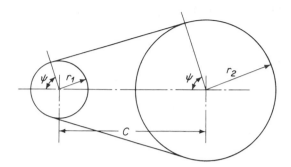

Figure 6-4 Layout for finding center distance.

distance for locating the pulleys. The situation is as shown in Fig. 6-4. Center distance can be found by the following equation.

$$c = \tfrac{1}{4}\left[H + \sqrt{H^2 - 8(r_2 - r_1)^2}\right] \qquad (7)$$

where $\qquad\qquad H = l - \pi(r_2 + r_1) \qquad\qquad\qquad (8)$

where l is the length of the belt.

Example 3. A belt has a pitch length of 70.9 in. and is to operate on pulleys of 10- and 14-in. diam. Calculate the center distance.

Solution. By Eq. (8): $H = l - \pi(r_2 + r_1) = 70.9 - \pi(7 + 5) = 33.20$ in.

By Eq. (7): $\qquad\qquad c = \tfrac{1}{4}\left[H + \sqrt{H^2 - 8(r_2 - r_1)^2}\right]$
$$= \tfrac{1}{4}\left[33.20 + \sqrt{33.20^2 - 8 \times 4}\right]$$
$$= \tfrac{1}{4}(33.20 + 32.72) = 16.48 \text{ in.}$$

4. Designing When Pulleys Are of Unequal Diameters

When the driving pulley is smaller than the driven pulley, the arc of contact between belt and pulley becomes less than a semicircle, and the T_1/T_2 ratio of 5 cannot be maintained. Adjustment for this condition can be made by use of the factors of Table 6-6.

The angle ψ, or the half-angle of contact in Fig. 6-4, can be found by the following equation.

$$\cos \psi = \frac{r_2 - r_1}{c} \qquad (9)$$

Let N_1 be the number of force peaks the belt could sustain with the smaller pulley. Each force peak then consumes $1/N_1$ of the belt life. Similarly, let N_2 be the number of force peaks the belt could sustain with the larger pulley. Each force peak will consume $1/N_2$ of the belt life. The proportion of belt life consumed in one passage of the belt on account of both pulleys is

Table 6-6   RATIO T_1/T_2 FOR V-BELTS FOR VARIOUS VALUES OF ANGLE OF CONTACT

Angle of Contact 2ψ	$\dfrac{T_1}{T_2}$	Angle of Contact 2ψ	$\dfrac{T_1}{T_2}$	Angle of Contact 2ψ	$\dfrac{T_1}{T_2}$	Angle of Contact 2ψ	$\dfrac{T_1}{T_2}$
180°	5.00	155°	4.00	130°	3.20	105°	2.56
175	4.78	150	3.82	125	3.06	100	2.44
170	4.57	145	3.66	120	2.92	95	2.34
165	4.37	140	3.50	115	2.80	90	2.24
160	4.18	135	3.34	110	2.67		

$$\frac{1}{N'} = \frac{1}{N_1} + \frac{1}{N_2} \tag{10}$$

where N' is the number of belt passes determined by the combined effects of both pulleys.[2]

Example 4. A C-section belt is 70.9 in. long and operates on pulleys 10 and 14 in. in diameter. Speed of the smaller pulley is 1,160 rpm. Load horsepower is 9, but a service factor of 1.6 must be used. Find the expected life in hours.

Solution. Note that data are the same as for Example 1 except for pulley size. For the small pulley:

By Example 1: $V = 3,037$ fpm

By Example 3: $c = 16.48$ in.

By Eq. (9): $\cos \psi = \dfrac{r_2 - r_1}{c} = \dfrac{7 - 5}{16.48} = 0.1214$

$$\psi = 83° \quad \text{and} \quad 2\psi = 166°$$

By Table 6-6: $\dfrac{T_1}{T_2} = 4.41 \quad \text{or} \quad T_2 = 0.227T_1$

By Eq. (2): $T_1 - T_2 = \dfrac{33,000 \text{ hp}}{V}$

$$T_1 - 0.227T_1 = 0.773T_1 = \frac{33,000 \times 9 \times 1.6}{3,037}$$

Then: $T_1 = 202.4$ lb

By Example 1: $T_{b1} = 160.5$ lb

By Example 1: $T_c = \underline{15.8}$ lb

Peak force: $F = 378.7$ lb

By Eq. (6) and Table 6-4:

$$N_1 = \left(\frac{Q}{F}\right)^x = \left(\frac{2,038}{378.7}\right)^x = 5.3816^{11.173}$$

$$\log N_1 = 11.173 \log 5.3887 = 11.173 \times 0.73149 = 8.17294$$

[2]This is an application of the so-called Miner equation.

$$N_1 = 148,900,000 \text{ force peaks (small pulley)}$$

For the large pulley: $T_1 = 202.4 \text{ lb}$

By Eq. (3): $T_{b2} = \dfrac{K_b}{d} = \dfrac{1,600}{14} = 114.3 \text{ lb}$

By Example 1: $T_c = 15.8 \text{ lb}$

Peak force: $F = 332.5 \text{ lb}$

By Eq. (6) and Table 6-4:

$$N_2 = \left(\frac{Q}{F}\right)^x = \left(\frac{2,038}{332.5}\right)^x = 6.1293^{11.173}$$

$$\log N_2 = 11.173 \log 6.1293 = 11.173 \times 0.78741 = 8.79773$$

$$N_2 = 627,700,000 \text{ force peaks (large pulley)}$$

By Eq. (10): $\dfrac{1}{N'} = \dfrac{1}{N_1} + \dfrac{1}{N_2}$

$$= \frac{1}{1,000,000}\left(\frac{1}{148.9} + \frac{1}{627.7}\right) = \frac{0.008309}{1,000,000}$$

$$N' = 120,400,000 \text{ belt passes}$$

By Example 1:

$$\frac{\text{belt passes}}{\text{min}} = 514$$

$$\text{Expected life} = \frac{120,400,000}{514 \times 60} = 3,900 \text{ hr}$$

This example shows that when a pulley is increased in diameter the fatigue effects are less and the life is lengthened accordingly. In tact when the diameter of the larger pulley is sufficiently large, the bending force becomes low and the peak force may be so small that the fatigue effects can be neglected entirely for that pulley.

5. Other Types of Belts

Other types of belts are available for power transmission purposes. Power bands, consisting of V-belts joined along their sides, as in Fig. 6-5, are of assistance in solving the length-matching problem when a number of V-belts must be run side by side.

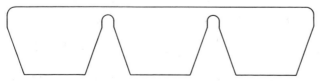

Figure 6-5 Joined V-belts.

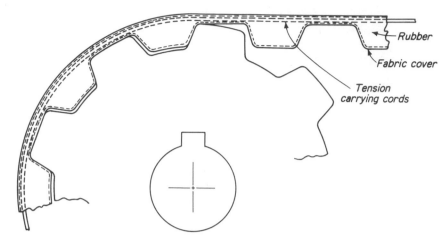

Figure 6-6 Timing belt.

The teeth of the so-called timing belt, Fig. 6-6, will keep the shafts completely synchronized.

As the pulley size becomes smaller, the belt tension increases for a given horsepower output. This increased force on the overhanging pulley at the motor causes increased loads in the motor bearings. The recommended

Table 6-7 RECOMMENDED MINIMUM SHEAVE DIAMETERS, INCHES, FOR GENERAL-PURPOSE ELECTRIC MOTORS

Motor hp	Motor rpm					
	575	690	870	1,160	1,750	3,450
$1/2$			2.2			
$3/4$			2.4	2.2		
1	3.0	2.5	2.4	2.4	2.2	
$1^1/2$	3.0	3.0	2.4	2.4	2.4	2.2
2	3.8	3.0	3.0	2.4	2.4	2.4
3	4.5	3.8	3.0	3.0	2.4	2.4
5	4.5	4.5	3.8	3.0	3.0	2.4
$7^1/2$	5.2	4.5	4.4	3.8	3.0	3.0
10	6.0	5.2	4.4	4.4	3.8	3.0
15	6.8	6.0	5.2	4.4	4.4	3.8
20	8.2	6.8	6.0	5.2	4.4	4.4
25	9	8.2	6.8	6.0	4.4	4.4
30	10	9	6.8	6.8	5.2	
40	10	10	8.2	6.8	6.0	
50	11	10	8.4	8.2	6.8	

See *National Electrical Manufacturers Association Standard MG-1-14.43a* and *MG-1-14.43*, Jan. 1968.

minimum sheave diameters for use with electric motors are shown in Table 6-7.

Allowance should be made in the design if the belt is to operate under adverse conditions such as may be caused by oil, high temperatures, or abrasive atmosphere.

Idler pulleys should be avoided because the additional flexing in passing around the pulley contributes to the fatigue effects. It is much better to maintain the tension as the belts stretch and wear by means of a take-up, which increases the center distance between the sheaves.

When designing a V-belt drive it is a good plan to calculate the cost of two or three different layouts of belts and pulleys to determine which has the smallest overall cost. The catalogs of the various manufacturers of V-belts contain much practical information.[3]

6. Disk Clutch

The plate clutch,[4] shown diagrammatically in Fig. 6-7, is used in both automotive and industrial service. Let it be assumed that the parts are sufficient-

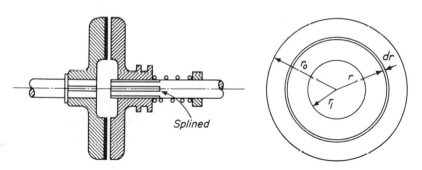

Figure 6-7 Plate clutch.

ly rigid to give uniform wear over the lining, and that the wear is proportional to the product of velocity and the pressure. Since the velocity is proportional to the radius r to the element, the following equation can be written:

$$\delta = Kpr \qquad\qquad (a)$$

[3]The following references should be consulted.

Worley, W. S., "Design of V-Belt Drives for Mass Produced Machines," *Prod. Eng.*, **24**, Sept. 1953, p. 154.

Marco, S. M., W. L. Starkey, and K. G. Hornung, "A Quantitative Investigation of the Factors Which Influence the Fatigue Life of a V-Belt," *Trans. ASME*, **82**, Feb. 1960, *J. Eng. Ind.*, p. 47

[4]See Jania, Z. J., "Friction-Clutch Transmissions," *Machine Design*, **30**, 1958, issues for Nov. 13, Nov. 27, Dec. 11, and Dec. 25.

where δ is the wear, and K is a constant. Since δ is constant for the entire face, the maximum pressure will occur at the smallest or inner radius r_i. Hence,

$$\delta = Kp_{max}r_i \tag{b}$$

Elimination of δ and K from the equations above gives

$$p = \frac{p_{max}r_i}{r} \tag{c}$$

The total normal force F_n, which must be exerted by the actuating spring, is found by multiplying the element of area $2\pi r\, dr$ by the pressure and integrating over the surface.

$$F_n = \int_{r_i}^{r_o} p\, dA = \int_{r_i}^{r_o} \frac{p_{max}r_i}{r} \times 2\pi r\, dr = 2\pi p_{max}r_i(r_o - r_i) \tag{11}$$

The torque is found by multiplying the force on the element by the coefficient of friction μ and the radius, and integrating over the area.

$$T = \int_{r_i}^{r_o} \mu p r\, dA = \int_{r_i}^{r_o} \mu \frac{p_{max}r_i}{r} \times 2\pi r^2\, dr$$

$$= \pi \mu p_{max}r_i(r_o^2 - r_i^2) = \tfrac{1}{2}\mu(r_o + r_i)F_n \tag{12}$$

Equations (11) and (12) are applicable when the assumption of uniform wear can be made.

Multiple-disk clutches have friction lining on both sides of alternate plates. Since Eq. (12) gives the torque for a single face, this quantity must be multiplied by the number of faces to find the torque for the entire clutch.

Equations (11) and (12) are applicable only when the parts of the clutch are very rigid so that the wear is uniform over the lining. Suppose an alternate type of construction is used in which the clutch lining is attached to flexible plates so that the pressure p is uniform over the entire surface. The total axial force F_n required for operation is

$$F_n = \pi p(r_o^2 - r_i^2) \tag{13}$$

The torque exerted by the clutch is

$$T = \mu p \int r\, dA = 2\pi \mu p \int_{r_i}^{r_o} r^2\, dr = 2\pi \mu p \left(\frac{r^3}{3}\right)_{r_i}^{r_o}$$

$$= \frac{2}{3}\pi \mu p(r_o^3 - r_i^3) = \frac{2\mu(r_o^3 - r_i^3)F_n}{3(r_o^2 - r_i^2)} \tag{14}$$

Equations (13) and (14) are applicable when the assumption of uniform pressure can be made.

Example 5. A plate clutch with a single friction surface is 10 in. OD and 4 in. ID. $\mu = 0.2$.

(a) If the uniform wear theory is valid, find the required axial force for $p_{max} = 100$ psi. Find the torque for the clutch.

(b) Do the same for a similar clutch where the uniform pressure theory is valid for $p = 100$ psi.

(c) If the uniform wear theory is valid, find the torque the clutch will carry for $F_n = 5,000$ lb, and the value of p_{max}.

(d) Do the same for a clutch where the uniform pressure theory is valid.

Solution. (a) By Eq. (11): $F_n = 2\pi 100 \times 2 \times 3 = 3,770$ lb
By Eq. (12): $T = \pi 0.2 \times 100 \times 2(25 - 4) = 2,639$ in. lb
   (b) By Eq. (13): $F_n = \pi 100(25 - 4) = 6,597$ lb
By Eq. (14): $T = \frac{2}{3}\pi 0.2 \times 100(125 - 8) = 4,900$ in. lb

   (c) By Eq. (11): $p_{max} = \dfrac{F_n}{2\pi r_i(r_o - r_i)} = \dfrac{5,000}{2\pi 2 \times 3} = 132.6$ psi

By Eq. (12): $T = \dfrac{0.2 \times 7 \times 5,000}{2} = 3,500$ in. lb

   (d) By Eq. (13): $p = \dfrac{F_n}{\pi(r_o^2 - r_i^2)} = \dfrac{5,000}{\pi(25 - 4)} = 75.8$ psi

By Eq. (14): $T = \dfrac{2 \times 0.2(125 - 8) \times 5,000}{3(25 - 4)} = 3,714$ in. lb

The equations for the disk clutch apply equally well to a disk brake. In general, instead of an annular area of friction material, the lining may consist of a small area on either side of the disk. Sometimes the area is so shaped that it can be approximated as a circular sector, and the equations applied over the angular extent of the lining. Servo action in general is not obtainable in disk brakes.

7. Cone Clutch

The cone clutch utilizes the wedging action of the parts for increasing the normal force on the lining; thus an increase in the tangential friction force and the torque results. The assumption is again made that the normal wear is proportional to the product of the normal pressure and the radius.

$$\delta_n = Kp_n r \qquad\qquad (a)$$

As shown in Fig. 6-8(b), the normal wear is uniform for all points on the cone. The maximum pressure thus occurs for the smallest value of r. Hence,

$$\delta_n = Kp_{max} r_i \qquad\qquad (b)$$

Elimination of δ_n and K in the foregoing equations gives

$$p_n = \frac{p_{max} r_i}{r} \qquad\qquad (c)$$

Let the radius r in Fig. 6-8(b) locate the elemental strip running around the cone. The differential area is then equal to $2\pi r\, dr/\sin \alpha$. The total normal force F_n is equal to the product of the normal pressure and the element of area integrated over the conical surface. Hence,

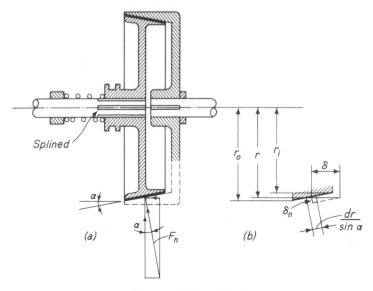

Figure 6-8 Cone clutch.

$$F_n = \int_{r_i}^{r_o} p_n \, dA = \int \frac{p_{max} r_i}{r} \cdot \frac{2\pi r \, dr}{\sin \alpha} = \frac{2\pi p_{max} r_i(r_o - r_i)}{\sin \alpha} \qquad (15)$$

As shown by Fig. 6-8(a), the reaction R, which the actuating spring must be capable of exerting, is equal to

$$R = F_n \sin \alpha = 2\pi p_{max} r_i(r_o - r_i) \qquad (16)$$

This value of R is required for steady operation after the load has been put in motion. For engaging the clutch while picking up the load, a greater spring force than indicated by Eq. (16) is required.

The torque is given by

$$T = \int_{r_i}^{r_o} \mu p_n r \, dA = \int \frac{\mu p_{max} r_i}{r} \cdot \frac{2\pi r^2 \, dr}{\sin \alpha}$$

$$= \frac{\pi \mu p_{max} r_i(r_o^2 - r_i^2)}{\sin \alpha} = \frac{\mu(r_o + r_i)}{2 \sin \alpha} R \qquad (17)$$

Values of the angle α range from a minimum of about 8° to larger values. Normal pressure on the lining may reach a value of 100 psi or more.

8. Band Brake

A simple band brake is shown schematically in Fig. 6-9. Sketch (b) shows, on an enlarged scale, the forces acting on an element of the lining lying between two axial planes a small angle $d\varphi$ apart. The tension on the tight side is equal to tension T of the slack side plus an increment dT. The

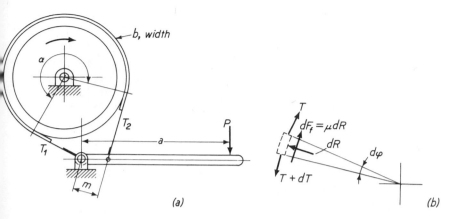

Figure 6-9 Simple band brake.

drum is pressing outwardly on the element with a reactive force dR. The friction force μdR is equal to dT, the difference between the tangential forces in the band. Thus

$$dT = \mu dR$$

Since angle $d\varphi$ is very small, the inward component from the tangential forces is $Td\varphi$, which is equal to the reaction dR. Then

$$dT = \mu T d\varphi$$

or

$$\frac{dT}{T} = \mu d\varphi$$

Band force T varies from T_2 where φ is zero, to the value T_1 where φ has the value α. The above equation should be integrated as follows.

$$\int_{T_2}^{T_1} \frac{dT}{T} - \int_0^{\alpha} \mu \, d\varphi$$

Then

$$\ln T \,|_{T_2}^{T_1} = \mu \varphi \,|_0^{\alpha}$$

$$\ln T_1 - \ln T_2 = \ln \frac{T_1}{T_2} = \mu \alpha$$

$$\frac{T_1}{T_2} = e^{\mu \alpha} \tag{18}$$

Values for the exponential function $e^{\mu \alpha}$ are given in Table 6-8.

In Fig. 6-9, the brake is actuated by the application of force P at the end of the lever. A smaller force P is needed for operation when the tight side of the band is attached to the stationary support and the slack side is attached to the lever.

In the differential band brake, the friction force assists in applying the band. For the brake shown in Fig. 6-10 the following equation can be written:

$$Pa = T_2 m_2 - T_1 m_1 = T_2(m_2 - m_1 e^{\mu \alpha}) \tag{19}$$

Table 6-8   VALUES OF EXPONENTIAL $e^{\mu\alpha}$

$\mu\alpha$	$e^{\mu\alpha}$	$\mu\alpha$	$e^{\mu\alpha}$	$\mu\alpha$	$e^{\mu\alpha}$	$\mu\alpha$	$e^{\mu\alpha}$	$\mu\alpha$	$e^{\mu\alpha}$	$\mu\alpha$	$e^{\mu\alpha}$	$\mu\alpha$	$e^{\mu\alpha}$
0.30	1.350	0.45	1.568	0.60	1.822	0.75	2.117	0.90	2.460	1.05	2.858	1.20	3.320
0.31	1.363	0.46	1.584	0.61	1.840	0.76	2.138	0.91	2.484	1.06	2.886	1.21	3.354
0.32	1.377	0.47	1.600	0.62	1.859	0.77	2.160	0.92	2.509	1.07	2.915	1.22	3.387
0.33	1.391	0.48	1.616	0.63	1.878	0.78	2.181	0.93	2.535	1.08	2.945	1.23	3.421
0.34	1.405	0.49	1.632	0.64	1.896	0.79	2.203	0.94	2.560	1.09	2.974	1.24	3.456
0.35	1.419	0.50	1.649	0.65	1.916	0.80	2.226	0.95	2.586	1.10	3.004	1.25	3.490
0.36	1.433	0.51	1.665	0.66	1.935	0.81	2.248	0.96	2.612	1.11	3.034	1.26	3.525
0.37	1.448	0.52	1.682	0.67	1.954	0.82	2.270	0.97	2.638	1.12	3.065	1.27	3.561
0.38	1.462	0.53	1.699	0.68	1.974	0.83	2.293	0.98	2.664	1.13	3.096	1.28	3.597
0.39	1.477	0.54	1.716	0.69	1.994	0.84	2.316	0.99	2.691	1.14	3.127	1.29	3.633
0.40	1.492	0.55	1.733	0.70	2.014	0.85	2.340	1.00	2.718	1.15	3.158	1.30	3.669
0.41	1.507	0.56	1.751	0.71	2.034	0.86	2.363	1.01	2.746	1.16	3.190	1.31	3.706
0.42	1.522	0.57	1.768	0.72	2.054	0.87	2.387	1.02	2.773	1.17	3.222	1.32	3.743
0.43	1.537	0.58	1.786	0.73	2.075	0.88	2.411	1.03	2.801	1.18	3.254	1.33	3.781
0.44	1.553	0.59	1.804	0.74	2.096	0.89	2.435	1.04	2.829	1.19	3.287	1.34	3.819

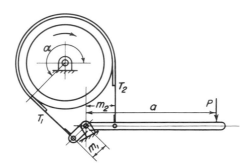

Figure 6-10 Differential band brake.

Should the product $m_1 e^{\mu\alpha}$ be greater than m_2, the brake will grab or be self-locking. Sometimes this feature is desirable, and is utilized in back stop devices for preventing rotation in the reverse direction. For counterclockwise rotation, shown in Fig. 6-10, the friction tends to loosen the band, and the drum revolves freely. Should the rotation reverse and become clockwise, the friction will apply the brake, and if it is self-locking, the rotation will be stopped.

Let the analysis of Fig. 6-9(b) be applied to the band at the point of tangency for T_1. Reaction R is equal to $brp_{max}\,d\psi$, where p_{max} is the pressure between drum and lining. The inward components of the band forces are equal to $T_1\,d\psi$. These two forces are equal to each other, and give the following useful equation.

$$T_1 = brp_{max} \qquad (20)$$

A similar equation can be written for the slack side.

Example 6. A band brake similar to Fig. 6-9 has a 16-in.-diam drum and a width of lining equal to 3 in. Speed is 200 rpm, $a = 10$ in., $m = 3$ in., $\alpha = 270°$, and $\mu = 0.2$. Find the torque and horsepower if the maximum lining pressure is 70 psi.

Solution. By Eq. (20): $T_1 = brp_{max} = 3 \times 8 \times 70 = 1{,}680$ lb

$$\mu\alpha = 0.2 \times 1.5\pi = 0.9425$$

By Table 6-8: $e^{\mu\alpha} = 2.566$

By Eq. (18): $T_2 = \dfrac{T_1}{e^{\mu\alpha}} = \dfrac{1{,}680}{2.566} = 655$ lb

Force: $P = \dfrac{T_2 m}{a} = \dfrac{655 \times 3}{10} = 196$ lb

Torque: $T = r(T_1 - T_2) = 8(1{,}680 - 655) = 8{,}200$ in. lb

Horsepower: $hp = \dfrac{Tn}{63{,}025} = \dfrac{8{,}200 \times 200}{63{,}025} = 26.0$

Band brakes are capable of exerting large torques. Careful attention must be paid to details so that the band seats properly on the drum when absorbing energy, and remains free and clear when not in use.

9. Block Brake with Short Shoe

The brake shown in Fig. 6-11 consists of a short block, which is pressed against the revolving drum by means of a lever. When the block is relatively short, it may be assumed that the normal force F_n between drum and shoe is concentrated at point B. If the lever is taken as the free body, a moment equation can be written about A as the center, and forces F_n and μF_n can be determined. It should be noted that if the direction of rotation of the drum is reversed, the direction of friction force μF_n is reversed also.

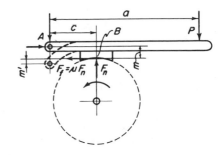

When the rotation is as shown in Fig. 6-11, the moment of the friction force aids in applying the shoe to the drum. Care must be taken that the friction moment is not so large that the brake is applied without the assistance of force P; the brake will then seize or grab, and unsatisfactory or dangerous operation results. When the pivot is located on the

Figure 6-11 Block brake with short shoe.

other side of the line of action of μF_n, as shown by the dotted outline in Fig. 6-11, the friction force tends to unseat the shoe. When the shoe is long, equations that take account of the variation in the pressure along the shoe must be used.

10. Pivoted Block Brake with Long Shoe

When the shoe is relatively long,[5] as shown in Fig. 6-12(a), the normal pressure p_n varies with different values of the angle φ. Frictional wear in engineering equipment is usually assumed to be proportional to the product of the velocity and the pressure. Since the velocity is the same for all points of the shoe, the wear of the lining is proportional to the pressure. Hence,

$$\delta_n = K p_n \tag{a}$$

where δ_n is the wear in the direction perpendicular to the lining, and K is a constant.

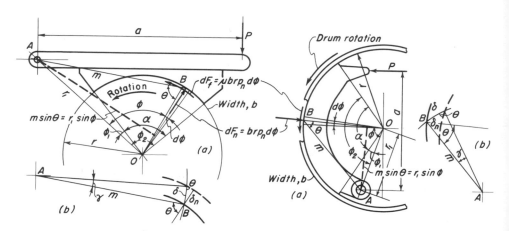

Figure 6-12 Block brake with long shoe.   Figure 6-13 Automotive type brake.

Figure 6-12(b) shows the motion δ of a point on the shoe as the lining wears and as the shoe rotates about point A through angle γ. Hence,

$$\delta = \gamma m \tag{b}$$

The wear δ_n in the direction normal to the drum is then equal to

$$\delta_n = \delta \sin \theta = \gamma m \sin \theta \tag{c}$$

The perpendicular dropped from A to radius OB has a length equal to $r_1 \sin \varphi$ or $m \sin \theta$. Substitution in Eq. (c) gives

$$\delta_n = \gamma r_1 \sin \varphi \tag{d}$$

Substitution into Eq. (a) gives

$$K p_n = \gamma r_1 \sin \varphi \tag{e}$$

[5] See Fazekas, G. A. G., "Graphical Shoe Brake Analysis," *Trans. ASME*, **79**, 1957, p. 1322. See also *J. Appl. Mech.*, **80**, 1958, p. 7.

One of the limiting factors in the design of a brake is the maximum pressure p_{max} between lining and drum. This occurs at the location having the maximum value of $\sin \varphi$ in Eq. (e). Hence,

$$Kp_{max} = \gamma r_1 (\sin \varphi)_{max} \tag{f}$$

Elimination of $\gamma r_1 / K$ between Eqs. (e) and (f) gives

$$p_n = \frac{p_{max}}{(\sin \varphi)_{max}} \sin \varphi \tag{g}$$

Should φ_2 in Fig. 6-12 be greater than 90°, then $(\sin \varphi)_{max}$ is equal to unity. For smaller values of φ_2 use the value of $\sin \varphi_2$ for the denominator.

If b is the width of the lining, the area of a small element, cut by two radii an angle $d\varphi$ apart, is equal to $br \, d\varphi$. When this area is multiplied by the normal force p_n and the arm $r_1 \sin \varphi$, and integrated over the entire shoe, the moment M_n of the normal forces about point A results.

$$M_n = \int_{\varphi_1}^{\varphi_2} brr_1 p_n \sin \varphi \, d\varphi = \frac{brr_1 p_{max}}{(\sin \varphi)_{mnx}} \int_{\varphi_1}^{\varphi_2} \sin^2 \varphi \, d\varphi$$

$$= \frac{brr_1 p_{max}}{4(\sin \varphi)_{max}} (2\alpha - \sin 2\varphi_2 + \sin 2\varphi_1) \tag{21}$$

Normal force dF_n on the element of Fig. 6-12(a) causes a counterclockwise moment about point A. If numerical substitution into Eq. (21) for the total moment gives a positive result, moment M_n is therefore counterclockwise. A negative result means that M_n is clockwise about A.

The moment of the friction forces M_f about A is equal to

$$M_f = \int_{\varphi_1}^{\varphi_2} \mu p_n (r - r_1 \cos \varphi) br \, d\varphi$$

$$= \frac{\mu brp_{max}}{(\sin \varphi)_{max}} \int \left[r \sin \varphi - \frac{1}{2} r_1 \sin 2\varphi \right] d\varphi$$

$$= \frac{\mu brp_{max}}{4(\sin \varphi)_{max}} [r_1 (\cos 2\varphi_2 - \cos 2\varphi_1) - 4r(\cos \varphi_2 - \cos \varphi_1)] \tag{22}$$

A positive value for M_f in Eq. (22) indicates a counterclockwise moment about A for the friction forces of Fig. 6-12, and a negative result indicates a clockwise moment. If the direction of rotation is reversed, these statements have the converse meanings for M_f.

In Fig. 6-12(a) the moment of the external force is equal to Pa. The shoe and lever are in equilibrium under the action of moments Pa, M_n, and M_f. A moment equation should be written for these terms with the proper signs. These signs depend on the location of the pivot with respect to the shoe and on the direction of rotation of the drum. When the friction moment assists in applying the shoe, the brake will be self-locking if M_f exceeds M_n. In practice, the ratio M_f / M_n should not be greater than about 0.7.

The torque exerted by the brake is found by taking the moment of the friction forces about the center of the drum O. Hence,

$$
\begin{aligned}
T &= \int_{\varphi_1}^{\varphi_2} \mu p_n b r^2 \, d\varphi \\
&= \frac{\mu b r^2 p_{max}}{(\sin \varphi)_{max}} \int \sin \varphi \, d\varphi \\
&= \frac{\mu b r^2 p_{max}}{(\sin \varphi)_{max}} (\cos \varphi_1 - \cos \varphi_2)
\end{aligned}
\qquad (23)
$$

Example 7. (a) Calculate the value of p_{max}, the torque, and the horsepower for the brake of Fig. 6-14. The coefficient of friction is equal to 0.2.

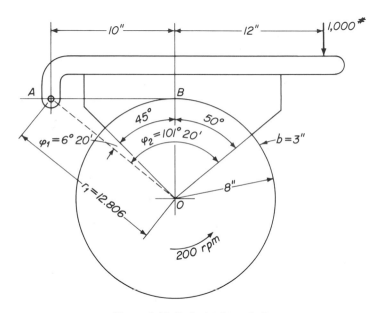

Figure 6-14 Brake for Example 7.

(b) Make the same calculations on the assumption that the total normal and friction forces are concentrated at point B and compare results with the more exact results of part (a).

Solution. (a) tan angle $AOB = \frac{10}{8} = 1.25$

angle $AOB = 51°20'$

Hence:

$$\varphi_1 = 51°20' - 45°0' = 6°20', \quad 2\varphi_1 = 12°40'$$

$$\varphi_2 = 6°20' + 95°0' = 101°20', \quad 2\varphi_2 = 202°40'$$

$$\alpha = 95° = 1.65806 \text{ rad}$$

$$r_1 = \sqrt{8^2 + 10^2} = 12.806 \text{ in.}$$

$$\cos \varphi_1 = 0.99390 \qquad\qquad \sin 2\varphi_1 = 0.21928$$
$$\cos \varphi_2 = -0.19652 \qquad\qquad \cos 2\varphi_1 = 0.97566$$
$$\sin 2\varphi_2 = -0.38537$$
$$\cos 2\varphi_2 = -0.92276$$

Since φ_2 is greater than $90°$: $(\sin \varphi)_{max} = 1$

By Eq. (21): $M_n = \frac{1}{4} \times 3 \times 8 \times 12.806 p_{max}(2 \times 1.65808 + 0.38357$
$$+ \, 0.21928) = 301.26 p_{max}$$

This moment is counterclockwise about point A.

By Eq. (22): $M_f = \frac{1}{4} \times 0.2 \times 3 \times 8 p_{max}[12.806(-0.92276 - 0.97566)$
$$- \, 4 \times 8(-0.19652 - 0.99390)] = 16.54 p_{max}$$

This moment is counterclockwise about point A.
Externally applied moment:

$$Pa = 1,000 \times 22 = 22,000 \text{ in. lb}, \qquad \text{clockwise}$$

Equilibrium of moments on shoe: $(301.26 + 16.54)p_{max} = 22,000$

$$p_{max} = 69.2 \text{ psi}$$

By Eq. (23): $\qquad\qquad T = 0.2 \times 3 \times 8^2 \times 69.2(0.99390 + 0.19652)$
$$= 3,164 \text{ in. lb}$$

$$\text{hp} = \frac{Tn}{63,000} = \frac{3,164 \times 200}{63,000} = 10.04$$

(b) Moments about A: $F_n = \dfrac{1,000 \times 22}{10} = 2,200 \text{ lb}$

Projected area of shoe: $A = 3(8 \sin 45° + 8 \sin 50°) = 35.36 \text{ in.}^2$

Average pressure: $p_{av} = \dfrac{2,200}{35.36} = 62.2 \text{ psi}$

Torque: $T = \mu F_n r = 0.2 \times 2,200 \times 8 = 3,520 \text{ in. lb}$

$$\text{hp} = \frac{Tn}{63,000} = \frac{3,520 \times 200}{63,000} = 11.17$$

As indicated by this example, the approximate equations frequently give a smaller pressure than actually occurs on the lining as well as a larger horsepower than can actually be developed.

Figure 6-13 shows an internal brake of a type widely used in automotive service. The equation for the moment M_n of the normal forces on the shoe is the same as Eq. (21), but a positive result indicates a clockwise moment about A. The equations for friction moment M_f and torque T are the same as for the brake of Fig. 6-12. A positive or negative result for M_f should be interpreted in the same way as for a brake with an external shoe.

11. Brake with Pivoted Symmetrical Shoe

In Fig. 6-15 the brake shoe is supported by a symmetrically located pin at A. Let it be assumed that distance r_1 is of such magnitude that the

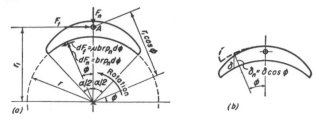

Figure 6-15 Block brake with pivoted symmetrical shoe.

friction forces dF_f have no resultant moment about A. The normal forces must then be symmetrical with respect to the vertical center line, and also have zero moment about A. Thus, no rotation of the shoe on the pin can occur, and the vertical component of the wear is uniform for the entire shoe. Assume as before that normal wear is proportional to normal pressure.

$$\delta_n = Kp_n \tag{a}$$

As shown in Fig. 6-15(b) the normal wear is equal to $\delta \cos \varphi$. Hence,

$$\delta \cos \varphi = Kp_n \tag{b}$$

The maximum normal pressure occurs at the location for the maximum value of $\cos \varphi$, that is, for $\varphi = 0$, giving

$$\delta = Kp_{max}$$

Substitution into Eq. (b) gives

$$p_n = p_{max} \cos \varphi \tag{c}$$

The equation for the moment of the friction forces about A is found by multiplying the tangential force on an element of area by the arm $(r_1 \cos \varphi - r)$ and integrating over the entire shoe. By the foregoing assumption the result is equal to zero.

$$M_f = 2 \int_0^{\alpha/2} \mu brp_n(r_1 \cos \varphi - r) \, d\varphi = 0$$

Substitution of the value for p_n from Eq. (c) gives

$$2\mu brp_{max} \int_0^{\alpha/2} (r_1 \cos^2 \varphi - r \cos \varphi) \, d\varphi = 0$$

Integrating and solving for r_1,

$$[r_1(\tfrac{1}{2}\varphi + \tfrac{1}{4} \sin 2\varphi) - r \sin \varphi]_0^{\alpha/2} = 0$$

$$r_1 = \frac{4r \sin (\alpha/2)}{\alpha + \sin \alpha} \tag{24}$$

The torque of the friction forces about O as a center is given by the following equation:

$$T = 2 \int_0^{\alpha/2} \mu br^2 p_n \, d\varphi$$

$$= 2\mu b r^2 p_{max} \int_0^{\alpha/2} \cos \varphi \, d\varphi$$

$$= 2\mu b r^2 p_{max} \sin \frac{\alpha}{2} \tag{25}$$

The pin reaction F_n can be found as the summation of the vertical components of the normal pressures.

$$F_n = 2 \int_0^{\alpha/2} b r p_n \cos \varphi \, d\varphi$$

$$= 2 b r p_{max} \int_0^{\alpha/2} \cos^2 \varphi \, d\varphi$$

$$= \tfrac{1}{2} b r p_{max}(\alpha + \sin \alpha) \tag{26}$$

The pin reaction F_t is found by summing the horizontal components of the frictional forces.

$$F_t = 2 \int_0^{\alpha/2} \mu b r p_n \cos \varphi \, d\varphi = \mu F_n \tag{27}$$

12. Lining Pressures

Wooden blocks are sometimes used as the friction elements for industrial brakes. Cast-iron shoes bearing on cast-iron or steel wheels are in wide use for brakes in railroad service. However, most brake linings depend upon asbestos as the basic friction material because of its ability to resist the effects of heat. Asbestos may be spun into yarn, woven and infused with a binder, and then consolidated by heat and pressure. Molded blocks and linings are also made directly from asbestos and binder without weaving.

Data on friction materials are shown in Table 6-9. The correct value to be used depends on the conditions of service for the particular application.

Table 6-9 TYPICAL FRICTION MATERIALS FOR BRAKE AND CLUTCH SERVICE

	Asbestos			Sintered Metal (Copper or Iron Base)
	Brake Blocks	Molded Linings	Woven Linings	
Coefficient of running friction	0.40–0.60	0.40–0.45	0.20–0.45	0.10–0.50
Max rubbing speed, fpm	7,500	5,000	3,000–7,500	
Max drum temperature, °F	750	500	500	Up to 1,000
Max working pressure, psi	150	100	100	Up to 500
Rate of wear	E, G	G	G, M	E

E, excellent; G, good; M, moderate.

Properties can vary over a considerable range depending on composition and method of fabrication.

The proper value should be obtained from the manufacturer's catalog for the particular type of lining considered. The coefficient of friction for brake materials may not remain constant in service but will vary because of conditions of temperature, or the presence of moisture and grease.[6]

13. Heating of Brakes

The capacity of a brake or clutch to absorb power is largely determined by its ability to dispose of the frictional heat. The heat-dissipating quality is determined by factors such as the size, shape, and condition of the surface of the various parts. If the clutch or brake is not enclosed in a housing, or if the surrounding air is in motion, the brake can be more readily·cooled. For service in excavating machinery, the following values for power absorption have been found to be satisfactory:

For open, exposed band, and cone clutches:

power absorption = 0.25–0.40 hp/in.² of contact surface

For open, exposed band brakes:

power absorption = 0.20–0.30 hp/in.² of contact surface

Although clutches and brakes in this type of service operate intermittently, they generally do not have sufficient time for cooling between applications, and the operating temperature may therefore be rather high. Excessive temperatures may be damaging to the lining.[7] The coefficient of friction will usually decrease with increase in temperature. However, certain compositions of lining show an increase in the coefficient for moderate rises of temperature. Wear of the lining is an important factor in the maintenance cost of a brake. If the drum is ground to a high degree of smoothness, the wear will be less than if it is merely turned.

Experience has shown that the product of the pressure p in pounds per square inch and the velocity V in feet per minute must be kept within certain limits. On this basis, the following values can be considered as an additional criterion for the design of brakes.

$pV \leqq 55,000$ for intermittent applications of the load, comparatively long periods of rest, and poor dissipation of heat.

$pV \leq 28,000$ for continuous application of the load, as in lowering operations, and poor dissipation of heat.

$pV \leqq 83,000$ for continuous application of the load and good dissipation of heat, into an oil bath, for instance.

[6]See Parker, R. C., "The Frictional Behavior of Engineering Materials," *Engineering*, 1949, pp. 193 and 217. See also Courtel, R., and L. M. Tichvinsky, "A Brief History of Friction," *Mech. Eng.*, **85**, Sept., 1963, p. 55; Oct., p. 33.

[7]See Rasmussen, A. C., "Heat-Radiating Capacity of Clutches and Brakes," *Prod. Eng.*, **2**, 1931, p. 529, and **3**, 1932, p. 282.

The tangential force F on an area A sq in. and a lining pressure of p psi is

$$F = \mu A p$$

Substitution in the horsepower equation gives

$$\text{hp} = \frac{FV}{33,000} = \frac{\mu A p V}{33,000} \tag{28}$$

Example 8. Find the horsepower per square inch absorbed by a brake if $\mu = 0.2$ and $pV = 55,000$.

Solution. In Eq. (28). $\dfrac{\text{hp}}{A} = \dfrac{\mu p V}{33,000} = \dfrac{0.2 \times 55,000}{33,000} = 0.33 \,\dfrac{\text{hp}}{\text{in.}^2}$

14. Roller Chains

A roller chain provides a readily available and efficient method for transmitting power between parallel shafts. Details of construction are shown in Fig. 6-16. The rollers turn on bushings that are press-fitted to the

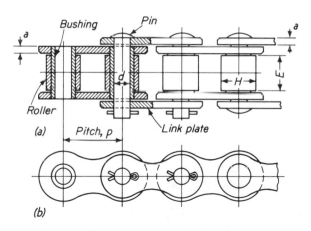

Figure 6-16 Construction of typical roller chain.

inner link plates. The pins are prevented from rotating in the outer link plates by the press-fit assembly. Dimensions for standard sizes are given in Table 6-10. The range in sizes is large so that chains can be used for both large and small amounts of power. Chains can be used for shaft spacings too great for gears. A large reduction in speed can be obtained if desired. The tolerances for a chain drive are greater than for gears, and the installation is relatively easy.

Chains present no fire hazard and are unaffected by relatively high temperatures and the presence of oil or grease. Chains are, however, more noisy

Table 6-10 DIMENSIONS OF STANDARD ROLLER TRANSMISSION CHAINS

Chain No.	Pitch p, in.	Roller		Pin Diameter d, in.	Link Plate Thickness a	Average Ultimate Strength, lb	Weight per foot, lb
		Diameter H, in.	Width E, in.				
25*	1/4	0.130*	1/8	0.0905	0.030	875	0.084
35*	3/8	0.200*	3/16	0.141	0.050	2,100	0.21
41†	1/2	0.306	1/4	0.141	0.050	2,000	0.28
40	1/2	5/16	5/16	0.156	0.060	3,700	0.41
50	5/8	0.400	3/8	0.200	0.080	6,100	0.68
60	3/4	15/32	1/2	0.234	0.094	8,500	1.00
80	1	5/8	5/8	0.312	0.125	14,500	1.69
100	1-1/4	3/4	3/4	0.375	0.156	24,000	2.49
120	1-1/2	7/8	1	0.437	0.187	34,000	3.67
140	1-3/4	1	1	0.500	0.219	46,000	4.93
160	2	1-1/8	1-1/4	0.562	0.250	58,000	6.43
180	2-1/4	1-13/32	1-13/32	0.687	0.281	76,000	8.70
200	2-1/2	1-9/16	1-1/2	0.781	0.312	95.000	10.51
240	3	1-7/8	1-7/8	0.937	0.375	130,000	16.90

*Without rollers.
†Lightweight chain.

than belts. A small adjustable idler sprocket should be provided to remove excessive slack from the chain as it wears. It should be located on the outside of the chain on the slack side and near the smaller sprocket.

15. Horsepower Capacity of Roller Chains

At lower speeds, the horsepower capacity is determined by the fatigue life of the link plates. The following equation applies.[8]

$$\text{hp} = 0.004 N_1^{1.08} n_1^{0.9} p^{3.0-0.07p} \tag{29}$$

where

N_1 = number of teeth in the smaller sprocket
n_1 = speed, rpm, of the smaller sprocket
p = chain pitch, inches

Constant 0.004 becomes 0.0022 for the No. 41 chain.

At higher speeds the horsepower is determined by the roller bushing fatigue life. The equation is

[8]See *Design Manual, Roller and Silent Chain Drives*, Park Ridge, Illinois: American Sprocket Chain Manufacturers Assn., 1968, p. 46.

$$\text{hp} = \frac{1{,}000 K N_1^{1.5} p^{0.8}}{n_1^{1.5}} \tag{30}$$

where

$$K = 29 \text{ for chains Nos. 25 and 35}$$
$$= 3.4 \text{ for chain No. 41}$$
$$= 17 \text{ for chains Nos. 40 to 240}$$

The data given in Tables 6-11 and 6-12 will be helpful in making numer-

Table 6-11   CONSTANTS USED IN ROLLER CHAIN DESIGN

N_1	$N_1^{1.08}$	$N_1^{1.5}$	N_1	$N_1^{1.08}$	$N_1^{1.5}$	p	$p^{0.8}$	$p^{3.0-0.07p}$
11	13.33	36.48	23	29.55	110.3	1/4	0.3299	0.01601
12	14.64	41.57	24	30.95	117.6	3/8	0.4563	0.05411
13	15.96	46.87	25	32.34	125.0	1/2	0.5744	0.1281
14	17.29	52.38	28	36.55	148.2	5/8	0.6866	0.2492
15	18.63	58.10	30	39.38	164.3	3/4	0.7944	0.4283
16	19.97	64.00	32	42.22	181.0	1	1	1
17	21.32	70.09	35	46.52	207.1	1-1/4	1.195	1.915
18	22.68	76.37	40	53.73	253.0	1-1/2	1.383	3.234
19	24.05	82.82	45	61.02	301.9	1-3/4	1.565	5.004
20	25.42	89.44	50	68.37	353.6	2	1.741	7.260
21	26.79	96.23	55	75.78	407.9	2-1/4	1.913	10.025
22	28.17	103.2	60	83.25	464.8	2-1/2	2.081	13.310
						3	2.408	21.437

Table 6-12   SPEED n_1, RPM, OF SMALL SPROCKET RAISED TO POWERS

n_1	$n_1^{0.9}$	$n_1^{1.5}$	n_1	$n_1^{0.9}$	$n_1^{1.5}$	n_1	$n_1^{0.9}$	$n_1^{1.5}$
5	4.257	11.18	175	104.4	2,315	1,200	590.6	41,570
10	7.943	31.62	200	117.7	2,828	1,400	678.4	52,380
15	11.44	58.10	250	143.9	3,953	1,600	765.1	64,000
20	14.82	89.44	300	169.6	5,196	1,800	850.6	76,370
25	18.12	125.0	350	194.8	6,548	2,000	935.3	89,440
30	21.35	164.3	400	219.7	8,000	2,400	1,102	117,600
40	27.66	253.0	450	244.3	9,546	2,800	1,266	148,200
50	33.81	353.6	500	268.6	11,180	3,200	1,428	181,000
60	39.84	464.8	600	316.5	14,700	3,600	1,587	216,000
80	51.62	715.6	700	363.6	18,520	4,000	1,745	253,000
100	63.10	1,000	800	410.0	22,630	5,000	2,133	353,600
125	77.13	1,398	900	455.8	27,000	6,000	2,514	464,800
150	90.88	1,837	1,000	501.2	31,620	7,000	2,888	585,700

ical calculations. In many cases it is necessary to apply both Eqs. (29) and (30) and to use the smaller of the two results.

Example 9. For a single-strand No. 60 chain, $p = \frac{3}{4}$ in., and the smaller sprocket has N_1 equal to 15 teeth. Smooth loading.

(a) Find the horsepower capacity at n_1 equal to 900 rpm for the smaller sprocket.

(b) Find the horsepower capacity at n_1 equal to 1,400 rpm for the smaller sprocket.

Solution. Numerical values should be substituted from Tables 6-11 and 6-12.

(a) Link plate fatigue,

$$\text{hp} = 0.004 \times 15^{1.08} \times 900^{0.9} \times 0.75^{3.0-0.07\times0.75}$$
$$= 0.004 \times 18.63 \times 455.8 \times 0.4283 = 14.60$$

Roller bushing fatigue,

$$\text{hp} = \frac{17,000 \times 15^{1.5} \times 0.75^{0.8}}{900^{1.5}} = \frac{17,000 \times 58.10 \times 0.7944}{27,000}$$
$$= 29.06$$

Link plate fatigue controls and the horsepower is 14.6.

(b) Link plate fatigue,

$$\text{hp} = 0.004 \times 15^{1.08} \times 1,400^{0.9} \times 0.75^{3.0-0.07\times0.75}$$
$$= 0.004 \times 18.63 \times 678.4 \times 0.4283 = 21.65$$

Roller bushing fatigue,

$$\text{hp} = \frac{17,000 \times 15^{1.5} \times 0.75^{0.8}}{1,400^{1.5}} = \frac{17,000 \times 58.10 \times 0.7944}{52,380}$$
$$= 14.98$$

At this speed, roller bushing fatigue controls, and the horsepower is 14.98.

The horsepower at different speeds for a single-strand No. 60 chain is plotted in Fig. 6-17. This figure illustrates graphically the need of using both Eqs. (29) and (30) when making calculations for the horsepower. The catalogs of the chain manufacturers contain extensive tables, which give the horsepower for different speeds and sprocket sizes.[9]

Chains have a considerable degree of elastic flexibility, which is helpful in prolonging their life. A service factor, however, must be applied to the nominal load under impact conditions to obtain the design horsepower carried by the chain. Factors are given by Table 6-13.

An increase in capacity can be obtained by the use of multiple-strand roller chains. These are essentially an assembly of single-strand chains placed side by side. The pins extend through the entire width of the chain and maintain the alignment of the different strands. Factors for multiplying the capac-

[9] Loading tables are also given in the March 1966 supplement to ASA B29.1, same address as footnote 8.

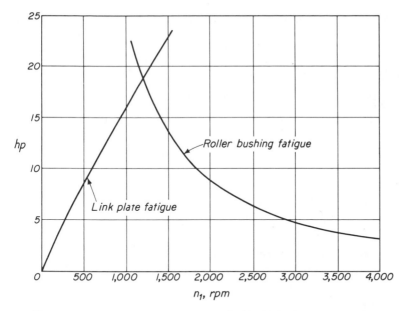

Figure 6-17 Horsepower vs. speed curves for No. 60 single strand roller chain, pitch, p = 3/4 in., operating on 15 tooth sprocket.

Table 6-13   SERVICE FACTORS AND MULTIPLE-STRAND FACTORS

Type of Driven Load	Type of Input Power			No. of Strands	Strand Factor
	I.C. Engine, Hydraulic Drive	Electric motor or Turbine	I.C. Engine, Mechanical Drive		
Smooth	1.0	1.0	1.2	2	1.7
Moderate shock	1.2	1.3	1.4	3	2.5
Heavy shock	1.4	1.5	1.7	4	3.3

ity of a single-strand chain for double, triple, and quadruple chains are given in Table 6-13. Overhanging shafts should be avoided when using multiple-strand chains unless the shaft is very rigid with a very small deflection.

16. Galling of Roller Chains

When the load is heavy and the speed is high, especially for pitches of 1 in. and greater, there is danger of galling or the welding together of the high spots of the contacting surfaces. Although such minute welds are immediately broken, rapid wear of the metal results. The maximum speed of operation to avoid galling for different chains and sprocket sizes is shown in

Table 6-14 MAXIMUM PERMISSIBLE SPEED, RPM, TO AVOID GALLING

N_1	Chain Number and Pitch							
	No. 80, 1 in.	No. 100, 1-1/4 in.	No. 120, 1-1/2 in.	No. 140, 1-3/4 in.	No. 160, 2 in.	No. 180, 2-1/4 in.	No. 200, 2-1/2 in.	No. 240, 3 in.
11	2,800	1,800	1,500	1,100	800	600	300	175
12	2,800	1,800	1,500	1,100	800	400	300	175
13	2,800	1,800	1,500	1,100	800	400	300	150
14	2,800	1,800	1,400	1,100	800	400	300	150
15	2,700	1,800	1,400	1,000	750	400	300	150
16	2,700	1,800	1,400	1,000	700	400	300	150
17	2,600	1,600	1,400	1,000	700	400	250	150
18	2,600	1,600	1,300	1,000	700	400	250	150
19	2,600	1,600	1,300	1,000	700	400	250	150
20	2,400	1,600	1,300	1,000	650	400	250	150
21	2,400	1,600	1,300	900	600	400	250	125
22	2,400	1,600	1,200	900	600	400	250	125
23	2,400	1,600	1,200	900	500	350	250	125
24	2,400	1,600	1,200	900	450	350	250	125
25	2,200	1,400	1,200	800	450	350	250	125
28	2,200	1,400	1,100	800	400	300	250	125
30	2,200	1,400	1,100	700	400	300	250	125
32	2,000	1,400	1,000	600	350	300	250	100
35	2,000	1,200	1,000	600	350	300	200	100
40	1,600	1,200	700	550	350	300	200	100
45	1,600	1,000	700	550	350	250	200	100
50	1,400	900	600	550	350	250	200	80
55	1,400	900	600	550	350	250	150	80
60	1,200	800	600	550	350	250	150	80

Table 6-14. Operation at speeds higher than those shown in the table is permissible provided that a suitable reduction in the load is made. The calculations are involved and the maker of the chain should be consulted for a suitable recommendation.[10]

17. Lubrication of Roller Chains

Lubrication is a very important feature in the successful design of a roller chain drive. In general a chain should have a sheet metal casing for protection from atmospheric dust and to facilitate lubrication.

[10]See Horner, W., and W. Kilgard, "Speed Limits for Chain Drives," *Machine Design*, **38**, Apr. 14, 1966, p. 183.

There are four types of lubrication for chain drives. The maximum chain speed for the use of each type is indicated by Table 6-15.

A chain should be washed in kerosene periodically to remove gummed lubricant and wear particles. It should then be soaked in oil to restore the internal lubricant.

Table 6-15 PERMISSIBLE CHAIN SPEED, FPM, FOR DIFFERENT TYPES OF LUBRICATION

Type	*Number and Pitch of Chain, in.*												
	25	35	40–41	50	60	80	100	120	140	160	180	200	240
	1/4	*3/8*	*1/2*	*5/8*	*3/4*	*1*	*1-1/4*	*1-1/2*	*1-3/4*	*2*	*2-1/4*	*2-1/2*	*3*
I	500	370	300	250	220	170	150	130	115	100	95	85	75
II	2,500	1,700	1,300	1,000	850	650	520	430	370	330	300	260	220
III	3,500	2,800	2,300	2,000	1,800	1,500	1,300	1,200	1,100	1,000	950	900	800
IV	Up to maximum permissible speed												

Type   I, Oil supplied periodically with brush or spoutcan.
   II, Drip lubrication. Oil applied between link plate edges from a drip lubricator.
   III, Oil bath or slinger. Oil level maintained in casing at predetermined height.
   IV, Oil stream. Oil supplied by circulating pump inside chain loop on lower strand.

18. Polygonal Action

When the number of teeth in the sprockets is small, the driven shaft of a roller chain drive may be given a pulsating or jerky motion. In Fig. 6-18, the shaft centers are located not an integral number of pitches apart, but some number plus one-half pitches apart. If the driver is assumed to be rotating uniformly, the driven shaft has the angular velocities indicated in the figures. In (a) the driven sprocket is rotating at a higher speed than the driver, whereas in (b) the driven shaft is moving slower. The equations for ω_2 indicate, for example, that for sprockets having 10 teeth, the velocity of the driven shaft will vary from 5 per cent above to 5 per cent below that of the driver. The chain velocity also varies. Sometimes long-link conveyor chains are operated on sprockets having as few as 5 or 6 teeth. The variation in velocity will then be most undesirable, and there will be an increase in the stresses in the chain and connected parts.[11]

19. Silent Chain

The silent or inverted tooth chain, shown in Fig. 6-19, is in wide use in the power transmission field. Different details of joint construction are used

[11]See Bouillon, G., and G. V. Tordion, "On Polygonal Action in Roller Chain Drives," *Trans. ASME*, **87**, 1965, *J. Eng. Ind.*, p. 243.

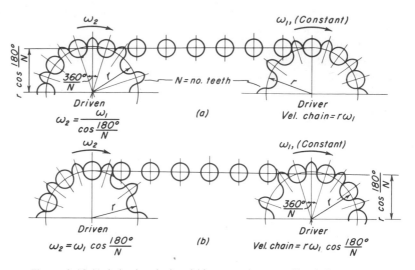

$$\omega_2 = \frac{\omega_1}{\cos \frac{180°}{N}}$$ (a)   Vel. chain = $r\omega_1$

$$\omega_2 = \omega_1 \cos \frac{180°}{N}$$ (b)   Vel. chain = $r\omega_1 \cos \frac{180°}{N}$

Figure 6-18 Variation in velocity of driven sprocket caused by shaft centers no being an integral number of pitches apart.

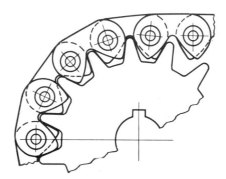

Figure 6-19 Silent chain.

with the object of increasing chain life. Careful attention to lubrication is required. Enclosures for the chain are usually necessary. Service factors should be employed in the design.

Standard pitches, or the distance center to center of pins, are $\frac{3}{8}$, $\frac{1}{2}$, $\frac{5}{8}$, $\frac{3}{4}$, 1, $1\frac{1}{4}$, $1\frac{1}{2}$, and 2 in. Sprockets may have 21–150 teeth. Center distance adjustment is required to compensate for wear. Greater power capacity can be obtained by using chains of greater width. Horsepower capacities are given in the manufacturer's catalog.

REFERENCES

1. Black, Peter, *Mechanics of Machines*, New York: Pergamon Press, Inc., 1967, Chapter 3.

2. Rothbart, H. S., Ed., *Mechanical Design and Systems Handbook*, New York: McGraw-Hill Book Company, 1964, Sections 28 and 29.

PROBLEMS

1. A 5V high-capacity V-belt carries a load horsepower of 12.5, but a service factor of 1.5 must be used. Pulley diameters are 9.75 in. and turn 1,750 rpm. Find the expected life in hours for belt lengths of 90, 95, and 100 in.

Ans. 18,800 hr; 19,900 hr; 20,900 hr.

2. A D-section V-belt has a length of 240.8 in. and operates on 16-in.-diam pulleys. Speed is 870 rpm. For an expected life of 20,000 hr, find the number of belts required to transmit 100 hp. *Ans.* 4 belts.

3. An 8V high-capacity V-belt is 280 in. long and operates on 21.2-in.-diam pulleys. Speed is 870 rpm. For an expected life of 20,000 hr, find the number of belts required to transmit 200 hp. *Ans.* 4 belts.

4. A C-section V-belt is 146.9 in. long and operates on pulleys 11 in. in diameter. Speed is 1,160 rpm. For a life of 20,000 hr, find the number of belts required to transmit 50 hp. *Ans.* 4 belts.

5. A B-section V-belt is 129.8 in. long and operates on pulleys of diameters of 6 and 25 in. Speed of the small pulley is 1,750 rpm. Expected life is 10,000 hr. Find the horsepower transmitted by the belt. Assume that the large pulley does not contribute to the fatigue effects. Then show that this assumption is permissible.

Ans. hp = 6.76.

6. A 5V high-capacity V-belt carries a load horsepower of 15, but a service factor of 1.5 must be used. Pulley diameters are 10.9 and 15 in. Speed of the small pulley is 1,750 rpm. Length of the belt is 100 in. Find the center distance and the expected life of the belt. *Ans.* c = 29.6 in.; life = 18,900 hr.

7. A 5V high-capacity V-belt is 71 in. long and carries a load horsepower of 9, but a service factor of 1.6 must be used. Pulley diameters are 10 and 14 in. Speed of the small pulley is 1,160 rpm. Find the expected life in hours.

Ans. life = 20,400 hr.

8. A 24-in.-OD plate clutch has a maximum lining pressure of 50 psi. One hundred and eighty horsepower are to be carried at 400 rpm. If the coefficient of friction is 0.30, find the inside diameter and the spring force required to keep the clutch engaged. *Ans.* d_i = 17.22 in.; F_n = 9,170 lb and
d_i = 10.20 in.; F_n = 11,060 lb.

9. Work Problem 8 but with the outside diameter 18 in. The maximum pressure is 75 psi, hp = 150, $\mu = 0.25$, and rpm = 600.

Ans. $d_i = 12.18$ in.; $F_n = 8,351$ lb and
$d_i = 8.50$ in.; $F_n = 9,513$ lb.

10. Find the torque that a cone clutch of mean radius 6 in. and angle α of 8° can carry if the force R for steady running is 500 lb; $\mu = 0.2$.   *Ans.* $T = 4,311$ in. lb.

11. A cone clutch of inclination 10° is to carry 40 hp at 600 rpm. The width of the lining along an element of the cone is 2 in. Maximum lining pressure is to be 50 psi. The coefficient of friction is 0.2. Find suitable values for r_o and r_i.

Ans. $r_o = 6.04$ in.; $r_i = 5.70$ in.

12. A cone clutch has a mean radius of 8 in. and a cone angle of 8°. Maximum lining pressure is 100 psi and the coefficient of friction is 0.2. Find the torque the clutch can exert and the engaging force required for steady operation. The lining measures 3 in. along an element of the cone. What is the horsepower for a speed of 600 rpm?   *Ans.* $R = 2,046$ lb;hp = 227.

13. Work Problem 12 but with the cone angle of 12°.

Ans. $R = 3,014$ lb; hp = 221.

14. Find the horsepower that the brake of Fig. 6-20 can absorb, and the length of arm a.   *Ans.* hp = 5.82; $a = 4.85$ in.

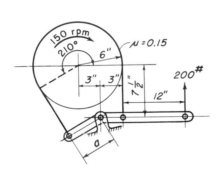

Figure 6-20 Problem 14.

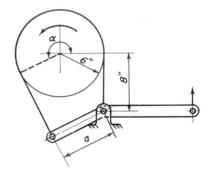

Figure 6-21 Problem 15.

15. The band brake in Fig. 6-21 is to absorb 6 hp at 150 rpm. The maximum pressure between lining and drum is 100 psi. The width of the band is 2 in. and $\mu = 0.12$. Find the angle of wrap α and the distance a.

Ans. $\alpha = 205.7°$; $a = 7.94$ in.

16. Find the horsepower that the brake of Fig. 6-22 can absorb, and the length of arm a.   *Ans.* hp = 13.33; $a = 1.46$ in.

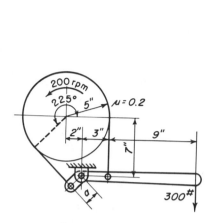

Figure 6-22 Problem 16.

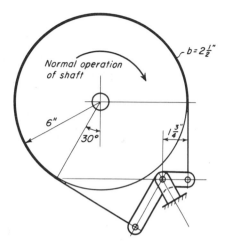

Figure 6-23 Problem 17.

17. Find the value of the torque that the backstop in Fig. 6-23 can resist if the maximum pressure between lining and drum is 200 psi. What must be the minimum value of the coefficient of friction to ensure the brake's holding the load?

Ans. $T = 9,310$ in. lb; $\mu = 0.174$.

18. In Fig. 6-24 the maximum lining pressure for any point on the brake is to be limited to 100 psi; $\mu = 0.2$. Find the torque capacity of the brake and the value of force P.

Ans. $T = 9,080$ in. lb; $P = 487$ lb.

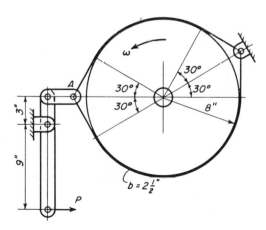

Figure 6-24 Problem 18.

19. In Fig. 6-25, the maximum lining pressure for any point on the brake is limited to 100 psi; $\mu = 0.2$. Find the torque capacity of the brake.

Ans. $T = 9,820$ in. lb.

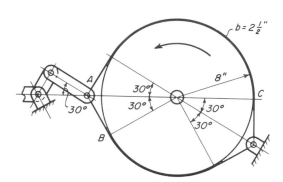

Figure 6-25 Problem 19.

20. (a) Write the expression for torque and the maximum lining pressure in terms of T_2 for the brake shown in Fig. 6-26; $\mu = 0.3$.

(b) Do the same for the brake in sketch (b). Note that for the same value of the actuating force T_2, the brake in (b) can exert approximately 3 times the torque as the brake in (a), but the lining pressure is almost twice as great.

Ans. (a) $T = 0.87T_2r; p = 1.874T_2/br$.
(b) $T = 2.51T_2r; p = 3.51T_2/br$.

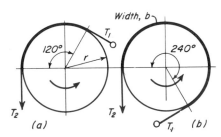

Figure 6-26 Problem 20.

21. A band brake exerts a torque of 15,000 in. lb. The drum is 2 in. wide and 10 in. in radius. If the maximum pressure between the lining and drum is 100 psi and the coefficient of friction is 0.3, find the angle of contact between lining and drum.

Ans. $\alpha = 264°46'$.

22. Work Problem 21 but with the width made equal to $2\frac{1}{2}$ in.

23. Work Problem 21 with a torque of 13,000 in. lb and μ equal to 0.25.

24. Find the value of dimension c in Fig. 6-27 that will cause the friction forces neither to help nor resist the application of the shoe to the drum.

Ans. $c = 1.414r$.

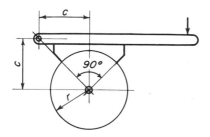

Figure 6-27 Problem 24.

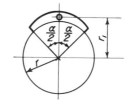

Figure 6-28 Problem 25.

25. (a) For $\alpha/2 = 45°$, find the value of r_1 in Fig. 6-28 that will cause the shoe to be free from moment about its pivot.

(b) Work (a) using $\alpha/2 = 60°$. *Ans.* (a) $r_1 = 1.10r$; (b) $r_1 = 1.17r$.

26. If the permissible value for p_{max} in Fig. 6-29 is 100 psi, find (a) the corresponding value of P, and (b) the horsepower that the brake will absorb; $\mu = 0.25$.

Ans. (a) $P = 777$ lb; (b) hp $= 28.6$.

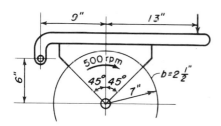

Figure 6-29 Problem 26.

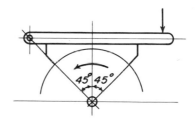

Figure 6-30 Problem 27.

27. (a) Find the resultant vertical component of the normal and friction forces for the brake of Fig. 6-30.

(b) Find the resultant horizontal component of the normal and friction forces.

Ans. (a) $F_v = \dfrac{\sqrt{2}}{8} brp_{max} [2(1 - \mu) + \pi(1 + \mu)]$.

(b) $F_h = \dfrac{\sqrt{2}}{8} brp_{max} [2(1 + \mu) - \pi(1 - \mu)]$.

28. Find the force P required if the maximum pressure of the lining in Fig. 6-31 is equal to 100 psi; $\mu = 0.25$. The width of the lining is 2 in. Find the torque about center O.   *Ans.* $P = 723$ lb; $T = 1,531$ in. lb.

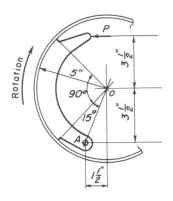

Figure 6-31 Problem 28.

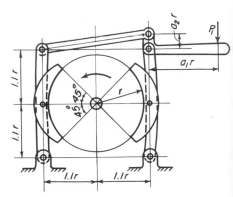

Figure 6-32 Problem 29.

29. (a) Make an exploded view of the double-block brake in Fig. 6-32. Draw each part, including the drum, separately, and show all forces and moments necessary for equilibrium. Assume that the normal and friction forces for the shoes are concentrated at points B.

(b) Compute the value of the maximum shoe pressure and the horsepower absorbed by the brake. Find the horsepower absorbed per unit area, and the pV value.   *Ans.* $p_{max} = 93.4$ psi; hp $= 20.5$; hp/in.² $= 0.34$; $pV = 58,700$.

30. (a) Make an exploded view of the double-block brake of Fig. 6-33 and show all forces and moments required for equilibrium of each part. $\mu = 0.2$; width $b = 4$ in.

(b) Find the value of p_{max} for each shoe and the value of the torque exerted by the brake. Also find the horsepower.

Ans.   (b) $p_{max} = 23.3$ and 20.7 psi; hp $= 47.4$.

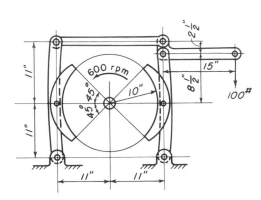

Figure 6-33 Problem 30

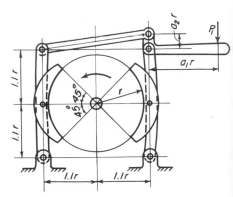

Figure 6-34 Problem 31.

31. Make an exploded view of the brake in Fig. 6-34 showing the forces required for the equilibrium of each part. Write the expression for the torque in terms of μ, P_1, r, a_1, and a_2.   *Ans.* $T = 4.4\mu P_1 r a_1/a_2$.

32. A crane brake is lowering the load shown in Fig. 6-35 at a speed of 18 fpm. The motor shaft is connected to the cable drum shaft by a gear train with a 130:1 ratio. Assume that the gears and bearings are frictionless. The brake drum rotates at one-half motor speed and is 16 in. in diameter; the lining is 3 in. wide; μ is equal to 0.2. The shoes and levers are arranged as in Problem 31. Find the value of p_{max} and the value of P_1 for $a_1 = 1$ and $a_2 = \frac{1}{3}$. Also find the pV value.   *Ans.* $p_{max} = 24.3$ psi; $P_1 = 125$ lb; $pV = 39,600$.

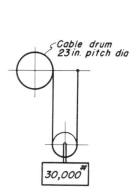

Figure 6-35 Problem 32.

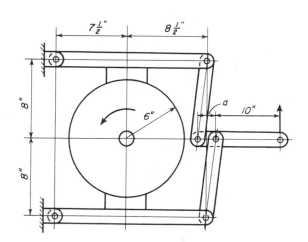

Figure 6-36 Problem 33.

33. Assume the brake in Fig. 6-36 has short shoes. The coefficient of friction is 0.3. Find the value of distance a that will cause the wear to be the same for both shoes.   *Ans.* $a = 1.74$ in.

34. A block brake with a short shoe is to be designed for a pV value of 55,000; μ is equal to 0.2; the area of the shoe is 18 in.^{2}; and the diameter of the brake drum is 16 in. A cable drum 12 in. in diameter is connected to the brake drum by means of gearing. The brake drum revolves 3 times as fast as the cable drum. See Fig. 6-37. Find the uniform velocity at which a 1,000-lb weight at the end of the cable is being lowered. Find the value of the pressure of the shoe.

Ans.　$V = 791$ fpm; $p = 69.5$ psi.

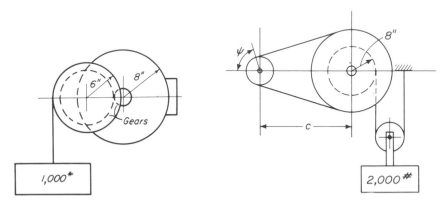

Figure 6-37 Problem 34.　　　　　Figure 6-38 Problem 36.

35. A 3V high-capacity V-belt carries a load horsepower of 1.25, but a service factor of 1.5 must be used. Pulley diameters are 3.15 and 6.00 in. Length of the belt is 63 in. Speed of the small pulley is 1,160 rpm. Find the center distance and the expected life in hours.　　*Ans.*　$c = 24.3$ in.; life $= 19,600$ hrs.

36. The weight in Fig. 6-38 must be raised at the rate of 10 fps. Motor pulley is 9.25 in. in diameter, and turns 870 rpm. Length of the belts is 125 in. Expected life is 10,000 hr. Neglect fatigue effects at large pulley. Assume a service factor of 1.6 will care for impact effects and friction of the system. Find the center distance, diameter of the large pulley, and number of belts required.

Ans.　$c = 31.8$ in.

No. of belts $= 5$.

37. Work Problem 8 but with horsepower equal to 195.

38. Find the horsepower absorbed and force P in Fig. 6-39 if the maximum lining pressure is 100 psi. $\mu = 0.2$.

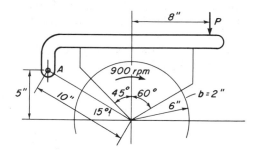

Figure 6-39 Problem 38.   **Figure 6-40** Problem 39.

39. In Fig. 6-40 the horsepower absorbed is 15; $\mu = 0.3$. Find the value of the actuating force P.

40. Find the value of force P for the brake in Fig. 6-41. $p_{max} = 100$ psi. $\mu = 0.25$.

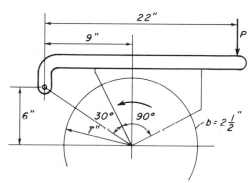

Figure 6-41 Problem 40.

41. Work Problem 33 but with the dimensions of $7\frac{1}{2}$, 8, and $8\frac{1}{2}$ in. in Fig. 6-36 each increased by $\frac{1}{2}$ in.

42. Work Problem 30 but with short shoes rigidly attached to the two uprights.

43. An automobile of weight W with four-wheel brakes has a wheel base of l inches. The center of gravity is c in. forward of the rear axle, and is h in. above the ground. If the coefficient of friction between the pavement and the tires is μ, show that the load carried by the two front tires during braking with the motor disconnected is equal to $W(c + \mu h)/l$.

44. At the time the clutch is engaged in Fig. 6-42, each shaft is rotating at the velocity shown. If Ω_1 is greater than Ω_2, the system on the right will be accelerated when the two shafts are clutched together. Upon engagement assume that the clutch slips, but applies a constant torque of T in. lb. Show that the time required for the two shafts to be rotating at the same velocity is given by $(\Omega_1 - \Omega_2)I_1 I_2/T(I_1 + I_2)$. Neglect any effects due to torsional vibration.

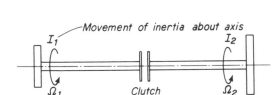

Figure 6-42 Problem 44.   Figure 6-43 Problem 45.

45. The body of weight W_1 in Fig. 6-43 has a downward velocity of v_o when a constant retarding torque T from a brake is applied. Total mass moment of inertia of the rotating body is I lb in. sec². Neglect friction and the inertia of the rope and lower pulley. Show that the velocity of W after t sec have elapsed is given by $v = v_o - (T - \frac{1}{2}Wr)t/[(2I/r) + (Wr/2g)]$.

46. (a) Find the horsepower capacity for a No. 40 chain at a speed of 900 rpm for the small sprocket, which has 12 teeth.

(b) Find the horsepower for a speed of 2,400 rpm.

Ans. (a) hp $= 3.42$; (b) hp $= 3.45$.

47. (a) Find the horsepower capacity for a No. 120 chain at a speed of 500 rpm for the small sprocket, which has 20 teeth.

(b) Find the horsepower for a speed of 900 rpm.

Ans. (a) hp $= 88.3$; (b) hp $= 77.9$.

48. A No. 60 chain operates with a small sprocket of 15 teeth. Calculate and plot the curves for horsepower vs. n_1 for link plate fatigue and roller bushing fatigue and compare with Fig. 6-17.

49. A No. 80 roller chain operates on a small sprocket of 20 teeth. Calculate and plot the curves for horsepower vs. speed from 0 to 2,400 rpm.

50. A No. 120 roller chain operates on a small sprocket of 12 teeth. Calculate and plot the curves for horsepower vs. speed from 0 to 1,400 rpm.

51. A No. 60 roller chain operates on a small sprocket of 16 teeth. Calculate and plot the curves for horsepower vs. speed from 0 to 4,000 rpm.

7 Welded and Riveted Connections

A weld is a union between metal surfaces brought about by the localized application of heat. Welding has assumed an important place in speeding the manufacture of component parts and in the assembly of these parts into engineering structures. Forge or pressure welding of wrought iron has been practiced for centuries; methods such as torch, arc, or resistance welding have appeared only in recent years. The field of usefulness of these latter methods is being rapidly expanded.

Rivets, also, can be used for forming the joints and connections between the parts of a structure. Although welding has replaced riveting to a considerable extent, rivets are customarily employed for certain types of joints. Long experience with this method of fastening has given confidence in the reliability of riveted joints.

A, throat area
FS, factor of safety
h, size of weld
I, moment of inertia
J, polar moment of inertia
K, stress concentration factor
l, length of weld

M, moment
p, pressure
P, load
r, radius
r_1, distance from center of gravity of joint to center of weld or rivet
s, stress, tension, or compression

s_{av}, average stress
s_e, endurance limit stress
s_r, range stress
s_s, shearing stress

s_{yp}, yield point stress
s_{syp}, yield point stress in shear
s_{ult}, ultimate stress
T, torque

1. Fabrication by Welding

Because of lower initial cost, many structural parts of machinery formerly made by casting are now fabricated by welding. The components can be sheared or flame cut from hot-rolled steel plate and then welded together. Figure 7-1 shows a number of typical welded assemblies. Sometimes the intricate portion of the body can be cast or stamped. The flat areas, made of plates, then can be attached by welding.

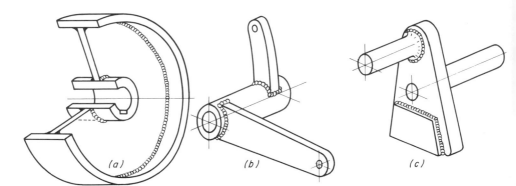

Figure 7-1 Machine parts fabricated by fusion welding.

Welded assemblies usually provide greater strength at a reduction in weight—an important advantage for moving parts of machines and transport equipment. In a welded design it is usually necessary to do a smaller amount of machining than for an equivalent casting. The design must provide accessibility to the welds so they can be properly made and inspected.

2. Fusion Welding

In the fusion process, heat is obtained from an oxyacetylene flame or from an electric arc passing between an electrode and the work. The edges of the parts are heated to the fusion temperature and joined together with the addition of molten filler material from a welding rod.

In metallic arc welding, the electrode is composed of suitable filler material, which is melted and fed into the joint as the weld is progressively formed. Shielded arc welding uses an electrode with a heavy coating of fluxing materials. These are consumed as the rod is melted, and perform the usual functions of a flux, as shown in Fig. 7-2. When the oxyacetylene torch is used, the molten metal is protected from the atmosphere by the outer envelope of the flame. The flame is generally adjusted until it is neutral or slightly reducing. A flux is used in the gas welding of some metals to float out any impurities that may be present and thus aid in forming a sound weld.

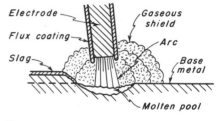

Figure 7-2 Arc welding with coated electrode.

Either direct or alternating current can be used with the metallic arc process. When the weld is larger than about $\frac{3}{8}$ in. minimum thickness, it is usually made in successive layers. Deposited weld metal frequently has the coarse structure characteristic of cast metals.

3. Strength of Fusion Welds

Several different types of welds with the equations for the stresses[1] arising from the given loadings are shown in Fig. 7-3. The height h for a butt weld

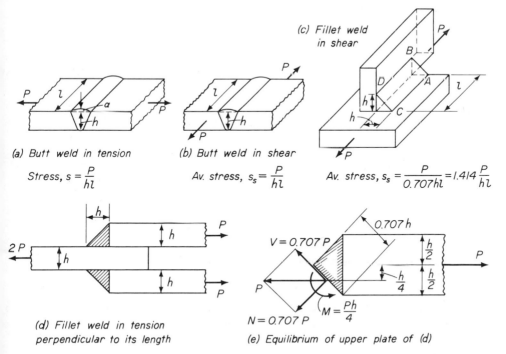

(a) Butt weld in tension

Stress, $s = \dfrac{P}{hl}$

(b) Butt weld in shear

Av. stress, $s_s = \dfrac{P}{hl}$

(c) Fillet weld in shear

Av. stress, $s_s = \dfrac{P}{0.707hl} = 1.414\dfrac{P}{hl}$

(d) Fillet weld in tension perpendicular to its length

$V = 0.707\,P$

$N = 0.707\,P$

$M = \dfrac{Ph}{4}$

(e) Equilibrium of upper plate of (d)

Figure 7-3 Various types of welds and loadings.

[1]See Jennings, C. H., "Welding Design," *Trans. ASME,* **58**, 1936, p. 497, and **59**, 1937, p. 462.

does not include the bulge or reinforcement *a*. Plates that are $\frac{1}{4}$ in. thick and heavier should be beveled before welding, as shown.

The fillet weld loaded parallel to its length is shown in Fig. 7-3(c). The throat area is 0.707 *hl*, where *h* is the length of the leg. The shearing stress on the throat area is

$$s_s = \frac{P}{0.707hl} = 1.414\frac{P}{hl} \tag{1}$$

The usual equation for the factor of safety for static loads can be employed.

$$FS = \frac{s_{syp}}{s_s} = \frac{0.5s_{yp}}{s_s} \tag{2}$$

Example 1. A $\frac{1}{4}$-in. fillet weld 2 in. long carries a steady load of 3,000 lb parallel to the weld. The weld metal has a yield strength of 50,000 psi. Find the value of the factor of safety.

Solution. By Eq. (1): $s_s = 1.414 \times \dfrac{3,000}{(1/4) \times 2} = 8,490$ psi

By Eq. (2): $FS = \dfrac{25,000}{8,490} = 2.95$

The fillet weld loaded in a direction perpendicular to its length, Fig. 7-3(d), has the advantage that precise fit-up of the plates is not required. The equilibrium diagram of the upper plate is shown in sketch (e). The weld has been cut through the throat, and force *P* and moment *M* necessary for equilibrium have been applied. Force *P* has been resolved into normal component *N* and shear component *V*.

When only two plates are present, the joint is eccentrically loaded, and is subjected to an additional moment something like that shown in Fig. 7-14 for the single-riveted lap joint. Because of the shape of the weld, the stress situation on the throat arising from loads *N*, *V*, and *M* is very complex, and exact equations suitable for design calculations are not available. It is common practice to consider the shearing stress in the throat of a transverse fillet weld as being merely the load divided by the throat area, which results again in Eq. (1).

4. Eccentrically Loaded Welds—Static Loads

When the load on a welded joint is applied eccentrically, the effect of the torque or moment must be taken into account as well as the direct load. The state of stress in such a joint is complicated, and it is necessary to make simplifying assumptions.

When a joint consists of a number of welds, it is customary to assume that the moment stress at any point is proportional to the distance from the center of gravity of the group of welds. Let the weld shown in Fig. 7-4 be one of a group forming a joint with the center of gravity of all the weld areas at O. The moment stress s acts perpendicularly to radius r on element dA of the weld. The external moment or torque T is equal to the moment from stress s integrated over all the welds of the joint.

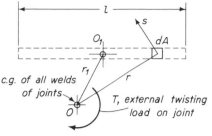

$$T = \int sr\, dA = \int \frac{s}{r} r^2\, dA = \frac{s}{r} \int r^2\, dA = \frac{sJ}{r}$$

$$s = \frac{Tr}{J} \qquad (3)$$

Figure 7-4 Stress on element of eccentrically loaded welded joint.

Ratio s/r is a constant since the stress is assumed to vary directly with r. The integral $\int r^2\, dA$ in Eq. (3) has been replaced by J, the polar moment of inertia about O for the group of welds. For the maximum torsional stress, the value of r to the point furthest removed from the center of gravity O must be used. The stress from the direct load must be added vectorially to the moment stress in order to obtain the resultant stress. For static loads, it is usual practice to assume that the direct stress in a weld is uniformly distributed throughtout its area.

The parallel axis equation can be used for finding the value for J for a weld about the center O. For the weld in Fig. 7-4 this equation would be written

$$J = J_o + Ar_1^2 \qquad (4)$$

The area A in this equation refers to the throat area of the weld. The fact that the throat for a fillet weld is inclined at 45° to the plane of the joint has no effect on the value of J. Radius r_1 extends from the center O_1 of the weld to the center of gravity O for the group. Symbol J_o represents the moment of inertia of the single-weld area about its own center O_1. This value can be found from the following equation.

$$J_o = \frac{Al^2}{12} \qquad (5)$$

where A is again the throat area and l is the length of the weld.

When Eq. (5) is substituted into Eq. (4), the result is

$$J = A\left(\frac{l^2}{12} + r_1^2\right) \qquad (6)$$

The value of J for each weld about O should be computed by Eq. (6); the results are added to obtain the moment of inertia of the entire joint.

Example 2. An eccentrically loaded bracket is welded to its support as shown in Fig. 7-5. If the load is steady, find the value of the maximum stress in the weld. Find the factor of safety if the yield strength of the weld metal is 50,000 psi.

Solution. By symmetry the center of gravity of the welds in the x-direction

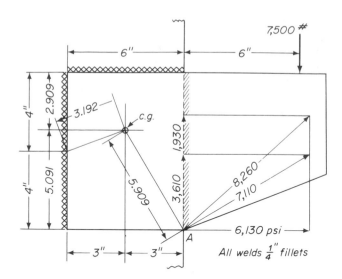

Figure 7-5 Example 2.

is halfway between the vertical welds. The center of gravity in the vertical direction is found by taking moments about the top weld.

$$22\bar{y} = 2 \times 8 \times 4$$

$$\bar{y} = 2.909 \text{ in. from top}$$

Width of throat: $b = 0.707h = 0.707 \times \frac{1}{4} = 0.1768 \text{ in.}$

For vertical weld: $A = 0.177 \times 8 = 1.414 \text{ in.}^2$

$$r_1 = \sqrt{1.091^2 + 3^2} = \sqrt{10.1903} = 3.192 \text{ in.}$$

By Eq. (6): $J = 1.4142\left(\frac{8^2}{12} + 3.192^2\right) = 21.953 \text{ in.}^4$

For top weld: $A = 0.1768 \times 6 = 1.061 \text{ in.}^2$

By Eq. (6): $J = 1.061\left(\frac{6^2}{12} + 2.909^2\right) = 12.159 \text{ in.}^4$

Total J: $J = 2 \times 21.953 + 12.159 = 56.065 \text{ in.}^4$

Total area: $A = 2 \times 1.414 + 1.061 = 3.889 \text{ in.}^2$

Direct stress: $s_s = \dfrac{7,500}{3.889} = 1,930 \text{ psi}$

Torsional stress: The maximum stress will occur at A.

Radius OA: $r = \sqrt{5.091^2 + 3^2} = 5.909 \text{ in.}$

$$s_s = \frac{Tr}{J} = \frac{7,500 \times 9 \times 5.909}{56.065} = 7,110 \text{ psi}$$

This stress is directed perpendicular to OA. It is now resolved into the components shown. The total vertical component is $3,610 + 1,930$ or $5,540$ psi.

Resultant stress: $s_s = \sqrt{5{,}540^2 + 6{,}130^2} = 8{,}260$ psi

By Eq. (2): $FS = \dfrac{0.5 s_{yp}}{s_s} = \dfrac{25{,}000}{8{,}260} = 3.03$

Example 3. Find the value of static force P in Fig. 7-6 if electrode E6010 is used at a factor of safety equal to 2.

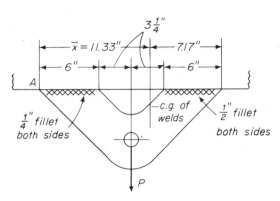

Figure 7-6 Example 3.

(a) All welds are $\frac{1}{4}$-in. fillets.

(b) Welds on the left side are $\frac{1}{4}$ in. and on the right side are $\frac{1}{2}$ in.

Solution. (a) By Table 7-3, for rod E6010, $s_{yp} = 50{,}000$ psi.

Throat area of all welds: $A = 4 \times \frac{1}{4} \times 6 \times 0.707 = 4.242$ in.^{2}

Then: $P = s_s A = \dfrac{25{,}000}{2} \times 4.242 = 53{,}020$ lb

    (b) Left, throat area: $A = 2 \times \frac{1}{4} \times 6 \times 0.707 = 2.121$ in.^{2}
                                              (both welds)

       Right, throat area: $A = 2 \times \frac{1}{2} \times 6 \times 0.707 = 4.242$ in.^{2}
                                              (both welds)

       Total area: $A = 6.363$ in.^{2}

Take moments at left end: $\bar{x} = \dfrac{2.121 \times 3 + 4.242 \times 15.5}{6.363}$

                                       $= 11.33$ in.

For left welds: Eq. (6): $J = 2.121\left(\dfrac{6^2}{12} + 8.33^2\right) = 153.65$ in.^{4}

For right welds: Eq. (6): $J = 4.242\left(\dfrac{6^2}{12} + 4.17^2\right) = 86.37$ in.^{4}

Total: $J = 153.65 + 86.37 = 240.02$ in.^{4}

Direct stress: $s_s = \dfrac{P}{6.363} = 0.1572P$

Moment arm of the load is 11.33 − 9.25 or 2.08 in.

Moment stress at A: $\qquad\qquad\qquad s_s = \dfrac{Tr}{J} = \dfrac{P2.08 \times 11.33}{240.02} = 0.0984P$

Total stress at A: $\qquad\qquad\qquad s_s = 0.1572P + 0.0984P = 0.2556P$

By Eq. (2): $\qquad\qquad\qquad 0.2556P = \dfrac{25,000}{2} \quad$ or $\quad P = 48,900\ \text{lb}$

It should be noted that, although the welds of part (b) are larger, because of the eccentricity, the carrying capacity of the joint has been actually reduced. It is generally advantageous to maintain symmetry in the design of a welded joint.

The method explained in this section cannot be considered an exact analysis of weld stresses, but should be looked upon simply as a reasonable effort to take into account the fact that the capacity of a joint to resist moment loads is increased by locating the welds further from the center. The theory also assumes that the weld stresses are within the elastic limit of the weld material, and that any effects from the deflection of the welded parts can be neglected.

5. Stress Concentration in Welds

Since abrupt changes in form occur in welds, stress concentrations are present as the force passes from one portion of the assembly to the other. Stress concentration effects are usually ignored for static or steady loads, and are applied only to the variable or range component in accordance with general design practice for fluctuating loads.

Care must be exercised that the weld metal and the plates at the base of a butt weld be thoroughly fused together. With insufficient fusion, sharp-cornered notches extend inward, as at A in Fig. 7-7(a), and a serious stress

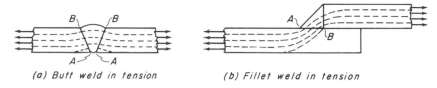

(a) Butt weld in tension   (b) Fillet weld in tension

Figure 7-7 Stress concentration in welds.

concentration results. Stress concentration also occurs at points B, where the force spreads into the reinforcement. A fillet weld has concentrations at the toe and heel, points A and B in Fig. 7-7(b), where the force is passing from one plate to the other through the weld. Values of stress concentration factors are given in Table 7-1.

Table 7-1   STRESS CONCENTRATION FACTORS
K FOR WELDS

Location	K
Reinforced butt weld	1.2
Toe of transverse fillet weld	1.5
End of parallel fillet weld	2.7
T-butt joint with sharp corners	2.0

A fillet weld loaded by parallel forces as in Fig. 7-3(c) has stress concentrations at each end caused by the unequal elongations of the plates. The upper plate at point B has the maximum elongation, because here it is carrying the entire load P. The lower plate at A has a small elongation, because it is carrying very little load at this point. Since the weld joins the two plates together, it is subjected to a greater deformation than the average for the weld as a whole, and an increase or concentration of stress occurs at the end of the weld. Similar reasoning applied to the other end, where the lower plate at C has a large elongation and the upper plate at D has a small elongation. Similar stress concentration effects are present at the ends of the butt weld in shear, shown in Fig. 7-3(b).

6. Residual Stress—Weldability

Residual stresses of considerable magnitude can result from contraction of the weld metal upon cooling. Such stresses are usually greatest in the transverse direction and are thus more harmful when the weld is subjected to tensile loads. The presence of residual stress is particularly dangerous if the weld is subjected to repeated loading or impact. Removal of residual stresses can be accomplished by annealing, or by application of an overload, which stresses the entire weld to the yield point value.

It is important that both the weld metal and adjacent parent metal be ductile and free from brittleness. The properties of the weld metal depend on the composition of the welding rod. However, for steel parent metal with a carbon content higher than about 0.15 per cent, there is danger of air hardening upon rapid cooling from the welding temperature. The quenching effect of the cold metal surrounding a weld can be reduced if the parts are preheated to 600°–1500°F before the welding is done. An annealing treatment after welding may be required to restore the original ductility of the parent metal. The fusion welding characteristics of some commonly used metals and alloys are given in Table 7-2.

Residual welding stresses in gray and alloy cast iron can be eliminated by preheating before welding, followed by slow cooling after welding. Plain-carbon and alloy steel castings are ordinarily welded by the same procedure

Table 7-2 WELDABILITY OF VARIOUS METALS AND ALLOYS

Metal or Alloy	Gas	Arc	Metal or Alloy	Gas	Arc
Carbon steels			Magnesium alloys	A	No
1. Low- and medium-carbon	A	A	Copper and copper alloys		
2. High-carbon	B	A	1. Deoxidized copper	A	B
3. Tool steel	B	B	2. Pitch, electrolytic, and lake	B	A
Cast steel, plain-carbon	A	A	3. Commercial bronze, red brass, low brass, and ounce metal	A	B
Gray and alloy cast iron	A	B	4. Spring, admiralty, yellow, and commercial brass	A	B
Malleable iron	B	B	5. Muntz metal, tobin bronze, naval brass, and managenese bronze	A	B
Low-alloy, high-tensile steels					
1. Ni-Cr-Mo and Ni-Mo	B	B	6. Nickel silver	A	B
2. All other usual compositions	A	A	7. Phosphor bronze, bell metal, and bearing bronze	A	A
			8. Aluminum bronze	B	A
Stainless steels			9. Beryllium copper	—	A
1. Chromium	B	A			
2. Chromium-Nickel	A	A			
Aluminum			Nickel and nickel alloys	A	A
1. Commercially pure Al	A	A			
2. Al-Mn alloy	A	A	Lead	A	No
3. Al-Mg-Mn and Al-Si-Mg alloy	B	A			
4. Al-Cu-Mg-Mn alloys	No	B			

A—commonly used. B—occasionally used under favorable conditions.

used for rolled steel of similar composition. Air hardening occurs if the carbon or alloy content is sufficiently high.

For the copper and aluminum alloys, strips of the parent metal are frequently used as filler material where an exact color match is required.

Fluxes are necessary in welding aluminum to remove the oxide coating from both the parent metal and filler rods. When the welding is completed, the parts must be thoroughly cleaned of the flux to avoid corrosion. Since the welding temperature of aluminum and magnesium is below the visible light range, it is difficult for the operator to determine when the welding temperature is being approached. At high temperatures these alloys are very weak, and there is a tendency for the member to collapse unless positive support is provided during the welding operation.[2]

[2] For the welding of aluminum, see "How to Weld Aluminum," *Iron Age*, **157**, June 20, 1946, p. 27, and **158**, July 4, 1946, p. 11.

7. Welding Electrodes

Many different types of electrodes have been standardized to fit the various conditions found in the welding of machinery and structures. The strength and ductility properties of several of these are given in Table 7-3.

Table 7-3 TENSILE STRENGTH, YIELD POINT, AND ELONGATION REQUIREMENTS FOR ALL-WELD-METAL TENSION TEST IN THE AS-WELDED CONDITION, ASTM DESIGNATION A233-64T

	E60 Series				E70 Series		
AWS-ASTM Classification	*Tensile Strength, min psi*	*Yield Point, min psi*	*Elongation in 2 in., min per cent*	*AWS-ASTM Classification*	*Tensile Strength, min psi*	*Yield Point, min psi*	*Elongation in 2 in., min per cent*
E6010	62,000	50,000	22	E7014	72,000	60,000	17
E6011	62,000	50,000	22	E7015	72,000	60,000	22
E6012	67,000	55,000	17	E7016	72,000	60,000	22
E6013	67,000	55,000	17	E7018	72,000	60,000	22
E6020	62,000	50,000	25	E7024	72,000	60,000	17
E6027	62,000	50,000	25	E7028	72,000	60,000	22

The numbering system is based on the use of an "E" prefix followed by four digits. The last digit indicates a group of welding technique variables, such as current supply and application. The next to the last digit indicates a welding position number as 1 for all positions—flat, horizontal, vertical, and overhead; 2 for flat and horizontal fillet welding; and 3 for flat only. The two digits on the left indicate the approximate tensile strength in thousands of

Table 7-4 ELECTRODE CLASSIFICATION

Classification	*Penetration*	*Basic Application*
E6010 E6011	Deep	Good mechanical properties, especially where multipass welds are employed, as in buildings, bridges, pressure vessels, and piping.
E6012	Medium	Good for single-pass high-speed horizontal fillet welds. Easy to handle. Especially adapted to cases of poor fit-up.
E6013	Medium	Designed for obtaining good-quality welds in thin metal.
E6020	Medium Deep	A high-production electrode for horizontal fillet welds in heavy sections.
E6027	Medium	Iron powder electrode. Fast and easy to handle.

pounds per square inch. Welding electrodes are available in diameters from $\frac{1}{16}$ to $\frac{5}{16}$ in.

The application properties for a number of electrodes are shown in Table 7-4. The choice of the proper electrode for a particular application should be left to a man with long experience in the field of welding.

8. Design for Fluctuating Loads

When the loading fluctuates, design calculations can be made by the methods explained in Chapter 2, as illustrated by the following examples.

Example 4. The load on a butt weld fluctuates continuously between 10,000 and 40,000 lbs. Plates are 1 in. thick. Factor of safety equals 2.5. Use an E6010 welding rod. Let the endurance limit for the weld be equivalent to that of an as-forged surface. Find the required length of weld.

Solution. By Table 7-3, for E6010:

$$s_{ult} = 62,000 \text{ psi}$$

$$s_{yp} = 50,000 \text{ psi}$$

By Fig. 2-16: $s_e = 16,000 \text{ psi}$

By Table 7-1: $K = 1.2$

Given loading: $P_{av} = 25,000 \text{ lb}, \qquad s_{av} = \dfrac{25,000}{1 \times l}$

$$P_r = 15,000 \text{ lb}, \qquad s_r = \dfrac{15,000}{1 \times l}$$

By Eq. (11), Chapter 2: $\dfrac{s_{yp}}{FS} = s_{av} + \dfrac{Ks_{yp}}{s_e} s_r$

$$\frac{50,000}{2.5} = \frac{25,000}{l} + \frac{1.2 \times 50}{16} \times \frac{15,000}{l}$$

From which: $l = 4.06 \text{ in.}$

Example 5. The moment in Fig. 7-8 fluctuates continuously between the values

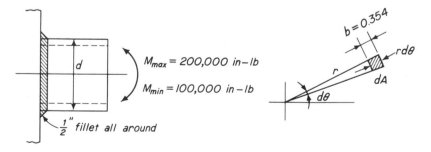

Figure 7-8 Example 5.

shown. Let an E6010 electrode be used. Factor of safety is 2. Let the endurance limit for the weld material be equivalent to that of an as-forged surface. Find the required diameter d to carry the load.

Solution. Moment of inertia I for the throat area will be found by first getting J for the throat and then taking one-half for I.

Width of throat:
$$b = 0.707h = 0.707 \times 0.5 = 0.354 \text{ in.}$$

By Fig. 7-8:
$$J = \int_0^{2\pi} r^2 \, dA = \int_0^{2\pi} r^2 \, brd\theta = br^3\theta \Big|_0^{2\pi} = 2\pi br^3$$

Then:
$$I = \tfrac{1}{2}J = \pi br^3 = \pi 0.354 r^3 = 1.111 r^3$$

$$M_{av} = 150,000 \text{ in. lb}, \qquad M_r = 50,000 \text{ in. lb}$$

$$S_{sav} = \frac{M_{av}c}{I} = \frac{150,000r}{1.111r^3} = \frac{135,030}{r^2}$$

$$S_{sr} = \frac{M_r c}{I} = \frac{50,000r}{1.111r^3} = \frac{45,010}{r^2}$$

By Table 7-3:
$$S_{ult} = 62,000 \text{ psi}$$

$$S_{yp} = 50,000 \text{ psi}, \qquad S_{syp} = \tfrac{1}{2} \times 50,000 = 25,000 \text{ psi}$$

By Fig. 2-16:
$$S_e = 16,000 \text{ psi}$$

By Table 7-1:
$$K = 1.5$$

By Eq. (11), Chapter 2:

$$S_s = S_{sav} + \frac{K s_{yp}}{S_e} S_{sr}$$

$$- \frac{135,030}{r^2} + \frac{1.5 \times 50}{16} \times \frac{45,010}{r^2} - \frac{346,000}{r^2}$$

By Eq. (2):
$$\frac{346,000}{r^2} = \frac{S_{syp}}{FS} = \frac{25,000}{2}$$

Or:
$$r^2 = \frac{2 \times 346,000}{25,000} = 27.681$$

Then:
$$r = 5.26 \text{ in.} \quad \text{and} \quad d = 10.52 \text{ in.}$$

9. Resistance Welding

In resistance welding, a heavy electric current is passed through the parts at the place where the weld is desired. The resistance of the metals to the current causes the temperature to rise quickly and the metal to become plastic. The weld is completed by mechanical pressure from the electrodes. This pressure brings the parts into complete union. Alternating current is used from a suitable transformer, which must be capable of supplying a large current at low voltage. The copper electrodes are faced with harder alloys at the contact end. To confine the heat in the proper region, the resistance between electrode and part must be less than the resistance between the surfaces to be joined.

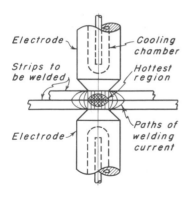

Figure 7-9 Schematic diagram of spot welding.

Spot welding is a type of resistance welding in which cylindrical electrodes are used that have a contact area approximately equal to the size of the desired weld. As shown in Fig. 7-9, a chamber is provided for cooling water, which prevents overheating and prolongs the life of the electrode. The squeezing pressure must be properly adjusted to the thickness and strength of the parts being welded. A chart showing the spot welding possibilities of various combinations of metals and alloys is given in Fig. 7-10. The metal surfaces to be welded must be clean if sound welds are expected. It is especially necessary that the oxide coating on aluminum be removed by mechanical or chemical means before spot welding.

Seam welding is similar to spot welding except that copper alloy disks about 6–9 in. in diameter are used. The work is rolled between the electrodes, and evenly spaced spot welds are obtained by periodic application of the

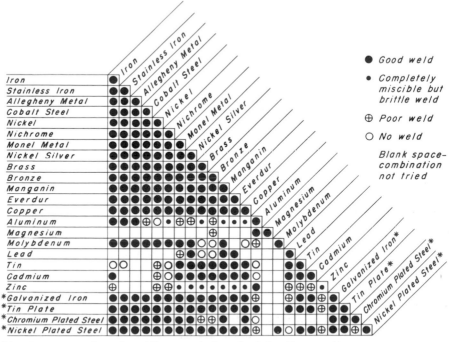

Figure 7-10 Spot welding chart for various combinations of metals.

current. It is possible to overlap the welds by close spacing and thus form a fluid-tight joint.

10. Other Kinds of Welding

Cold welding. Some metals, notably aluminum and copper, can be welded cold[3] if they are subjected to very high localized pressure. There must be sufficient lateral displacement to break down the normal oxide coating and to bring pure metals into contact with each other.

Atomic hydrogen welding. When molecular hydrogen is passed through the welding arc it is dissociated into atomic hydrogen. When the gas in this condition strikes the colder work it recombines into molecular hydrogen with a large evolution of heat, which is used in forming the weld.

Electron beam welding.[4] Here a beam of high-speed electrons moving at about one-half the velocity of light is focused on a small spot on the workpiece. The workpiece is quickly brought to the melting temperature and the parts joined together. The process must be conducted in a vacuum.

Ultrasonic welding. Here the workpieces are clamped together and given a high-frequency vibration in the direction of the interface. Local plastic deformation occurs; oxide and other surface films are broken and dispersed and nascent metallic contact is achieved.[5]

Friction welding. A cylindrical workpiece is attached to the end of a flywheel shaft and rotated at high speed and the power is shut off. The other part is stationary but is pressed firmly against the moving member. The friction raises the interface to the welding temperature as the flywheel coasts down. Upon cooling, the parts are welded together.[6]

Hard facing. This is the process of depositing by welding of an alloy material on metallic parts to build up a worn area or to form a protective surface to resist abrasion, impact, heat, or corrosion.

11. Soldering and Brazing

Many cast and wrought metals can be united by soldering and brazing. Soft solders are tin-lead alloys of low melting points; hard solders comprise the silver solders and brazing alloys of different compositions and melting

[3]See p. 21 of Reference 9, end of chapter.

[4]See Solomon, J. L., "Electron Beam Welding," *Mech. Eng.*, **87**, Jan. 1965, p. 28. See also Miller, K. J., and L. M. Reese, "Electron Beam Welding," *Machine Design*, **36**, Apr. 23, 1964, p. 218, and May 7, 1964, p. 165.

[5]See *Machine Design*. **36**, Apr. 9, 1964, p. 130.

[6]See Alm, G. V., "Grown Together Joints Grow Up," *Machine Design*, **40**, Jan. 4, 1968, p. 100. See also Kiwalle, J., "Designing for Inertia Welding," *Machine Design*, **40**, Nov. 7, 1968, p. 161.

points. In any soldering or brazing process, the parts are heated above the melting point of the solder but below that of the parts. The wetting action of the solder brings it into intimate contact with the surfaces to be joined. The solder, after cooling, serves as an adhesive to bind the two parts together, even though in some cases the solder forms an alloy with the metal of the part. In general, the strength of the joint is improved as the wetting action becomes more perfect.

The surfaces must be clean and covered with a flux, which is liquid at the soldering temperature. The flux dissolves any oxides present as well as preventing oxidation while the parts are being heated. The flux is drawn into the joint by capillary attraction. When the soldering temperature is reached, the solder is also drawn in by capillarity and displaces the flux. For this action, the surfaces must be closely spaced with respect to each other. The capillarity is greater for a tightly fitted joint, but higher temperatures are required to secure sufficient fluidity of the solder. In practice, clearances of 0.003 or 0.004 in. have been found satisfactory.

The strength of a soldered joint depends on many factors, such as the quality of the solder, thickness of the joint, smoothness of the surfaces, kind of materials soldered, soldering temperature, and duration of contact between solder and part at the soldering temperature. Optimum stress values in shear for soft-soldered joints can be taken as 6,000 psi for copper, 5,000 psi for mild steel, and 4,000 psi for brass. These are ultimate values.

The properties of soft-solder metals are given in Table 7-5. Melting of the alloy begins at the melting point shown, and is completed at the temperature of the flow point.

The silver solders have higher melting temperatures, and the soldering operation is carried on while the parts are at a red heat. Cast-iron, wrought-iron, and carbon steels can be joined to each other or to brass, copper, nickel, silver, monel, and other nonferrous alloys. Silver solders are malleable and ductile and are suitable for vibration and impact loads. They are also resistant to corrosion. The tensile strength of cast-silver solder varies from 40,000 to 60,000 psi. Joint preparation and clearances are the same as for soft solder. Heating can be done with a torch or with a high-frequency induction heating coil.

Brazing is a form of soldering at high temperatures that uses alloys of copper and zinc. The properties of several brazing alloys are given in Table 7-5. Tobin bronze is in wide use as a welding rod for general repair service and for building up defects in iron castings. The tensile strength is approximately 50,000 psi. Manganese bronze is similarly used, especially where the deposited metal is called upon to resist abrasion. Brazing alloys for aluminum are available with a melting point below that of the aluminum parts. In fusion welding, the high temperature permits impurities to be floated away by the flux. However, for brazing, the oxides must be removed and the joint cleaned by mechanical means.

Table 7-5 COMPOSITIONS AND USES FOR SOLDERS AND BRAZING ALLOYS

Soft Solders

Sn	Pb	Melting Point, F	Flow Point, F	Uses
60	40	361	372	High-grade solder. Has low flow-point temperature.
50	50	361	421	Widely used general-purpose solder.
40	60	361	453	For wiped joints. For automobile radiators and heating units.
30	70	361	486⎱	Low-grade solder. For filling dents and seams in
20	80	361	523⎰	automobile bodies.

Silver Solders

Ag	Cu	Zn	Melting Point, F	Flow Point, F	Uses
10	52	38	1,510	1,600	Solders of low silver give best results in stainless steel. Application of pressure aids in forming good joint.
20	45	35	1,430	1,500	Solder of minimum silver for good joints in brass.
50	15.5	16.5	1,160	1,175	Cd 18, general-purpose solder. Suitable for joining unlike metals.*
65	20	15	1,280	1,325	Solders of high silver are preferred for copper, especially for thin sections.

Miscellaneous Brazing Alloys

Alloy	Zn	Sn	Cu	Melting Point, F	Uses
Brazing alloy	rem		50–53	1,595–1,620	General-purpose brazing alloy.
Tobin bronze	38–43	0.50–1.50	rem	1,625	Strong oxyacetylene welds on steel, cast iron, copper, and nickel alloys.
Manganese bronze	38–43	0–1.5	rem	1,600	Mn 0.50–0.75. Hard and wear resistant.
Cu-Ni-Zn alloy	43		47	1,700	Ni 10. Strong, general-purpose alloy.

*Proprietary alloy, "Easy-Flo," Handy and Harman.

12. Furnace Brazing

Furnace or copper brazing[7] is a very satisfactory method for joining parts with pure copper. The component parts are assembled together, a

[7]See Webber, H. M., "Furnace Brazing of Machine Parts," *Mech. Eng.*, **72**, 1950, pp. 863 and 969. See also *Iron Age*, **142**, 1938, Sept. 8, 15, and 22; Nov. 3, 10 and 24; and Dec. 8, and 29; **143**, 1939, Feb. 2, Mar. 16, and Apr. 6.

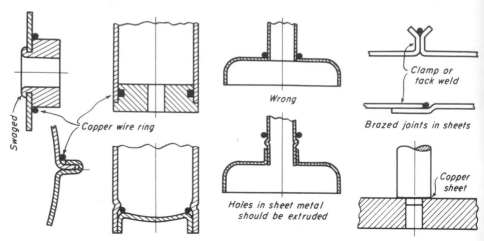

Figure 7-11 Typical joints fabricated by furnace welding.

ring of pure copper wire is placed adjacent to the joint as shown in Fig. 7-11, and the assembly is passed through a furnace. The copper melts at 1980°F and is drawn into the joint by capillary attraction. The strength of the assembly increases with the tightness of the fit. For steel, the clearance should run from about 0.001 in. loose to a light press fit.

Castings, forgings, and parts machined from solid stock can sometimes be redesigned and fabricated from stampings, screw machine parts, and pieces of tubing brazed together. Machining operations can be minimized and substantial economies effected. Parts must be held in their proper relationship with each other in their trip through the furnace. It is advantageous to have self-locking joints and thus avoid the use of holding fixtures. When the heating is done in an electric furnace with a reducing atmosphere, no flux is required, and the parts are delivered with smooth bright surfaces.

13. Riveted Joint with Central Load

Rivets are in wide use as fastenings for joints in buildings, bridges, boilers, tanks, ships, and miscellaneous frameworks. For centrally applied loads, it is customary to assume that all the rivets in a joint are equally stressed.[8] This assumption, however, is only approximately valid. As a simple example, consider the two bars in Fig. 7-12, which are held together by a single row of rivets. The material at B and C has large elongations, because it is carrying almost the entire load P. The material at A and D has small elongations because the bars at these points are carrying only small

[8]See Jenkins, E. S., "Rational Design of Fastenings," *Soc. Automotive Eng. J.* **52**, Sept. 1944, p. 421. See also *Proc. Soc. Exp. Stress Anal.*, **2**, No. 2, 1944, p. 28, and **7**, No. 2, 1949, p. 17.

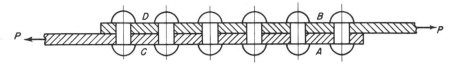

Figure 7-12 Bars carrying tensile forces joined by single row of rivets.

loads. Because of these unequal elongations, the end rivets carry excessive loads. The loads are smaller for the next rivets toward the center, until at the middle of the joint the rivets will be carrying less than the average load. For ductile materials, a more uniform distribution of the loading among the rivets occurs if the joint is first subjected to an overload, which causes the material in the highly stressed regions to yield in plastic flow.

In structural steel work, it is standard practice to use punched holes, which are $\frac{1}{16}$ in. larger than the diameter of the rivet. Although the rivet after driving may completely fill the hole, computations are made on the basis of the original rivet diameter. Since punching injures the metal around the circumference of the hole, work of higher quality is secured by subpunching the hole to a smaller diameter and then reaming it until it is large enough to admit the rivet. Boiler codes required the holes to be either drilled or subpunched and reamed. In boiler work, computations are made on the basis of the hole diameter, since the rivets must be tightly driven and must completely fill the holes.

Rivets must be spaced neither too close nor too far apart. The minimum rivet spacing, center to center, for structural steel work is usually taken as three rivet diameters. A somewhat closer spacing is often used in boilers. Rivets should not be spaced too far apart or buckling of the plates will take place. The maximum spacing is usually taken as 16 times the thickness of the outside plate. The edge distance, or distance from the rivet center to the edge of the plate, must not be less than a specified amount or there is danger of failure, as shown in Fig. 7-13(d) and (e).

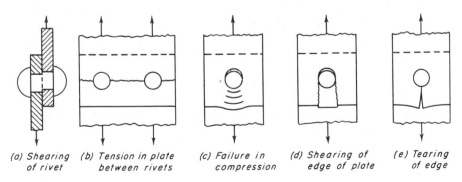

(a) Shearing   (b) Tension in plate   (c) Failure in   (d) Shearing of   (e) Tearing
   of rivet   between rivets   compression   edge of plate   of edge

Figure 7-13 Types of failure for riveted joints.

14. Stresses in Rivets

Power-driven hot rivets contract upon cooling and draw the plates tightly together so that the friction between the parts assists in the transfer of the load. Under certain conditions, the entire load might be carried in this way. The friction in a joint breaks down as the loading increases, and the joint can then fail in a number of different ways, as shown in Fig. 7-13. For fluctuating loads, it is especially desirable to prevent slip in the joint. The loading should therefore be conservative and the rivets tightly driven.

Despite the assistance given by the friction, it is customary to compute the strength of a riveted joint from the strength of the rivets in shear, the tension in the plates between the rivet holes, or the strength in compression or bearing for the rivets on the plates. Power-driven rivets are rivets driven by a riveting machine or by a hand-operated air hammer. When turned bolts are used, they must fit the holes with not over $\frac{1}{50}$-in. diametral clearance.

Figure 7-14 Eccentric load on rivet of single lap joint.

When rivets are loaded in single shear, as shown in Fig. 7-14, the stress situation is complicated by the effects of the eccentricity. When loaded in single shear, lower working-stress values must be used for shear and compression than when the rivet is in double shear. Proper provision must be made in riveted joints for the effects of fluctuating loads, impact, and extremes of temperature.

15. Stresses in a Cylindrical Shell

The stresses in the walls of a thin cylindrical shell are generally computed on the assumption that the stress is uniform throughout the wall thickness. Let the shell of Fig. 7-15 be loaded by the internal pressure p. If the length of the section perpendicular to the paper is taken as l, an element of area on the drum surface is equal to $rl\,d\theta$. The horizontal component of the force on the element is $prl\cos\theta\,d\theta$. When this force is integrated over a quarter-circle, the result is equal to the hoop force F. Thus,

$$F = \int_0^{\pi/2} prl\cos\theta\,d\theta = prl\sin\theta\Big|_0^{\pi/2} = prl \qquad (7)$$

The hoop or tangential stress s_1 is found by dividing force F by the area of the cross section tl, giving

$$s_1 = \frac{F}{tl} = \frac{pr}{t} \qquad (8)$$

In the longitudingal direction, the stress s_2 in the material must resist

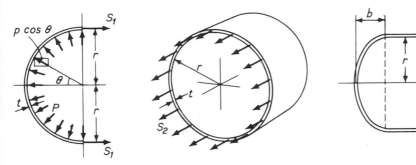

(a) Hoop stress in shell   (b) Longitudinal stress in shell   (c) Shell with eliptical
                                                                                    end closure

Figure 7-15 Stresses in thin cylindrical shell.

the force of the internal pressure acting on the end surface of the drum. Hence,

$$2\pi r t s_2 = \pi r^2 p$$

$$s_2 = \frac{pr}{2t} \tag{9}$$

The hoop stress is seen to be twice as great as the longitudinal stress. Equations (8) and (9) are valid only when thickness t is much less than radius r. From Eq. (7) the conclusion can be drawn that the bearing or compressive force on a rivet, Fig. 7-13(c), is equal to the stress in bearing times the projected area, or the diameter times the thickness of the plate.

Stresses at discontinuities in shells increase rapidly when the plate thickness is such that bending must be taken into account.[9] The theory becomes very complex.

A riveted boiler joint is weaker than the plate composing the shell because of the area lost in the rivet holes. Computations can be made for various assumed types of failure that will free the main plate of the joint in order to make a comparison with the strength of the unpunched plate and thus arrive at a value for the efficiency of the joint. The process is illustrated by the following example.

Example 6. Find the working force carried by an $8\frac{1}{4}$-in. width of the triple-riveted butt joint in Fig. 7-16. Make computations for the assumed types of failure shown in Table 7-6, and compute the efficiency of the joint. If this joint is used on a 60-in.-diam cylindrical tank, find the permissible internal pressure. Use the following stress values.

Tension: $s_t = 55,000$ psi ult for steel plate

[9]See Timoshenko, S., *Strength of Materials, Part II*, 2d ed., New York: Van Nostrand Reinhold Company, 1941, p. 164. See also Timoshenko, S., and S. Woinowsky-Krieger, *Theory of Plates and Shells*, 2d ed., New York: McGraw-Hill Book Company, 1959, p. 481.

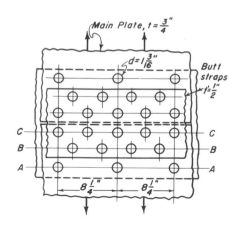

Figure 7-16 Triple riveted butt joint. Example 6.

Compression: $s_c = 95,000$ psi ult for steel plate

Shear: $s_s = 44,000$ psi ult for steel rivets

Factor of safety: $FS = 5$

Solution. (a) Tension in main plate at section *A-A*.

$$F = \frac{55,000}{5}(8.25 - 1.1875) \times 0.75 = 58,270 \text{ lb}$$

(b) Shear in rivets. There are nine surfaces in shear: four on *B-B*, four on *C-C*, and one on *A-A*. Area of rivet in shear:

$$A = \frac{\pi}{4} \times 1.1875^2 = 1.108 \text{ in.}^2$$

$$F = \frac{44,000}{5} \times 9 \times 1.108 = 87,710 \text{ lb}$$

(c) Bearing on rivets. On *B-B* and *C-C*, failure occurs in the main plate. On *A-A*, failure occurs in the butt plate.

Table 7-6 LOADS FOR JOINT OF FIGURE 7-16

No.	Type of Failure	Working Force, lb
(a)	Tension at *A-A*	58,270
(b)	Shear in rivets	87,710
(c)	Bearing on rivets	78,970
(d)	Tension at *B-B* Shear at *A-A*	58,220
(e)	Tension at *B-B* Bearing at *A-A*	59,750
(f)	Bearing at *B-B* and *C-C* Shear at *A-A*	77,430
(g)	Tension in gross plate	68,060
(h)	Efficiency	85.5%
(i)	Permissible pressure	235 psi

Area in bearing: $A = 4 \times 1.1875 \times 0.75 + 1.1875 \times 0.5 = 4.156 \text{ in.}^2$

$$F = \frac{95,000}{5} \times 4.156 = 78,970 \text{ lb}$$

(d) Tension at section *B-B*, and shear at section *A-A*.

Area in tension: $A = (8.25 - 2 \times 1.1875) \times 0.75 = 4.406 \text{ in.}^2$

$$F = \frac{55,000}{5} \times 4.406 + \frac{44,000}{5} \times 1.108$$

$$= 48,470 + 9,750 = 58,220 \text{ lb}$$

(e) Tension at section *B-B* and bearing at *A-A*.

$$F = 48{,}470 + \frac{95{,}000}{5} \times 1.1875 \times 0.5 = 59{,}750 \text{ lb}$$

(f) Bearing at sections *B-B* and *C-C*, and shear at section *A-A*.

$$F = \frac{95{,}000}{5} \times 4 \times 1.1875 \times 0.75 + 9{,}750 = 77{,}430 \text{ lb}$$

(g) Tension in unpunched plate.

$$F = \frac{55{,}000}{5} \times 8.25 \times 0.75 = 68{,}060 \text{ lb}$$

(h) Efficiency $= \dfrac{58{,}220}{68{,}060} \times 100 = 85.5\%$

(i) By Eq. (7): $p = \dfrac{F}{rl} = \dfrac{58{,}220}{30 \times 8.25}$

$$= 235 \text{ psi maximum permissible internal pressure}$$

16. Riveted Joint with Eccentric Load

When the load is applied eccentrically to a group of rivets forming a joint, the effect of the torque or moment must be taken into account, as well as the direct load. A typical example is shown in Fig. 7-17, where the

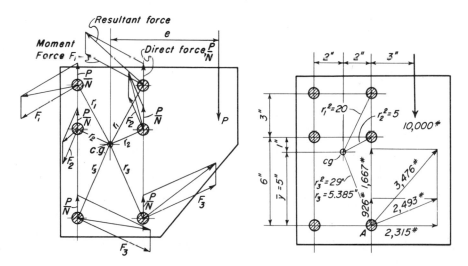

Figure 7-17 Riveted joint with eccentric load.   **Figure 7-18** Example 7.

joint of *N* rivets is subjected to a moment equal to *Pe*. Let it be assumed that the moment load on a rivet varies directly with the distance from the center of gravity *O* of the group of rivets, and is directed perpendicular to the radius to the center of gravity. Equations for moment forces F_1, F_2, and F_3 in Fig.

7-17 may then be written

$$F_1 = Cr_1; \qquad F_2 = Cr_2; \qquad F_3 = Cr_3 \tag{a}$$

where C is the constant of proportionality.

The externally applied moment is equal to the summation of the products of these forces and their arms to the center of gravity O. Hence,

$$\begin{aligned} T = Pe &= N_1 F_1 r_1 + N_2 F_2 r_2 + N_3 F_3 r_3 + \cdots \\ &= C(N_1 r_1^2 + N_2 r_2^2 + N_3 r_3^2 + \cdots) \end{aligned} \tag{10}$$

where N_1 is the number of rivets with radius r_1, N_2 is the number with radius r_2, and so on, until the entire joint has been taken care of. When the value of C has been determined from Eq. (10), the moment force for each rivet can be computed by multiplying by the appropriate r. It is customary to assume that the direct load P/N is the same for all rivets of the joint. The vectorial sum of moment force and direct force is the resultant force on the rivet.

Example 7. Find the value of the force carried by the most heavily loaded rivet for the joint of Fig. 7-18. Find the value of the shear stress for $\frac{3}{4}$-in. rivets and the value of the bearing stress if the plate is $\frac{5}{16}$ in. thick.

Solution. The plate is considered the free body.

Direct force on rivets:
$$F = \frac{10,000}{6} = 1,667 \text{ lb}$$

Take moments about the bottom row of rivets:
$$\hat{y} = \frac{2 \times 6 + 2 \times 9}{6} = 5 \text{ in,}$$

In Eq. (10):
$$10,000 \times 5 = C(2 \times 20 + 2 \times 5 + 2 \times 29)$$
$$C = 463$$

Rivet at A has the maximum force.

Moment force:
$$F_3 = 463 \times 5.385 = 2,493 \text{ lb}$$

This force has the horizontal and vertical components shown in Fig. 7-18.

Resultant force:
$$F_r = \sqrt{(926 + 1,667)^2 + 2,315^2} = 3,476 \text{ lb}$$

Area in shear:
$$A = \frac{\pi}{4} \times 0.75^2 = 0.4418 \text{ in.}^2$$

Shear stress in rivet:
$$s_s = \frac{3,476}{0.4418} = 7,870 \text{ psi}$$

Area in bearing:
$$A = \frac{5}{16} \times \frac{3}{4} = 0.2344 \text{ in.}^2$$

Bearing stress on rivet:
$$s = \frac{3,476}{0.2344} = 14,830 \text{ psi}$$

Example 8. Find the value of P for the two joints shown in Fig. 7-19 based on a working stress of 15,000 psi in shear.

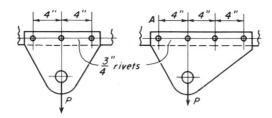

Figure 7-19 Example 8.

Solution. area in shear $= \dfrac{\pi}{4}\left(\dfrac{3}{4}\right)^2 = 0.4418$ in.^{2},   each rivet.

(a) $P = 3 \times 0.4418 \times 15,000 = 19,880$ lb

(b) Moment arm of the load is 2 in.

In Eq. (10): $2P = C(2 \times 2^2 + 2 \times 6^2)$, $C = 0.025P$

On the rivet at A: Moment force $= 6 \times 0.025P = 0.15P$

$$\text{Direct force} = \frac{P}{4} = 0.25P$$

$$\text{Total force} = 0.40P$$

$$\text{Stress} = \frac{0.4P}{0.4418} = 15,000$$

$$P = 16,570 \text{ lb}$$

Thus, although the joint (b) has more rivets and is more expensive to make, the carrying capacity is less than the symmetrical joint of (a).

17. Adhesives

A design technique of rapidly growing importance is metal-to-metal adhesive bonding. High-strength phenolic elastomers, vinyl-phenolics, and epoxy-resin adhesives are being used to an increasing extent for the structural joining of metals. Organic materials can be bonded as well. Adhesive bonding is in wide use in the automotive and aircraft industries.

The important matter of selecting the right adhesive for each type of service is not a simple procedure. There is a great variety of adhesive products, and there is little standardization of their properties. Adhesives seldom display their best properties when substituted directly for other fastenings. In general it is necessary to design the joint specifically for the use of adhesives. It is possible under certain conditions to form a successful joint by the use of adhesives that would not be possible by mechanical means.

The adhesive joint has the advantage of a larger contact area than other types of joints. Thick or thin materials of any shape can be joined. Any combination of similar or dissimilar materials can be employed. The adhesive layer minimizes or prevents electrochemical corrosion between dissimilar metals. Such joints in general show good resistance to fatigue or cyclic loading and aid in dampening vibration. The adhesive can sometimes be arranged to provide sealing for fluids and gases. The warping and residual stress that accompany welding are avoided.

The surfaces to be joined must be carefully cleaned and sometimes given special treatments. Heat and pressure may be required for curing and sometimes the curing time is a disadvantage. Some adhesives will harden only in the absence of air. Jigs and fixtures may also be required, and rigid process control may be necessary. Some adhesives may be subject to attack by bacteria, moisture, or certain solvents. The upper service temperature is limited to about 350°F. Some adhesives lose strength faster than others with increase of temperature. As a consequence of the differential strain between the parts, a severe stress concentration occurs at the ends of the overlap in an ordinary lap joint.

In mechanical bonding the adhesive infiltrates porous materials to give cementing of a mechanical nature. In chemical bonding the adhesive reacts with the parts or joins them largely through intermolecular attraction. For creep resistance the adhesive material should be rigid, but when the joint undergoes flexure in service, a more flexible material should be used.

Adhesive bonding is a rapidly growing field with excellent possibilities for future development. Simpler, cheaper, stronger, and more easy-to-use adhesives can be expected in the future.[10]

REFERENCES

1. Fish, G. D., *Arc Welded Steel Frame Structures*, New York: McGraw-Hill Book Company, 1941.

2. Koopman, K. H., "Elements of Joint Design for Welding," *Welding J.*, **37**, 1958, p. 579.

3. *Metals Handbook*, Novelty, Ohio: American Society for Metals, 1961, p. 337.

[10]See Goland, M., and E. Reissner, "The Stresses in Cemented Joints," *Trans. ASME*, **66**, 1944, p. A-17. See also the following articles.

Sharpe, L. H., "Assembling with Adhesives," *Machine Design*, **38**, Aug. 18, 1966, p. 178.

Scott, R. C., "Guide to Joint Sealants," *Prod. Eng.*, **35**, Mar. 16, 1964, p. 73.

Eickner, H. W., "Basic Shear Strength Properties of Metal-Bonding Adhesives . . . ," *Bull. 1850*, Madison, Wis.: U.S. Dept. of Agriculture, Forest Products Laboratory, Aug. 1955.

Kuenzi, E. W., "Determination of Mechanical Properties of Adhesives for Use in the Design of Bonded Joints." *Bull*; *1851*, Madison, Wis,: U. S. Dept. of Agriculture, Forest Products Laboratory, Jan, 1956.

4. *Procedure Handbook of Arc Welding—Design and Practice*, Cleveland: The Lincoln Electric Co., 1957.

5. Rossi, B. E., *Welding and Its Applications*, New York: McGraw-Hill Book Company, 1941.

6. Rothbart, H. A., Ed., *Mechanical Design and Systems Handbook*, New York: McGraw-Hill Book Company, Section 25.

7. Skeist, I., Ed., *Handbook of Adhesives*, New York: Van Nostrand Reinhold Company, 1962.

8. *Symposium on the Welding of Iron and Steel*, Vols. **1** and **2**, London: The Iron and Steel Institute, 1935.

9. Udin, H., E. R. Funk, and J. Wulff, *Welding for Engineers*, New York: John Wiley & Sons, Inc., 1954.

10. *Welding Handbook*, 4th ed., New York: American Welding Society, 1960.

PROBLEMS

1. Find the value of the maximum stress in the weld of Fig. 7-20. Assume the direct stress to be uniformly distributed over the throat area.

Ans. $s_s = 11,500$ psi.

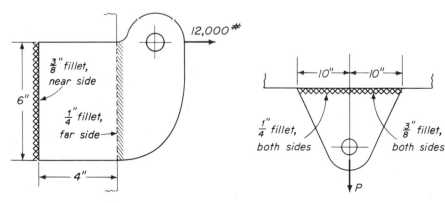

Figure 7-20 Problem 1.   Figure 7-21 Problem 2.

2. Find the permissible static load P in Fig. 7-21 if an E6010 electrode is used and the factor of safety by the maximum shear theory of failure is to be equal to 2.

Ans. $P = 82,400$ lb.

3. Work Problem 2 for the joint of Fig. 7-22.   *Ans.* $P = 29{,}600$ lb.

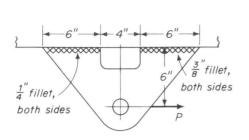

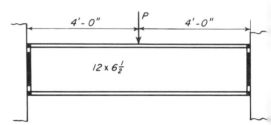

Figure 7-22 Problem 3.   Figure 7-23 Problem 4.

4. Work Problem 2 for the joint in Fig. 7-23.   *Ans.* $P = 49{,}700$ lb.

5. The ends of the channel of Fig. 7-24 lie on immovable supports. If all welds are of the same size, find the length of leg h of the fillet welds required to carry the given load. Use E6010 electrode. Factor of safety is equal to 3. The beam is to be fixed at the ends by the welds.   *Ans.* $h = 0.38$ in.

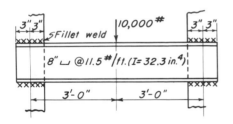

Figure 7-24 Problem 5.   Figure 7-25 Problem 6.

6. The I-beam in Fig. 7-25 is 11.87 in. deep with flange $6\frac{1}{2}$ in. wide and of a uniform thickness of 0.355 in. The beam is welded to immovable supports by butt welds. Assume the bending moment is carried by the flange welds, and the shear forces are carried by the web welds. Flange welds are of the same size as the flange of the beam. Find the value of force P for a bending stress of 12,000 psi in the flange welds.   *Ans.* $P = 25{,}860$ lb.

7. Find the required length of weld l in Fig. 7-26 if an E6010 electrode is used with an *FS* of 2.   *Ans.* $l = 10$ in.

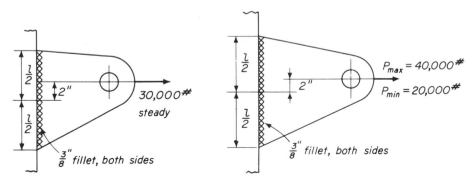

Figure 7-26 Problem 7.   **Figure 7-27** Problem 8.

8. Find the required length of weld l in Fig. 7-27 if an E6010 electrode is used with an *FS* of 2. Let the endurance limit for the weld metal be equivalent to that of an as-forged surface. Stress concentration factor is equal to 1.5.   *Ans.* $l = 18.9$ in.

9. Find the value of P_{max} for the joint of Fig. 7-28 if P_{min} is equal to zero. Other data is the same as for Problem 8.   *Ans.* $P_{max} = 7,460$ lb.

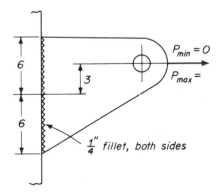

Figure 7-28 Problem 9.

10. Find the factor of safety for the most highly stressed point of the welds of Fig. 7-29. The weld metal has a yield strength of 50,000 psi.   *Ans.   FS = 2.95.*

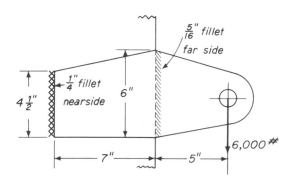

Figure 7-29 Problem 10.

11. Find the value of the maximum stress in the joint of Fig. 7-30. The weld extends across the top and along the vertical edge.   *Ans.   s = 10,680 psi.*

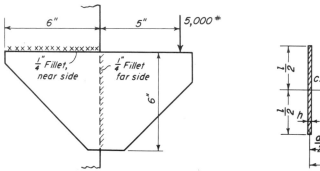

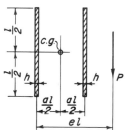

Figure 7-30 Problem 11.   **Figure 7-31** Problem 12.

12. Derive an equation for the value of the maximum stress in the welds of Fig. 7-31. Let A be the area of one weld.

$$Ans.   s = \frac{P}{2A}\sqrt{K^2 + (1 + aK)^2}, \quad \text{where} \quad K = \frac{3(2e - a)}{1 + 3a^2}$$

13. Same as Problem 12 but for the joint of Fig. 7-32.

$$Ans.   s = \frac{P}{2A}\sqrt{a^2K_1^2 + (1 + K_1)^2}, \quad \text{where} \quad K_1 = \frac{3(2e - 1)}{1 + 3a^2}$$

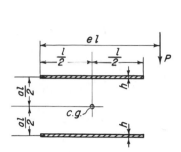

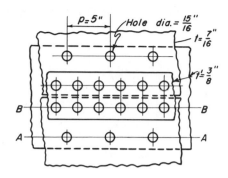

Figure 7-32 Problem 13.

Figure 7-33 Problem 14.

14. Compute and tabulate the working forces for a pitch length of 5 in. for the double-riveted butt joint of Fig. 7-33 for the following assumed methods of failure. Also find the efficiency of the joint. (a) Tension at section *A-A*. (b) Shear in the rivets. (c) Bearing on the rivets. (d) Tension at *B-B* and shear at *A-A*. (e) Tension at section *B-B* and bearing at *A-A*. (f) Tension in gross plate. Ultimate strength of steel plate in tension is 55,000 psi, and in compression, 95,000 psi. Ultimate shear strength for rivets is 44,000 psi. Use a factor of safety of 5.

Ans. Tension at *A-A*, 19,550 lb; eff = 81.3 per cent.

15. Make computations similar to those of Example 6, but for the triple-riveted butt joint in Fig. 7-34. Tank diam = 60 in. Use stresses from Example 6.

Ans. Tension at *B-B* and shear at *A-A*, 70,860 lb; eff = 84.1%.

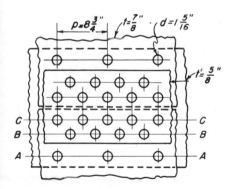

Figure 7-34 Problem 15.

Figure 7-35 Problem 16.

16. Find the resultant force on each rivet of the joint in Fig. 7-35.

17. Repeat Problem 16 for the joint of Fig. 7-36.

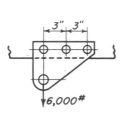

Figure 7-36 Problem 17.

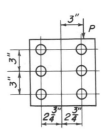

Figure 7-37 Problem 18.

18. Find the permissible load P for the riveted joint in Fig. 7-37 if the resultant shearing stress for the most highly stressed rivet is 15,000 psi. Rivets are 1 in. in diameter.   *Ans.* $P = 40,600$ lb.

19. Find the shearing stress in the most heavily loaded rivet in the group shown in Fig. 7-38. Rivets are $\frac{3}{4}$ in. in diameter.   *Ans.* $s_s = 12,680$ psi.

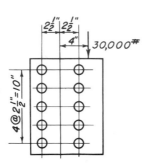

Figure 7-38 Problem 19.

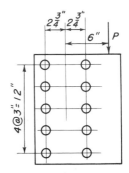

Figure 7-39 Problem 20.

20. Find the permissible load P for the joint in Fig. 7-39 if the maximum value of the shearing stress is 15,000 psi. Rivets are $\frac{3}{4}$ in. in diameter.

Ans $P = 30,600$ lb.

21. Find the distance b in Fig. 7-40 if the maximum shearing stress on the most heavily loaded rivet is to be 13,500 psi.   *Ans.* $b = 5.04$ in.

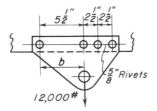

Figure 7-40 Problem 21.

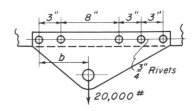

Figure 7-41 Problem 22.

22. Repeat Problem 21 for Fig. 7-41.   *Ans.* $b = 6.71$ in.

23. Find the value of the force on the most heavily loaded rivet in Fig. 7-42.

Ans.   1,270 lb.

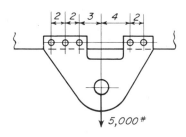

Figure 7-42 Problem 23.

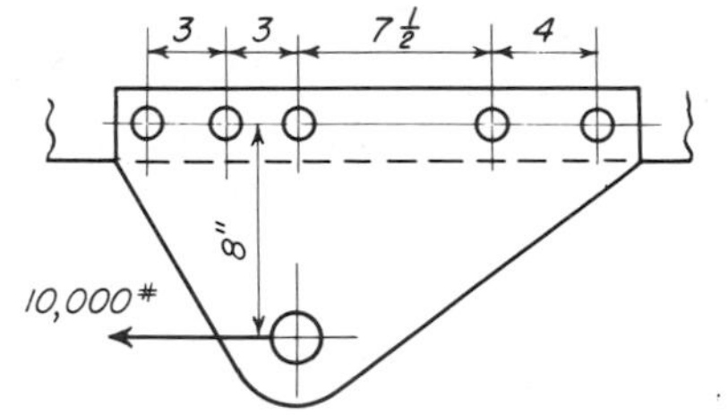

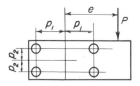

Figure 7-43 Problem 24.

24. Find the value of the force on the most heavily loaded rivet in Fig. 7-43.

Ans.   4,080 lb.

25. Find the value of the force on the most heavily loaded rivet in Fig. 7-44.

Ans.   4,300 lb.

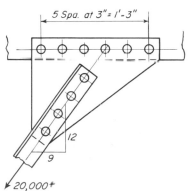

Figure 7-44 Problem 25.

Figure 7-45 Problem 26.

26. Derive an equation that will give the value of the maximum force on the right-hand rivets in Fig. 7-45.

Ans. $F = \dfrac{P}{4}\sqrt{\dfrac{(e + p_1)^2 + p_2^2}{p_1^2 + p_2^2}}.$

27. The lap joint of Fig. 7-46 is made of carbon steel plates and rivets. Find the efficiency of the joint based on shear in the rivets on *A-A* and tension in the plate on *B-B*. Use stress values from Example 6.

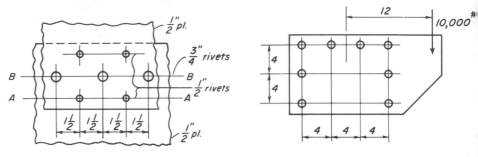

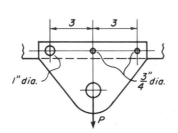

Figure 7-46 Problem 27.   **Figure 7-47** Problem 28.

28. Find the resultant force on the most heavily loaded rivet in Fig. 7-47.

29. Find the value of *P* in Fig. 7-48 if the shearing stress on the most heavily loaded rivet is 12,000 psi.

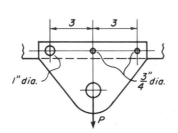

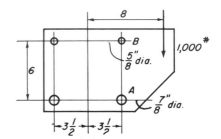

Figure 7-48 Problem 29.   **Figure 7-49** Problem 30.

30. Compute the value of the shearing stress at the center of rivets *A* and *B* in Fig. 7-49.

8 Lubrication

A lubricant is used to reduce the friction of bearings and sliding surfaces in machines and thus diminish the wear, heat, and possibility of seizure of the parts. Although a layer of oil will eliminate the excessive friction of metal-to-metal contact, the friction within the oil film must be taken into account. The study of lubrication and the design of bearings is therefore concerned mainly with phenomena related to the oil film between the moving parts. The literature on lubrication is very extensive, and a mathematical treatment of the subject is beyond the scope of this book. Fortunately it is possible to make design calculations from the graphs obtained by mathematical analysis.

A, area
A_c, cooling area of bearing housing
c, radial clearance of journal bearing
d, diameter of journal bearing
f, coefficient of friction
fpm, feet per minute
F, total friction force

F_1, tangential friction force per axial inch
h, film thickness
h_0, minimum film thickness
hp, horsepower
l, length in axial direction
n, revolutions per minute
p, oil pressure in film, psi
P, load per unit of projected area

psi, pounds per square inch	W, total load, pounds
r, radius of journal	W_1, load per axial inch
s_s, shearing stress, psi	Z, viscosity, centipoises
SUS, viscosity, Saybolt Universal Seconds	$sp.gr_{60}$, specific gravity at $60°F$
t, temperature, Fahrenheit	$sp.gr_t$, specific gravity at $t°F$
ΔT, rise in temperature of bearing housing	ϵ, (epsilon) eccentricity ratio
U, tangential velocity, inches per second	μ, (mu) viscosity, pounds seconds per square inch
VI, viscosity index	ρ, (rho) density, mass per unit volume

1. Viscosity and Newton's Law

The plate in Fig. 8-1 is resting on top of an oil film of thickness h, and is being moved with velocity U under the action of the force F. When the

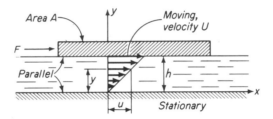

Figure 8-1 Flat plate moving on oil film.

plate moves, it does not slide along on top of the film. The oil adheres tightly to the plate, and motion is accompanied by slip or shear between the oil particles throughout the entire height of the film. Thus, if the plate and contacting layer of oil move with velocity U, the velocity at intermediate heights is directly proportional to the distance from the fixed or bottom plate, as shown in Fig. 8-1.

According to Newton, the shearing stress in the oil film varies directly with the velocity U and inversely with the film thickness h. Hence

$$s_s = \frac{F}{A} = \mu \frac{U}{h} \tag{1}$$

The factor of proportionality μ is called the coefficient of viscosity, or simply the viscosity. This equation assumes that the area A is very large as compared to h, in order that disturbances around the edges may be neglected.

Viscosity is the measure of the ability of the lubricant to resist shearing stress. It is a molecular phenomenon, and the work done by force F is turned into heat, which raises the temperature of the oil and the surrounding parts.

2. Measurement of Viscosity

The unit of viscosity in the metric system is called the *poise*. Equation (1) indicates that its dimensions are dyne sec/cm². Thus, in Fig. 8-1, if a force F of 1 dyne is required to maintain a plate of 1-cm² area and 1-cm film thickness at a velocity of 1 cm/sec, the oil would have a viscosity of 1 poise. The viscosity of most lubricating oils is less than 1 poise, and the centipoise, or $\frac{1}{100}$ poise, represented by Z, is used in lubrication calculations. The viscosity of water at 68.4°F is equal to 1 centipoise. It can be easily shown[1] that the following equations transfer viscosity from dyne sec/cm² to the English system of lb sec/in.²

$$\text{viscosity } \mu, \qquad \frac{\text{lb sec}}{\text{in.}^2} = \frac{\text{viscosity, } Z \text{ centiposes}}{6,895,000} \qquad (2)$$

$$= \text{viscosity, } Z \text{ centiposes} \times 0.000000145 \qquad (3)$$

Another unit in use is the kinematic viscosity, ν (nu). It is called, in the metric system, the stoke, and is found by dividing the viscosity in poises by ρ, the density in mass per cubic centimeter. Since the dimensions of ρ are dyne sec²/cm⁴, the dimensions of ν are cm²/sec. A centistoke is equal to 0.01 stoke.

Because of experimental difficulties, these absolute units are not used commercially in designating the viscosity of lubricating oils. Terms are used that pertain to the type of viscosimeter used in making the test. One method in wide use is to specify viscosity by Saybolt Universal Seconds, SUS, Fig. 8-2, at a given temperature. The Saybolt viscosity is the time in seconds required for 60 cm³ of the oil to flow through a standardized capillary tube. The change to absolute units can be made with sufficient accuracy for most purposes by means of the empirical equation

$$Z = sp.gr_t\left(0.22\text{SUS} - \frac{180}{\text{SUS}}\right) \text{ centipoises} \qquad (4)$$

where Z is viscosity in centipoises at temperature t°F, $sp.gr_t$ is the specific gravity of the oil at t°F, and SUS is the viscosity, Saybolt Universal Seconds, also at t°F.

The specific gravity of the oil changes with the

[1] Dyne sec/cm² can be obtained from lb sec/in.² by replacing pounds by 980.7 × 1,000/2.205 dynes and inches by 2.54 cm.

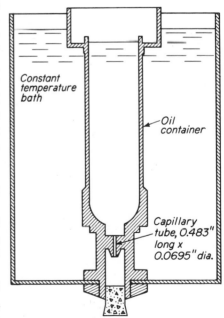

Constant temperature bath

Oil container

Capillary tube, 0.483" long x 0.0695" dia.

Figure 8-2 Saybolt Universal viscosimeter.

temperature, becoming smaller as the temperature rises. This property can be determined by means of the empirical equation

$$sp.gr_t = sp.gr_{60} - 0.00035(t - 60) \tag{5}$$

where $sp.gr_{60}$ is the specific gravity at 60°F, and t is the temperature in °F.

Viscosity is only one of the many properties of a satisfactory lubricating oil. It is, however, the only variable directly related to the lubricant that appears in the design equations. The SAE classification is widely used for specifying the viscosity of lubricating oils (see Table 8-1).[2] From this table, it can be

Table 8-1   VISCOSITY RANGE FOR OILS, SAYBOLT UNIVERSAL SECONDS, SUS

SAE Viscosity No.	at 0°F		at 210°F	
	Minimum	Maximum	Minimum	Maximum
5W	—	6,000	—	—
10W	6,000*	less than 12,000	—	—
20W	12,000†	48,000	—	—
20	—	—	45	less than 58
30	—	—	58	less than 70
40	—	—	70	less than 85
50	—	—	85	110

*Minimum viscosity at 0°F can be waived, provided viscosity at 210°F is not below 40 SUS.

†Minimum viscosity at 0°F can be waived provided viscosity at 210°F is not below 45 SUS.

seen that the viscosity for any given designation can vary over a considerable range. To make calculations, precise information is needed concerning the viscosity of the oil that is to be used. Fair average values[3] for the viscosities of SAE oils are shown by the curves[4] of Fig. 8-3 plotted on ASTM Chart D341-39. It is unique that when the viscosities for two temperatures, usually 100°F and 210°F, are plotted on this chart, a straight line between them will give the viscosity at other desired temperatures. Since the designer requires the viscosity in absolute units, the viscosities of the oils of Fig. 8-3 are shown in lb sec/in.² units in Fig. 8-4. These curves are easily drawn by reading the SUS values from Fig. 8-3 and then by transforming to the absolute units of Fig. 8-4 by application of Eqs. (5), (4), and (3).

[2]See *SAE Handbook*.

[3]These are from the booklet *Physical Properties of Lubricants*, published by the American Society of Lubrication Engineers.

[4]American Society for Testing and Materials, Philadelphia, Pa.

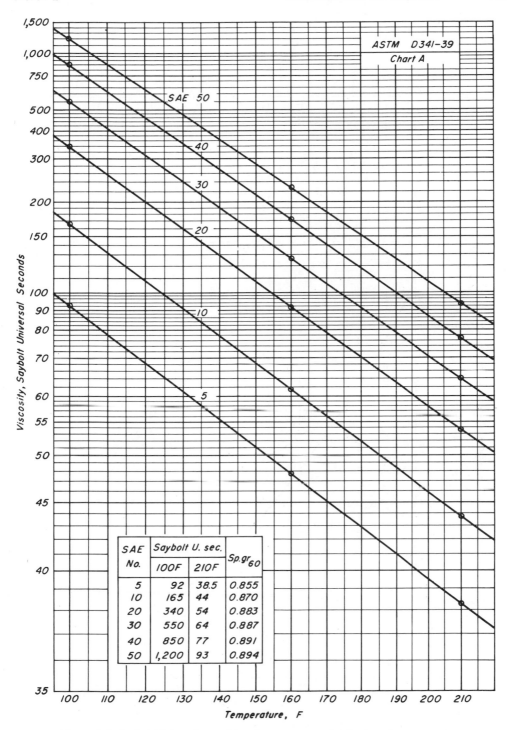

SAE No.	Saybolt U. sec.		Sp.gr 60
	100F	210F	
5	92	38.5	0.855
10	165	44	0.870
20	340	54	0.883
30	550	64	0.887
40	850	77	0.891
50	1,200	93	0.894

Figure 8-3 Saybolt Universal viscosities of some common lubricating oils.

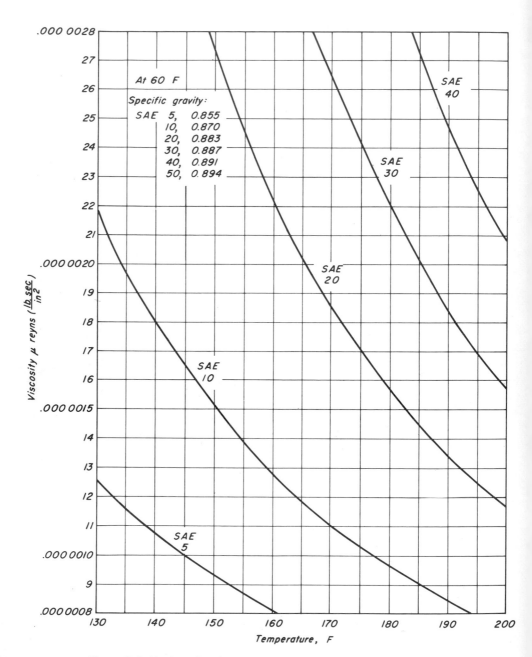

At 60 F

Specific gravity:

SAE 5,   0.855
      10,   0.870
      20,   0.883
      30,   0.887
      40,   0.891
      50,   0.894

SAE 40

SAE 30

SAE 20

SAE 10

SAE 5

Viscosity μ reyns $\left(\dfrac{lb\ sec}{in^2}\right)$

Temperature, F

Figure 8-4 Absolute viscosity vs. temperature for the SAE oils of Fig. 8-3.

Example 1. An oil has a viscosity of 400 SUS at 100°F and 55 SUS at 210°F. Specific gravity at 60°F is equal to 0.93. Find the viscosity in absolute units at 170°F.

Solution. When the given viscosities are plotted on Fig. 8-3, the Saybolt viscosity at 170°F is found to be 84 sec.

The specific gravity at 170°F is found by substituting as follows.

In Eq. (5): $sp.gr_t = 0.93 - 0.00035(170 - 60) = 0.892$

In Eq. (4): $Z = 0.892\left(0.22 \times 84 - \dfrac{180}{84}\right) = 14.57$ centipoises

In Eq. (3): $\mu = 14.57 \times 0.000000145 = 0.00000211$ lb sec/in.^{2}

Example 2. An oil has an SUS value of 50.5 at 210°F. Find the kinematic viscosity for this temperature.

Solution. In Eq. (4): $Z = sp.gr_t\left(0.22 \times 50.5 - \dfrac{180}{50.5}\right) = 7.55 sp.gr_t$ centipoises

In the metric system, the density ρ, grams per cubic centimeters, has the same numerical value as the specific gravity $sp.gr_t$. Hence,

$$v = \frac{Z}{\rho} = \frac{7.55 sp.gr_t}{sp.gr_t} = 7.55 \text{ centistokes}$$

3. Viscosity Index

A widely used method for specifying the rate of change of viscosity with temperature is known as the *viscosity index*.[5] The method was originally devised by testing samples of oils made from two extreme types of crudes. The series of oils having a small change of viscosity with temperature were given a viscosity index of 100, and the oils with a large change of viscosity with temperature were given a viscosity index of 0.

To find the viscosity index, *VI*, of a sample of oil, its SUS values are first determined for 210°F and 100°F and plotted on the ASTM chart as *x* and *y*, respectively, as shown in Fig. 8-5. Among the $VI = 100$ oils, there will be one whose SUS value at 210°F will be *x*, the same value as that of the sample. Let the SUS at 100°F for this oil be designated by *H*, as shown in the figure. Similarly, the viscosity at 100°F for the *VI* = 0 oil, which has the same SUS at 210°F as the sample, is represented by *L*.

By definition, the viscosity index is found by means of the following equation:

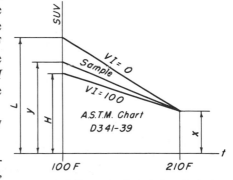

Figure 8-5 Determination of viscosity index.

[5]See Dean, E. W., and G. H. B. Davis, "Viscosity Variation of Oils with Temperature," *Chem. Met. Eng.*, **36**, 1929, p. 618. See also *Ind. Eng. Chem.*, **32**, 1940, p. 102.

$$VI = \frac{L - y}{L - H} \times 100 \text{ per cent} \tag{6}$$

The viscosity index thus designates the rate of change of viscosity of an oil as compared to oils with very small and very large rates of change of viscosity. It should be noted that the VI tells nothing about the value of the viscosity at any given temperature. Values of VI below 0 and above 100 are recognized.

When the SUS value at 210°F for the sample lies between 50 and 350, the values of H and L can be determined from the following equations.

$$H = 0.0408x^2 + 12.568x - 475.4 \tag{7}$$

$$L = 0.2160x^2 + 12.070x - 721.2 \tag{8}$$

Example 3. Find the viscosity index for SAE 20 oil in Fig. 8-3.

Solution. $x = 54$ and $y = 340$.

By Eq. (7): $H = 0.0408 \times 54^2 + 12.568 \times 54 - 475.4 = 322.2$

By Eq. (8): $L = 0.216 \times 54^2 + 12.07 \times 54 - 721.2 = 560.4$

By Eq. (6): $VI = \dfrac{L - y}{L - H} \times 100 = \dfrac{560.4 - 340}{560.4 - 322.2} \times 100 = 92.5 \text{ per cent}$

4. The Zn/P Curve

The phenomenon that takes place when a shaft is rotating in a bearing in the presence of a lubricant can be represented by the experimentally determined curve shown in Fig. 8-6. Here the abscissa is taken as the nondimensional group Zn/P, where Z is the viscosity in centipoises, n is the speed in revolutions per minute, and P is the load in pounds per square inch of the

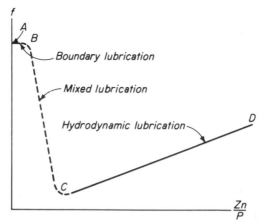

Figure 8-6 Plot of operating characteristic vs. coefficient of friction for journal bearing.

projected journal area. The ordinate is taken as the coefficient of friction f, where f is the ratio of the tangential friction force to the load carried by the bearing.

Nondimensional coordinates such as these are very convenient for engineering problems where many variables are present. Experiments can be made by changing the value of the nondimensional group without the need for changing each individual variable, and much labor can be saved. The curve can be divided into three more or less distinct parts.

(1) Operation in the AB portion occurs for very heavy loads and slow speeds, and generally with an insufficient supply of oil. The film thickness is smaller than the combined heights of the asperities on the opposing surfaces. Interlocking of the asperities may occur, and gross seizure is prevented by the coating of lubricant on the surfaces. This is called boundary lubrication.

(2) In the BC region the separation between the parts is greater and the load is carried partly on the asperities and partly on localized patches of lubricant, which form a load-supporting film. This is called mixed lubrication.

(3) From C to D the viscosity Z and the speed n are relatively large and the unit load P is relatively small so that the film is relatively thick if a copious supply of lubricant is present. The friction depends on the viscosity of the lubricant and is independent of the smoothness of the surfaces and the kinds of materials used for the bearing. It is only in this region that the hydro-dynamic theories, as discussed in the following sections, are valid. To avoid danger of seizure, the operating value of Zn/P should be at least 5 or 6 times that of the minimum point of the curve. It can be noted from the CD portion of the curve that the coefficient of friction for lubricated surfaces increases with increase of velocity. This is in contrast with dry surfaces where the coefficient of friction is relatively independent of the velocity.

Example 4. Let the curve for Zn/P and f for a 3 in. diam $\times$ 6 in. long bearing be a straight line passing through points $Zn/P = 75$, $f = 0.002$, and $Zn/P = 375$, $f = 0.0065$. SAE 10 oil like that in Fig. 8-3 is used. The total load on the bearing is 1,350 lb, and the rpm $= 1,200$. What is the horsepower friction loss for this bearing if the oil film has a temperature of 170°F?

Solution. In Eq. (5): $sp.gr_t = 0.870 - 0.00035(170 - 60)$

$$= 0.832 \text{ at } 170°F$$

By Fig. 8-3, at 170°F: SUS $= 56.3$ sec

In Eq. (4): $$Z = 0.832\left(0.22 \times 56.3 - \frac{180}{56.3}\right)$$

$$= 7.64 \text{ centipoises at } 170°F$$

$$P = \frac{1,350}{18} = 75 \text{ psi of projected journal area}$$

$$\frac{Zn}{P} = \frac{7.64 \times 1,200}{75} = 122$$

Draw the given Zn/P and f curve, and read $f = 0.0027$ for $Zn/P = 122$.

$$F = fW = 0.0027 \times 1,350$$
$$= 3.645 \text{ lb tangential friction force}$$
$$\text{friction hp} = \frac{Tn}{63,000} = \frac{3.645 \times 1.5 \times 1,200}{63,000} = 0.104$$

5. Petroff's Bearing Equation

Equation (1) for the flat plate can be easily adapted to the cylindrical or journal bearing, provided that the speed and viscosity are high and that the load is very light so that the journal is in a central position in the bearing. Thus the plate in Fig. 8-1 is assumed to be wrapped into the cylindrical shaft in Fig. 8-7. If $2r$ or d is the diameter of the journal, and l its length in the axial direction, the developed journal area A is $2\pi rl$.

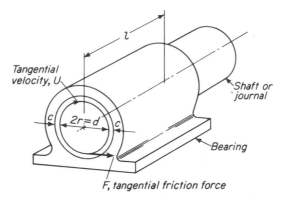

Figure 8-7 Very lightly loaded journal centered in bearing.

The thickness h becomes the radial clearance c, or the difference between the radius of the bearing, and the radius of the shaft. Substitution for A and h in Eq. (1) gives $F = 2\pi\mu Url/c$. However, if F_1 is taken as the tangential friction force per inch of axial length, then $F = F_1 l$, and the foregoing equation becomes

$$\frac{F_1}{\mu U}\left(\frac{c}{r}\right) = 2\pi \tag{9}$$

This is known as Petroff's equation. It is valid only for the hypothetical case of zero load and centrally located journal.

The tangential velocity of the journal is given by

$$U = \frac{\pi dn}{60}\text{in./sec} \tag{10}$$

The friction horsepower is given by

$$\text{hp} = \frac{F_1 l U}{12 \times 550} \tag{11}$$

Example 5. Find the friction horsepower of a very lightly loaded 360° journal bearing. The bearing is 3 in. in diameter and 5 in. long. Radial clearance is 0.0025 in. Use SAE 20 oil of Fig. 8-3. Rpm = 1,800. Oil film temperature is 150°F.

Solution. From Fig. 8-4, at 150°F:

$$\mu = 0.00000273 \frac{\text{lb sec}}{\text{in.}^2}$$

In Eq. (10): $U = \dfrac{3\pi \times 1{,}800}{60} = 282.7 \dfrac{\text{in.}}{\text{sec}}$

In Eq. (9): $F_1 = \dfrac{2\pi \times 0.00000273 \times 282.7 \times 1.5}{0.0025} = 2.91 \dfrac{\text{lb}}{\text{in.}}$

In Eq. (11):   friction hp $= \dfrac{2.91 \times 5 \times 282.7}{12 \times 550} = 0.62$

6. Load Carrying Journal Bearing

The flat plate in Fig. 8-1 is incapable of supporting a vertical load because no pressure exists in the oil film. If a vertical load were placed on the plate, the oil would squeeze out around the edges, and the plate would approach the surface beneath, resulting in metal-to-metal contact. If, however, the plates are tipped at a small angle to each other, as shown in Fig. 8-8, the oil

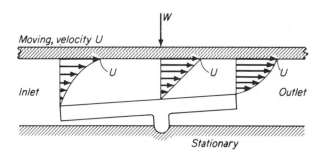

Figure 8-8 Load on moving plate supported on wedge-shaped film.

will be drawn into the wedge-shaped opening, and sufficient pressure will be generated in the film to support the load *W*. This fact is utilized in the thrust bearings for hydraulic turbines and propeller shafts of ships, as well as in the conventional journal bearing with flooded-oil lubrication.

Since the same quantity of oil is drawn past each cross section, the velocity distribution throughout the height of the film is no longer linear as it is for parallel plates. The velocity curve will be concave at the wider cross sections and convex at the narrower, as indicated in Fig. 8-8.

If the shaft in Fig. 8-7 carried a downward load, the journal would no longer remain central with the bearing. For the classical case of the infinitely long bearing, the shift of the journal center is not downward, but is theoretically to one side as shown in Fig. 8-9. The shift of the journal center is usually expressed as a proportion of the radial clearance, c. Thus in Fig. 8-9 the shaft has shifted ϵc inches, where the eccentricity ratio ϵ is a pure number less than unity. A copious supply of oil is assumed, which fills the entire clearance space. The oil clings to the journal and is drawn around into the region of decreasing film thickness, and thereby builds up sufficient pressure to support the load W. The bearing automatically adjusts itself to the imposed load. Thus, should W be increased, the shift to the left would become greater, and the resulting decrease of the film thickness would cause higher oil pressures, which would take care of the larger load.

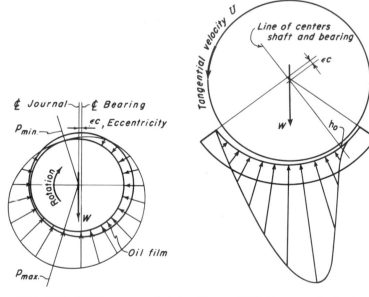

Figure 8-9 Oil-film pressure in full journal bearing.

Figure 8-10 Pressure distribution in partial bearing.

7. Load and Friction Curves for Journal Bearings

In actual bearings, the full continuous film of Fig. 8-9 does not exist. The film ruptures and the load is carried by a partial film located beneath the journal. Even in partial bearings, as in Fig. 8-10, the ruptured film leaves a portion of the arc at the trailing edge with zero pressure.

A computer solution for journal bearings for different bearing arcs and length-diameter ratios has been made[6] and is given by the curves in Figs. 8-11, 8-12, and 8-13. The ordinates are the nondimensional load variable $(W_1/\mu U)(c/r)^2$ and friction variable $(F_1/\mu U)(c/r)$, where W_1 is the load per axial inch of bearing, and F_1 is the tangential friction force per axial inch. The abscissa is taken as the eccentricity ratio ϵ.

It is seen from Fig. 8-10 that the minimum film thickness h_0 is

$$h_0 = c(1 - \epsilon) \tag{12}$$

Example 6. A 4 in. diam $\times$ 6 in. long 120° central partial bearing has a minimum film thickness of 0.001 in. Radial clearance is 0.002 in. SAE 10 oil like that in Fig. 8-4 is used. The bearing carries a load of 82 psi of projected journal area at 900 rpm. Find the temperature of the film and the friction horsepower.

Solution.
$$U = \frac{2\pi rn}{60} = \frac{2\pi 2 \times 900}{60} = 188.5 \text{ in./sec}$$

By Eq. (12):
$$\epsilon = 1 - \frac{h_0}{c} = 1 - \frac{0.001}{0.002} = 0.5$$

$$\frac{c}{r} = \frac{0.002}{2} = \frac{1}{1,000}$$

From Fig. 8-11(a):
$$\frac{W_1}{\mu U}\left(\frac{c}{r}\right)^2 = 1.69$$

From Fig. 8-11(b):
$$\frac{F_1}{\mu U}\left(\frac{c}{r}\right) = 3.60$$

$$\mu = \frac{W_1}{1.69U}\left(\frac{c}{r}\right)^2 = \frac{82 \times 4}{1.69 \times 188.5 \times 1,000^2}$$
$$= 0.000001028 \text{ lb sec/in.}^2$$

From Fig. 8-4, the oil film temperature is 175° F:

$$F_1 = 3.60\mu U\left(\frac{r}{c}\right)$$
$$= 3.60 \times 0.000001028 \times 188.5 \times 1,000$$
$$= 0.697 \text{ lb/in.}$$

By Eq. (11): friction hp $= \dfrac{0.697 \times 6 \times 188.5}{12 \times 550} = 0.120$

Other variables are in use for presenting lubrication data. Thus, $W_1 = Pd$, where P is the pressure, psi, on the projected area of the bearing, and $U = \pi dN'$, where N' is the speed of rotation, revolutions per second, rps. Then

$$\frac{W_1}{\mu U}\left(\frac{c}{r}\right)^2 = \frac{P}{\pi\mu N'}\left(\frac{c}{r}\right)^2 = \frac{1}{\pi S} \tag{13}$$

[6]See Raimondi, A. A., and J. Boyd, *A Solution for the Finite Journal Bearing and Its Application to Analysis and Design*, Report III, **1**, No. 1, American Society of Lubrication Engineers, New York: Pergamon Press, Inc., p. 194.

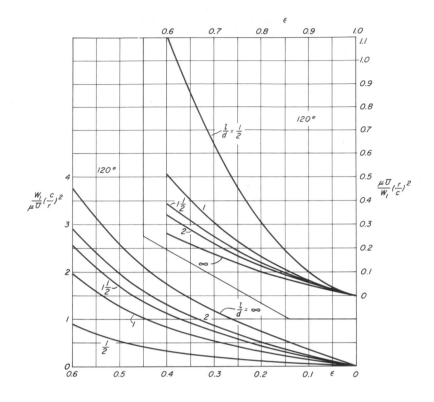

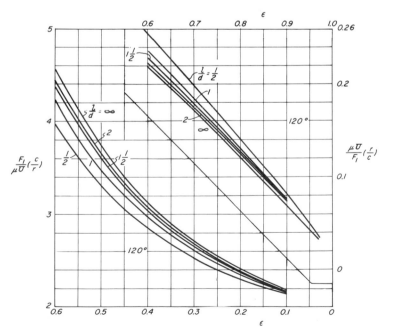

Figure 8-11 Load and friction characteristics of 120° central partial bearing.

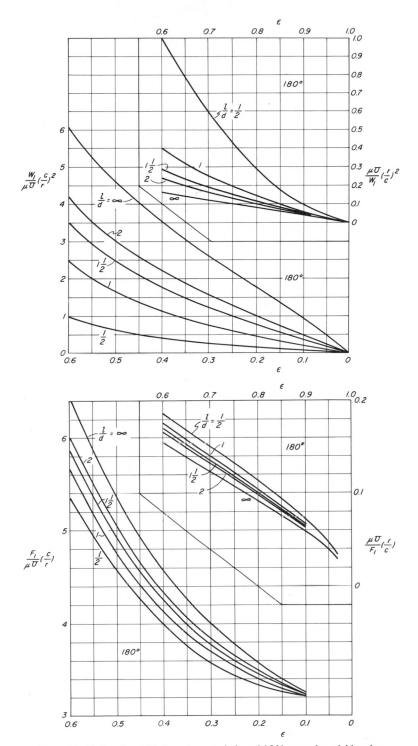

Figure 8-12 Load and friction characteristics of 180° central partial bearing.

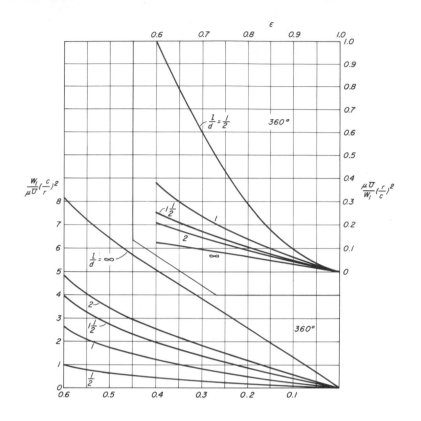

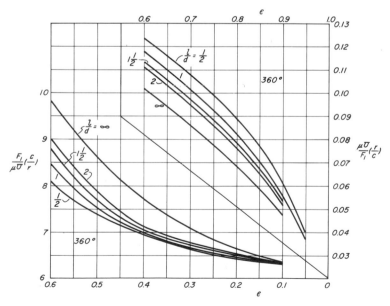

Figure 8-13 Load and friction characteristics of 360° journal bearing.

where

$$S = \frac{\mu N'}{P}\left(\frac{r}{c}\right)^2 \qquad (14)$$

Here S is the so-called "Sommerfeld number."

For journal bearings, the coefficient of friction, f, is defined in the usual way as the ratio of the tangential force to the normal force. Then

$$f = \frac{F}{W} = \frac{F_1}{W_1} \qquad (15)$$

8. Heat Balance of Bearings

The frictional loss of energy in a bearing is turned into heat, which raises the temperature of the lubricant and the adjacent parts.[7] The rate at which such heat is produced will now be determined. Let P be the load on the bearing in pounds per square inch of projected journal area ld. Since the total load on the bearing is Pld, the tangential friction force is $fPld$, where f is the coefficient of friction. The friction moment is $\frac{1}{2}fPld^2$. The number of radians turned by the shaft per minute is $2\pi n$, where n is the revolutions per minute. The work of friction per minute, or power, then is $\pi fPld^2 n$ in. lb/min. Since 1 Btu is equal to 778×12 in. lb of mechanical work, the heat produced is

$$H = \frac{\pi fPld^2 n}{778 \times 12} \text{Btu/min} \qquad (16)$$

Under steady operating conditions, the heat is removed from a bearing at the same rate that it is produced. The heat, however, must be dissipated at a sufficiently rapid rate to maintain an operating temperature low enough so that no harm will come to the oil. Sometimes cooling coils are provided for large and important bearings. In forced-feed systems the excess heat is carried away by the lubricant, which is cooled before being returned to the bearing.

However, the majority of bearings in use depend upon the surrounding air as the cooling medium. Many bearings are constructed in the form of a self-contained unit with an oil reservoir in the base or pedestal,[8] as shown in Fig. 8-14 or

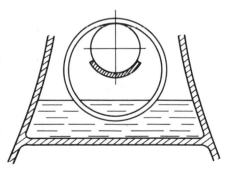

Figure 8-14 Ring-oiled partial journal bearing.

[7]See Chapter 5, Reference 9, end of chapter.

[8]See Karelitz, G. B., "Heat Dissipation in Self-Contained Bearings," *Trans. ASME*, **64**, 1942, p. 463. Also Trans ASME, **52** (1), 1930, p, APM 57. See also Lemmon, D. C., and E. R. Booser, "Bearing Oil-Ring Performance," *Trans. ASME*, **82**, 1960, *J. Basic Eng.*, p. 327.

8-18(d). An oil ring or chain, rotating with the shaft, dips into the lubricant and carries a supply of oil to the top of the shaft where it spreads over the surface.

For self-cooled bearings of moderate rises in temperature, Newton's law of cooling is found to be sufficiently accurate. This law states that

$$H = \frac{C_1}{144 \times 60} A_c \, \Delta T \text{ Btu/min} \tag{17}$$

where H is the heat lost, Btu per minute; C_1 is the heat lost per square foot of heated surface per hour, per °F; A_c is the area of the surface in square inches which is losing heat; and ΔT is the difference between the temperature of the surface and the surrounding air.

The constant C_1 depends upon the shape and the condition of the surface of the bearing housing, and also on whether or not the surrounding air is in motion. Thus it is difficult to estimate C_1 accurately. A commonly used value for the rate of losing heat in still air is to take C_1 as 2 Btu/hr/ft²/°F.

Since the area of the housing A_c is more or less under the control of the designer, fairly accurate estimates may be made for this quantity. It is usually convenient to express A_c in terms of the developed journal area πdl. Thus,

$$A_c = C_2 \pi dl \tag{18}$$

where C_2 is a constant.

Since the heat produced under steady operating conditions is equal to the heat lost, Eqs. (16) and (17) can be set equal to each other. When this is done, and Eq. (18) substituted, the following result will be obtained.

$$fPdn = 1.0805 C_1 C_2 \, \Delta T \tag{19}$$

By substituting $f = F_1/W_1$ and $P = W_1/d$, Eq. (19) becomes

$$F_1 n = 1.0805 C_1 C_2 \, \Delta T \tag{20}$$

The oil film itself is at a considerably higher temperature than the exterior surface of the housing. Just how much higher, it may be difficult to say; the designer is again confronted with a situation of considerable uncertainty. However, on the basis of experiments it can be assumed that the housing rises in temperature one-half as much as the oil film. Hence,

$$\text{oil film temp.} = \text{air temp.} + 2 \, \Delta T \tag{21}$$

Example 7. Let the shaft diameter of a 120° central partial bearing be $3\frac{1}{2}$ in., the axial length $5\frac{1}{4}$ in., and the rpm 900. The oil film temperature is not to exceed 80°F above room temperature of 100°F. The r/c ratio is 1,000, and the minimum film thickness is 0.00075 in. The housing area is 8 times the developed journal area ($C_2 = 8$). Assume that the rate of cooling C_1 is 2 Btu/hr/ft²/°F and the the housing rises in temperature one-half as much as the oil film.

Find the load P psi of the projected journal area that the bearing will carry.

Solution. $$c = \frac{r}{1,000} = 0.00175 \text{ in. radial clearance}$$

By Eq. (12)   Eccentricity ratio $\epsilon = 1 - \dfrac{h_0}{c} = 1 - \dfrac{0.00075}{0.00175} = 0.571$

From Fig. 8-11(a): $\dfrac{W_1}{\mu U}\left(\dfrac{c}{r}\right)^2 = 2.27$ (a)

From Fig. 8-11(b): $\dfrac{F_1}{\mu U}\left(\dfrac{c}{r}\right) = 4.13$ (b)

Divide Eq. (b) by Eq. (a): $f\left(\dfrac{r}{c}\right) = \dfrac{4.13}{2.27} = 1.82$ or $f = 0.00182$

In Eq. (21): $$\Delta T = \frac{1}{2}(180 - 100) = 40°F$$

In Eq. (19): $$P = \frac{1.0805 \times 2 \times 8 \times 40}{0.00182 \times 3.5 \times 900}$$

$$= 120.7 \text{ psi projected journal area}$$

The maximum temperature at which a bearing should operate depends upon the physical properties of the lubricant used. For usual industrial applications, an average film temperature in the range 160°F–180°F is considered satisfactory.

9. Designing for Film Temperature and Minimum Film Thickness

The temperature of the oil film and the minimum film thickness at which the bearing operates are usually the important factors that the designer attempts to predict when he makes the computations for a new design. The curves of Figs. 8-11, 8-12, and 8-13 permit these quantities to be found, although not by direct calculation. The load and film thickness for a number of assumed values of the film temperature can be found and curves plotted. From such curves, the temperature and film thickness can be read for any given load that the bearing might be carrying. The procedure can be carried out conveniently in tabular form as shown by the following example.

Example 8. A 3 in. diam × 3 in. long 120° central partial oil ring bearing operates at 1,200 rpm with SAE 10 oil of Fig. 8-4. Room temperature is 80°F; assume that housing temperature rises one-half as much as the film temperature. Assume the cooling rate to be 2 Btu/hr/ft²/°F, and that the cooling area is 16 times the developed journal area. The r/c ratio is 1,000. Plot the curves for P vs. temperature and P vs. h_0 for the range 155°F–165°F at 2.5 deg. intervals.

Solution. $U = \dfrac{\pi dn}{60} = \dfrac{\pi 3 \times 1,200}{60} = 188.5 \text{ in./sec}$

In Eq. (20): $F_1 = \dfrac{1.0805 C_1 C_2 \Delta T}{n} = \dfrac{1.0805 \times 2 \times 16 \Delta T}{1,200} = 0.0288 \Delta T$

The following table should be constructed. The values in columns 2–5 can be found by computation. Column 6 is determined from Fig. 8-11(b), and column 7 from Fig. 8-11(a). The values in the remaining columns can then be computed.

1	2	3	4	5	6	7	8	9	10
Film t_{av}	ΔT	$F_1 = 0.0288\,\Delta T$	μ	$\frac{\mu U}{F_1}\left(\frac{r}{c}\right)$	ϵ	$\frac{\mu U}{W_1}\left(\frac{r}{c}\right)^2$	W_1	P	h_0
155	37.5	1.0805	0.000 001390	4.124*	0.585	1.85*	485	161.6	0.00062
157.5	38.75	1.1165	1332	0.225	0.622	0.460	546	182.0	0.00057
160	40	1.1525	1278	0.209	0.653	0.397	607	202.3	0.00052
162.5	41.25	1.1885	1228	0.195	0.680	0.347	667	222.4	0.00048
165	42.5	1.2246	1184	0.182	0.704	0.305	733	244.2	0.00044

*Indicates inverse of quantity at column heading.

The desired curves are plotted in Fig. 8-15.

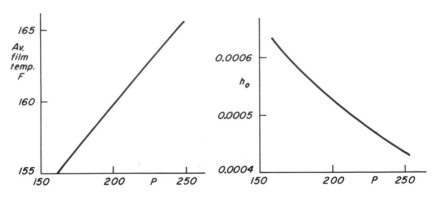

Figure 8-15 Film temperature and minimum film thickness vs. load for Example 8.

Now suppose the probable film temperature and minimum film thickness are desired when the bearing is carrying a load of 200 psi of projected journal are. Reference to Fig. 8-15 shows that the film temperature, for the assumed conditions, is 160°F, and the minimum film thickness is 0.00053 in.

The permissible film thickness is largely dependent on the surface roughness of the parts, as discussed in Section 14 and Fig. 8-19.

10. Pressure Lubricated Bearing

Sometimes a 360° bearing is lubricated by a pressurized inlet line feeding the oil to a circumferential groove at the center of the bearing, Fig. 8-16.

The oil flows both ways from the central groove and is extruded at the sides of the bearing. The clearance space can be adjusted to provide sufficient oil flow for the proper cooling of the bearing.

The quantity Q of oil flowing through one side of the bearing is given by the equation

$$Q = \frac{\pi rc^3 p_i}{6\mu l'}(1 + 1.5\epsilon^2)\frac{\text{in.}^3}{\text{sec}} \qquad (22)$$

where p_i is the inlet pressure in the central groove, and l' is the axial length for one side.

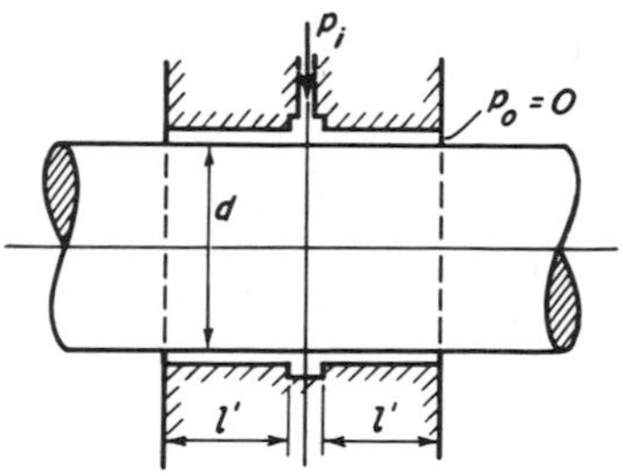

Figure 8-16 Pressure lubricated journal bearing.

The heat removed by the flow Q for an inlet temperature of t_i and an outlet temperature of t_o is equal to the following equation when the specific heat of the lubricant is taken as $\frac{1}{2}$ Btu/lb/°F.

$$\frac{\text{Btu}}{\text{sec}} = \frac{1}{2}Q\gamma(t_o - t_i) \qquad (23)$$

where γ is the weight per cubic inch of the oil. This can be equated to the heat produced as represented by Eq. (16).

$$\frac{\pi fPl'd^2n}{778 \times 12 \times 60} = \frac{1}{2}Q\gamma(t_o - t_i)$$

In this equation, F_1/W_1 should be substituted for f, W_1/d should be substituted for P, Eq. (22) should be substituted for Q, and the result should then be solved for p_i.

$$p_i = \frac{F_1}{Bc^3(1 + 1.5\epsilon^2)} \qquad (24)$$

where

$$B = \frac{23{,}340\gamma(t_o - t_i)}{\mu l'^2 n} \qquad (25)$$

Example 9. A 2 in. diam × 4 in. long ($l' = 2$ in.) 360° bearing has a central groove. It is lubricated with SAE 20 oil. Speed is 900 rpm, $r/c = 500$, and $P = 200$ psi projected area. Find the value of the inlet pressure if the inlet and outlet temperatures of the oil are 160°F and 175°F, respectively.

Solution, Average oil temperature is 167.5°F.

By Fig. 8-4: $\mu_{av} = 0.00000194$ lb sec/in.^{2}

By Eq. (5): $sp.gr_{167.5} = 0.883 - 0.00035 \times 107.5 = 0.8454$

Let the weight of water be taken as 62.4 lb/ft^3.

$$\gamma = \frac{62.4}{1{,}728} \times 0.8454 = 0.03053 \text{ lb/in.}^3$$

In Eq. (25): $$B = \frac{23{,}340 \times 0.03053 \times 15}{0.00000194 \times 2^2 \times 900} = 1{,}530{,}000$$

$$U = \frac{\pi dn}{60} = \frac{\pi 2 \times 900}{60} = 94.25 \text{ in./sec}$$

$$W_1 = 2 \times 200 = 400 \text{ lb/in.}$$

$$\frac{\mu U}{W_1}\left(\frac{r}{c}\right)^2 = \frac{0.00000194 \times 94.25 \times 500^2}{400} = 0.114$$

By Fig. 8-13(a): $\quad \epsilon = 0.83$

By Fig. 8-13(b): $\quad \dfrac{\mu U}{F_1}\left(\dfrac{r}{c}\right) = 0.076$

Then: $\qquad F_1 = \dfrac{0.00000194 \times 94.25 \times 500}{0.076} = 1.20$

$$c = \frac{1}{500} = 0.002 \text{ in.}$$

In Eq. (24): $\qquad p_i = \dfrac{1.20}{1{,}530{,}000 \times 0.002^3(1 + 1.5 \times 0.83^2)}$

$$= 48.3 \text{ psi}$$

11. Bearing Materials

Some of the qualities required in a material for a sleeve bearing are load-carrying capacity, thermal conductivity, low coefficient of friction, smoothness of surface, and resistance to wear, fatigue, and corrosion. No one material can possess all the desired characteristics to a sufficient degree,[9] and it is necessary to compromise in most designs. The chemical compositions of a number of typical bearing materials are given in Tables 8-2 and 8-3.

(a) *Tin- and Lead-Base Babbitts.* Babbitt bearings are in very wide use and are of two general types: tin base and lead base. They are quickly run in and assume very smooth surfaces. Application is usually made to a steel back as shown in Fig. 8-18(b) and (c). Thickness of the lining is usually about

Table 8-2   PER CENT COMPOSITION OF TYPICAL BEARING ALLOYS*

	Babbitt		Bronze		Copper-Lead
	Tin Base SAE 11	Lead Base SAE 13	SAE 791	SAE 794	SAE 48
Copper (Cu)	5.75		rem	73.5	70
Tin (Sn)	87.5	6	4	3.5	
Lead (Pb)		rem	4	23	30
Antimony (Sb)	6.75	10			
Zinc (Zn)			4		

*See *SAE Handbook.*

[9]See References 1 and 4, end of chapter.

Table 8-3   MAXIMUM DESIGN PRESSURES FOR ENGINE BEARINGS*

Bearing Material	Per Cent Composition of Material	Maximum Mean Bearing Pressure, psi
Lead-base babbitt	Pb (75–85), Sn (4–10), Sb (9–15)	600–800
Tin-base babbitt	Pb (0.35–0.6), Sn (86–90), Sb (4–9), Cu (4–6)	800–1,000
Cadmium-base alloy	Cu (0.4–0.75), Cd (97), Ni (1–1.5), Ag (0.5–1.0)	1,200–1,500
Copper-lead alloy	Pb (45), Cu (55)	2,000–3,000
Copper-lead alloy	Pb (25), Sn (3), Cu (72)	3,000–4,000
Silver	Ag (99), 0.5–1.0 lead on surface	5,000 up

*Pressures are based on fatigue life of 500 hr at a bearing temperature of 300°F. Bearing metal thicknesses range from 0.01 to 0.015 in. for lead, tin, and cadmium-base metals, and 0.025 in. for copper, lead, and silver, all on steel backs. At lower temperatures, the fatigue life will be greatly extended. Data from References of footnote 13.

0.015 in. Thin babbitt bearings with linings 0.002–0.005 in. thick are also used and can carry somewhat heavier loads. Babbitt bearings have good *conformability*, or the property of adjusting themselves to small misalignments or shaft deflections. They also make excellent bearings from the standpoint of *embeddability* because a reasonable amount of dirt or foreign matter in the lubricant can be absorbed by the soft bearing material and the shaft is thus protected against scoring. Journal material may be soft steel, hard steel, or cast iron.

(b) *Bronze Bearings.* Bronze bearings are suitable for high loads and slow speeds, but the alignment between shaft and bearing must be good. They are made in a great number of alloy compositions to obtain various physical properties. Standardized bushings, as shown in Fig. 8-18(a), can be purchased ready for use.

(c) *Copper-Lead Bearings.* These bearings are used where the loads are higher than can be carried by babbitt. The conformability, however, is less and consequently such bearings must be used where the shafts are rigid and the alignment is good.

(d) *Cast-Iron Bearings.* Cast iron is a bearing material widely used where the service is not too severe. Journals should have a hardness in the 150–250 Brinell range, and should have a smooth finish. Some of the good performance of cast iron is usually attributed to the graphite inclusions that are normally present. The alignment between journal and bearing, however, must be good. Cast iron runs well with phosphor bronze, and also with cast iron as evidenced by the cylinder block and pistons of internal combustion engines.

(e) *Porous Bearings.* The so-called "self-lubricating" or "porous" bearing is made by sintering powdered metal and then impregnating with oil.

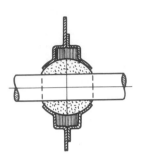

Figure 8-17 Self-aligning porous bearing with oil saturated felt washer.

Various compositions of bronze are in wide use. Iron is used to a lesser extent. Figure 8-17 shows a self-aligning porous-metal bearing applied to a small electric motor. The felt washer can frequently be omitted. Such bearings are available from stock in both cylindrical and spherical form.

Porous bearings are sometimes designed to a specified value of PV, where P is the load per square inch of projected area, and V is the surface velocity of the shaft in feet per minute.[10]

Example 10. Find the required length for a porous bearing $\frac{1}{2}$ in. in diameter turning at a speed of 1,750 rpm. Load on the bearing is 50 lb. $PV = 30,000$.

Solution. Let W be the load on the bearing.

$$PV = \frac{W}{ld} \times \frac{\pi dn}{12} = \frac{\pi Wn}{12l}$$

Then:

$$l = \frac{\pi Wn}{12(PV)} = \frac{\pi 50 \times 1,750}{12 \times 30,000} = 0.76 \text{ in.}$$

(f) *Carbon and Plastic Bearings.* For high-temperature service, or where conventional types of lubrication cannot be used, bearings of pure carbon have been satisfactorily employed. Graphite is a layer-lattice solid, and the low friction of such bearings is due to the low resistance to slip between the lattice planes.

Teflon has an extremely low coefficient of friction, and requires no oil lubrication. Bearings can be operated at high loads combined with slow or oscillating speeds. Wear resistance is increased by filling with glass fiber, powdered carbon, bronze, or metallic oxides. Bearings can be used under conditions of corrosion, abrasion, or stickiness. Bearings are frequently formed merely by the use of liners cut from Teflon sheet or strips.

(g) *Laminated Phenolic Bearings.* These are plastic bearings formed by impregnating paper or fabric with the phenolic resin, forming into shape, and then curing by heat and pressure. Since this is a thermosetting material subsequent heating will have no effect. Such bearings show good resistance to corrosion, fatigue, and shock. They have been successfully applied to large sizes and heavy loads, such as in the steel mill industry. Water is sometimes used as the lubricant.

(h) *Rubber and Wood.* Soft rubber, lubricated with water, has been used for many years in pumps and general marine service. Lignum vitae and wood are also used for bearings to a limited extent.

(i) *Bonded Coatings.* Molybdenum disulphide, MoS_2, which resembles

[10]See *ASTM Specifications B 438 and 439*, Vol. 7.

graphite in appearance, possesses very-low-friction characteristics. Like graphite, it has a laminar structure. To ensure retention of the lubricant on the surface where it is needed, it can be mixed with a suitable binder, usually a thermosetting plastic, spread on the surface, and then dried at a suitable temperature. Wear life in general is satisfactory. This lubricant has proved to be very effective under operating conditions not possible for fluid lubricants.[11]

Long experience has shown that certain combinations of materials will work well together.[12] Other combinations will not but will exhibit excessive wear.

Soft steel works well with cast iron, babbitt, and soft brass, but works poorly with soft steel, bronze, nylon, and thermoset resins.

Hard steel works well with soft bronze, phosphor bronze, brass, cast iron, nylon, and the thermoset resins, but poorly with the harder heat-treated bronzes. Hardened nickel steel does not work well with hardened nickel steel.

12. Bearing Loads

Bearing loads are determined mainly from experience with similar applications that have proved successful. The permissible loading is not a direct function of the particular material under consideration, but depends rather on the type of service. Table 8-4 gives a survey of current practice and can be

Table 8-4   CURRENT PRACTICES IN BEARING DESIGN PRESSURES[13]

Type of Bearing	Design Pressure, psi
Diesel engines, main bearings	800–1,500
Connecting rod bearings	1,000–2,000
Wrist pins	1,800–2,000
Electric motor bearings	100–200
Marine lineshaft bearings	25–35
Steam turbines and reduction gears	100–250
Automotive gasoline engine, main bearings	500–700
Connecting rod bearings	1,500–2,500
Aircraft engine connecting rod bearings	700–2,000
Centrifugal pumps	80–100
Roll neck bearings	1,500–2,500
Railway axle bearings	300–350
Light lineshaft bearings	15–25
Heavy lineshaft bearings	100–150

[11]See Campbell, W. F., p. 197 of Reference 2, end of chapter. See also Reference 3, end of chapter.

[12]See p. 277 of Rothbart, H. A., *Cams*, New York: John Wiley & Sons, Inc., 1956. See also Halsey, F. A., *Handbook for Machine Designers, Shop Men, and Draftsmen*, New York: McGraw-Hill Book Company, 2d ed., 1916, p. 18.

[13]See Fuller, D. D., "Design Analysis of Journal Bearings," *Machine Design*, **28**, Feb. 9, 1956, p. 119. See also p. 221, Reference 6, end of chapter.

used as a general guide when more specific information is lacking. Careful attention should be paid to the reliability of the oil supply for small values of Zn/P, which occur when the speed is slow and the load is large.

13. Construction of Bearings

A separate insert or bushing is usually provided to form the bearing surface for supporting the shaft in a sleeve bearing. It is made of a material that is known to have desirable qualities of high load capacity and low friction. The types and details of construction for sleeve bearings can vary over a considerable range.

Figure 8-18(a) shows a solid bronze bushing, a type of bushing that is in

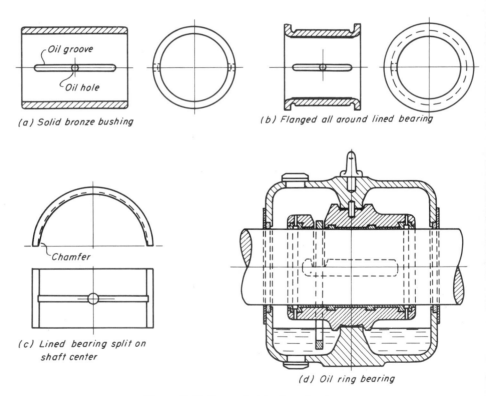

(a) Solid bronze bushing

(b) Flanged all around lined bearing

(c) Lined bearing split on shaft center

(d) Oil ring bearing

Figure 8-18 Types of plain sleeve bearings.

wide use. It can be cast, machined from bar stock, or formed from tubing or sheets. It is usually held in place by a light press fit. The bearings shown in sketches (b) and (c) consist of a steel or bronze back with a thin lining of babbitt or one of the numerous other bearing alloys. Such bearings can be made either all around or split on the centerline of the shaft. A wise variety of types and sizes of such bearings is available from stock ready to use. They may be obtained solid, split longitudinally or flanged, and with many kinds of oil grooves.

The oil ring bearing[14] of Fig. 8-18(d) has a cast-iron back with a babbitt liner cast in place. To obtain a better oil supply in slow-speed service, the ring is sometimes replaced by an endless loop of chain hanging on the shaft.

14. Clearance and Oil Grooves

In addition to choosing the material, the designer must specify the shaft clearance and the oil grooving for the bearing.

The clearance depends to some extent on the desired quality. Small clearances can be maintained with high-grade workmanship, but when costs must be reduced, the clearances usually are made larger. Table 8-5 gives a summary of clearances used in industrial applications.

For small clearances, the surface roughness of the shaft and bearing must be considered. These surfaces, when examined by a microscope, are usually far from being geometrically smooth. In many cases, the roughness consists of two orders of irregularities. Over most of the surface there are relatively small, low irregularities at close intervals. In addition, at relatively large intervals, there are high peaks and deep valleys. Roughness is usually measured from the tops of the high peaks, as shown in Fig. 8-19. To accen-

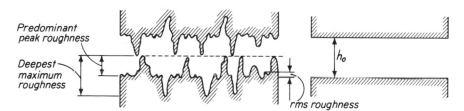

Figure 8-19 Lower limit at which hydrodynamic lubrication is assumed to be possible.

tuate the roughness, it is customary to prepare samples with a vertical scale that is 25 times as great as the horizontal scale. Profilometers give readings for either the root mean square or the arithmetic average of the roughness. The predominant peak roughness can then be had by multiplying by a suitable factor. For example, for ground surfaces, the arithmetic average roughness should be multiplied by 5 to obtain the predominant peak roughness.[15]

[14]Karelitz, G. B., "Performance of Oil Ring Bearings." *Trans. ASME*, **52**(1), 1930, p. APM 57.

[15]See Tarasov, L. P., "Relation of Surface Roughness Readings to Actual Surface Profile," *Trans. ASME*, **67**, 1945, p. 189. The rms average for a surface is usually about 10 per cent greater than the arithmetic average roughness.

Table 8-5   BEARING CLEARANCES IN INDUSTRIAL APPLICATIONS*

	Running Clearance, Thousandths of an Inch, for Shaft Diameter under				
	1/2 in.	1 in.	2 in.	3-1/2 in.	5-1/2 in.
Precision spindle practice—hardened and ground spindle lapped into the bronze bushing. Below 500 fpm and 500 psi	0.00025 to 0.00075	0.00075 to 0.0015	0.0015 to 0.0025	0.0025 to 0.0035	0.0035 to 0.005
Precision spindle practice—hardened and ground spindle lapped into bronze bushing. Above 500 fpm and 500 psi	0.0005 to 0.001	0.001 to 0.002	0.002 to 0.003	0.003 to 0.0045	0.0045 to 0.0065
Electric motor and generator practice— ground journal in broached or reamed bronze bushing or reamed babbitt bushing	0.0005 to 0.0015	0.001 to 0.002	0.0015 to 0.0035	0.002 to 0.004	0.003 to 0.006
General machine practice (continuous rotating motion)—turned steel or cold-rolled steel journals in bored and reamed bronze or poured and reamed babbitt bushings	0.002 to 0.004	0.0025 to 0.0045	0.003 to 0.005	0.004 to 0.007	0.005 to 0.008
General machine practice (oscillating motion)—journal and bearing material as above	0.0025 to 0.0045	0.0025 to 0.0045	0.003 to 0.005	0.004 to 0.007	0.005 to 0.008
Rough machine practice—turned steel or cold-rolled steel journals in poured babbitt bearings	0.003 to 0.006	0.005 to 0.009	0.008 to 0.012	0.011 to 0.016	0.014 to 0.020

*Data by Johnson Bronze Co.

Figure 8-19 shows the surfaces for journal and bearing located so that the high peaks are just in contact. It has been proposed[16] that the condition illustrated be taken as the limit at which hydrodynamic lubrication can be assumed to exist. Thus if each surface has an arithmetic average roughness of 30μ in., the minimum permissible film thickness in Fig. 8-19 would be $2 \times 30 \times 5$ or 0.0003 in. Surface finish should thus be specified and closely controlled if the design calculations indicate that the bearing will operate with a very thin film. Experience has indicated, however, that other factors, such as misalignment, deformation, foreign matter in the oil, and fatigue due to dynamic loads, are the cause of most bearing failures.

Oil grooves with carefully rounded edges should be provided to distribute the oil over the entire surface. It is very important, however, that no grooves

[16] See Ocvirk, F. W., and G. B. Dubois, "Surface Finish and Clearance Effects on Journal-Bearing Load Capacity and Friction," *Trans. ASME*, **81**, 1959, *J. Basic Eng.*, p. 245.

be placed in the load-carrying portion of the lining. When so placed, such grooves merely provide an easy exit for the oil and prevent the formation of the film and the high pressures necessary to support the load. A good rule is to use the least possible amount of grooving. Sometimes the chamfer between the bearing and cap provides a channel for the lateral flow of the lubricant. Figure 8-20 shows the reduction in load capacity (for equal minimum film thickness) due to an incorrectly located oil groove. If the load on the bearing with the middle groove is not reduced, operation may take place with a dangerously thin oil film. If a belt pulley is adjacent to a bearing, care should be taken that the forces from the belt do not pull the shaft toward the side where oil grooves may be located and where a satisfactory oil film cannot be formed.

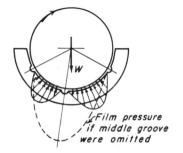
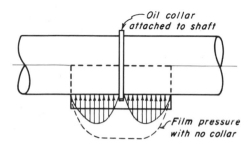

Figure 8-20 Loss of capacity from incorrect grooving.   **Figure 8-21** Loss of capacity caused by oil collar.

In a similar manner, a circumferential groove reduces the load-carrying qualities, as shown in Fig. 8-21. An oil collar is a very effective means of distributing the lubricant. However, an oil collar practically cuts the bearing in two, and the sum of the loads carried by each part is less than that carried by a single bearing of the total width. Forced-feed-lubricated bearings frequently are constructed without oil grooves.

The most important point to be considered when choosing the method of lubricating a bearing is the reliability of the oil supply. Because of low first cost, many bearings are lubricated by hand oilcans, drip or wick oilers, oil-soaked waste, or grease. As previously mentioned, such bearings usually operate in the region of boundary lubrication, with its attendant high friction and danger of seizure, should the scanty oil supply become temporarily deranged.

Safety lies in having a copious oil supply at all times. Such a supply can be provided by an oil ring and reservoir, or by piping the oil to the bearing under pressure from a central pumping station. Whatever method is used, close attention to the mechanical details is required if satisfactory operation is to be expected.

Means for preventing leakage and loss of oil at the ends of the bearing must often be provided. Felt gaskets or patented oil seals are in wide use. Sometimes the outer housing is enlarged to accommodate a small disk fastened to the shaft. The oil that has leaked past the side of the bearing is thrown off the disk by centrifugal force and caught by the housing. Often a circumferential groove at the end of the bearing is provided with drain holes to return the oil to the reservoir.

Grease is widely used for exposed locations because it is easier to retain in the bearing. Investigations have shown that the action can be hydrodynamic with a pressure distribution somewhat like that for oil.

15. Elastic Matching

When the bearings are some distance apart and the shaft carries a lateral load, the inclination of the shaft at the ends may be great enough to give rise to the misalignment shown in Fig. 8-22(a). Obviously such a bearing is

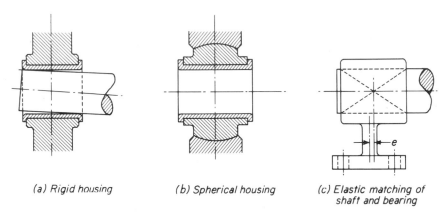

(a) Rigid housing   *(b) Spherical housing*   *(c) Elastic matching of shaft and bearing*

Figure 8-22 Types of bearing supports.

subjected to rapid wear and is not able to form a satisfactory load-carrying film. The difficulty can be reduced by setting the bearings closer together and using a larger-diameter shaft or a smaller value for the l/d ratio. The spherical seat shown in sketch (b) gives self-alignment from all causes, but is expensive, and increases the size of the bearing.

Sometimes the bearing can be located eccentrically on an elastic support in such a way that it tilts slightly under load and assumes the same inclination as the shaft. See Fig. 8-22(c) and Problem 116, Chapter 1. The oil film then has a uniform thickness throughout the axial length of the bearing, and the difficulty shown in Fig. 8-22(a) is overcome.[17]

[17]See Baker, J. G., "Elastic Matching Improves Bearing Life and Performance," *Machine Design*, **20**, May 1948, p. 106.

16. Dry Friction

Dry friction or the friction between surfaces without intentional lubrication is a very complex phenomenon. Despite more than a century of study, a satisfactory explanation of the mechanism of dry friction does not as yet exist. The so-called "classical rules of dry friction" are usually stated as follows.

(1) The friction force is directly proportional to the force pressing the bodies together.

(2) The friction force is independent of the area of the contacting surfaces.

(3) The friction force is largely independent of the velocity of sliding.

(4) The friction force depends on the nature of the sliding surfaces.

The coefficient of static friction for bodies at rest, or with motion impending, is somewhat greater than for kinetic friction, which prevails after the sliding member is set in motion.

According to Coulomb, 1785, the friction force is due to the interlocking of the asperities of the two surfaces and the necessity of deforming the irregularities as the two bodies pass by each other. Within rather wide limits the magnitude of the friction force is independent of the degree of roughness of the surfaces. The asperities that are deformed are very minute so that it apparently makes no difference whether they are located on a smooth surface or consist of protuberances located upon larger irregularities.

For very rough surfaces, the protuberances may interlock to the extent that parts of them are torn off as the bodies move with respect to each other. This is a wear phenomenon and is usually not considered to be friction in the usual meaning of the term. Such action is especially likely to occur when one of the bodies is much harder than the other.

Although the nature of the sliding surfaces determines the friction force, surfaces of naked metal do not exist in everyday life. Exposure to the atmosphere will contaminate the surface with oxide films, moisture, absorbed gases, grease, dirt, and so on. These films cling very tightly to a surface and can be removed only with difficulty. Ordinary cleaning does not suffice. Even in the laboratory, unless the body is kept in an inert atmosphere, an oxide film quickly forms. Thus when two ordinary bodies are sliding together, it is the surface films that are touching each other and not the metals themselves.

Reference handbooks usually make no attempt to give exact specifications for the surfaces under consideration, and this may partially account for the wide variations in the published values for the coefficient of friction.[18]

When bodies with very smooth surfaces are brought into intimate contact, adhesive or cohesive forces come into action and hold the bodies tightly together. Thus when gage blocks or optical flats are wrung together,

[18]See Schmidt, A. O., and E. J. Weiter, "Coefficients of Flat Surface Friction," *Mech. Eng.*, **79**, 1957, p. 1130. See also Bikerman, J. J., "Surface Roughness and Sliding Friction," *Rev. Mod. Phys.*, **16**, 1944, p. 53.

a relatively large force may be required for tangential motion even though the normal force pressing them together is very small. The classical rules of friction do not apply under such conditions.

A surface of naked metal can be obtained by removal of the oxide and surface film and degasing in the laboratory. When two metallic surfaces of this kind are brought together in a vacuum or inert atmosphere, the metals weld together at the high spots where contact occurs between them. When one body is drawn along the other the welds will be made and broken continuously. The resulting motion is somewhat jerky or discontinuous and has been given the name of *stick slip*. The force required to produce motion under such conditions is relatively very large and may be as much as 20 times the value for corresponding contaminated surfaces. The force also depends upon the velocity of sliding and is thus not in accord with the classical rules of friction.

When two bodies with contaminated surfaces are brought together, actual contact occurs only at the high spots of the two surfaces. The actual area of touching may be only a very small fraction of the nominal area. The contact stress at the isolated points where the bodies touch reaches very high values. Such stresses are required to deform the asperities sufficiently to permit motion to take place. Under such conditions of extreme stress, the surface film may rupture and the welding phenomenon described above may take place. Such welding, together with any adhesive forces that may be present, serves to further complicate the situation.

The wide scatter in the published values for the coefficient of friction is undoubtedly due to the lack of precise information on the conditions of the actual surfaces being tested. ". . . in view of the extremely high friction found with really clean metals it is fortunate, for engineering, that metals are not found with perfectly clean surfaces in practise."

17. Boundary or Thin-Film Lubrication

Machine parts coated with a lubricant and rubbing together without benefit of a fluid film are a common occurrence in machinery. Such operation can also occur in a journal bearing for low values of the viscosity or speed or for high values for the load as indicated by the left end of the curve in Fig. 8-6. Such action, known as boundary lubrication, can also occur during starting, stopping, or the reversing of journal bearings. It can also occur during the running-in period of a new machine. The separation of the surfaces is of molecular dimensions.

The rules previously stated for dry friction apply also to boundary lubrication with the following additions.

(5) The friction force depends on the composition of the lubricant and its reaction with the contacting surfaces.

(6) The friction force is influenced by the temperature and surface roughness of the bodies.

(7) The friction force is independent of the viscosity of the lubricant.

The ordinary rules of hydrodynamic lubrication do not apply to boundary lubrication, and design equations for this type of service are not available.[19]

Fatty acids, as found in certain animal and vegetable oils, can be added in small quantities to the lubricant to give a friction reducing quality distinct from the viscosity of the lubricant. These react with the surfaces to form very tough and durable coatings, and at the same time provide a plane of low shearing strength on which the relative motion between the parts takes place. Massive seizure between the parts is thus prevented. Should the film be ruptured when two peaks come together exposing pure metals, instantaneous welding will occur. Such welds of the high spots are immediately broken by the relative movement between the bodies resulting in wear and increased friction force.

Boundary lubrication breaks down at elevated temperatures resulting in a large increase in the coefficient of friction accompanied by wear and surface damage.

True boundary lubrication is difficult to achieve even in the laboratory. Usually there will be sufficient lubricant present to form localized patches of film, which carry part of the load while the remainder of the surfaces are rubbing without complete separation of the peaks of the two surfaces. The result is the so-called mixed lubrication.

18. Mixed or Semifluid Lubrication

Mixed lubrication, as shown by the BC section of Fig. 8-6, represents a partial breakdown of the full hydrodynamic film as might occur under heavy loads, slow speed, and low viscosity. An insufficient supply of lubricant as given by drip cups or wick oilers may be a contributing factor.[20] Bearing and journal should be made of materials that experience has shown will operate together satisfactorily. Mixed lubrication can also occur when starting a heavily loaded bearing that has been inactive for some time.

For larger values of Zn/P, operation is in the CD region. The lubrication is hydrodynamic and follows the laws given in the earlier sections of the chapter. Such operation is largely independent of the smoothness of the surfaces or the kinds of materials used for the moving parts.

Many other means are available for achieving lubrication in mechanical equipment. Some of these are slider bearings, hydrostatic bearings, and gas bearings. Information can be obtained from the appended list of reference works.

[19]See Reference 2, end of chapter.

[20]See Larsen, R. G., and G. L. Perry, "Investigation of Friction and Wear under Quasi-Hydrodynamic Conditions," *Trans. ASME,* **67**, 1945, p. 45.

REFERENCES

1. Bassett, H. N., *Bearing Metals and Alloys*, New York: Longmans, Green & Co., Inc., 1937.

2. *Boundary Lubrication, An Appraisal of World Literature*, New York: American Society of Mechanical Engineers, 1969.

3. Braithwaite, E. R., *Solid Lubricants and Surfaces*, Elmsford, N. Y.: Pergamon Press, Inc.

4. Dayton, R. W., Ed., *Sleeve Bearing Materials*, Cleveland: American Society for Metals, 1949.

5. Fuller, D. D., "*Design Analysis of Journal Bearings*," *Machine Design*, **28**, Feb. 9, 1956, p. 119.

6. Fuller, D. D., *Theory and Practice of Lubrication*, New York: John Wiley & Sons, Inc., 1956.

7. *General Discussion on Lubrication*, London: Institution of Mechanical Engineers, 1938.

8. Gross, W. A., *Gas Film Lubrication*, New York: John Wiley & Sons, Inc., 1962.

9. Hersey, M. D., *Theory of Lubrication*, New York: John Wiley & Sons, Inc., 1938.

10. Norton, A. E., *Lubrication*, New York: McGraw-Hill Book Company, 1942.

11. Radzimovsky, E. I., *Lubrication of Bearings*, New York: the Ronald Press Company 1959.

12. Shaw, M. C., and E. F. Macks, *Analysis and Lubrication of Bearings*, New York: McGraw-Hill Book Company, 1949.

13. Slaymaker, R. R., *Bearing Lubrication Analysis*, New York: John Wiley & Sons, Inc., 1955.

14. Trumpler, P. R., *Design of Film Bearings*, New York: The Macmillan Company, 1966.

PROBLEMS

1. Plot the curve for SAE 10 oil in Fig. 8-4. For this type of problem it is convenient to carry out the work in tabular form. A separate column heading should be used for each successive step in passing from the SUS values of Fig. 8-3 to μ in. lb sec/in.^{2} units.

2. An oil has a *VI* of 60 per cent and a Saybolt viscosity at 100°F of 400 sec. Find the Saybolt viscosity for this oil at 180°F.   *Ans.*   SUS = 70.5 sec.

3. An oil has a *VI* of 90 per cent and a Saybolt viscosity at 100°F of 300 sec. Find the Saybolt viscosity of this oil at 160°F. *Ans.* SUS = 84 sec.

4. A very lightly loaded. 360° bearing, 6 in. in diameter and 9 in. long, consumes 2 hp in friction when running at 1,200 rpm. Radial clearance is 0.003 in. Find the temperature of the oil film using SAE 10 oil in Fig. 8-4.

Ans. $t = 145°F.$

5. A 2.5 in. diam × 3.75 in. long 360° bearing turns at 1,200 rpm and consumes 0.1 hp in friction. SAE 10 oil in Fig. 8-4 is used; $r/c = 1,000$. Find the total load for the bearing if the film is at a temperature of 175°F.

Ans. $W = 1,030$ lb.

6. A 3 in. diam × 3 in. long 360° bearing turns at 1,200 rpm; $c/r = 0.001$; $h_0 = 0.001$ in. SAE 20 oil is used. The film temperature is 180°F. Find the total load *W*. *Ans.* $W = 830$ lb.

7. A 3 in. diam × 6 in. long, 120° central partial bearing is operating at a value for ϵ equal to 0.6. Coefficient of friction is equal to 0.0025. Find the value of the minimum film thickness. *Ans.* $h_o = 0.00098$ in.

8. A bearing 10 in. in diameter and 10 in. long operates at 1,500 rpm, carrying 25,000 lb. The c/r ratio is 0.0015. Let the viscosity of the lubricant be taken as 30 centipoises. Find the friction horsepower loss in accordance with Petroff's equation, 360° loaded bearing, and for the 120° central partial bearing.

Ans. 17.03; 20.17; and 10.38 hp.

9. Use data as for Example 7, except that the oil film temperature and cooling area are not known. For this bearing, *P* is 125 psi of projected journal area, and SAE 10 oil in Fig. 8-4 is used. Find the required cooling area of the housing.

Ans. 4.0 ft.²

10. A 120° central partial bearing 3.5 in. in diameter and 5.25 in. long carries 125 psi projected journal area at 600 rpm. Room temperature is 100°F and housing is at 140°F. $C_1 = 2$; $C_2 = 8$; $\epsilon = 0.52$. What is the minimum film thickness?

Ans. $h_0 = 0.0011$ in.

11. A 120° central partial bearing 4 in. in diameter and 5 in. long carries 90 psi of projected journal area. The r/c ratio is 1,000; h_0 is equal to 0.001 in. SAE 20 oil in Fig. 8-4 is used. Room temperature is 100°F. Rpm = 600. Assume the rate of cooling is 2.5 Btu/hr/ft²/°F. Assume housing temperature rises one-half as much as oil film. Find the oil film temperature and the required area of housing.

Ans. $t = 168°$ F; $A_c = 2.44$ ft.²

12. A 120° central partial bearing 2.5 in. in diameter and 4.0 in. long has a minimum film thickness of 0.0006 in. SAE 20 oil in Fig. 8-4 is used; $n = 860$ rpm and $r/c = 1,000$. If the oil film is at 165°F, find the load the bearing is carrying.

Ans. $W = 1,740$ lb.

13. Find the required area of housing in order to maintain the film temperature of a 3 in. diam × 4.5 in. long, 120° central partial bearing at 165°F; rpm = 900 and $r/c = 1,000$. Room temperature is 100°F. Assume housing temperature rises one-half as much as the oil film. Cooling rate is 2 Btu/hr/ft²°F. SAE 10 oil is used. Load is 100 psi of projected journal area. *Ans.* $A_c = 2.33$ ft².

14. A 2 in. diam × 4 in. long, 120° central partial bearing turns 1,350 rpm. SAE 10 oil is used. Film temperature is 175°F and room temperature is 100°F. Assume housing rises one-half as much as the film, and that the cooling rate is 2 Btu/hr/ft²/°F. Let the area of the housing be 8 times the developed journal area.

If the shaft and bearing are dimensioned as shown in Fig. 8-23, determine the load and minimum film thickness for an assembly at the loosest permissible fit, and also for an assembly at the tightest permissible fit.

Ans.   Loosest, $W = 880$ lb; tightest, $W = 500$ lb.

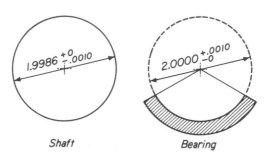

Shaft   Bearing

Figure 8-23 Problem 14.

15. A 120° central partial bearing is 2 in. in diameter and 3 in. long. $r/c = 1,000$; $n = 900$ rpm. SAE 10 oil is used. Bearing is self-cooled with $C_1 = 2$ Btu/hr/ft²/°F and $C_2 = 8$. Room temperature is 100°F. Assume the housing surface temperature rises one-half as much as the average film temperature.

(a) Plot the curve for average film temperature vs. P.

(b) Plot the curve for h_o vs. P.

Use temperature range from 145°F to 170°F inclusive at 5 deg. intervals.

16. A 360° pressure-lubricated bearing with a central groove is 4 in. diam × 4 in. long ($l' = 2$ in.). $r/c = 1,000$; $P = 300$ psi projected area. $n = 900$ rpm. SAE 20 oil is used.

For an inlet temperature of 100°F and an outlet temperature of 220°F, find the value of the inlet pressure p_i if all the frictional heat is carried away by the oil.

Ans. $p_i = 27.2$ psi.

17. A 120° central partial bearing is 2 in. in diameter × 3 in. long with a radial clearance of 0.001 in. The shaft turns at 900 rpm. Loading is 150 psi of projected area. Room temperature is 100°F. SAE 10 oil is used. Heat transfer coefficient C_1 equals 2 Btu/hr/ft²/°F. For a minimum film thickness of 0.0004 in. find the average temperature in the film and the required value for constant C_2.

Ans. $t_{av} = 162°$ F; $C_2 = 6.86$.

18. (a) Use the data in Example 8 and calculate the values of P for the given temperature range for values of r/c of 600, 800, 1,200, and 1,400. Calculations for r/c of 1,000 have already been made in Example 8. Plot the curves for P vs. film temperature.

(b) From the curves read and tabulate the temperature values for loads P of 180, 200, and 220 psi of projected area for each of the values above of r/c. Plot the curve for film temperature vs. r/c for $P = 180$ psi. Do the same for $P = 200$ psi and for $P = 220$ psi. Note that there is an optimum value for r/c at which the film temperature is a minimum.

(c) Plot the ϵ values vs. P for the above r/c ratios. From the curves read and tabulate the ϵ values for loads P of 180, 200, and 220 psi for each r/c curve. Calculate the corresponding h_0 values and plot vs. r/c for $P = 180$ psi. Do the same for $P = 200$ psi and for $P = 220$ psi. Note that there is an optimum r/c at which h_0 is a maximum for any given pressure P.

19. A 360° pressure-fed bearing with a central groove is 2 in. in diameter by 4 in. long ($l' = 2$ in.) and turns 900 rpm. $r/c = 500$. SAE 20 oil is used. Inlet temperature is 160°F. Loading is 200 psi projected area. Plot curve for outlet oil temperature t_o vs. inlet oil pressure p_i for the outlet temperature range 175°F–190°F inclusive at 5 deg. intervals.

20. A journal bearing 3 in. in diameter and 6 in. long has the same Zn/P vs. f curve as for Example 4. Assume the specific gravity at the film temperature to be 0.84. Total load is 1,800 lb, and rpm = 1,500. Find the SUS value at the operating temperature if the power loss for the bearing is 0.15 hp.   *Ans.*   SUS = 51 sec.

21. A journal bearing 3 in. in diam and 6 in. long has the same Zn/P vs. f curve as that in Example 4. The oil film is at 180°F. The oil has a SUS value of 52 sec at 210°F, a VI of 80 per cent, and a $sp.gr_{60}$ of 0.9. Total load is 1,500 lb and rpm = 1,200. Find the horsepower loss for this bearing.   *Ans.*   hp = 0.133.

22. A 120° central partial bearing is 2.5 in. in diameter and 4.0 in. long. Rpm = 860; SAE 20 oil is used, and the r/c ratio is 1,000. For an oil temperature of 160°F and minimum film thickness of 0.0006 in., find the load on the bearing.
Ans. $W = 1,940$ lb.

23. Three bearings have respective film arcs of 120°, 180°, and 360°. The diameter is 2.5 in. and length is 4 in., Rpm = 1,200, r/c ratio is 1,000, and load is 200 psi of projected journal area. SAE 10 oil. Find the film temperature for each bearing if the minimum film thickness is 0.0006 in.

This problem illustrates the fact that, when the angle is reduced, the bearing must be operated at a lower temperature if the film thickness is to be maintained.
Ans.   144°F; 167°F; 175°F.

24. Let the film arcs for a series of bearings be 120°, 180°, and 360°. The diameter is 2.5 in. and the length is 4.0 in. Rpm = 1,200, r/c ratio is 1,000, and film temperature is 170°F. SAE 10 oil is used and minimum film thickness is 0.0006 in. Find the load P on the projected journal area that each bearing is carrying.

This problem illustrates the fact that the load must be reduced for smaller values of the bearing arc in order to maintain a given film thickness and oil temperature.
Ans.   132, 191, and 213 psi.

25. Let the film arcs for three bearings be 120°, 180°, and 360°, respectively. Diameter is 3.5 in. and length is 4.5 in. The load on the bearings is 1,800 lb. Rpm = 900 and film temperature is 165°F. SAE 10 oil is used; r/c ratio is 800. Find the minimum film thickness and the friction horsepower.

This problem shows that a reduction in film arc will give a reduction in the friction horsepower, but that the film thickness will also be reduced.
Ans.   Friction hp: 0.090, 0.106, 0.149.

9 Ball and Roller Bearings

Ball and roller bearings have been brought to their present state of perfection only after a long period of research and development. The benefits of such specialized research can be obtained when it is possible to use a standardized bearing of the proper size and type. Ball bearings are used in almost every kind of machine and device with rotating parts. However, such bearings cannot be used indiscriminately without a careful study of the loads and operating conditions. In addition, the bearing must be provided with adequate mounting, lubrication, and sealing.

Pillow blocks, incorporating ball or roller bearings, are available from stock, and are very convenient as the mounting and sealing of the bearing has already been done.

1. Construction and Types of Ball Bearings

A ball bearing usually consists of four parts: an inner ring, an outer ring, the balls, and the cage or separator. To increase the contact area and permit larger loads to be carried, the balls run in curvilinear grooves in the rings. The radius of the groove is slightly larger than the radius of the ball and a very slight amount of radial play must be provided. The bearing is thus permitted to adjust itself to small amounts of angular misalignment in the assembled shaft and mounting. The separator keeps the balls evenly spaced and prevents them from touching each other on the sides where their relative velocities are the greatest.

Ball bearings are made in a wide variety of types and sizes. Single-row radial bearings are made in four series, extra light, light, medium, and heavy, for each bore, as illustrated in Fig. 9-1(a), (b), and (c). The heavy series of

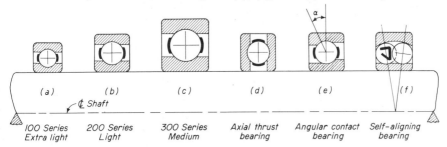

(a)	(b)	(c)	(d)	(e)	(f)
100 Series Extra light	200 Series Light	300 Series Medium	Axial thrust bearing	Angular contact bearing	Self-aligning bearing

Figure 9-1 Types of ball bearings.

bearings is designated by 400. Most manufacturers use a numbering system so devised that if the last two digits are multiplied by 5, the result will be the bore in millimeters. The digit in the third place from the right indicates the series number. Thus, bearing 307 signifies a medium-series bearing of 35-mm bore. Additional digits, which may be present in the catalog number of a bearing, refer to manufacturer's details. Some makers list deep groove bearings and bearings with two rows of balls.

The radial bearing is able to carry a considerable amount of axial thrust. However, when the load is directed entirely along the axis, the thrust type of bearing should be used. The angular contact bearing will take care of both radial and axial loads. The self-aligning ball bearing will take care of large amounts of angular misalignment. An increase in radial capacity may be secured by using rings with deep grooves, or by employing a double-row radial bearing.

Radial bearings are divided into two general classes, depending on the method of assembly. These are the Conrad, or nonfilling-notch type, and the maximum, or filling-notch type. In the Conrad bearing, the balls are placed between the rings as shown in Fig. 9-2(a). Then they are evenly spaced and the separator is riveted in place. In the maximum-type bearing, the balls are inserted through a filling notch ground into each ring, as shown in Fig. 9-2(b). Because more balls can be placed in such bearings, their load capacity is greater than that of the Conrad type. However, the presence of the notches limits the load-carrying ability of these bearings in the axial direction.

High-carbon chromium steel, 52100, is used for balls and rings. It is heat treated to high strength and hardness, and the surfaces are smoothly

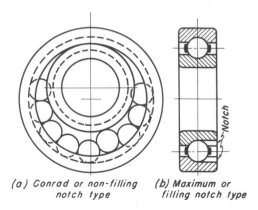

(a) Conrad or non-filling
notch type

(b) Maximum or
filling notch type

Figure 9-2 Methods of assembly for ball bearings.

ground and polished. The dimensional tolerances are very small; the balls must be very uniform in size. The stresses are extremely high because of the small contact areas, and the yield point of the material may be exceeded at certain points. Because of the high values of the fluctuating stresses, antifriction bearings are not designed for unlimited life, but for some finite period of service determined by the fatigue strength of the materials. A specified speed and number of hours of expected service must therefore accompany the given load values for these bearings.

2. Selection of Ball Bearings

Ball bearings were formerly rated on the basis of the compressive stress in the most heavily loaded ball. Except for static loads, experience has shown that the actual cause of failure is fatigue. Fatigue characteristics are thus used for load rating and are dependent to a large extent on experimental results.

Experience has indicated that the life of an individual bearing cannot be precisely predicted. It is also impossible to predict the minimum life that any one of a group of apparently identical bearings under test will exceed. Because of these uncertainties, definitions of terms relating to the life of bearings must be carefully noted.

"The life of an individual ball bearing is defined as the number of revolutions (or hours at some given constant speed) which the bearing runs before the first evidence of fatigue develops in the material of either ring or of any of the rolling elements.[1]

"The *rating life* of a group of apparently identical ball bearings is defined as the number of revolutions (or hours at some given constant speed) that 90 per cent of a group of bearings will complete or exceed before the first

[1] *Method of Evaluating Load Ratings of Ball Bearings*, Sec. 9 New York: Anti-Friction Bearing Mfg. Assn., Inc., 1960.

evidence of fatigue develops. As presently determined, the life which 50 per cent of the group of ball bearings will complete or exceed is approximately *five times* this rating life.

"The *basic load rating* is that constant stationary radial load which a group of apparently identical ball bearings with stationary outer ring can endure for a rating life of one million revolutions of the inner ring."

The basic load rating C for a rating life of one million revolutions for radial and angular contact ball bearings, except filling slot bearings, with balls not larger than 1 in. diam is[2] given by the equation

$$C = f_c (i \cos \alpha)^{0.7} Z^{2/3} D^{1.8} \tag{1}$$

where f_c = a constant from Table 9-1, as determined by the value of $(D \cos \alpha)/d_m$

Table 9-1 CONSTANTS FOR SINGLE-ROW RADIAL CONTACT GROOVE BALL BEARINGS

$\dfrac{D \cos \alpha}{d_m}$	f_c	$\dfrac{D \cos \alpha}{d_m}$	f_c	$\dfrac{F_a}{iZD^2}$	X	Y
0.05	3,550	0.22	4,530	25	0.56	2.30
0.06	3,730	0.24	4,480	50		1.99
0.07	3,880	0.26	4,420	100		1.71
0.08	4,020	0.28	4,340	150		1.55
0.09	4,130	0.30	4,250	200		1.45
0.10	4,220	0.32	4,160	300		1.31
0.12	4,370	0.34	4,050	500		1.15
0.14	4,470	0.36	3,930	750		1.04
0.16	4,530	0.38	3,800	1,000		1.00
0.18	4,550	0.40	3,660			
0.20	4,550					

f_c is also valid for single- and double-row angular contact groove ball bearings. X and Y are also valid for double-row radial contact groove ball bearings.

i = number of rows of balls in the bearing

α = nominal angle of contact (angle between line of action of ball load and plane perpendicular to bearing axis)

Z = number of balls per row

D = ball diameter

d_m = pitch diameter of ball races

Table 9-2 shows dimensions for typical ball bearings, and the basic load rating C.

Example 1. Find the value of C for a 207 radial bearing.

Solution. By Table 9-2: $d_m = \frac{1}{2}(2.8346 + 1.3780) = 2.1063$ in.

$$\frac{D \cos \alpha}{d_m} = \frac{0.4375}{2.1063} = 0.208$$

[2]For balls larger than 1 in. diam, the exponent for D is 1.4.

Table 9-2 DIMENSIONS AND BASIC LOAD RATINGS FOR CONRAD-TYPE SINGLE-ROW RADIAL BALL BEARINGS

Bearing No.	Bore		Outside Diameter		Width		Balls		Capacity	
	mm	in.	mm	in.	mm	in.	No.	diam	Dynamic C	Static P_{st}
102	15	0.5906	32	1.2598	9	0.3543	9	3/16	965	560
202			35	1.3780	11	0.4331	7	1/4	1,340	780
302			42	1.6535	13	0.5118	8	17/64	1,660	1,000
103	17	0.6693	35	1.3780	10	0.3937	10	3/16	1,040	630
203			40	1.5748	12	0.4724	7	5/16	1,960	1,220
303			47	1.8504	14	0.5512	6	3/8	2,400	1,500
104	20	0.7874	42	1.6535	12	0.4724	9	1/4	1,620	1,000
204			47	1.8504	14	0.5512	8	5/16	2,210	1,390
304			52	2.0472	15	0.5906	7	3/8	2,760	1,750
105	25	0.9843	47	1.8504	12	0.4724	10	1/4	1,740	1,110
205			52	2.0472	15	0.5906	9	5/16	2,420	1,560
305			62	2.4409	17	0.6693	8	13/32	3,550	2,350
106	30	1.1811	55	2.1654	13	0.5118	11	9/32	2,290	1,550
206			62	2.4409	16	0.6299	9	3/8	3,360	2,250
306			72	2.8346	19	0.7480	8	1/2	5,120	3,560
107	35	1.3780	62	2.4409	14	0.5512	11	5/16	2,760	1,910
207			72	2.8346	17	0.6693	9	7/16	4,440	3,070
307			80	3.1496	21	0.8268	8	17/32	5,750	4,020
108	40	1.5748	68	2.6772	15	0.5906	13	5/16	3,060	2,260
208			80	3.1496	18	0.7087	9	1/2	5,640	4,000
308			90	3.5433	23	0.9055	8	5/8	7,670	5,560
109	45	1.7717	75	2.9528	16	0.6299	13	11/32	3,630	2,730
209			85	3.3465	19	0.7480	9	1/2	5,660	4,000
309			100	3.9370	25	0.9843	8	11/16	9,120	6,730
110	50	1.9685	80	3.1496	16	0.6299	14	11/32	3,770	2,940
210			90	3.5433	20	0.7874	10	1/2	6,070	4,450
310			110	4.3307	27	1.0630	8	3/4	10,680	8,010
111	55	2.1654	90	3.5433	18	0.7087	13	13/32	4,890	3,820
211			100	3.9370	21	0.8268	10	9/16	7,500	5,630
311			120	4.7244	29	1.1417	8	13/16	12,350	9,400
112	60	2.3622	95	3.7402	18	0.7087	14	13/32	5,090	4,110
212			110	4.3307	22	0.8661	10	5/8	9,070	6,950
312			130	5.1181	31	1.2205	8	7/8	14,130	10,900
113	65	2.5591	100	3.9370	18	0.7087	15	13/32	5,280	4,410
213			120	4.7244	23	0.9055	10	11/16	10,770	8,410
313			140	5.5118	33	1.2992	8	15/16	16,010	12,520
114	70	2.7559	110	4.3307	20	0.7874	14	15/32	6,580	5,480
214			125	4.9213	24	0.9449	10	11/16	10,760	8,410
314			150	5.9055	35	1.3780	8	1	18,000	14,240

By Table 9-1: $f_c = 4,550$

By Table 9-2: $D = \frac{7}{16} = 0.4375$ in.

$$\log D = 9.64098 - 10$$

$$1.8 \log D = 9.35376 - 10$$

$$D^{1.8} = 0.2258$$

$$Z = 9, \qquad Z^{2/3} = \sqrt[3]{9^2} = 4.327$$

By Eq. (1): $C = 4,550 \times 4.327 \times 0.2258 = 4,440$ lb,

                                             load for one million revolutions

If two groups of identical bearings are run with different loads P_1 and P_2, their rating lives N_1 and N_2 revolutions are found to be inversely proportional to the cubes of their loads.

$$\frac{N_1}{N_2} = \frac{P_2^3}{P_1^3} \tag{2}$$

This equation can be generalized as follows.

$$N_1 P_1^3 = N_2 P_2^3 = N_3 P_3^3 = \cdots = 1,000,000 C^3, \qquad \text{a constant} \tag{3}$$

The terms in this equation are constants and are all equal to one another. They are also equal to $10^6 C^3$ as previously defined. If average life is to be used, the entire equation can be multiplied by a constant, usually 5. Care must be exercised, however, in the use of catalogs, because some makers use other factors. Life N in revolutions in all equations can be replaced if desired by

$$N = 60nL \tag{4}$$

where

$$n = \text{speed, rpm}$$

$$L = \text{life, hr}$$

Example 2. For the 207 bearing of Example 1, find the radial load for a rating life of 500 hr, or an expected average life of 2,500 hr at 1,500 rpm.

Solution.

By Eqs. (3) and (4): $P_1^3 = \dfrac{10^6 C^3}{N_1} = \dfrac{10^6 C^3}{60nL}$

Or: $= \dfrac{10^6 \times 4,440^3}{60 \times 1,500 \times 500} = 1,945,000,000$

From which: $P_1 = 1,250$ lb

3. Effect of Axial Load

When an axial component of load is present in addition to the radial, the *equivalent radial load* P_e is the larger of the values given by the two following equations.[3]

[3]See footnote 1.

$$P_e = V_1 F_r \tag{5}$$

$$P_e = X V_1 F_r + Y F_a \tag{6}$$

where F_r = radial component of load

F_a = axial component of load

X = radial factor from Table 9-1

Y = axial or thrust factor from Table 9-1 as determined from value of F_a/iZD^2

V_1 = race rotation factor, equal to unity for inner ring rotation, and 1.2 for outer ring rotation

Equations (5) and (6) apply to radial and angular contact bearings, but not to filling-slot bearings.

A service factor C_1 can be inserted into Eqs. (5) and (6) to care for any shock and impact conditions to which the bearing may be subjected.

$$P_e = C_1 V_1 F_r \tag{7}$$

$$P_e = C_1(X V_1 F_r + Y F_a) \tag{8}$$

Values to be used for C_1 depend on the judgment and experience of the designer, but Table 9-3 may serve as a guide.

Table 9-3   SHOCK AND IMPACT FACTORS

Type of Load	C_1
Constant or steady	1.0
Light shocks	1.5
Moderate shocks	2.0
Heavy shocks	3.0 and up

The impact load on a bearing should not exceed the static capacity as given by Table 9-2 or Eq. (11) or the race may be damaged by Brinelling from the balls. This load may be exceeded somewhat if the bearing is rotating and the duration of the load is sufficient for the bearing to make one or more complete revolutions while the load is acting.

Example 3. Suppose the bearing in Example 1 carries a combined load of 400 lb radially and 300 lb axially at 1,200 rpm. The outer ring rotates, and the bearing is subjected to moderate shock. Find the average expected life of this bearing in hours.

Solution.
$$\frac{F_a}{iZD^2} = \frac{300}{9 \times 0.4375^2} = 174$$

By Table 9-1: $Y = 1.50$

By Table 9-3: $C_1 = 2$

By Eq. (8): $P_e = 2(0.56 \times 1.2 \times 400 + 1.5 \times 300)$

$\qquad\qquad\qquad = 1{,}440$ lb equivalent radial load

By Eqs. (3) and (4): $N = \dfrac{10^6 C^3}{P^3} = 60nL$

Expected rating life: $L = \dfrac{10^6 C^3}{60n P^3} = \dfrac{10^6 \times 4{,}440^3}{60 \times 1{,}200 \times 1{,}440^3} = 410 \text{ hr}$

Expected average life $= 410 \times 5 = 2{,}050 \text{ hr}$

Example 4. What change in the loading of a ball bearing will cause the expected life to be doubled?

Solution. Let N_1 and P_1 be the original life and load for the bearing. Let N_2 and P_2 be the new life and load.

Then: $N_2 = 2N_1$

By Eq. (3): $P_2^3 = \dfrac{N_1 P_1^3}{N_2} = \dfrac{N_1 P_1^3}{2N_1} = 0.5\, P_1^3$

Or: $P_2 = 0.794 P_1$

Hence a reduction of the load to 79 per cent of its original value will cause a doubling of the expected life of a ball bearing.

4.   Design for Variable Loading

Ball bearings frequently operate under conditions of variable load and speed. Design calculations should take into account all portions of the work cycle and should not be based solely on the most severe operating conditions. The work cycle should be divided into a number of portions in each of which the speed and load can be considered as constant.

Suppose $P_1, P_2, \ldots$ are the loads on the bearing for successive intervals of the work cycle. Let N_1 be the life of the bearing revolutions, if operated under the constant load of P_1; N_2 is the life if the load is constant at P_2; and so on. Then, by Miner's equation as derived in Chapter 2,

$$\frac{\alpha_1}{N_1} + \frac{\alpha_2}{N_2} + \cdots = \frac{1}{N} \tag{9}$$

where α_1 represents the proportion of the revolutions turned in the work cycle for load P_1, α_2 represents the proportion at load P_2, and so on. By Eq. (3),

$$N_1 = \frac{10^6 C^3}{P_1^3}, \qquad N_2 = \frac{10^6 C^3}{P_2^3}$$

and so on. Substitution should be made into Eq. (9).

$$\frac{\alpha_1 P_1^3}{10^6 C^3} + \frac{\alpha_2 P_2^3}{10^6 C^3} + \cdots = \frac{1}{N}$$

or

$$\alpha_1 P_1^3 + \alpha_2 P_2^3 + \cdots = \frac{10^6 C^3}{N} \tag{10}$$

It is obvious that $\alpha_1 + \alpha_2 + \cdots$ must equal unity. The application of this equation will be demonstrated by the following examples.

Example 5. A 207 bearing is to operate on the following work cycle.

Radial load of 1,400 lb at 200 rpm for 25 per cent of the time
Radial load of 2,000 lb at 500 rpm for 20 per cent of the time
Radial load of   800 lb at 400 rpm for 55 per cent of the time

The inner ring rotates. The loads are steady. Find the expected average life of this bearing in hours.

Solution. The equation will be applied on the basis of a work cycle of 1 min. The following table should be constructed.

	Assumed Interval, min	rpm	In Assumed Interval, rev.
$P_1 = 1,400$ lb	0.25	200	50
$P_2 = 2,000$ lb	0.20	500	100
$P_3 = \ \ 800$ lb	0.55	400	220
	1.00		370 rpm

Then: $\qquad \alpha_1 = \dfrac{50}{370}, \qquad \alpha_2 = \dfrac{100}{370}, \qquad \alpha_3 = \dfrac{220}{370}$

In Eq. (10): $\dfrac{5}{37} \times 1,400^3 + \dfrac{10}{37} \times 2,000^3 + \dfrac{22}{37} \times 800^3 = \dfrac{10^6 \times 4,440^3}{N}$

$$\frac{5}{37} \times 2.744 + \frac{10}{37} \times 8 + \frac{22}{37} \times 0.512 = \frac{10^6 \times 4.440^3}{N}$$

$$0.3708 + 2.1622 + 0.3044 = \frac{10^6 \times 87.528}{N}$$

Or: $\qquad\qquad\qquad\qquad\qquad\qquad\qquad N = 30,850,000$ rev.

In Eq. (4): $\qquad\qquad L = \dfrac{N}{60n} = \dfrac{30,850,000}{60 \times 370} = 1,390$ hr $\qquad$ rating life

average life $= 5 \times 1,390 = 6,950$ hr

Example 6. A 306 radial ball bearing with inner ring rotation has a 10-sec work cycle as follows.

For 2 sec	For 8 sec
$F_r = 800$ lb	$F_r = 600$ lb
$F_a = 400$ lb	$F_a = 0$
$n = 900$ rpm	$n = 1,200$ rpm
Light shock	*Steady load*

Find the expected average life of this bearing.

Solution. Equation (10) will be applied to the work cycle of 10 sec. For 2 sec:

By Eq. (7): $\qquad\qquad\qquad P_e = C_1 V_1 F_r = 1.5 \times 1 \times 800 = 1,200$ lb

Data from Table 9-2: $\dfrac{F_a}{iZD^2} = \dfrac{400}{8 \times 0.5^2} = 200$

By Table 9-1: $Y = 1.45$

In Eq. (8): $P_e = 1.5(0.56 \times 1 \times 800 + 1.45 \times 400)$

                         $= 1,540\,\text{lb},$ equivalent radial load

                                                  (will be used)

No. of revolutions $= 2 \times \dfrac{900}{60} = 30,$ in 2 sec

For 8 sec: $P_e = F_r = 600\,\text{lb}$

No. of revolutions $= 8 \times \dfrac{1,200}{60} = 160,$ in 8 sec

For 10 sec: No. of revolutions $= 30 + 160 = 190$

Then: $\alpha_1 = \dfrac{30}{190}, \quad \alpha_2 = \dfrac{160}{190}$

By Eq. (10): $\dfrac{3}{19} \times 1,540^3 + \dfrac{16}{19} \times 600^3 = \dfrac{1\,0^6 \times 5,120^3}{N}$

$\dfrac{3}{19} \times 3.6665 + \dfrac{16}{19} \times 0.216 = \dfrac{10^6 \times 5.12^3}{N}$

$0.5789 + 0.1819 = \dfrac{10^6 \times 134.218}{N}$

$N = \dfrac{10^6 \times 134.218}{0.7608} = 176,420,000\,\text{rev.}$

Average speed. $n = \dfrac{190}{10} \times 60 = 1,140\,\text{rpm}$

By Eq. (4): $L = \dfrac{N}{60n} = \dfrac{176,420,000}{60 \times 1,140} = 2,580\,\text{hr},$

                                                  rating life

average life $= 5 \times 2,580 = 12,900\,\text{hr}$

5. Static Capacity

Unless there is relative motion between the rings of a ball bearing, the depressions of the balls into the races will gradually enlarge, and permanent indentations will remain. The static capacity is ordinarily defined as the maximum allowable static load that does not impair the running characteristics of the bearing to make it unusable.

For single-row radial ball bearings the equation[4] usually used for static capacity is

$$P_{st} = 1,780ZD^2 \tag{11}$$

where, as before, Z is the number of balls in the row, and D is the ball diameter.

[4]See footnote 1.

Example 7. Calculate the static capacity for the 306 bearing of Table 9-2.

Solution. By Table 9-2, $Z = 8$ balls of diameter $D = \frac{1}{2}$ in.
By Eq. (11): $P_{st} = 1,780 \times 8 \times (\frac{1}{2})^2 = 3,560$ lb
Equation (11) was used in determining the static capacities of Table 9-2.

6. Friction and Lubrication of Ball Bearings

Rolling friction prevails to a large extent in ball and roller bearings. Ball bearings, in general, have slightly less frictional resistance than high-grade partial journal bearings operating with flooded lubrication. Ball bearings have definitely less friction than journal bearings operating with scanty oil supply.

Average values for the coefficient of friction[5] are as follows:

Self-aligning ball bearings	0.0010
Cylindrical roller bearings with flange guided short rollers	0.0011
Thrust ball bearings	0.0013
Single-row deep-groove ball bearings	0.0015
Tapered and spherical roller bearings with flange-guided rollers	0.0018
For cageless needle bearings, in a similar manner, is obtained	0.0045

The amount and viscosity of the lubricating oil used determine to a large extent the friction of a ball bearing. Tests have shown that drip feed, which supplies a drop of oil every 2–4 hr, gives much lower friction values than a more copious oil supply. In fact, for flooded lubrication, the power lost in churning the oil may be greater than the friction of the bearing alone. A very light or thin oil will give lower friction than one whose viscosity is high.

A light coating of oil or grease is all that is required to maintain an oil film between balls and races. When oil lubrication is used, more or less elaborate seals are needed to retain the lubricant. Lubrication by oil mist has proved successful for very high-speed applications.

Grease is essentially a soap impregnated with lubricating oil. For rolling bearing service, the oil has viscosity properties similar to SAE 20 or SAE 30.

Grease tends to remain in the bearing and protect the surfaces. Therefore it does not require such elaborate retainers. At low temperatures, the balls cut a channel through the grease, but enough oil usually sweats off to provide lubricant. Prelubricated sealed bearings can frequently be mounted at lower cost because sealing parts and grease fittings are eliminated. The bearings are filled at the factory with the proper quantity of grease. They have been known to run for years without servicing. For temperatures above 250°F, special high-temperature greases must be used.

[5] See Reference 6, p. 41, end of chapter. See also Styri, H., "Friction Torque in Ball and Roller Bearings," *Mech. Eng.*, **62**, 1940, p. 886.

Reliability of the oil supply is of the utmost importance. Lack of lubricant can cause local heating, expansion, and loss of radial play. The load on the balls may be increased to the extent that spalling and early bearing failure occur.

Back and forth rotation of the shaft through small angles can cause early failure of bearings unless the load is very light. Lubrication is difficult because the oil or grease may not be replenished back of a ball or roller before the motion is reversed.

The lubricant in a ball bearing serves not only to reduce the friction, but to prevent foreign matter, which would injure the surfaces, from entering the bearing. Every effort must be made to protect the highly polished surfaces from grit, water, acids, or anything that will cause scratches or corrosion. Corrosion fatigue is particularly injurious in caus-ing early fatigue failure of the bearing. Fatigue failures are caused by the bearing surface breaking down; small particles of the metal come out and leave pits or spalls, as illustrated in Fig. 9-3. This breakdown is preceded by minute surface cracks, which are developed by repeated stress applications until they become sufficiently large to form zones of local weakness.

Figure 9-3 Failure of bearing race by pit-ting or spalling. (*Courtesy of New Depar-ture, Division of General Motors.*)

7. Bearing Materials and Surface Finish

As previously mentioned, high-carbon chromium steel 52100 is widely used as a ball bearing material. This is a through hardening steel, and a commonly accepted minimum hardness for bearing components is 58 Rockwell C. This material is not suitable for temperatures over 350°F. For higher temperatures, steels especially developed for high-temperature service should be used.

In contrast to ball bearings, roller bearings are usually made of case-hardened steels. The carburized case or exterior should have a hardness of 58–63 R_c. The core is softer with a hardness of 25–40 R_c. Certain plain-carbon and alloy steels have been found suitable for roller bearing service. The maximum temperature is limited to about 350°F.

The separator, cage, or retainer for conventional bearings is usually a stamping of low-carbon steel. For higher speeds or precision service, the separator is machined from a suitable copper alloy, such as bronze. Cages are also made of a solid lubricant material for use where conventional types of lubrication cannot be used.

The life of a rolling element bearing depends to a large extent on the smoothness of the contacting surfaces—the balls, rollers, and races. Typical surface roughness dimensions for production bearings are as follows.[6] These are in terms of microinches or millionths of an inch, usually written μ in.

Balls	2–3 μin. rms
Ball races	6–10 μin. rms
Rollers	8–12 μin. rms
Roller races	10–20 μin. rms

Surface finish values vary considerably from manufacturer to manufacturer. For critical applications it may be necessary to obtain the actual value of the surface finish for the bearing under consideration.

8. Mounting of Ball Bearings

For a rotating shaft, relative rotation between shaft and bearing is usually prevented by mounting the inner ring with a press fit and securing it with a nut threaded on the shaft. Excessive interference of metal must be avoided in press fits, or the stretching of the inner ring may decrease the small but requisite internal looseness of the bearing.

Although the outer ring, when the shaft rotates, is mounted more loosely than the inner ring, rotational creep between the ring and housing should be prevented. When two bearings are mounted on the same shaft, the outer ring of one of them should be permitted to shift axially to care for any differential expansion between shaft and housing. Several examples of typical mounting details with oil retainers are shown in Fig. 9-4. The catalogs of the various manufacturers contain useful illustrations of this kind, as well as other practical information.

Shafts or spindles in machine tools and precision equipment that must rotate without play or clearance in either the radial or axial directions can be mounted on preloaded ball bearings. The preloading, which removes all play from the bearing, can be secured in a number of different ways. For example, suppose the outer rings of the bearings at A in Fig. 9-5 project a small but controlled amount beyond the inner rings. When the inner rings are brought into contact at B by means of the locknut, the balls will be displaced in the rings an amount sufficient to remove all looseness from the bearing. Close attention must be paid to dimensions and tolerances to secure just enough projection of the ring to remove the play, but not so much as to induce excessive pressure or binding of the balls. The bearing at the other end of the shaft must be arranged for free axial movement of the outer ring. The bearings in Fig. 9-5 can be separated if desired with one bearing at each end of the shaft.

[6]For a discussion of root mean square (rms), average, see Chapter 13.

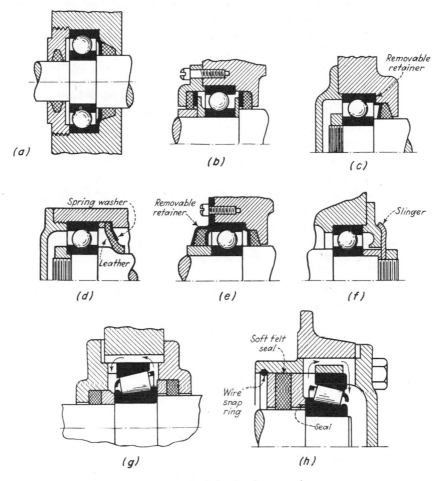

Figure 9-4 Details of typical anti-friction bearing mountings.

Although this arrangement will remove the looseness from both ends of the shaft, serious stresses may be induced by a temperature difference between shaft and housing. Preloaded, double-row radial bearings are made by some manufacturers.

9. Permissible Misalignment

Misalignment of ball and roller bearings can have a serious effect on their life and load-carrying capacity. Misalignment can occur from shaft

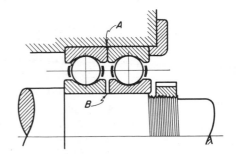

Figure 9-5 Method for obtaining preloading in ball bearings.

deflections, from loading, or from errors in machining the supports for the rings. Sometimes the support for a bearing will deflect elastically and the design should be arranged so that the housing deflection offsets the shaft deflection rather than opposing it.

Antifriction bearings are manufactured with various amounts of internal clearance. If the misalignment does no more than remove the internal clearance of a ball bearing, in general no shortening of the life will occur. Pressing a ring on a shaft may remove a portion of the original clearance. Permissible misalignment is also influenced by such factors as types of loading, race curvature and depth, diameter and number of balls, method of lubrication, retainer geometry, and minor dimensional details of the various makes.

Roller bearings are more sensitive to the effects of misalignment so that any slope of the shaft has a bad effect on the life and load capacity. A slight crown on the rolls has proved to be beneficial in overcoming the effects of misalignment. In needle bearings the rollers depend on each other for guidance so that some resultant clearance is necessary for satisfactory operation.

Recommended misalignment limits are given in Table 9-4. In this table

Table 9-4 PERMISSIBLE ANGULAR MISALIGNMENT BETWEEN BEARING AND SHAFT

Single-row radial ball bearing:	
Slow speed and loose fit	0.0020–0.0040 in./in.
High speed	0.0010 in./in.
Cylindrical roller bearing	0.0003–0.0015 in./in.
Needle bearing	0.0003–0.0010 in./in.
Tapered roller bearing	0.0005 in./in.

some makers recommend the lower ranges, while others permit the higher values. In case of doubt, the safest plan is to obtain the assistance of the maker of the particular bearing that is under consideration.[7]

10. Unground Ball Bearings

The foregoing discussion has referred to ball bearings of the highest quality of materials and workmanship. Other bearings of lower quality can be purchased for installations requiring less accuracy or where cost is the controlling factor. The rings are made on automatic screw machines and are hardened but not ground.[8]

[7]See Smith, R. J., "Rules for Application of Needle Bearings," *Machine Design*, **30**, Jan. 9, 1958, p. 116. See also Korff, W. H., "Twelve General Considerations for Needle Bearing Application," *Prod. Eng.*, **15**, 1944, p. 389.

[8]See Agnoff, C., "Multipurpose Bearings," *Machine Design*, **39**, Aug. 31, 1967, p. 98. See also Recknagel, F. W., "Construction and Characteristics of Low-Cost Ball Bearings," *Prod. Eng.*, **21**, Jan. 1950, p. 106.

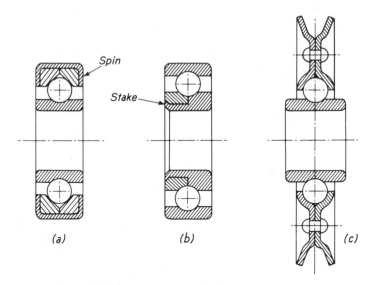

Figure 9-6 Types of unground ball bearings.

Different types of construction are in use. The bearing of Fig. 9-6(a) has the outer ring split by a plane perpendicular to the axis. The bearing is assembled by spinning the edges of the bushing, which is slipped over the outer rings. The bearing of Fig. 9-6(b) has a split inner ring, and is made by staking the bore of the inner ring as shown. Various additional features, such as pulleys, gears, castor wheels, and so on, can be incorporated as an integral part of the outer ring. Sketch (c) shows a sheave-idler in which the outer ring is formed by the stampings comprising the sheave. Unground ball bearings are frequently cheaper than an equivalent plain bushing.

11. Relative Advantages of Rolling Element Bearings

Some of the advantages of rolling element are as follows.

(1) Starting friction is low—a desirable feature for intermittent service or for starting at low temperatures.

(2) Loads can be inclined at any angle in the transverse plane.

(3) Thrust components can be carried.

(4) Maintenance costs are low.

(5) Bearings are easily replaced when worn out.

(6) Less axial space is required than for journal bearings. Shafts are thus shorter, and may even be smaller in diameter.

Some of the advantages of fluid film bearings are as follows.

(1) First cost is usually lower.

(2) Less radial space is required than for ball bearings.

(3) They are better suited to overload and shock conditions.

(4) Operation is quieter than with ball bearings, especially after wear has taken place.

(5) There is less difficulty with fatigue.

(6) They are less easily injured by foreign matter.

12. Roller Bearings

When shock and impact loads are present, or when a large bearing is needed, cylindrical and tapered roller bearings are usually used. A roller bearing in general consists of the same four elements as a ball bearing: the two rings, the cage, and the rollers. Some typical examples of roller bearings are shown in Fig. 9-7.

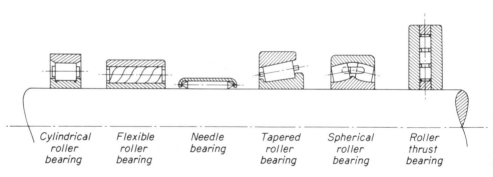

| Cylindrical roller bearing | Flexible roller bearing | Needle bearing | Tapered roller bearing | Spherical roller bearing | Roller thrust bearing |

Figure 9-7 Types of roller bearings.

In the cylindrical roller bearing, the flanges on the rings serve to guide the rollers in the proper direction. When the flanges are omitted from one of the rings, as shown in Fig. 9-7, the rings can then be displaced axially with respect to each other, and no thrust component can be carried.

In addition to the radial load, the tapered roller bearing can carry a large axial component whose magnitude depends on the angularity of the rollers. The radial load will also produce a thrust component. The outer ring is separable from the remainder of the bearing. In this type of bearing, it is possible to make adjustment for the radial clearance. Two bearings are usually mounted opposed to each other, and the clearance is controlled by adjusting one bearing against the other. Double-row tapered roller bearings are also available.

Roller bearings in general can be applied only where the angular misalignment caused by shaft deflection is very slight. This deficiency is not present in the spherical roller bearing. It has excellent load capacity and can carry a thrust component in either direction.

In the flexible roller bearing, the rollers are wound from strips of spring steel, and afterwards are hardened and ground to size. If desired, the rollers can bear directly on the shaft without an inner ring, particularly if the shaft surface has been locally hardened. This bearing has been successfully applied under conditions of dirty environment.

The needle bearing has rollers that are very long as compared to their diameters. Cages are frequently not used, and the inner ring may or may not be present. The outer ring may consist of hardened thin-walled metal as shown in Fig. 9-7; the housing in which the bearing is mounted must have sufficient thickness to give adequate support. The friction of needle bearings is several times as great as for ordinary cylindrical roller bearings. Because of the tendency of the unguided rollers to skew, needle bearings are particularly adapted to oscillating loads, as in wrist pins, rocker arms, and universal joints. For continuous rotation, needle bearings are usually suitable where the loading is intermittent and variable so that the needles will be frequently unloaded and thus tend to return to their proper locations. When the application involves angular misalignment of the shaft, two short bearings end to end usually are better than one bearing with long rollers. The needle bearing is low-priced and requires very little radial space.

Spherical roller bearings, Figs. 9-7 and 9-8, can be used when the shaft has angular misalignment.

Thrust bearings can be constructed by the use of straight or tapered rollers.

Roller bearings are selected by a process similar to that used for ball bearings. They must be chosen, however, in accordance with the recommendations given in the catalog of the manufacturer of the particular type of bearing under consideration.

Figure 9-8 Self-aligning spherical roller bearing for radial and thrust loads. (*Courtesy McGill Mfg. Co.*)

13. Contact Stress Between Cylinders

When two bodies having surfaces of different curvatures are pressed together, an area of contact develops, because of course there is no such thing as line or point contact. This area is usually very small and the resulting stress can attain a very high value. In addition to rolling contact bearings, the problem is of importance in the design of cams and gear teeth.

The elastic analysis is very complex, but Fig. 9-9 gives some results for

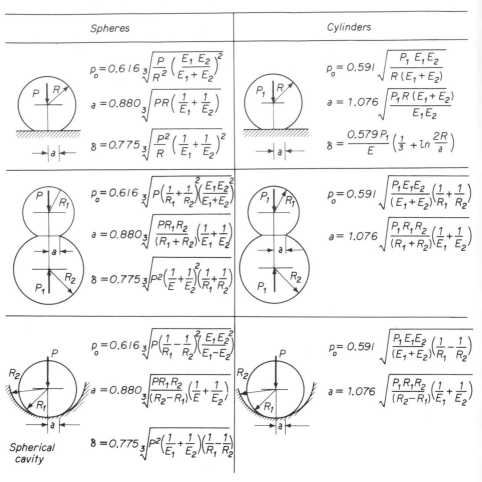

| | Spheres | | Cylinders |

$$p_o = 0.616 \sqrt[3]{\frac{P}{R^2}\left(\frac{E_1 E_2}{E_1 + E_2}\right)^2}$$

$$a = 0.880 \sqrt[3]{PR\left(\frac{1}{E_1} + \frac{1}{E_2}\right)}$$

$$\delta = 0.775 \sqrt[3]{\frac{P^2}{R}\left(\frac{1}{E_1} + \frac{1}{E_2}\right)^2}$$

$$p_o = 0.591 \sqrt{\frac{P_1 E_1 E_2}{R(E_1 + E_2)}}$$

$$a = 1.076 \sqrt{\frac{P_1 R(E_1 + E_2)}{E_1 E_2}}$$

$$\delta = \frac{0.579 P_1}{E}\left(\frac{1}{3} + \ln\frac{2R}{a}\right)$$

$$p_o = 0.616 \sqrt[3]{P\left(\frac{1}{R_1} + \frac{1}{R_2}\right)^2\left(\frac{E_1 E_2}{E_1 + E_2}\right)^2}$$

$$a = 0.880 \sqrt[3]{\frac{PR_1 R_2}{(R_1 + R_2)}\left(\frac{1}{E_1} + \frac{1}{E_2}\right)}$$

$$\delta = 0.775 \sqrt[3]{P^2\left(\frac{1}{E} + \frac{1}{E_2}\right)^2\left(\frac{1}{R_1} + \frac{1}{R_2}\right)}$$

$$p_o = 0.591 \sqrt{\frac{P_1 E_1 E_2}{(E_1 + E_2)}\left(\frac{1}{R_1} + \frac{1}{R_2}\right)}$$

$$a = 1.076 \sqrt{\frac{P_1 R_1 R_2}{(R_1 + R_2)}\left(\frac{1}{E_1} + \frac{1}{E_2}\right)}$$

$$p_o = 0.616 \sqrt[3]{P\left(\frac{1}{R_1} - \frac{1}{R_2}\right)^2\left(\frac{E_1 E_2}{E_1 - E_2}\right)^2}$$

$$a = 0.880 \sqrt[3]{\frac{PR_1 R_2}{(R_2 - R_1)}\left(\frac{1}{E} + \frac{1}{E_2}\right)}$$

$$\delta = 0.775 \sqrt[3]{P^2\left(\frac{1}{E_1} + \frac{1}{E_2}\right)\left(\frac{1}{R_1} - \frac{1}{R_2}\right)}$$

$$p_o = 0.591 \sqrt{\frac{P_1 E_1 E_2}{(E_1 + E_2)}\left(\frac{1}{R_1} - \frac{1}{R_2}\right)}$$

$$a = 1.076 \sqrt{\frac{P_1 R_1 R_2}{(R_2 - R_1)}\left(\frac{1}{E_1} + \frac{1}{E_2}\right)}$$

Spherical cavity

Figure 9-9 Contact stress equations. p_o, maximum compressive stress; a, half width of contact zone; P, total load on spheres; P_1, load per axial inch on cylinders; $\mu = 0.3$; δ, normal approach towards each other for points remote from contact zone; $\ln x = \log_e x = 2.3026 \log_{10} x$.

both spheres and cylinders. For spheres, a is the radius of the contact circle, but for rectangles, a represents the half-width of the contact rectangle. The total force pressing two spheres together is P, while P_1 represents the load per axial inch for cylinders. The maximum compressive stress in the zone of contact is represented by p_o. Poisson's ratio in these equations was taken as 0.3. Normal approach toward each other for points remote from the contact zone is given by δ. As usually written, $\ln x = \log_e x = 2.3026 \log_{10} x$.

Example 8. A $\frac{1}{2}$-in.-diam steel roller runs on the inside of a steel ring of inside diameter 3.2 in. If the load on the roller is 1,000 lb/axial in., find the half-width a of the contact zone, and the value of the maximum compressive stress p_o.

Solution. Use the equations in the lower-right corner of Fig. 9-9.

$$p_o = 0.591 \sqrt{\frac{1,000 \times 30,000,000^2}{2 \times 30,000,000}\left(\frac{1}{0.25} - \frac{1}{1.6}\right)}$$

$$= 0.591\sqrt{500 \times 30,000,000(4 - 0.625)}$$

$$= 0.591 \times 225,000 = 133,000 \text{ psi}$$

$$a = 1.076 \sqrt{\frac{1,000 \times 0.25 \times 1.6}{1.6 - 0.25}\left(\frac{2 \times 30,000,000}{30,000,000^2}\right)}$$

$$= 1.076 \sqrt{296.3 \times \frac{2}{30,000,000}}$$

$$= 1.076 \times 0.00444 = 0.0049 \text{ in.}$$

14. Elasto-hydrodynamic Lubrication

Elasto-hydrodynamic lubrication deals with the lubrication of highly loaded contacts, where the elastic deformation of the parts must be taken into account as well as the increase in viscosity of the lubricant due to the high pressure.[9] This small elastic flattening of the parts, together with the increase in viscosity, provides a film, although very thin, that is much thicker than would prevail with completely rigid parts. The theory is complex and beyond the scope of this book.

REFERENCES

1. Allan, R. K., *Rolling Bearings*, London: Sir Isaac Pitman & Sons Ltd., 1945.

2. *Bearings Reference Issue, Machine Design,* **42**, June 18, 1970.

3. *Boundary Lubrication—An Appraisal of World Literature*, New York: American Society of Mechanical Engineers, 1969.

4. Dowson, D., and G. R. Higginson, *Elasto-hydrodynamic Lubrication*, New York: Pergamon Press, Inc., 1966.

5. Harris, T. A., *Rolling Bearing Analysis*, New York: John Wiley & Sons, Inc., 1966.

6. Palmgren, A., *Ball and Roller Bearings*, Philadelphia: SKF Industries, Inc., 3d Ed., 1959.

[9]See Zaretsky, E. V., and W. J. Anderson, "EHD Lubrication," *Machine Design*, **40**, Nov. 7, 1968, p. 167. See also the following.
Burton, R. A, "An Analytical Investigation of Visco-elastic Effects in the Lubrication of Rolling Contact," *Trans. ASME*, 3, 1960, p. 1.
Cheng, H. S., "A Numerical Solution of the Elastohydrodynamic Film Thickness in an Elliptical Contact," *Trans. ASME*, **92**, Series F, *Jour. of Lubrication Tech.* Jan. 1970, p.155.

7. Rothbart, H. A., Ed., *Mechanical Design and Systems Handbook*, New York: McGraw-Hill Book Company, 1964, Sec. 13.

8. Shaw, M. C., and E. F. Macks, *Analysis and Lubrication of Bearings*, New York: McGraw-Hill Book Company, 1949, Chapter 10.

PROBLEMS

In the following problems, it is assumed that the bearing has but one row of balls.

1. If a 206 bearing is installed at the left reaction in Fig. 9-10, find the expected average life in hours. Speed is 1,800 rpm and the bearing is subjected to light shock. Inner rings are pressed on the shaft; outer rings have free fits.   *Ans.*   1,320 hr.

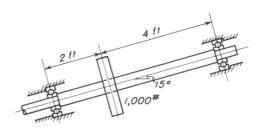

Figure 9-10 Problem 1.

2. Use Fig. 9-10 but the rise is 5 in. vertically for 12 in. horizontally. A 110 bearing is installed at the left end. The shaft rotates 1,800 rpm for 60 per cent of the time and 750 rpm for the remainder of the time. Loads are steady. Make a diagram and show the reaction at each end. Compute the expected average life of the left bearing.   *Ans.*   4,700 hr.

3. A cross section through an industrial car, and the loading for one axle, are shown schematically in Fig. 9-11. The inner rings of the bearings are fixed on the shaft; the outer rings have free fits. Wheels are press fitted on the axle.

Calculate and show reactions at the rails. Now separate the system at the bearings and show all forces acting at the points of separation for each portion. Consider the forces from bearings and wheels to be concentrated at their respective centers. Draw and dimension the bending moment diagram for the axle.

Based on the loading conditions of the left bearing, find the expected average life of a 310 bearing at a speed of 300 rpm. Assume the bearing is subjected to moderate shock.   *Ans.*   15,150 hr.

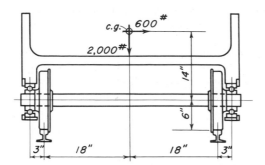

Figure 9-11 Problem 3.

4. Work Problem 3, except the downward load is 2,400 lb and the sidewise load is 450 lb. *Ans.* 16,910 hr.

5. In Fig. 3-14, let the loads be steady and the rpm equal to 900. Find the expected average life for a 212 bearing installed at the left reaction. *Ans.* 4,740 hr.

6. A 205 bearing carries a 1,000-lb radial and a 250-lb axial load for one-sixth of the work cycle, and a 500-lb radial for the remainder of the cycle. Speed is uniform at 1,200 rpm. The inner ring rotates and loads are steady. Find the expected average life for this bearing. *Ans.* 3,630 hr.

7. A 210 bearing has a work cycle with 1,000 rpm for one-third of the time, 2,000 rpm for one-third of the time, and 4,000 rpm for one-third of the time. The outer ring rotates. Assume light-shock conditions. The radial load is 800 lb and the axial load is 300 lb. Find the expected average life for this bearing. *Ans.* 1,740 hr.

8. A 206 bearing carries a radial load of 2,000 lb at 500 rpm for one-half the time, and 600 lb radial at 3,600 rpm for the remaining half of the time. The inner ring rotates, and the loads are steady. Find the expected average life. *Ans.* 1,325 hr.

9. A 206 bearing is subjected to the following work cycle.

> Radial load of 1,000 lb at 150 rpm for 30 per cent of the time
> Radial load of 1,500 lb at 600 rpm for 10 per cent of the time
> Radial load of 500 lb at 300 rpm for 60 per cent of the time

The inner ring rotates; loads are steady. What is the expected average life of this bearing? *Ans.* 11,700 hr.

10. What change in the loading of a ball bearing will cause the expected life to be halved?

11. A 104 bearing operates under the following schedule of loads and speeds. The inner ring rotates; loads are steady. Find the expected average life for this bearing.

> Radial load of 740 lb at 2,000 rpm for 5 per cent of the time
> Radial load of 510 lb at 3,300 rpm for 15 per cent of the time
> Radial load of 250 lb at 1,750 rpm for 35 per cent of the time
> Radial load of 200 lb at 2,200 rpm for 45 per cent of the time

12. A 307 bearing operates under the following schedule of loads and speeds. The inner ring rotates; loads are steady. Find the expected average life for this bearing.

> Radial load of 900 lb at 1,000 rpm for 30 per cent of the time
> Radial load of 2,000 lb at 300 rpm for 15 per cent of the time
> Radial load of 4,000 lb at 100 rpm for 55 per cent of the time

13. The work cycle for a 304 ball bearing is as follows.

> 1,000-lb radial and 250-lb axial at 300 rpm for one-half the cycle
> 1,200-lb radial at 500 rpm for one-quarter of the cycle
> 700-lb radial at 1,000 rpm for one-quarter of the cycle

Loads are steady; the outer ring rotates. Find the expected average life of this bearing.

14. Find the expected rating life of a 108 bearing that is subjected to a steady radial load of 615 lb, and an axial load of 350 lb. The inner ring rotates. The speed is 3,600 rpm.

15. An electric motor for a drill press has 102 bearings mounted as in Fig. 9-10 except that the shaft is vertical. Bearings are 10 in. center to center. The center of the V-belt pulley is located 2.5 in. from the lower motor bearing. During two-thirds of the work cycle the steady belt forces equal 128 lb at 900 rpm. During the remainder of the cycle the steady belt forces equal 88 lb at 1770 rpm. The motor armature weighs 80 lb. Find the expected rating life of the most heavily loaded bearing in hours.

10 Spur Gears

The designer is frequently confronted with the problem of transferring power from one shaft to another while maintaining a definite ratio between the velocities of rotation of the shafts. Various types of gearing have been developed for this purpose, which will operate quietly and with very low friction losses. Smooth and vibrationless action is secured by giving the proper geometric shape to the outline of the teeth. The proportions of the gear tooth, as well as the sizes of the teeth, have been standardized. This procedure has simplified design calculations and has reduced the required number of cutting tools to a minimum. The proper material must be selected to obtain satisfactory strength, fatigue, and wear properties. Ease of manufacture and ease of inspection are necessary if production costs are to be kept at their lowest level. All these problems must be taken into account by the designer.

a, addendum
b, face width
BHN, Brinell Hardness Number
c, center distance
d, pitch diameter

e, error; shift in position of generating rack
E, modulus of elasticity
f, clearance
F_b, bending capacity of tooth

375

F_d, dynamic load

F_t, horsepower or transmitted load

F_w, wear capacity of tooth

$g = 386$ in/sec², acceleration due to gravity

hp, horsepower

I, mass moment of inertia

k, spring constant of tooth

l, lead of screw

m, mass

n, revolutions per minute, rpm

N, number of teeth

p, circular pitch

p_b, base pitch

P_d, diametral pitch

r, pitch radius

rpm, revolutions per minute

R, radius of curvature of involute

s, bending stress

s_{ec}, surface endurance stress

t, time; tooth thickness

T, torque

V, velocity, feet per minute

y, form factor for circular pitch

Y form factor for diametral pitch

z, length of contact for meshing teeth

γ, (gamma) weight per cubic inch

φ, (phi) pressure angle

ω, (omega) angular velocity, radians per second

1. Introduction

In the friction cylinders of Fig. 10-1, the tangential velocities of the surfaces are equal if it is assumed that no slip occurs at the point of contact O. Hence,

$$r_1\omega_1 = r_2\omega_2 \tag{1}$$

where r_1 and r_2 refer to the radii, and ω_1 and ω_2 are the angular velocities in radians per second. These cylinders can be transformed into spur gears by placing teeth on them that run parallel to the axes of the cylinders. The circles in Fig. 10-1 are then called the pitch circles; their diameters are called the pitch diameters of the gears. The teeth are arranged to extend both outside and inside the pitch circles. The names of some of the more important parts

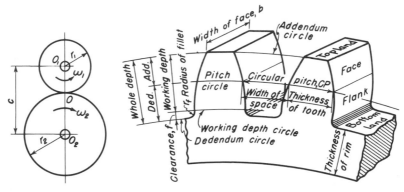

Figure 10-1 Friction cylinders.

Figure 10-2 Principal parts of gear teeth.

of spur gear teeth are shown in Fig. 10-2. When operating together, the teeth of one gear extend to the working-depth circle of the other. Clearance f is required to prevent the end of the tooth of one gear from riding on the bottomland of the mating gear.

For positive transmission of motion, the teeth need not be of any particular shape. However, for quiet and vibrationless operation, the velocities of the pitch circles of the two gears must be the same at all times. This statement refers especially to the short interval of time during which two particular teeth are in contact. If the pitch circle of the driver is moving with constant velocity, the shape of the teeth must be such that the velocity of the pitch circle of the driven gear is neither increased nor decreased at any instant while the two teeth are touching. When this condition is satisfied, the gears are said to fulfill the fundamental law of toothed gearing.

2. Fundamental Law of Toothed Gearing

Portions of two gears having centers at O_1 and O_2 are shown in Fig. 10-3. The gears have angular velocities of ω_1 and ω_2, respectively. The teeth are in contact at point K_1, but no particular shape has as yet been specified for the outline of the teeth. Lines NN and TT represent respectively the normal and tangent drawn to the tooth surfaces at K_1. Normal NN intersects the line of centers at point O. The vector K_1M_1 represents the velocity of K_1 considered as a point of gear 1. The vector is perpendicular to radius O_1K_1. Similarly, K_1M_2 represents the velocity of K_1 when K_1 is considered as a point on gear 2. Since the teeth remain in contact, the projection K_1N_1 of the velocity vectors on the common normal must be the same for both gears.

From the figure, $K_1M_1 = \omega_1 \times O_1K_1$, and $K_1M_2 = \omega_2 \times O_2K_1$, so that

$$\frac{\omega_2}{\omega_1} = \frac{O_1K_1}{K_1M_1} \times \frac{K_1M_2}{O_2K_1} \qquad (a)$$

From sides of similar triangles, the following relationships are secured from the figure.

$$\frac{O_1K_1}{K_1M_1} = \frac{O_1A}{K_1N_1} \qquad \text{and} \qquad \frac{K_1M_2}{O_2K_1} = \frac{K_1N_1}{O_2B} \qquad (b)$$

Substitution in Eq. (a) gives

$$\frac{\omega_2}{\omega_1} = \frac{O_1A}{O_2B} \qquad (c)$$

Also, from similar triangles,

$$\frac{O_1A}{O_2B} = \frac{O_1O}{O_2O} \qquad (d)$$

Substitution in Eq. (c) gives

$$\omega_1 \times O_1O = \omega_2 \times O_2O \qquad (2)$$

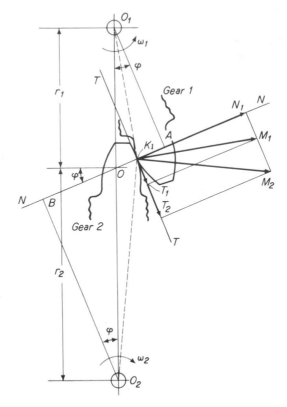

Figure 10-3 Component of velocity normal to tooth surface at point of contact is same for both gears.

The ratio ω_1/ω_2 from Eq. (2) must remain constant at all times if there is to be no change in the velocity ratio of the two gears. Equations (1) and (2) should be solved simultaneously, making use of the relationship

$$r_1 + r_2 = O_1O + O_2O$$

The result is

$$O_1O = r_1$$
$$O_2O = r_2 \qquad \text{(e)}$$

Point O is therefore a fixed point through which the pitch circles are drawn. Hence, in order to fulfill the fundamental law of toothed gearing, the sides of the teeth must be so shaped that the normal drawn through the point of contact will at all times pass through the pitch point O.

The difference $K_1T_2 - K_1T_1$ between the tangential vectors is equal to the velocity of sliding of one tooth on the other. The velocity of sliding is not constant, but varies with the location of the contact point K_1. When

K_1 coincides with O, the sliding velocity is zero, and pure rolling contact exists between the teeth for this point.

It can be shown that gear teeth composed of involutes or cycloids fulfill the fundamental law. For involute gearing, the normal NN not only passes through point O at all times, but maintains a constant inclination φ with respect to the common tangent to the pitch circles. For cycloidal gearing, point O remains fixed, but angle φ varies as the location of the point of contact K_1 changes.

3. Kinematics of Involute Gear Teeth

Assume that the required velocity ratio for a pair of gears is secured by pitch circles of radii O_1O and O_2O, respectively, in Fig. 10-4(a). Draw the line AB at an angle φ to the common tangent to the circles. In practice, φ is usually made either $14\frac{1}{2}°$ or $20°$. Circles are drawn tangent to AB and

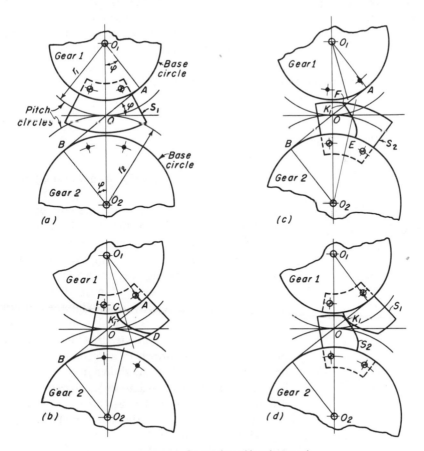

Figure 10-4 Generation of involute teeth.

are called the base circles of the gears. Let cylinders be made equal to the base circles, and let a string be passed around the base circle of gear 1, then from A to B, and then around the base circle of gear 2 somewhat in the manner of one strand of a crossed belt.

In Fig. 10-4(a), let a smooth plate S_1 be attached to gear 1, and let a scribing point be fastened to the cord at location O. Keeping the string taut, give the cylinders a small rotation in both directions, thus drawing the curve CD shown in Fig. 10-4(b). For all locations of the gear, the distance K_1A is equal to the arc AC. The curve CD is therefore an involute and is at all times normal to the string. A similar procedure should be followed on gear 2 after first removing S_1 and attaching surface S_2 to gear 2. A corresponding curve EF will be obtained as shown in Fig. 10-4(c).

If certain portions of S_1 and S_2 are now removed, both plates may be attached to the gear simultaneously, as in Fig. 10-4(d). The tooth outlines are in contact at K_1, and one gear may be turned by a pressure supplied by the other at this point. The string is therefore not required for producing motion and may be removed. The common normal to the tooth outlines coincided with the string and therefore will always cross the line of centers at the fixed point O. The rule of gearing is accordingly fulfilled. The resulting velocity ratio for the gears will be the same as the velocity ratio of the given pitch circles.

In the similar triangles O_1AO and O_2BO,

$$\frac{O_2B}{O_1A} = \frac{O_2O}{O_1O} \tag{3}$$

Substitution in Eq. (2) gives

$$\omega_1 \times O_1A = \omega_2 \times O_2B \tag{4}$$

Thus the linear velocities of the base circles of the two gears are also equal to each other.

In involute gearing, at least one pair of teeth must always be in contact with each other along the line AB. This line is accordingly called the line of action, or the pressure line, and the angle φ is usually referred to as the pressure angle.

A large-scale drawing of gear teeth is frequently helpful or necessary in studying the form or kinematic action of a proposed gear design. In laying out the involute on the drawing board, the pressure line is first drawn in accordance with the desired pressure angle φ. As shown in Fig. 10-5, the base circle is drawn tangent to the pressure line. Tangents are drawn to the base circle to represent various positions of the string as it unwinds.

The starting point E for the involute is found by making arc EB equal to tangent OB by taking

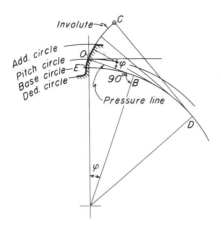

Figure 10-5 Drawing board layout for the involute.

short steps along these curves with a divider. Other points on the involute are found similarly. For example, arc *ED* must be equal to tangent *CD*. It should be noted that the involute curve becomes a straight line for a rack or a gear of infinite radius.

4. Cycloidal Gear Teeth

Gear teeth, when formed of portions of epicycloids and hypocycloids, also fulfill the law of tooth gearing. These curves are traced by a point on the circumference of the generating circle as it rolls without slipping along the inside and outside of the pitch circle of the gear as shown in Fig. 10-6.

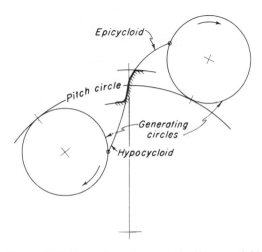

Figure 10-6 Generation of epicycloid and hypocycloid.

Cycloidal gearing was formerly in extensive use for cast-iron gears with cast teeth. Cycloidal gearing has been superseded by the involute tooth form, principally because the cutting tools required for manufacturing the latter are simpler in form. Variations in center distance have no effect on involute teeth, whereas for cycloidal gearing, tooth action is not correct unless the center distance is accurately maintained. The chief interest at the present time in cycloidal gearing lies in the fact that the outline for the $14\frac{1}{2}°$ composite system is partially composed of cycloids. The rotors of the Root positive blower are also cycloidal in shape.

5. Pitches of Gear Teeth

For gear calculations, the following different pitches are in use.
(a) *Circular pitch* is defined as the distance in inches from a point on

the pitch circle of one tooth to the corresponding point on the adjacent tooth measured along the pitch circle. If circular pitch is represented by p, pitch diameter by d, and the number of teeth in the gear by N, then

$$p = \frac{\pi d}{N} \tag{5}$$

(b) *Diametral pitch* is defined as the number of teeth in the gear per inch of pitch diameter. If diametral pitch is represented by P_d, then

$$P_d = \frac{N}{d} = \frac{N}{2r} \tag{6}$$

From Eq. (6), the following equation can be easily derived for center distance c.

$$c = r_1 + r_2 = \frac{N_1 + N_2}{2P_d} \tag{7}$$

The smaller of two meshing gears is usually called the *pinion*. Combination of Eqs. (5) and (6) gives the useful relationship

$$pP_d = \pi \tag{8}$$

Figure 10-7 shows a number of gear teeth of different diametral pitches in actual size.

(c) *Base pitch* is defined as the distance from a point on one tooth to the corresponding point on the adjacent tooth measured around the base circle. As shown in Fig. 10-8, the base pitch is also the distance from tooth to tooth measured on a tangent to the base circle. Since the radius of the base circle is $r \cos \varphi$, where r is the pitch radius, the base pitch is equal to the circumference of the base circle divided by the number of teeth in the gear. Thus,

$$p_b = \frac{2\pi r \cos \varphi}{N} = p \cos \varphi = \frac{\pi \cos \varphi}{P_d} \tag{9}$$

For $\varphi = 14\frac{1}{2}°$,

$$p_b = \frac{3.0415}{P_d} \tag{10}$$

For $\varphi = 20°$,

$$p_b = \frac{2.9521}{P_d} \tag{11}$$

Example 1. What is the thickness of a 4P tooth measured along the pitch circle?

Solution. By Eq. (8): $p = \frac{\pi}{P_d} = \frac{\pi}{4} = 0.7854$ in.

The thickness of the tooth along the pitch circle can be taken as one-half the value of the p if clearance or backlash is neglected. Hence

tooth thickness = $\frac{1}{2} \times 0.7854 = 0.3927$ in.

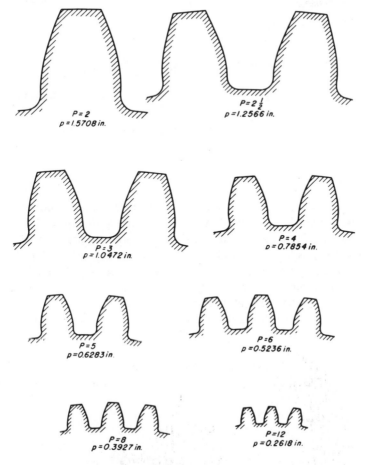

P=2
p=1.5708 in.
P=2 1/2
p=1.2566 in.
P=3
p=1.0472 in.
P=4
p=0.7854 in.
P=5
p=0.6283 in.
P=6
p=0.5236 in.
P=8
p=0.3927 in.
P=12
p=0.2618 in.

Figure 10-7 Actual sizes of gear teeth of various diametral pitches.

Example 2. A 20-tooth 5P gear meshes with a 63-tooth gear. Find the value of the standard center distance.

p_b
p_b
Pitch circle
Base circle

Figure 10-8 Base pitch.

Solution. By Eq. (6):

Pitch radius: $$r_1 = \frac{N_1}{2P_d} = \frac{20}{2 \times 5} = 2.0 \text{ in.}$$

Pitch radius: $$r_2 = \frac{N_2}{2P_d} = \frac{63}{2 \times 5} = 6.3 \text{ in.}$$

Center distance: $c = r_1 + r_2 = 2.0 + 6.3 = 8.3$ in.

Example 3. Two 8P gears are to be mounted on a center distance of 16 in. The speed ratio is to be 7: 9. Find the number of teeth in each gear.

Solution. By the given conditions:

$$\frac{N_1}{N_2} = \frac{7}{9} \quad \text{or} \quad N_1 = \frac{7}{9}N_2$$

In Eq. (7): $2P_dc = N_1 + N_2 = \frac{7}{9}N_2 + N_2 = \frac{16}{9}N_2$

$$N_2 = \frac{9}{16} \times 2P_dc = \frac{9}{16} \times 2 \times 8 \times 16 = 144 \text{ teeth}$$

$$N_1 = \frac{7}{9} \times 144 = 112 \text{ teeth}$$

6. Standard Systems of Gearing

The American Gear Manufacturers Association has recommended that the basic pressure angle φ be either 20° or 25° with full-depth addendums equal to $1/P_d$. Standards are retained for the $14\frac{1}{2}$° full-depth involute system, the $14\frac{1}{2}$° composite system, and the 20° involute stub-tooth system. These latter systems are required mainly for the replacement of gears already in service.

As the number of teeth in a gear is increased, the involute tooth outline becomes straighter until at the limit, when the radius is infinite, the gear becomes a rack with straight-sided teeth. Such racks have been standardized by the AGMA as follows.

(a) 20° *Full-Depth Involute.* These gears have a 20° pressure angle and the basic rack illustrated in Fig. 10-9. This is a widely used system of gearing. The rack for φ equal to 25° is similar except for the change in the angle.

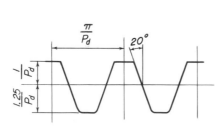
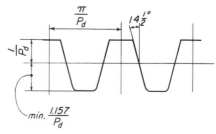

Figure 10-9 Basic rack for 20° full-depth involute system. 25° system is similar except for change in angle.

Figure 10-10 Basic rack for $14\frac{1}{2}$° full-depth involute system.

(b) $14\frac{1}{2}$° *Full-Depth Involute.* This is an involute tooth form whose basic rack has straight-sided teeth, as shown in Fig. 10-10. This system of gearing is very satisfactory, provided there is a considerable number of teeth in the gears. As will be explained, when the number of teeth is small, these gears, if made by one of the generating processes, are subject to undercutting, which may reduce the duration of contact between the teeth. They cannot be operated interchangeably with gears of the $14\frac{1}{2}°$ composite system. The tooth outlines are different, and true gear action cannot take place.

(c) $14\frac{1}{2}°$ *Composite System.* In this system, the tooth curve is an involute for a short distance each side of the pitch line, but is a cycloid for the inner and outer portions of the outline. Such gears are usually cut with the formed milling cutter shown in Fig. 10-13. The proportions for the basic rack for the system are given in Fig. 10-11. The least number of teeth for satisfactory action is twelve. This system is sometimes erroneously called the "standard $14\frac{1}{2}°$ involute system." However, as mentioned above, only a portion of the tooth is of involute form.

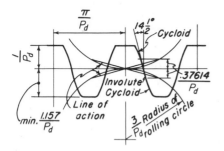

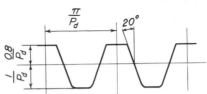

Figure 10-11 Basic rack for $14\frac{1}{4}°$ composite system.

Figure 10-12 Basic rack for 20° stub-tooth involute system.

(d) 20° *Stub-Tooth Involute.* Gears in this system operate on a pressure angle of 20°, and have shorter addenda and dedenda than the full-depth systems. The basic rack is shown in Fig. 10-12. Although undercutting has been lessened, the short addendum reduces the duration of contact. Vibration may occur, especially in gears of few teeth, because of insufficient overlap in the gear action, as is explained in Section 14.

(e) *Fellows Gear Shaper System.* This system uses a pressure angle of 20° and two diametral pitches, such as $\frac{4}{5}$, $\frac{6}{8}$, and so forth. The numerator indicates the pitch, which determines the thickness of the tooth and the pitch diameter. The denominator is used for determining the addendum in the usual manner. A gear having shortened or stub teeth is thus produced. Figures 10-9 to 10-12, inclusive, illustrate the basic racks for the designated systems of gearing. The actual cutting tools, however, must have the tip of the teeth extended in order to cut the clearance f on the mating gear. This clearance is usually made equal to $0.157/P_d$ or $0.25/P_d$.

7. Methods of Manufacture

Gear teeth are formed by the following milling or generating processes.

(a) *Milling Cutter.* Spur gears may be made from a blank by removing

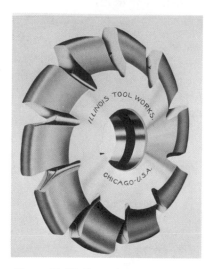

Figure 10-13 Formed circular cutter for gear teeth. (*Courtesy Illinois Tool Works.*)

the material between the teeth on a milling machine that uses the formed cutter shown in Fig. 10-13.

Gears of the $14\frac{1}{2}°$ composite system are usually made by this method. Since the geometric curves forming the sides of the teeth vary with the number of teeth in the gear, eight cutters are required for each pitch if gears of all sizes are to be cut. The eight cutters and the range of each are as follows.

No. 1:   135–rack
No. 2:   55–134 teeth
No. 3:   35–54 teeth
No. 4:   26–34 teeth
No. 5:   21–25 teeth
No. 6:   17–20 teeth
No. 7:   14–16 teeth
No. 8:   12 and 13 teeth

Cutters are theoretically correct only for the lowest number of teeth in each range. If, when gears are being cut near the higher end of the range, a more accurate tooth form is desired, cutters in half-numbers suitable for gears with such numbers of teeth are also available.

The following pitches are considered standard.

P_d 1–P_d 2 varying by $\frac{1}{4}$ diametral pitch
P_d 2–P_d 3 varying by $\frac{1}{2}$ diametral pitch
P_d 3–P_d 12 varying by 1 diametral pitch
P_d 12–P_d 32 varying by 2 diametral pitch

Gears with large-size teeth are frequently made by formed cutters because of the lack of suitable generating machines and tools.

Example 4. Two spur gears have a 3:1 ratio. The center distance is to be 11.600 in. Determine whether it is possible to use standard gears of $P_d = 4$ for this train.

Solution.
$$N_2 = 3N_1$$

In Eq. (7):
$$N_1 + N_2 = N_1 + 3N_1 = 2P_d c$$
$$4N_1 = 2 \times 11.6P_d$$
$$N_1 = 5.8P_d$$

When $P_d = 4$ is substituted in this equation, the right side will not be an integer. Gears with $P_d = 4$ therefore cannot be used for this train.

(b) *Rack Generation.* Gear tooth forms may be produced by another method, known as generating. Since a rack may be considered a gear of infinite radius, a tool of this shape may be constructed of hardened steel with cutting edges around the boundaries of the teeth. The tool is given a reciprocating motion parallel to the gear axis. At the same time, the gear blank is slowly rotated, and the rack is given a lateral motion equal to the pitch-line velocity of the gear. The material between the gear teeth is cut away and involute teeth are generated, as shown in Fig. 10-14. Only one tool for each pitch will be required to cut gears of any number of teeth.

(c) *Hobbing.* The hobbing process generates teeth from a straight-sided tool as shown in Fig. 10-15. The hob may be considered a cylinder around which a thread of the same cross section as a rack tooth has been helically wound. The resulting worm is gashed, the edges are relieved, and are then hardened and ground. The hob is located to give the proper depth of cut and is then rotated. Kinematically, the action of the hob on the blank is equiva-

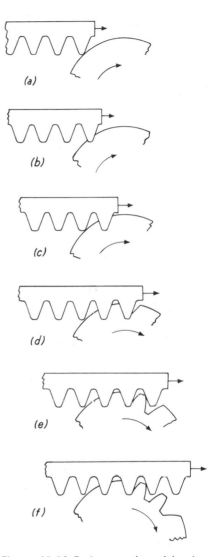

Figure 10-14 Rack generation of involute teeth.

Figure 10-15 Hob with partially cut gear blank.

lent to that of a rack cutter. The lead of the helix of the hob as it rotates simulates the lateral motion of the rack. As the cutting progresses, the hob is fed axially along the blank until the teeth extend across the entire width of face. The hobbing process accounts for the major portion of gears made in quantity production.

(d) *Fellows Gear Shaper Method.* This is a generating process using a cutter that resembles a hardened gear with properly relieved edges. Cutter and blank are mounted on parallel axes and are slowly rotated; the cutter is given an additional reciprocating motion on its axis. Teeth are generated in the blank as shown in Fig. 10-16. At the beginning, the cutter is fed radially into

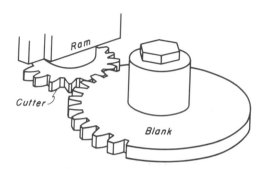

Figure 10-16 Fellows gear shaper method of forming gear teeth.

the blank a distance equal to the depth of the tooth. The Fellows method must be used for cutting internal gears. The method is also suitable for shoulder gears where space is restricted at one end of the teeth, as in the cluster gears of automobile transmissions. Fellows cutters are made to cut gears of all systems.

Hobbed and shaped teeth are usually provided with a small amount of tip relief. The cutting tools are modified to cut the tooth slightly narrower in the outer portion. The oncoming pair of teeth pick up the load more gradually, and errors in tooth form and spacing have less effect in producing noise and vibration.

8. Gear Finishing Methods

For high speeds and heavy loads, a finishing operation, subsequent to cutting, may be required to bring the tooth outline to a sufficient degree of accuracy and surface finish. The following methods are in use.

(a) *Gear Shaving*. The gear is run with a specially made cutter resembling a gear or rack. The cutter tooth has an axial component of motion on the tooth surface, which removes fine hair-like chips. Shaving is done prior to hardening, and sufficient stock must be allowed for this operation. Shaving is a widely used finishing method for gears made in large quantities.

(b) *Honing*. Here the hardened gear is run with an abrasive impregnated plastic helical gear-shaped tool, which will make minor tooth-form corrections and improve the smoothness of the surface.

(c) *Lapping*. The gear is run with a gear-shaped lapping tool in an abrasive-containing medium. Sometimes two mating gears are similarly run. An additional relative motion in the axial direction is required for spur and helical teeth. Excessive lapping can be detrimental to the involute shape.

(d) *Grinding*. Grinding can be used to give the final form to the teeth after heat treatment. Errors resulting from distortion in hardening can be corrected. When cutting, sufficient stock must be allowed on the tooth surface. Some fine-pitch gears are ground from solid blanks.

(e) *Burnishing*. This is a plastic smearing process resulting from rubbing pressure, which flattens and spreads minute surface irregularities. A special hardened burnishing tool is used. The process will not correct errors due to improper cutting.

9. Transmitted or Horsepower Load

With a pair of gears, power is transmitted by the force that the tooth of one gear exerts on the tooth of the other. This force is directed along the line of action or pressure line, as shown in Fig. 10-4. If the gears are transmitting power at a constant rate and are turning at a constant rpm, the force along the pressure line must be a constant, also. The velocity along the pressure line is equal to the tangential velocity of the base circles.

The tangential velocity of the pitch circle is given by

$$V = \frac{\pi dn}{12} \text{ fpm} \qquad (12)$$

where d is the pitch diameter in inches, and n is the rpm.

A principle of mechanics states that a force can be considered as acting at any point along its line of action. In Fig. 10-17 let the force F_n between the teeth be considered as acting at the pitch point O. This force has the value

$$F_n = \frac{33,000 \text{ hp}}{V_b} \qquad (a)$$

where V_b is the tangential velocity of the base circle in feet per minute. It is equal to

$$V_b = \frac{\pi dn \cos \varphi}{12} = V \cos \varphi$$

From Fig. 10-17,

$$F_n = \frac{F_t}{\cos \varphi}$$

When these are substituted into Eq. (a), the result is

$$F_t = \frac{33,000 \text{ hp}}{V} \qquad (13)$$

The radial component F_r of force F_n is equal to $F_t \tan \varphi$.

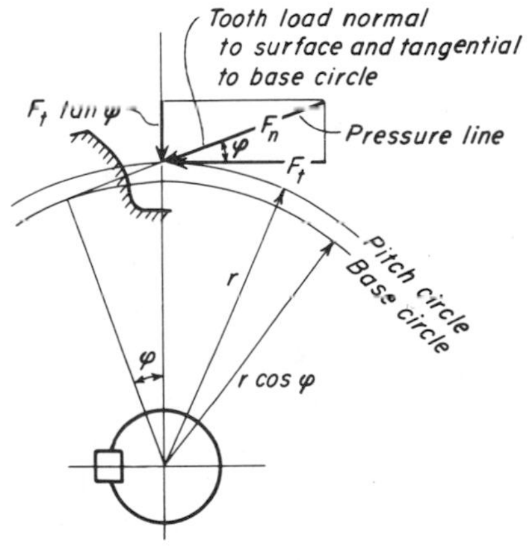

Figure 10-17 Transmitted or horsepower force.

10. Bending Capacity of Spur Gear Teeth

As the teeth rotate through the loading zone, the number of pairs of teeth in contact simultaneously varies from one to two or more. It is customary, however, to determine the ability of a single tooth to carry a bending load when the load is acting through the corner or the most unfavorable point as shown in Fig. 10-18.

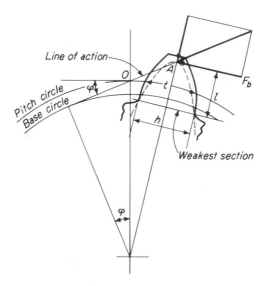

Figure 10-18 Beam strength of gear tooth.

The force along the pressure line in Fig. 10-18 is considered as being applied at the centerline of the tooth, where it is divided into radial and tangential components. The radial component causes a uniform compressive stress over the cross section, but it is customary to neglect this force when making stress calculations. The tangential component F_b produces a bending moment $F_b l$ at the base or narrowest portion of the tooth.

It is customary to compute the bending stress on the assumption that the tooth is a cantilever beam. If the elementary equation, $s = 6M/bh^2$, for bending stress is used, the result at best is only approximate. Accuracy cannot be secured because the tooth is short and thick and nonuniform in cross section. It must be remembered that the derivation of the equation for bending stress assumed a long thin beam of constant cross section. In addition, for concentrated loads, the equation is valid only at points a considerable distance away from the point of application of the force. Nevertheless it is customary to use the equation, and to make application in the following manner.

$$s = \frac{6M}{bh^2} = \frac{F_b}{b} \times \frac{6l}{h^2} \tag{14}$$

Here b is the width of the tooth in the axial direction, and l and h are the height and thickness, as shown in Fig. 10-18.

The factor $h^2/6l$ is a purely geometrical property of the size and shape of the tooth and may be written as a function of the circular pitch. Therefore, let

$$py = \frac{h^2}{6l} \quad \text{or} \quad y = \frac{h^2}{6lp} \tag{15}$$

The term y is a pure number and is called the form or Lewis factor. It depends on the number of teeth in the gear and the system of gearing used. Dimensions h and l must be for the cross section, which makes $h^2/6l$ a minimum.

Substitution of Eq. (15) into Eq. (14) gives

$$F_b = sbyp \tag{16}$$

Equation (16) gives the tangential load that the tooth can carry in beam action. Values of y for gears of different numbers of teeth are given in Table 10-1. Sometimes the form factor is expressed as Y, which includes the factor π. Thus,

$$Y = \pi y \tag{17}$$

Since $p = \pi/P_d$, Eq. (16) may be written

BENDING CAPACITY

$$F_b = sb\frac{Y}{P_d} \tag{18}$$

The distinction between Eqs. (13) and (16) should be noted. Equation (13) gives the values of the tangential force the gears are carrying. Equation (16) gives merely the capacity of the tooth to resist a bending load and makes no reference to the actual load the teeth may be carrying. It may be neces-

Table 10-1   FORM OR LEWIS FACTOR y FOR SPUR GEARS WITH LOAD AT TIP OF TOOTH

No. of Teeth	14-1/2° Full Depth	20° Full Depth	20° Stub	No. of Teeth	14-1/2° Full Depth	20° Full Depth	20° Stub	No. of Teeth	14-1/2° Full Depth	20° Full Depth	20° Stub
10	0.056	0.064	0.083	19	0.088	0.100	0.123	43	0.108	0.126	0.147
11	0.061	0.072	0.092	20	0.090	0.102	0.125	50	0.110	0.130	0.151
12	0.067	0.078	0.099	21	0.092	0.104	0.127	60	0.113	0.134	0.154
13	0.071	0.083	0.103	23	0.094	0.106	0.130	75	0.115	0.138	0.158
14	0.075	0.088	0.108	25	0.097	0.108	0.133	100	0.117	0.142	0.161
15	0.078	0.092	0.111	27	0.099	0.111	0.136	150	0.119	0.146	0.165
16	0.081	0.094	0.115	30	0.101	0.114	0.139	300	0.122	0.150	0.170
17	0.084	0.096	0.117	34	0.104	0.118	0.142				
18	0.086	0.098	0.120	38	0.106	0.122	0.145	rack	0.124	0.154	0.175

sary to adjust the right side of Eqs. (16) and (18) to take care of the effects of stress concentration.

11. Form or Lewis Factors for Spur Teeth

The form or Lewis factor is named for the American engineer who first made application of the bending equation to gear teeth.[1] Although gear materials and methods of manufacture have changed greatly since the values in Table 10-1 were published in 1893, they have remained in use to the present day with but minor changes.

Values of l and h, from which y could be computed by Eq. (15), are somewhat uncertain since they are influenced to a large extent by the size of the fillet radius r_f. For teeth made by milling cutters, the radius may sometimes be as small as $0.05p$. For radii of this size, the actual values of y are smaller than those shown in the table, and the teeth are then not as strong as Eq. (16), using tabular values, would indicate. In fact, fillets of rather generous size are required to give the y values of Table 10-1.

For teeth that increase in thickness all the way to the base, the length of the moment arm l may be found by inscribing a parabola within the tooth outline. It should be tangent to the fillets on either side, and the vertex should be at A, as in Fig. 10-18. The bending stresses are then computed for the cross section passing through the points of tangency.

Since the tensile strength of materials used for gears increases with the hardness,[2] working stress values for bending can be found in Table 10-2.

Table 10-2 ALLOWABLE WORKING STRESSES, PSI, AND VALUES OF K FOR GEAR MATERIALS

Material	Hardness		Bending, s		Com-pression, s_{ec}	K, 14-$1/2°$	K, $20°$
	BHN	R_c	Spur, Helical	Bevel			
Cast iron	160–200		5,000	3,000	50,000	56	76
Cast iron	210–245		7,000	4,000	60,000	80	110
Steel	160–200		20,000	10,000	60,000	43	59
Steel	210–245		22,000	11,000	70,000	58	80
Steel	302–351	33–38	32,000	15,000	100,000	119	163
Flame or induction hardened		48–53	35,000	15,000	160,000	305	417
Carburized or case hardened		58–63	55,000	30,000	200,000	477	651

Values for K are calculated for both gears of the same material.

[1]See Lewis, W., "Investigation of the Strength of Gear Teeth," *Proc. Engr's Club Phila.*, **10**, 1893, p. 16.

[2]See pp. 130 and 134 of Reference 4, end of Chapter.

Example 5. Two $20°$ full-depth gears, $P_d = 4$, are made of steel, BHN $= 350$. Number of teeth are 24 and 48. Width of the face is 3 in. Find the bending capacity of the teeth.

Solution. By Eq. (8): $p = \dfrac{\pi}{P_d} = \dfrac{\pi}{4}$

By Table 10-2: $s = 32{,}000$ psi,   in bending

The pinion is the weaker. By Table 10-1, $y = 0.107$ for $N_1 = 24$ teeth.

By Eq. (16): $F_b = sbyp = 32{,}000 \times 3 \times 0.107 \times \dfrac{\pi}{4} = 8{,}070$ lb

Stress concentration may indicate a reduction in the bending capacity.

12. Limit Load for Wear

The gear teeth must also be sufficiently strong to carry the load F_w arising from the contact or compressive stress between the teeth. Calculations are made by considering the teeth as two parallel cylinders in contact.[3] The radii of the cylinders are taken as the radii of curvature for the involutes when the teeth are making contact at the pitch point O, Fig. 10-19. Approximating the teeth as cylinders is satisfactory, for although the compressive

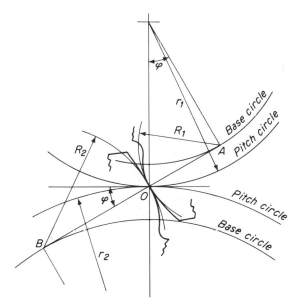

Figure 10-19 Radii of curvature R_1 and R_2 for tooth surfaces at pitch point O.

[3]See Buckingham, E., *Dynamic Loads on Gear Teeth*, New York: American Society of Mechanical Engineers, 1931, p. 41.

stresses are very high, they decrease rapidly at locations removed from the area of contact.

In Fig. 9-9 of Chapter 9, the maximum compressive stress p_o in the contact zone is shown as

$$p_o = 0.591 \sqrt{\frac{P_1 E_1 E_2}{E_1 + E_2}\left(\frac{1}{R_1} + \frac{1}{R_2}\right)} \tag{a}$$

where P_1 is the load per axial inch pressing the cylinders together, E_1 and E_2 are the moduli of elasticity for the materials composing the gears, and R_1 and R_2 are the radii of curvature OA and OB, respectively, for the two gears.

When contact occurs at the pitch point, the radii of curvature for the involutes are

$$R_1 = \frac{d_1}{2} \sin \varphi \tag{b}$$

$$R_2 = \frac{d_2}{2} \sin \varphi = \frac{N_2 d_1}{2N_1} \sin \varphi \tag{c}$$

The last form of Eq. (c) arises from the relationship $d_1/N_1 = d_2/N_2$.

Let p_o in Eq. (a) be called s_{ec}, the surface endurance limit in compression for the gear materials. Substitution of Eqs. (b) and (c) gives

$$s_{ec}^2 = 0.35 \frac{P_1 E_1 E_2}{(E_1 + E_2)} \cdot \frac{2}{d_1 \sin \varphi}\left(1 + \frac{N_1}{N_2}\right)$$

Both sides of this equation are multiplied by the face width b, and the total load $P_1 b$ is given the name *wear load* F_w. The equation, when solved for F_w, gives

$$F_w = \frac{s_{ec}^2 b d_1 \sin \varphi}{1.40}\left(\frac{1}{E_1} + \frac{1}{E_2}\right)\left(\frac{2N_2}{N_1 + N_2}\right) \tag{19}$$

$$= d_1 b Q K \tag{20}$$

where

$$Q = \frac{2N_2}{N_1 + N_2} \tag{21}$$

and

$$K = \frac{s_{ec}^2 \sin \varphi}{1.4}\left(\frac{1}{E_1} + \frac{1}{E_2}\right) \tag{22}$$

Values for s_{ec} and K for materials of different hardnesses are shown in Table 10-2.

Example 6. Find the limit load for wear for the gears of Example 5.

Solution. By Eq. (21): $Q = \dfrac{2N_2}{N_1 + N_2} = \dfrac{96}{72} = \dfrac{4}{3}$

By Table 10-2: $K = 163$ for $\text{BHN} = 350$

By Eq. (20): $F_w = d_1 b Q K = 6 \times 3 \times \dfrac{4}{3} \times 163 = 3{,}910 \text{ lb}$

13. Dynamic Load

The tangential horsepower force F_t transmitted from one gear to another can be easily determined by Eq. (13). However, this is not the entire force that acts between the teeth. Inaccuracies of tooth form and spacing, combined with the inertia of the rotating masses, produce dynamic effects, which also act upon the teeth.

Since the accuracy of a gear cannot be fully determined in advance, the calculation of the dynamic load may introduce a considerable amount of uncertainty in the design of gears. Based on certain approximations, it is possible to derive the following equation[4] for the dynamic load F_d arising from errors of tooth form and spacing.

$$F_d = \frac{2e}{t}\sqrt{km_e} \tag{23}$$

Here e is the sum of the errors, inches, of the two mating teeth, t is the time, seconds, during which the error is acting, k is the spring constant, pounds per inch, of the two mating teeth, and m_e is the effective mass, lb sec²/in., of the two gears.

14. Errors in Gears

Errors of tooth form and spacing in the cutting of the teeth are due to various causes and are difficult to control.[5] Permissible tolerances for gear teeth can be found in the publication *Gear Classification Manual 390.02*, 1964, of the American Gear Manufacturers Association. It is convenient to use such tolerances as the expected error in gear teeth. Table 10-3 gives average values for the combined tooth-to-tooth spacing tolerances and profile tolerances.[6]

In the table quality No. 8 is the lowest and No. 12 is the highest. A gear hobbing machine in good condition with a properly sharpened hob can be expected to give a quality of about No. 8. This quality is widely used in the general field of power transmission. Higher-quality gears may require special finishing operations, and will be more expensive. Automotive gears, for example, have qualities of No. 10 or 11. The designer aims to use the lowest-quality number that the intended service will permit.

A conservative estimate for t in Eq. (23) would be to say that the error

[4]See Tuplin, W. A., "Dynamic Loads on Gear Teeth," *Machine Design*, **25**, Oct. 1953, p. 203. See also the author's *Mechanical Design Analysis*, Englewood Cliffs, N.J.: Prentice-Hall, Inc., 1964, p. 181. See Tucker, A. I., "Gear Design-Dynamic Loads," *Mechanical Engineering*, **93**, Oct. 1971, p. 29. Also see Spotts, M. F., "Dynamic Loads on Gear Teeth," *Design News*, June 7, 1971, p. 71.

[5]See *Right and Wrong of Hob Sharpening*, Chicago: Illinois Tool Works, 1942. See also pp. 9–28, 22–15, and 22–24 of Reference 3, end of chapter.

[6]These values are found by squaring the tooth-spacing tolerances and the profile tolerances, adding together, and taking the square root.

Table 10-3   TOTAL TOOTH-TO-TOOTH SPACING AND PROFILE ERRORS, *e*, INCHES

AGMA Quality No.	Diametral Pitch, P_d	Pitch Diameter					
		1-½ in.	*3 in.*	*6 in.*	*12 in.*	*25 in.*	*50 in.*
8	2			0.0048	0.0050	0.0052	0.0055
	4		0.0032	0.0034	0.0036	0.0037	0.0040
	8	0.0022	0.0023	0.0024	0.0025	0.0028	0.0029
	16–19.99	0.0017	0.0018	0.0019	0.0020	0.0021	0.0023
9	2			0.0033	0.0034	0.0037	0.0039
	4		0.0022	0.0023	0.0025	0.0027	0.0027
	8	0.0016	0.0017	0.0018	0.0019	0.0019	0.0020
	16–19.99	0.0013	0.0013	0.0014	0.0014	0.0015	0.0016
10	2			0.0024	0.0025	0.0026	0.0027
	4		0.0016	0.0017	0.0018	0.0019	0.0020
	8	0.0012	0.0012	0.0012	0.0013	0.0014	0.0015
	16–19.99	0.0009	0.0010	0.0010	0.0010	0.0011	0.0011
11	4		0.0012	0.0012	0.0013	0.0014	0.0015
	8	0.0008	0.0009	0.0009	0.0009	0.0010	0.0011
	16–19.99	0.0006	0.0007	0.0007	0.0007	0.0008	0.0008
12	4		0.0009	0.0009	0.0009	0.0010	0.0010
	8	0.0006	0.0006	0.0007	0.0007	0.0007	0.0008
	16–19.99	0.0005	0.0005	0.0005	0.0005	0.0006	0.0006

e is acting during the time required for the rotation of a single pitch. The number of pitches turned per second is $n_1 N_1/60$, where n_1 is the rpm speed of the pinion or smaller gear, and N_1 is the number of teeth in this gear. The time *t* for the passage of a single pitch, then is

$$t = \frac{60}{n_1 N_1} \tag{24}$$

15. Spring Constant and Effective Mass

It was shown in Example 10, Chapter 1, that if the spring constant for a cantilever is calculated, then any other geometrically similar cantilever, but of the same breadth *b*, will have an equal value for the spring constant. The same is true for gear teeth since if the only change is in the pitch, the teeth are geometrically similar. It can be shown[7] that the average value of the spring constant for a pair of meshing teeth can be represented by

[7]See Niemann, G., and H. Rettig, "Error-Induced Dynamic Gear Tooth Loads," *Proc. Intern. Conf. Gearing*, London: Institution of Mechanical Engineers, 1958, p. 31. See also p. 432 of Reference 1, end of chapter.

$$k = \frac{b}{9}\left(\frac{E_1 E_2}{E_1 + E_2}\right) \tag{25}$$

where E_1 and E_2 are the moduli of elasticity for the materials of the two gears.

When both gears are steel, $E_1 = E_2 = 30{,}000{,}000$ psi, the value of k for a meshing pair becomes

$$k = 1{,}670{,}000b \text{ lbs/in.} \tag{26}$$

When one gear is steel, $E_1 = 30{,}000{,}000$ psi and the other gear is cast iron or bronze, $E_2 = 16{,}000{,}000$ psi, then

$$k = 1{,}160{,}000b \text{ lbs/in.} \tag{27}$$

Here b is the width of face of the gears.

The moment of inertia of a gear can be found by approximating it as a solid circular cylinder of diameter equal to the pitch diameter of the gear, and axial length equal to the face width of the teeth. The mass moment of inertia I, Fig. 10-20(a), then is

$$I = m\frac{r^2}{2} \tag{a}$$

where m is the mass.

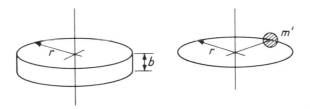

Figure 10-20 Moment of inertia of solid circular cylinder.

A concentrated mass m', located at the pitch circle, Fig. 10-20(b), has a moment of inertia I' about the axis of

$$I' = m'r^2 \tag{b}$$

Suppose m' were of such size as to make I and I' equal. Then

$$m' = \tfrac{1}{2}m \tag{c}$$

The dynamical system of the two gears is considered as the masses m'_1 and m'_2 concentrated at the pitch circles connected by a spring comprising the two teeth. For such a system, the effective mass m_e is given by the following equation.

$$\frac{1}{m_e} = \frac{1}{m'_1} + \frac{1}{m'_2} \tag{28}$$

Mass m_1' of the pinion is equal to

$$m_1' = \frac{1}{2}m_1 = \frac{\pi r_1^2 b \gamma_1}{2g} \tag{29}$$

where γ_1 is the weight per cubic inch of the material.

Mass m_2' of the gear is given by a similar equation.

$$m_2' = \frac{1}{2}m_2 = \frac{\pi r_2^2 b \gamma_2}{2g} \tag{30}$$

where γ_2 is the weight per cubic inch of the material composing the gear. Substitution into Eq. (28) gives

$$\frac{1}{m_e} = \frac{2g}{\pi b}\left(\frac{1}{\gamma_1 r_1^2} + \frac{1}{\gamma_2 r_2^2}\right) \tag{31}$$

16. Design Equations for Dynamic Load

The spring constant for a pair of gears will change while a tooth is passing through the contact zone since one, two, or sometimes more, pairs may be in contact at different times. During the central portion of the passage, a single pair is in contact, which must carry the entire load. On a conservative basis the spring constant, as given by Eq. (26) or (27), will be used in arriving at a design equation.

(a) *Steel Pinion—Steel Gear.* Here $\gamma_1 = \gamma_2 = 0.283$ lb/in.³ and acceleration g due to gravity is taken as 386 in./sec². Equation (31) becomes

$$\frac{1}{m_e} = \frac{868}{b}\left(\frac{r_1^2 + r_2^2}{r_1^2 r_2^2}\right) \tag{a}$$

When this equation, together with $k = 1,670,000b$ lb/in. and $t = 60/n_1 N_1$ sec, are substituted into Eq. (23), the result is

$$F_d = \frac{1.46 e n_1 N_1 b r_1 r_2}{\sqrt{r_1^2 + r_2^2}} \tag{32}$$

(b) *Steel Pinion—Cast-Iron Gear.* Here $\gamma_1 = 0.283$ lb/in.³ and $\gamma_2 = 0.256$ lb/in.³ Equation (31) becomes, after values for g, γ_1, and γ_2 are substituted,

$$\frac{1}{m_e} = \frac{2g}{\pi b \gamma_2}\left[\frac{r_1^2 + (\gamma_2/\gamma_1)r_2^2}{r_1^2 r_2^2}\right] = \frac{960}{b}\left(\frac{r_1^2 + 0.9r_2^2}{r_1^2 r_2^2}\right)$$

This, together with $k = 1,160,00b$ lb/in. and $t = 60/n_1 N_1$ sec, gives

$$F_d = \frac{1.16 e n_1 N_1 b r_1 r_2}{\sqrt{r_1^2 + 0.90r_2^2}} \tag{33}$$

The load capacity of a pair of gears is based on either the bending or wear capacity, whichever is the smaller. The force F_t remaining for transmitting the desired horsepower is found by deducting the dynamic load F_d. The following equations result.

$$F_t = F_b - F_d \tag{34}$$

$$F_t = F_w - F_d \tag{35}$$

The pitch line velocity is given by Eq. (12), and the horsepower by Eq. (13).

Example 7. Let the quality number for the gears of Examples 5 and 6 be No. 8. The pinion is turning 860 rpm. Find the dynamic load for these gears and the horsepower they are transmitting.

Solution. By Table 10-3: $e = 0.0034 + 0.0036 = 0.0070$ in.

By Eq. (32): $$F_d = \frac{1.46 \times 0.0070 \times 860 \times 24 \times 3 \times 3 \times 6}{\sqrt{3^2 + 6^2}}$$

$$= 1,700 \text{ lb}$$

Pitch-line velocity: $$V = \frac{\pi d_1 n_1}{12} = \frac{\pi 6 \times 860}{12} = 1,351 \text{ fpm}$$

Horsepower will be based on the wear capacity of Example 6 since it is smaller than the bending capacity of Example 5.

By Eq. (35): $F_t = F_w - F_d = 3,910 - 1,700 = 2,210$ lb

$$\text{hp} = \frac{F_t V}{33,000} = \frac{2,210 \times 1,351}{33,000} = 90.6$$

Example 8. Two 20° full-depth gears are carburized and have an AGMA quality No. 8. Center distance is 4.5 in.; 2:1 reduction. Pinion turns 3,450 rpm. If the gears are transmitting 190 hp, find suitable values for P_d and b.

Solution. This type of problem is best solved by assuming a value for P_d and then testing to see whether it is satisfactory.

Let: $P_d = 6$

$$r_1 = \frac{1}{3} \times 4.5 = 1.5 \text{ in.,} \qquad d_1 = 3 \text{ in.,} \qquad N_1 = 18$$

$$r_2 = \frac{2}{3} \times 4.5 = 3.0 \text{ in.,} \qquad d_2 = 6 \text{ in.,} \qquad N_2 = 36$$

By Table 10-2: $s = 55,000$ psi,   in bending

By Eq. (16): $F_b = sbyp = 55,000b \times 0.098 \times \dfrac{\pi}{6} = 2,820b$

$$Q = \frac{2N_2}{N_1 + N_2} = \frac{72}{54} = \frac{4}{3}$$

By Table 10-2: $K = 651$

By Eq. (20): $F_w = d_1 bQK = 3b\dfrac{4}{3} \times 651 = 2,604b$ *F_W WILL GOVERN*

By Eq. (12): $V = \dfrac{\pi d_1 n_1}{12} = \dfrac{\pi 3 \times 3,450}{12} = 2,710 \text{ fpm}$

By Eq. (13): $F_t = \dfrac{33,000 \text{ hp}}{V} = \dfrac{33,000 \times 190}{2,710} = 2,310 \text{ lb}$

By Table 10-3: $e = 0.00275 + 0.0029 = 0.00565$ in.

By Eq. (32): $F_d = \dfrac{1.46 \times 0.00565 \times 3{,}450 \times 18b \times 1.5 \times 3.0}{\sqrt{1.5^2 + 3.0^2}}$

$= 687b$

Since $F_w < F_b$: $F_t = F_w - F_d$

Use Eq. (35): $2{,}310 = 2{,}604b - 687b = 1{,}917b$

Whence: $b = \dfrac{2{,}310}{1{,}917} = 1.21$ in.

The assumed value of $P_d = 6$ gives a reasonable value for b. A trial will indicate that a gear with larger teeth, say $P_d = 5$, will give a smaller value for b and is therefore a gear that will contain less material. However, when $P_d = 5$, the pinion will have only 15 teeth. This number may be too small for, as explained in Section 25, undercutting for a 20° gear will occur when the number of teeth is less than 17. The bending capacity of the tooth would of course be increased for $P_d = 5$.

The actual situation in regard to dynamic load is much more complex than the simple treatment given above. The equations given here for dynamic load should be applied only to gears operating with moderate speeds and loads. The equations indicate a dynamic load increasing linearly with speed. In high-speed systems, the equation may indicate a dynamic load too great for the capacity of the teeth, when in fact such gears may operate satisfactorily. When the teeth are heavily loaded, the deflections have the effect of additional error for which the equation takes no account.[8]

17. Reduction of the Effective Mass

As indicated by Eq. (23), a reduction in the dynamic load force can be achieved by making a reduction in the effective mass. This can be done by cutting the teeth in a thin rim as indicated by Fig. 10-21. The moment of inertia of the rim about the axis is equal to the moment of inertia of a cylinder of radius r_o minus the moment of inertia of a cylinder of radius r_i. Thus

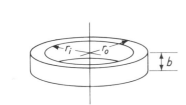

Figure 10-21 Moment of inertia of hollow circular cylinder.

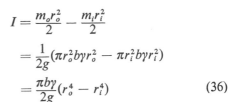

$$I = \frac{m_o r_o^2}{2} - \frac{m_i r_i^2}{2}$$

$$= \frac{1}{2g}(\pi r_o^2 b \gamma r_o^2 - \pi r_i^2 b \gamma r_i^2)$$

$$= \frac{\pi b \gamma}{2g}(r_o^4 - r_i^4) \tag{36}$$

As before, I is equated to the moment of inertia I', or $m'r_o^2$, of a mass con-

[8]Much effort has been expended on this difficult problem with no general agreement about what the equation for dynamic load really should be. The references of footnotes 4 and 7 should be consulted. See also Seireg, A., and D. R. Houser, "Evaluation of Dynamic Factors for Spur and Helical Gears," *Trans. ASME*, **92**, 1970, *Jour. of Engrg. for Industry*, p. 504.

centrated at the radius r_o. Then

$$m' = \frac{\pi b \gamma}{2 g r_o^2}(r_o^4 - r_i^4) \tag{37}$$

Values for m' for each gear are calculated, and the results substituted into Eq. (28).

The moment of inertia of the spokes and hub can usually be neglected. However, if the gear is made with a solid web, its moment of inertia should be included.

The effective mass of a mesh is sometimes increased when the gears are connected to other rotating bodies by stiff heavy shafts.

18. Number of Pairs of Teeth in Contact

Figure 10-22(a) to (d) shows the successive positions of a tooth as it passes through the zone of loading.

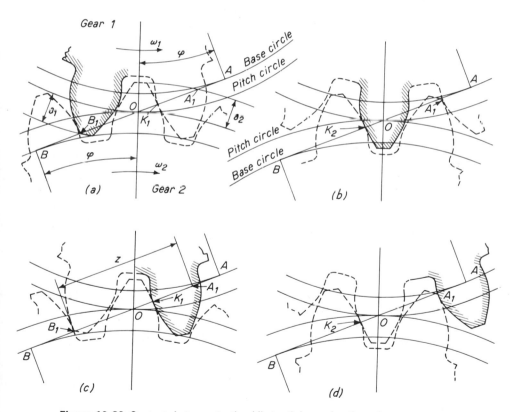

Figure 10-22 Contacts between teeth while tooth is passing through contact zone.

In sketch (a) the crosshatched tooth is just entering the contact zone at B_1 where the addendum circle crosses the line of action AB. Another pair is already in contact at K_1 so that the load is being carried by two pairs between sketches (a) and (b).

However, in (b), contact is about to be broken at A_1 so that between sketches (b) and (c) the load is carried by but a single pair.

At B_1 in sketch (c) a new pair is coming into contact so that this contact and the one at K_1 carry the load to sketch (d), where the crosshatched tooth will leave the contact zone as soon as the contact at A_1 is broken. The entire load will then be carried by K_2 until the process begins over again as in sketch (a).

It is obvious that length of contact A_1B_1 or z must be somewhat greater than a base pitch so that a new pair of teeth will come into contact before the pair that had been carrying the load separate. In fact, for smooth and vibrationless operation, there should be considerable overlap of gear action. The designer usually tries to obtain a minimum value of 1.4 for the quotient z/p_b. Distance z, or A_1B_1 can be computed as follows. Distance OA_1 is equal to $A_1B - OB$. Length A_1B can be found as the short leg of a right triangle, where $r_2 + a_2$ is the hypotenuse, and $r_2 \cos \varphi$ is the long leg. Length OB is merely $r_2 \sin \varphi$. Then

$$z = OA_1 + OB_1 \tag{38}$$

$$OA_1 = A_1B - OB = \sqrt{(r_2 + a_2)^2 - r_2^2 \cos^2 \varphi} - r_2 \sin \varphi \tag{39}$$

By similar reasoning,

$$OB_1 = AB_1 - OA = \sqrt{(r_1 + a_1)^2 - r_1^2 \cos^2 \varphi} - r_1 \sin \varphi \tag{40}$$

Example 9. Two 20° full-depth gears have a diametral pitch of 6 and operate on a center distance of 7.5 in. The speed ratio is 1.5 : 1. Find the number of base pitches in the length of contact.

Solution. $c = r_1 + r_2 = r_1 + 1.5r_1 = 2.5r_1 = 7.5$, $r_1 = 3$ in., $r_2 = 4.5$ in.

In Eq. (39): $OA_1 = \sqrt{\left(4.5 + \dfrac{1}{6}\right)^2 - 4.5^2 \times 0.93969^2}$

$$- 4.5 \times 0.34202 = 1.9740 - 1.5391 = 0.4349 \text{ in.}$$

In Eq. (40): $OB_1 = \sqrt{\left(3 + \dfrac{1}{6}\right)^2 - 3^2 \times 0.93969^2} - 3 \times 0.34202$

$$= 1.4424 - 1.0261 = 0.4163 \text{ in.}$$

By Eq. (11): $p_b = \dfrac{2.9521}{P_d} = \dfrac{2.9521}{6} = 0.4920$ in.

$$\text{number} = \frac{0.4349 + 0.4163}{0.4920} = 1.73 \text{ tooth intervals in length}$$

of contact

If the number of teeth in the gears is decreased, the radii become proportionally smaller and the addenda proportionally larger. Thus, point A_1 in Fig. 10-22 will eventually coincide with point A. Should gears be made with still fewer teeth, interference of metal will occur, unless the flanks are undercut, as will be explained later. The equations above for length of contact are applicable only to gears that are free from undercutting.

Points A and B in Fig. 10-22 are called the interference points for the gears. It is possible to have gear action all the way along the pressure line to the interference point. However, when the gears are highly stressed, it is good practice to have contact begin and end a considerable distance short of the interference points. Figure 10-5 shows how the radius of curvature of the involute curve becomes progressively smaller as the interference point, or the beginning of the involute, is approached. Contact stresses caused by the tooth load increase rapidly as the radius of curvature of the body is decreased. The stresses may become so high that pitting and surface failure will occur if the involute is too heavily loaded in the region of small radius of curvature.

The designer usually specifies teeth of the smallest size that will carry the load in order that the overlap of gear action will be as large as possible.

The stresses in spur gear teeth are very sensitive to the accuracy of mounting. If the two shafts are not parallel, the loading does not occur across the entire width of face, but is concentrated at one edge of the tooth, which can produce a large increase in the stresses. Sometimes the teeth are crowned, which means they are made slightly smaller toward the ends to ensure that the load is applied over the central portion of the width.

To assist the tooth in picking up its load gradually when coming into contact at B_1, Fig. 10-22(a), the width of the tooth is decreased slightly from the theoretical value near the tip. This is known as tip relief.[9]

19. Materials for Gears

Gears are made from a wide variety of materials, such as gray and alloy cast iron, cast and forged steel, brass and bronze, and impregnated fabric. Cast iron has good wearing properties but is weak in bending. It therefore requires the use of relatively large teeth. To secure a hard-wearing surface, carbon and alloy wrought steels must, in general, be given some sort of heat treatment, as, for example, quenching and tempering. If it is desired to harden only the surface, the steel may be heated by induction using high-frequency currents. The heated region is then quickly cooled by a water spray before the temperature of the interior has risen enough to be affected by the quenching. In some cases, the heating is done with a torch or a gas flame. A hard surface can also be secured by case hardening or carburizing. Carbon is

[9]See Walker, H., "Gear Tooth Deflection and Profile Modification," *The Engineer*, **166**, 1938, pp. 409 and 434; **170**, 1940, p. 102; **178**, 1944, pp. 484 and 502.

absorbed by the tooth surfaces while the gear is held at a red heat in a furnace. The core or interior of the gear will be strengthened, but it will retain its original ductility to a large extent. Another process for obtaining a very hard, though thin, surface layer is nitriding. The surface resistance of gears can also be improved by shot peening. A number of typical heat-treating steels used for gears are shown in Table 10-4. The composition and physical properties for these steels are given in the chapter on engineering materials.

Table 10-4   TYPICAL STEELS USED FOR HEAT-TREATED GEARS

Quenched and Tempered		Carburized	
SAE 1045	Plain-carbon	SAE 2315	Nickel
SAE 3140	Nickel-chromium	SAE 3115	Nickel-chromium
SAE 4140	Chromium-molybdenum	SAE 4615	Nickel-molybdenum
SAE 4640	Nickel-molybdenum	SAE 6115	Chromium-vanadium
SAE 5140	Chromium	AISI A8620	Chromium-nickel-molyb-
AISI A8640	Chromium-nickel-molyb-denum		denum

Warping or distortion of the gear in heat treatment is a serious matter. The load is then likely to be concentrated on a corner of the tooth instead of being evenly distributed across the face. Alloy steels retain their shape after heat treatment better than do plain-carbon steels. Accuracy of cutting and rigidity of mounting must also be given consideration so that a uniform load across the tooth may be obtained. Selection of a suitable material for a gear is frequently difficult and is usually based upon such considerations as cost, freedom from warping in heat treatment, good wearing properties, ability to sustain impact loads, and lack of sensitivity to stress concentrations. The tensile strength and chemical composition are thus of secondary importance except insofar as they contribute to the foregoing desirable qualities.

Distortion can be eliminated if the teeth are first rough cut in a blank that has been suitably prepared for machining by normalizing or annealing. The gear is then hardened to the upper limit for machining, about 300–350 Brinell, and the teeth are brought to final size by taking a finish cut. No subsequent heat treatment will be done. A free-machining steel of suitable physical properties is sometimes employed with the foregoing process, and a reduction in the machining costs can thereby be effected.

20. Gear Tooth Failures

At the present time, gear design is really a combination of art and science. Gear failures can best be avoided if practical field experience is available to assist in the design. Some gears run better and last longer than would be expected by the design equations, while others fail even when operated well within the limits imposed by the calculations.[10]

It is very important that a correct analysis be made as to why a pair of gears failed since merely making the set larger may not be a cure for the original cause of the failure. The three most common types of gear tooth failure are tooth breakage, surface pitting, and scoring.

(a) *Tooth breakage* may be caused by an unexpectedly heavy load being imposed on the teeth. A more common type of failure is due to bending fatigue, which results from the large number of repetitions of load imposed on the tooth as the gear rotates. A small value for the radius of the fillet may accentuate the bending fatigue effects.

(b) *Pitting* is a fatigue phenomenon caused by stresses exceeding the endurance limit of the surface material. After a sufficient number of repetitions of the loading cycle, bits of metal on the surface will fatigue and drop out. The process sometimes continues at an increasing rate since the remaining unpitted areas are less able to carry the load. Lubrication difficulties may contribute to pitting failures.

(c) *Scoring* can occur under heavy loads and inadequate lubrication. The oil film breaks down and metal-to-metal contact occurs. High temperatures result and the high spots of the two surfaces weld together. The welds are immediately broken, but the surfaces of the teeth undergo rapid wear.[11] Gearboxes should be broken in by preliminary operation at lower loads and smaller speeds until the tooth surfaces are highly polished. Misalignment of the shafts may shift the entire load to one edge of the tooth with resulting excessive stress and the probability of scoring. Fine-pitch teeth are largely immune to scoring. The diametral pitch should be at least 10 or greater. For coarse pitches, scoring is sometimes a greater hazard than breakage or pitting. At the present time there is no simple and reliable formula by means of which the designer can predict the possibility of scoring.

Foreign matter in the lubricant can cause a rapid increase in the rate of wear for a pair of gears. Unbalanced rotating masses and torsional vibration in a geared system can also cause unexpected loading, which the teeth are called upon to carry.

21. Determination of Tooth Loads

The tooth loads and bearing reactions for the various shafts of a gear train can be easily found by making suitable sketches. A typical case is shown in Fig. 10-23, the computations for which are carried out in the following example.

[10]See Chapter 7, Reference 4, and pp. 14–42 of Reference 3, end of Chapter.

[11]See Almen, J. O., Chapter 12 of *Mechanical Wear*, J. T. Burwell, Ed., Cleveland: American Society for Metals, 1950. See also pp. 52, 141, and 291 of Reference 4, end of chapter.

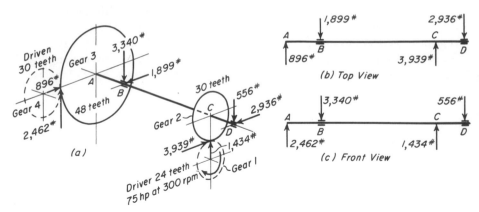

Figure 10-23 Shaft and gears of Example 10.

Example 10. Find the tooth loads and bearing reactions for the shaft shown in Fig. 10-23(a). All gears are 20° pressure angle with diametral pitch equal to 3. Make top and front views showing the loading for the shaft. Assume frictional losses to be negligible. $AB = 6$ in.; $\quad BC = 20$ in.; $\quad CD = 5$ in.

Solution. For gear 1: $\quad d = \dfrac{N}{P_d} = \dfrac{24}{3} = 8$ in.

For gears 1 and 2: $\quad V = \dfrac{\pi d n}{12} = \dfrac{\pi 8 \times 300}{12} = 628.3$ fpm

By Eq. (13): $\quad F_t = \dfrac{33{,}000\text{hp}}{V} = \dfrac{33{,}000 \times 75}{628.3} = 3{,}939$ lb,

tangential force on gear 2

Then: $\quad F_r = F_t \tan \varphi = 3{,}939 \times 0.36397 = 1{,}434$ lb

For gear 3: $\quad V = \dfrac{48}{30} \times 628.3 = 1{,}005.3$ fpm

$$F_t = \dfrac{33{,}000 \times 75}{1{,}005.3} = 2{,}462 \text{ lb},$$

tangential force on gear 3

Then: $\quad F_r = F_t \tan \varphi = 2{,}462 \times 0.36397 = 896$ lb

The tooth loads described above are shown in Fig. 10-23(a). The bearing reactions are found by simple statics. The top and front views for the shaft loads are shown in sketches (b) and (c), respectively. Bending moments and stresses for the shaft can be found in the usual manner.

22. Lubrication and Mounting of Gears

Gears operate under a diversity of conditions, and the methods of lubrication will vary accordingly. For unenclosed or exposed gearing, the lubricant is applied by an oilcan, a drip oiler, or a brush. Frequent applications of small amounts of lubricant are preferable to large volumes at longer intervals. If the gears are exposed to water or acids, a sticky lubricant that will adhere to the metal must be used.

When gears run in an enclosed casing, the larger gears may dip into a bath of oil, which will be carried to the wearing surfaces. Sometimes enclosed gearing is lubricated by spraying a jet of oil on the working surfaces as they revolve toward each other. When the contact pressure is very high, extreme-pressure EP lubricants must be used to prevent rupture of the oil film and the resulting metal-to-metal contact of the parts. Lubricants of the EP type contain additives that increase the load-carrying properties as well as prevent the squeezing out of the lubricant.

The mounting of gears is very important. Care must be exercised that shafts are parallel, or the entire load will be carried by the end of the tooth instead of across the entire width of each face. Improperly mounted gears are subjected to greater wear and danger of breakage; they are also the cause of noise and vibration.

In general, spur gears operate at high efficiencies. Good-quality commercial gears properly mounted and lubricated should not consume in friction more than 1 or 2 per cent of the power transmitted.

23. Backlash

When making gearing calculations, it is standard practice to assume that the thickness of the tooth measured around the pitch circle is exactly one-half the circular pitch. Because of unavoidable inaccuracies, it is necessary to cut the teeth slightly thinner to provide some clearance so that the gears will not bind but will roll together smoothly. This clearance is called *backlash*. It is illustrated in Fig. 10-24 and is defined as the amount by which a tooth

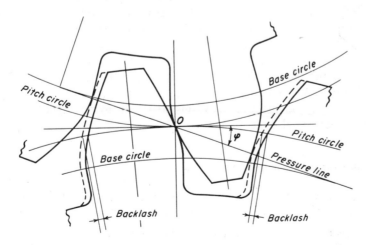

Figure 10-24 Backlash in gears.

space exceeds the thickness of the engaging tooth. The measurement is to be made on the pitch circle. If the measurement is made normal to the tooth profile, as by a feeler gage, the reading can be converted to a length taken along the pitch circle.

Backlash is usually obtained by setting the cutting tools deeper into the blank to produce a thinner tooth and wider space. Tooth thickness is usually reduced for each gear, although it is possible to obtain the backlash by making the teeth thinner on but one of the gears. Backlash can also be increased or decreased by giving a small variation to the center distance at which the gears are mounted. Recommended backlash values are given in Table 10-5. For gear trains in precision equipment and instruments it is frequently necessary to keep the backlash at a very low value. For involute teeth, backlash and variation in center distance have no effect on theoretically correct gear tooth action.

Table 10-5 RECOMMENDED BACKLASH AFTER ASSEMBLY

P_d	Backlash, in.	P_d	Backlash, in.
1	0.025–0.040	5	0.006–0.009
$1\frac{1}{2}$	0.018–0.027	6	0.005–0.008
2	0.014–0.020	7	0.004–0.007
$2\frac{1}{2}$	0.011–0.016	8–9	0.004–0.006
3	0.009–0.014	10–13	0.003–0.005
4	0.007–0.011	14–32	0.002–0.004

24. Dimensioning of Gears

The drawing for a gear should show the outside diameter and thickness of the blank with suitable tolerances. The number of teeth, diametral pitch, and pressure angle should be given. Backlash should be taken into account, and the tooth thickness, as measured around the theoretical pitch circle, should be shown. Tooth thickness should be expressed with a tolerance, but the pitch circle diameter should be given as a flat dimension. The addendum and whole depth should also be shown. It is usually helpful to include the center distance, number of teeth in the mating gear, and backlash after assembly. It is good practice, however, to avoid giving superfluous interdependent theoretical dimensions.

25. Undercutting in Gear Teeth

A gear having few teeth, if made by one of the generating processes, has undercut flanks[12] as illustrated in Fig. 10-25. For such gears, the end

[12]See Spotts, M. F., "Undercutting of Hobbed Spur Gear Teeth," *Machine Design*, **28**, Apr. 19, 1956, p. 123.

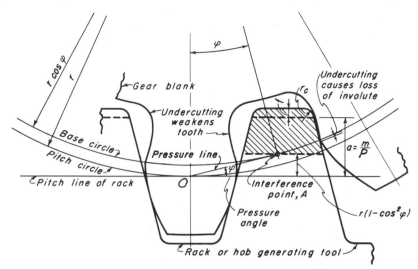

Figure 10-25 Undercutting in generated gear resulting from insufficient number of teeth.

of the cutting tool extends inside the interference point A or point of tangency of base circle and pressure line, and removes an excessive amount of metal. Undercutting not only weakens the tooth, but removes a small portion of the involute adjacent to the base circle; this loss of involute may cause a serious reduction in the length of contact. Gear action can take place only on the involute, which ends at the base circle. The portion of the mating tooth that projects inside the interference point is therefore not needed for the operation of the gears.

Undercutting can be prevented by using more and smaller teeth in the gear. No undercutting can occur if the addendum does not extend inside the interference point A. In Fig. 10-25 the height from the interference point to the center of the gear is $r \cos^2 \varphi$. The distance from the pitch line of the rack to the interference point, $r(1 - \cos^2 \varphi)$ or $r \sin^2 \varphi$, should be set equal to the addendum a or m/P_d of the cutting tool. Pitch radius r is equal to $N/2P_d$ so that addendum a is equal to

$$a = \frac{m}{P_d} = \frac{N}{2P_d} \sin^2 \varphi \qquad \text{or} \qquad N = \frac{2m}{\sin^2 \varphi} \qquad (41)$$

The minimum number of teeth that a gear must have if generated teeth are not to be undercut are as follows.

For $14\frac{1}{2}°$ full-depth gears: $m = 1$; $N = 32$
For $20°$ full-depth gears: $m = 1$; $N = 17$
For $20°$ stub teeth: $m = 0.8$; $N = 14$

Generated gears having fewer teeth than indicated above should be used only when the resulting loss of involute will not reduce the length of contact below a satisfactory value. The superiority of 20° gears, when the number of teeth is small, should be noted.

Should the gear in Fig. 10-25 be made by some method that would not undercut the flanks, there would be interference of metal, and the gear would probably neither mesh nor roll with another gear. Undercutting and interference can be avoided in gears of few teeth by using a system with a larger pressure angle, or by cutting the teeth on the long and short addendum system described in the following section. Cycloidal gear teeth neither undercut nor interfere. This advantage is retained by the $14\frac{1}{2}°$ composite system.

Involutes are of course produced by all the standard generating systems. However, in order to determine what the teeth actually look like, the method of production and the particular tool that will be used must be taken into account. Such information can be secured by making a large-scale layout of the teeth of both gears. It is especially needed when the number of teeth is small, and it is necessary to determine the extent of the undercutting. A magnified cutter should be made from a piece of tracing paper and rolled over the circle representing the blank; successive positions should be traced until the tooth outline is generated. The teeth of both gears can then be meshed together, and definite information can be obtained regarding undercutting and loss of involute, radius of fillet, length of contact, overlap, length of addendum, and depth of cut.

26. Long and Short Addendum Gearing

A very successful method of avoiding undercutting is to use the so-called "long and short addendum gearing." The addendum on the pinion or smaller gear is made longer than standard, and the addendum of the larger gear is made shorter by an identical amount. The gears are generated by the same rack or hob as for gears with standard addenda. The pressure angle during operation is unchanged.

The outside diameter of the pinion is increased, and since the hob penetrates less deeply into the blank, undercutting is reduced or eliminated. The increase of radius for the blank at which undercutting can be just avoided is equal to the value of e, as shown in Fig. 10-26. Here the end of the rack tooth, not counting clearance f, passes through the interference point. From the figure,

$$e = \frac{1}{P_d} - r(1 - \cos^2 \varphi) = \frac{1}{P_d} - r \sin^2 \varphi \qquad (42)$$

Example 11. Find the shift in the position of the rack necessary to eliminate undercutting in a $14\frac{1}{2}°$ full-depth gear of 18 teeth. $P_d = 1$.

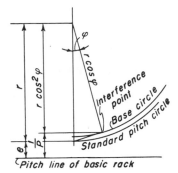

Figure 10-26 Shift in position of generating rack to cause addendum to pass through interference point of gear. $P_d = 1$.

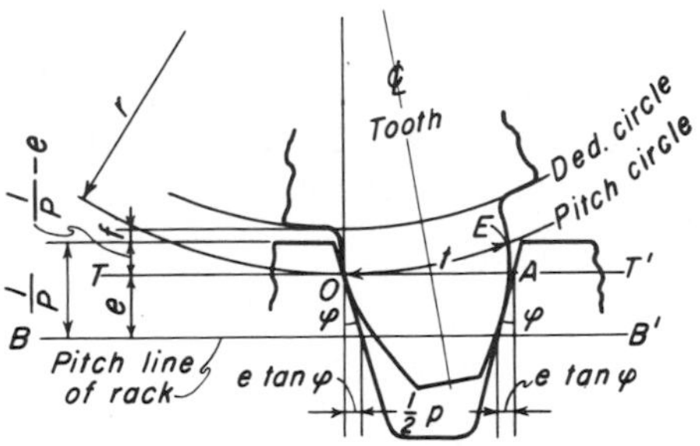

Figure 10-27 Change in tooth thickness resulting from non-standard setting of generating rack.

Solution. $r = \dfrac{N}{2P_d} = \dfrac{18}{2} = 9$ in.

In Eq. (42): $e = \dfrac{1}{1} - 9(1 - 0.96815^2) = 1 - 0.5642 = 0.4358$ in.

Although the value of e can be selected at the discretion of the designer, it is frequently rounded off to a value somewhat larger than that given by Eq. (42).

The teeth in Fig. 10-26 are thicker than those of a standard gear, and the spaces are thinner. Tooth thickness can be easily determined from Fig. 10-27, which shows an oversize blank with teeth cut by a rack located a distance e from the pitch circle of the gear. The pitch circle rolls without slipping along pitch line TT' of the rack. Segment PE equals arc OE, and the tooth thickness t on the pitch circle for the chosen value of e is

$$t_1 = 2e \tan \varphi + \tfrac{1}{2}p \tag{43}$$

Figure 10-27 illustrates the cutting of an oversize pinion. When the undersize gear blank is cut, the rack is located a distance e inside the pitch circle, and the tooth thickness is reduced to

$$t_2 = -2e \tan \varphi + \tfrac{1}{2}p \tag{44}$$

Example 12. Find the thickness of the tooth on the pitch circle for $14\tfrac{1}{2}°$ gears, if e is taken as 0.5 in. for $P_d = 1$.

Solution. $p = \dfrac{\pi}{P_d} = 3.1416$ in.

In Eq. (43), for pinion: $t_1 = 2 \times 0.5 \times 0.25862 + 0.5 \times 3.1416$
$$= 0.2586 + 1.5708 = 1.8294 \text{ in.}$$

In Eq. (44), for gear: $t_2 = -2 \times 0.5 \times 0.25862 + 0.5 \times 3.1416$
$$= -0.2586 + 1.5708 = 1.3122 \text{ in.}$$

The teeth for this example, for $N_1 = 18$ and $N_2 = 62$, are illustrated in Fig. 10-28. It can be seen that the pinion tooth, which normally would be badly undercut, is not.

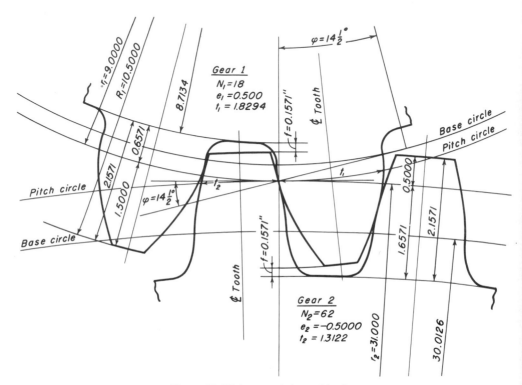

Figure 10-28 Long and short addendum gears.

Because the larger gear is undersize and the hob has been shifted inwardly, undercutting can be avoided only by having a sufficiently large number of teeth. The larger gear should have at least as many teeth in excess of the minimum number given in Section 25 as the smaller gear has fewer than the minimum.

Because of the change in tooth thickness, the y-factors of Table 10-1 cannot be used with long and short addendum gears.

When both mating gears have few teeth, undercutting in both can be prevented by the use of positive values for e. Assembly, when both gears have thicker teeth, can be made only at an increased center distance. The calculations become involved and are beyond the scope of this book.

27. Internal or Annular Gears

A gear can also be made as a ring (see Fig. 10-29) with teeth on the inside that are somewhat like the spaces between the teeth of an external spur gear. The assembly of an internal and external gear is very compact, with a reduced distance between the shaft centers.

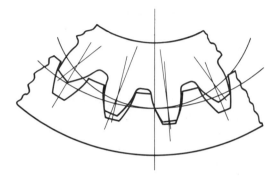

Figure 10-29 Internal or annual gear.

Internal gears are cut on the Fellows gear shaper. To prevent the corners on the inner ends of the teeth from being cut away, the gear must have considerably more teeth than the cutter. To obtain proper tooth action without undue modification of tooth shape, the smallest permissible difference between the number of teeth in the pinion and internal gear must be as follows.

    8   teeth for 20° stub-tooth form
   10   teeth for 20° full-depth teeth
   12   teeth for 14½° full-depth teeth

If the spokes or web of the internal gear are integral with the rim, a groove for cutter relief must be provided at the inner end of the teeth. This groove must be at least $\frac{3}{32}$ in. in width.

The teeth of internal gears are stronger than those of corresponding spur gears. Operation is smooth and quiet because a greater number of teeth are in contact.

28. Speed Ratios of Gear Trains

The speed ratio of a gear train can be easily found if the number of teeth in each gear is known. Thus in Fig. 10-30 the ratio between n_4, the rpm of the shaft for gear 4, and n_1, the rpm of the shaft for gear 1, is given by the following equation.

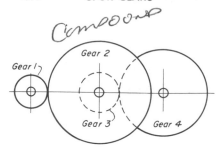

Gear 2

Gear 1

Gear 3   Gear 4

Figure 10-30 Gear train.

$$\frac{n_4}{n_1} = \frac{N_1}{N_2} \cdot \frac{N_3}{N_4} \tag{45}$$

Additional ratios can be inserted into Eq. (45) if the train consists of a larger number of gears.

The inverse problem, that of finding the number of teeth each gear should have if the ratio of the train is to equal an arbitrarily chosen value, is much more difficult. Such problems arise whenever it is necessary to give a specified relative motion to two portions of a mechanism.

A typical example is shown in Fig. 10-31, where a thread of lead l_1 is being

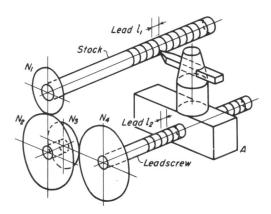

Figure 10-31 Thread being cut by lathe.

cut by an engine lathe whose lead screw has a lead of l_2. Let gear 1 make one complete revolution. Gear 4 makes $N_1 N_3/N_2 N_4$ revolutions, and carriage A moves a distance $N_1 N_3 l_2/N_2 N_4$ in. The distance must be equal to l_1. Hence

$$\frac{N_1 N_3}{N_2 N_4} = \frac{l_1}{l_2} \tag{46}$$

The problem is thus reduced to that of finding the numbers of teeth in the various gears.

In general, a given ratio cannot be exactly produced by a gear train because the individual members must necessarily contain integral numbers of teeth. It is possible, however, to approximate the desired ratio with a degree of accuracy that is sufficient for most applications. Although trial and error is involved, a systematic procedure helps reduce the labor to a minimum. A number of different methods can be used, of which the following is very effective.[13]

[13]See Spotts, M. F., "A Practical Method for Designing Gear Trains," *Prod. Eng.*, **24**, Feb. 1953, p. 211.

Let the given ratio between the first and last shafts be represented by G, and let u/w be a common fraction, u and w integers, whose value is close to that of G. Ratio G can be exactly represented by

$$G = \frac{uh - j}{wh} \tag{47}$$

where j is an integer, and h has the value obtained by solving the equation. Thus

$$h = \frac{j}{u - wG} \tag{48}$$

In Eq. (47), it is necessary that both numerator and denominator be integers factorable into terms suitable for use as the numbers of teeth in the gears of the train. However, the value of h, as given by Eq. (48), in general is not an integer. Then the numerator and denominator in Eq. (47) will not be integers either. However if an integer h', whose value is close to that of h, is used, an approximate value G' of the ratio is obtained that may be sufficiently close to the exact value. The equation for G' is

$$G' = \frac{uh' - j}{wh'} \tag{49}$$

The use of the above equations is illustrated by the following example.

Example 13. Suppose the shaft on the left in Fig. 10-30 is to make 2.54 revolutions for each revolution of the shaft on the right. Find suitable numbers of teeth for the gears.

Solution. $G = \dfrac{1}{2.54} = 0.3937008$

By trial it is found that fraction $\frac{24}{61}$ has a value close to G. Hence

$$\frac{u}{w} = \frac{24}{61}$$

Let: $j = 1$

In Eq. (48): $h = \dfrac{1}{24 - 61 \times 0.3937008} = -63.499$

Suppose: $h' = -63$

In Eq. (49): $G' = \dfrac{-24 \times 63 - 1}{-61 \times 63} = \dfrac{1,513}{3,843} = \dfrac{17 \times 89}{61 \times 63} = 0.3937028$

$$\text{error} = 0.3937028 - 0.3937008 = 0.0000020$$

If the solution above is sufficiently accurate, the numbers of teeth in the gears can be made $N_1 = 17$, $N_2 = 61$, $N_3 = 89$, and $N_4 = 63$.

Another solution can be obtained by taking

$$h' = -64$$

In Eq. (49): $G' = \dfrac{-24 \times 64 - 1}{-61 \times 64} = \dfrac{1,537}{3,904} = \dfrac{29 \times 53}{61 \times 64} = 0.3936988$

Another solution can be obtained as follows.

Let $j = 3$

$$h = \frac{3}{24 - 61 \times 0.3937008} = -190.498$$

Let: $h' = -190$

In Eq. (49): $G' = \dfrac{-24 \times 190 - 3}{-61 \times 190} = \dfrac{4,563}{11,590} = \dfrac{13 \times 13 \times 27}{10 \times 19 \times 61}$

$$= 0.3937015$$

$$\text{error} = 0.0000008$$

The accuracy has been improved, but a train of six gears is required.

If the numerator in Eq. (49) is not factorable, a different value of h' should be used. A factor table, such as that in *Machinery's Handbook*,[14] is very helpful. Because of the unavailability of large gears, it is frequently specified in change gear work that the largest gear should not exceed 100 or 120 teeth.

Suitable values for the fraction u/w can be found on the slide rule. Thus if the end of the C-scale is set at the value for G on the D-scale, numerous combinations for u/w can be found. Some additional values for the example above are $\frac{13}{33}$, $\frac{35}{89}$, $\frac{37}{94}$, $\frac{41}{104}$, $\frac{87}{221}$, $\frac{102}{259}$, $\frac{124}{315}$, and $\frac{137}{348}$.

Suitable values for u/w can also be found on a standard calculating machine. The given value for G should be placed on the keyboard, and successive values added until the sum is very close to an integer. Thus

$$wG \approx u \tag{50}$$

In the example above 61 times the value of G is a number very close to 24. Hence u/w can be taken as $\frac{24}{61}$.

29. Planetary Gear Trains

Planetary or epicyclic gear trains provide a compact arrangement suitable for speed reducers. In Fig. 10-32, the right-hand shaft is integral with arm 0. Gears 1 and 3 are keyed to a short length of shaft, which revolves in a bearing in arm 0. Gear 1 meshes with the fixed gear 2, and gear 3 meshes with gear 4, which in turn is keyed to the left-hand shaft.

Let N_1, N_2, N_3, and N_4 be the number of teeth in gears 1, 2, 3, and 4, respectively. As shown in the figure, arm 0, which was originally vertical, has been given an angular displacement α, which causes gear 1 to traverse arc AB on gear 2. Arc BC turned by gear 1 is equal to arc AB. Since angles are inversely proportional to the radii, or to the number of teeth, gears 1 and 3 are turned through angle $\alpha N_2/N_1$.

While the foregoing was taking place, gears 3 and 4 were rotating on

[14]New York: The Industrial Press, 1964. See also Reference 7, end of chapter.

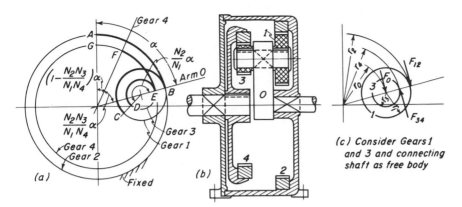

Figure 10-32 Planetary gear train.

each other through the equal arcs DE and EF. Gear 4 is turned in the reverse direction through angle $\alpha N_2 N_3 / N_1 N_4$.

The net effect of these two operations is to move the point of gear 4, which was originally vertical at G, over to location F. Gear 4 has thus been rotated clockwise through angle

$$\left[1 - \frac{N_2 N_3}{N_1 N_4}\right]\alpha$$

This latter value, when divided by α, gives the ratio of the rotations of shafts 4 and 0, respectively.

$$\frac{n_4}{n_0} = 1 - \frac{N_2 N_3}{N_1 N_4} \tag{51}$$

The equation above for the speed ratio can be derived in another way by giving consideration to the various torques. In Fig. 10-32(c) consider gears 1 and 3 and the connecting shaft as the free body. Let F_0 be the force exerted on the free body from arm 0, F_{12} be the force from gear 2 upon gear 1, and F_{34} be the force from gear 4 upon gear 3. If arm 0 is the driver, the input torque T_0 is equal to $F_0 r_0$. The output torque T_4 is equal to $F_{34} r_4$.

The moments of the forces about the center of the planet gears must be in equilibrium. Let r_1 be the radius of gear 1 and r_3 the radius of gear 3.

$$F_{12} r_1 = F_{34} r_3 \tag{a}$$

The moments about the main shaft centers must also be in equilibrium.

$$F_0 r_0 + F_0 r_0 = F_{34} r_4 \tag{b}$$

or

$$F_{12} r_2 = F_{34} r_4 - F_{12} r_2 \tag{c}$$

The value of F_{12} from Eq. (a) should now be substituted.

$$F_0 r_0 = F_{34} r_4 - \frac{F_{34} r_3 r_2}{r_1} = F_{34} r_4 \left[1 - \frac{r_2 r_3}{r_1 r_4}\right]$$

or
$$T_0 = T_4\left[1 - \frac{N_2 N_3}{N_1 N_4}\right]$$

If friction is neglected, the input and output energies for the gear train are equal. Hence

$$T_4 n_4 = T_0 n_0$$

or
$$\frac{n_4}{n_0} = \frac{T_0}{T_4} = 1 - \frac{N_2 N_3}{N_1 N_4} \tag{52}$$

The reaction or housing torque T_2 is equal to $F_{12} r_2$. Its value can be found by substitution for F_{34} from Eq. (a) into Eq. (c).

$$F_0 r_0 = \frac{F_{12} r_1 r_4}{r_3} - F_{12} r_2 = F_{12} r_2\left[\frac{r_1 r_4}{r_2 r_3} - 1\right]$$

$$T_0 = T_2\left[1 - \frac{N_2 N_3}{N_1 N_4}\right]\frac{N_1 N_4}{N_2 N_3}$$

or
$$T_2 = \frac{N_2 N_3 / N_1 N_4}{1 - N_2 N_3 / N_1 N_4} T_0 \tag{53}$$

The foregoing methods can be applied to all types of epicyclic systems, including those containing bevel gears. Either shaft 0 or 4 can be used as the driver.

Instead of a single arm 0, an actual gear train is usually constructed with either two or three equally spaced arms, each of which carries a pair of planet gears.

REFERENCES

1. Buckingham, E., *Analytical Mechanics of Gears*, New York: McGraw-Hill Book Company, 1949.

2. Buckingham, E., *Spur Gears*, New York: McGraw-Hill Book Company, 1928.

3. Dudley, D. W., Ed., *Gear Handbook*, New York: McGraw-Hill Book Company, 1962.

4. Dudley, D. W., *Practical Gear Design*, New York: McGraw-Hill Book Company, 1954.

5. Rothbart, H. A., Ed., *Mechanical Design and Systems Handbook*, New York: McGraw-Hill Book Company, 1964, Section 32.

6. *Standards of the American Gear Manufacturers Association*, Washington, D.C.

7. Stutz, C. C., *Formulas in Gearing*, Providence, R.I.: Brown and Sharpe Mfg. Co., 1938.

PROBLEMS

Assume that all gear blanks are solid circular cylinders unless specified otherwise.

1. From a piece of vellum, make a rack cutter similar to that in Fig. 10-33. $\varphi = 14\frac{1}{2}°$; $P = 1$. Use standard clearances at top and bottom of the tooth. Use a radius of the fillet $1\frac{1}{3}$ times the clearance. The fillet is tangent to both the sloping cutting edge and the clearance line. To preserve the strength of the tool, only a small portion of the paper along the cutting edges should be removed, as shown. The numbered lines at right angles to the pitch line are spaced $\frac{1}{8}p$ apart. Also, draw Fig. 10-34 to simulate the blank of a 12-tooth gear. The numbered radial lines are also spaced $\frac{1}{8}p$ apart around the pitch circle.

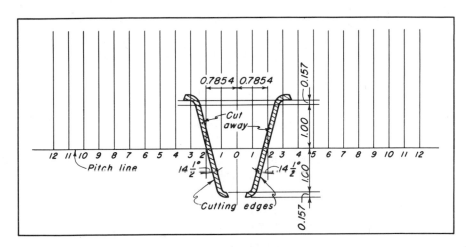

Figure 10-33 Problem 1.

Tool and blank are synchronized by placing the pitch line of the rack on the pitch circle of the gear, and having the correspondingly numbered lines coincide. A sharp drafting pencil should be drawn along the cutting edges for each position of the rack; an involute tooth is thus generated in the manner illustrated by Fig. 10-14.

Note the large amount of undercutting when a gear of few teeth is made by this process. Draw the base circle on the gear and estimate the percentage of involute lying between the base and pitch circles that has been lost by undercutting. If greater accuracy is desired, this problem can be worked out at twice the scale mentioned above.

2. For the gears of Problem 1, assume that the undercutting, as shown in Fig. 10-25, extends one-third of the way from the base to the pitch circles. Make a full-size layout and compute the number of tooth intervals in the remaining length of

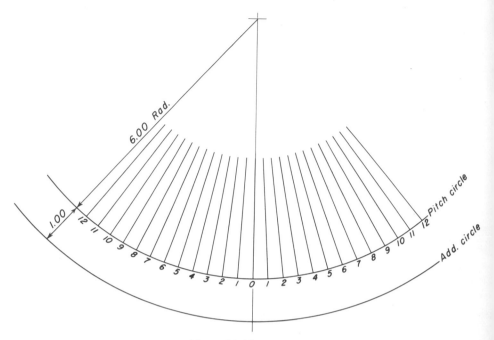

Figure 10-34 Problem 1.

contact if two of these gears are meshed together. Note the seriousness of the loss of even a small portion of the involute curve.

Ans.   No. of tooth intervals = 0.42.

3. How many teeth must there be in a $14\frac{1}{2}°$ gear of full-depth involute teeth if the base and working-depth circles coincide?   Ans. $N = 62.8$.

4. Make a full-size layout of a 20° full-depth involute tooth. $N = 12$; $P = 1$. Make the flank of the tooth radial inside the base circle. Let the radius of the fillet be $0.05p$ on one side of the tooth and $0.15p$ on the other side. Assume that the load is located on the pressure line at the end of the tooth similar to Fig. 10-18. Length l should be measured from point A to the cross section of minimum depth h.

Measure the tooth and compute y by Eq. (15). Do the same for several nearby cross sections until the one is found that gives the smallest value for y. Two sets of measurements and computations are needed, one for the tooth of the small fillet and the other for the large. Note how the size of the fillet as actually produced on the tooth will cause changes in the value of the y-factor. Can you obtain the tabulated value of 0.078 for y? If $r_f = 0.05p$, will the tooth be safe when carrying the load indicated by Eq. (16)?

5. Two 20° full-depth gears are made of soft steel, BHN = 200. AGMA Quality No. 8. Pinion turns 172.5 rpm. $N_1 = 48$; $N_2 = 96$; $P_d = 4$, and $b = 3$ in. Find the horsepower the gears are transmitting, and the number of base pitches in the interval of contact.   Ans.   hp = 23.2; No. of intervals = 1.80.

6. Two 20° full-depth steel gears are heat treated to BHN = 350. AGMA Quality No. 8. Pinion turns 860 rpm. $N_1 = 30$; $N_2 = 120$; $P_d = 5$, and $b = 2.5$ in. Find the horsepower the gears are transmitting.   *Ans.*   hp = 86.1.

7. Two 20° full-depth steel gears are heat treated to BHN = 350. AGMA Quality No. 8. Pinion turns 860 rpm. $N_1 = 24$; $N_2 = 96$; $P_d = 4$, and $b = 2$ in. Find the horsepower the gears are transmitting.   *Ans.*   hp = 77.2.

8. Two 20° full-depth gears are made of induction-hardened steel. AGMA Quality No. 8. Pinion turns 215 rpm $N_1 = 36$; $N_2 = 72$; $P_d = 3$, and $b = 2.5$ in. Find the horsepower the gears are transmitting.   *Ans.*   hp = 198.1.

9. Two 20° stub-tooth steel gears are heat treated to BHN = 350. AGMA Quality No. 8. Pinion turns 860 rpm. $N_1 = 30$; $N_2 = 90$; $P_d = 5$, and $b = 2$ in. Find the horsepower the gears are carrying, and the number of base pitches in the interval of contact.   *Ans.*   hp = 62.2; No. of intervals = 1.43.

10. Two 20° full-depth gears are carrying 45 hp at 860 rpm for the pinion. Speed ratio is 2 : 1 with a center distance of 9 in. AGMA Quality No. 9. $P_d = 5$. Both gears hardened to BHN = 350. Calculate the necessary value for face width b.
        Ans. $b = 1.29$ in.

11. Two 20° full-depth gears are made of induction-hardened steel. Center distance is 7.5 in. Speed ratio is 4 : 1. Pinion turns 3,450 rpm. AGMA Quality No. 8. If the gears are carrying 120 hp, find suitable values for P_d and b.

12. Calculate the value of K for the wear equation for the combination of steel pinion and cast-iron gear, $E_1 = 30,000,000$ psi and $E_2 = 16,000,000$ psi, for φ equal to $14\frac{1}{2}°$ and 20°.
   (a) For $s_{ec} = 50,000$ psi for BHN 160–200 cast iron.
   (b) For $s_{ec} = 60,000$ psi for BHN 210–245 cast iron.
        Ans. $s_{ec} = 50,000$ psi; $K = 43$ for $14\frac{1}{2}°$; $K = 59$ for 20°.
        $s_{ec} = 60,000$ psi; $K = 62$ for $14\frac{1}{2}°$; $K = 84$ for 20°.

13. A 20° full-depth steel pinion meshes with a cast-iron gear, BHN = 220. Pinion turns 1,140 rpm. AGMA Quality No. 9. $N_1 = 20$; $P_d = 4$, and $b = 2$ in.
   (a) Find the horsepower if $N_2 = 40$.   *Ans.*   hp = 24.4.
   (b) Find the horsepower if $N_2 = 120$.   *Ans.*   hp = 40.6.

14. A 20° full-depth steel pinion meshes with a cast-iron gear, BHN = 220. Center distance is 8 in. Speed ratio is 3 : 1. Pinion turns 430 rpm. AGMA Quality No. 8. $P_d = 4$ and $b = 1.75$ in. Find the horsepower capacity of the set, and the number of base pitches in the interval of contact.
        Ans.   hp = 9.4; No. in contact = 1.62.

15. Two 20° full-depth gears have AGMA Quality No. 8. Pinion turns 430 rpm. $N_1 = 24$; $N_2 = 48$; $P_d = 4$, and $b = 2$ in. Find the horsepower transmitted for the following combination of materials.
   (a) Steel pinion and cast-iron gear, BHN = 220.   *Ans.*   hp = 17.9.
   (b) Both gears steel with BHN = 200.   *Ans.*   hp = 7.7.
   (c) Both gears steel with BHN = 350.   *Ans.*   hp = 41.8.
   (d) Both gears induction-hardened steel.   *Ans.*   hp = 108.8.
   (e) Both gears carburized steel.   *Ans.*   hp = 177.6.

16. Find the horsepower for the gears of Problem 7 if (a) the quality is made No. 9, and (b) if the quality is made No. 10.

Ans.   (a) hp $= 92.2$; (b) hp $= 102.3$.

17. Find the minimum value of the pressure angle for a system of gearing that will generate a 12-tooth full-depth pinion without undercutting.

Ans. $\varphi = 24°6'$.

18. A 72-tooth $14\frac{1}{2}°$ pinion meshes with a 144-tooth gear on a standard center distance of 18 in. The smaller gear has full-depth teeth, and the larger gear has standard stub teeth. Find the number of tooth intervals in the length of contact. $P = 6$.

Ans.   No. of intervals $= 2.10$.

19. In Fig. 10-22(a), distance OA_1 is equal to three-fourths of OA. Gear 1 has 20 teeth. If gear 2 has full-depth addenda, find the (theoretical) value of r_2. Make computations for $P_d = 1$, and $\varphi = 20°$

Ans. $r_2 = 22.74$ in.

20. Two $20°$ spur gears have 17 teeth each. What must be the value of the addendum if the length of contact extends two-thirds of the way from the pitch point to the interference point? $P_d = 1$. Find the number of teeth in contact.

Ans.   Add. $= 0.842$ in.; No. $= 1.31$.

21. The length of contact on a $20°$ full-depth gear of 18 teeth extends two-thirds of the way from the pitch point to the interference point. Find the length of the addendum of the mating gear if it has 72 teeth. $P_d = 1$. Find the number of teeth in contact.

Ans.   Add. $= 0.752$; No. $= 1.46$.

22. In Fig. 10-35, consider shaft B together with gears 2 and 3 as the free body. Determine the tooth forces for these gears based on the horsepower. Draw a top view of shaft B and show all horizontal loads and reactions. Draw an elevation of shaft B and show all vertical loads and reactions. $P_d = 6$; $\varphi = 20°$.

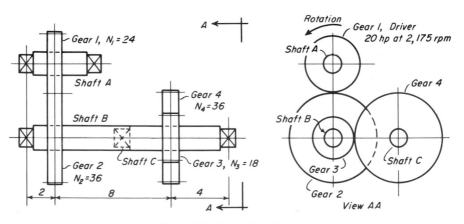

Figure 10-35 Problem 22.

23. Work Problem 22 but with view AA as given by Fig. 10-36.

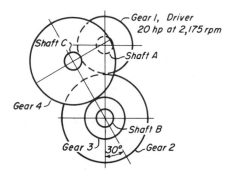

Figure 10-36 Problem 23.

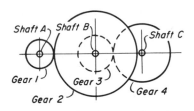

Figure 10-37 Problems 24 and 25.

24. In Fig. 10-37, gear 1 has 15 teeth of $P_d = 5$, and turns 2,150 rpm. Gear 4 has 43 teeth of $P = 4$, and turns 450 rpm. The distance between shafts A and C is 13 in. Find the number of teeth in gears 2 and 3.   *Ans.* $N_2 = 35$; $N_3 = 21$.

25. In Fig. 10-37, gear 1 has 24 teeth and gear 2 has 70 teeth of $P_d = 4$. Gear 3 has 20 teeth and gear 4 has 50 teeth of $P_d = 3$.

(a) Find the value of the center distance between shafts A and C.

(b) Find the rpm of gear 4 if gear 1 turns 1,200 rpm.

(c) Find the torque at shaft C if the input horsepower at shaft A is 50. The efficiency of each pair of gears is 98 per cent.

Ans.   (a) 23.417 in.; (b) $n_4 = 164.6$ rpm;

(c) $T = 18,390$ in. lb.

26. In Fig. 10-38, $N_1 = 32$, $N_2 = 75$, and $N_3 = 24$. Find the value of the p for gear 3 if the rack is to be moved 0.5 ft for each revolution of gear 1.

Ans. $p = 0.586$ in.

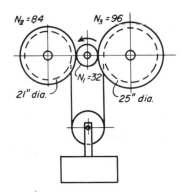

Figure 10-39 Problem 27.

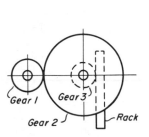

Figure 10-38 Problem 26.

27. In Fig. 10-39, gear 1 is located on the motor shaft and turns 1,200 rpm. Find the speed at which the load is being raised.   *Ans.*   52.36 fpm.

28. Determine some additional solutions for a gear train of ratio 2.54:1.

> *Ans.* $(29 \times 53)/(61 \times 64)$;
> $(11 \times 67)/(36 \times 52)$;
> $(13 \times 19 \times 71)/(16 \times 32 \times 87)$, etc.

29. Determine the suitable numbers of teeth for the gears of a train of ratio 251:93.   *Ans.* $(41 \times 61)/(75 \times 90)$;
> $(29 \times 67)/(57 \times 92)$;
> $(15 \times 53 \times 79)/(43 \times 54 \times 73)$, etc.

30. Make a front view of the reducer shown in Fig. 10-40 and show the relative displacements of all gears for a given rotation α of arm 0. Mark the values of all angles used in deriving the equation for the speed ratio.   *Ans.* $\dfrac{n_1}{n_0} = 1 + \dfrac{N_3}{N_1}$.

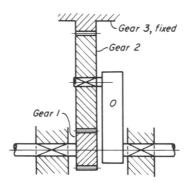

Gear 3, fixed
Gear 2
Gear 1
O

Figure 10-40 Problem 30.

31. Make a front view of the epicyclic gear train shown in Fig. 10-41. By a process similar to that illustrated in Section 29, derive the equation for the speed

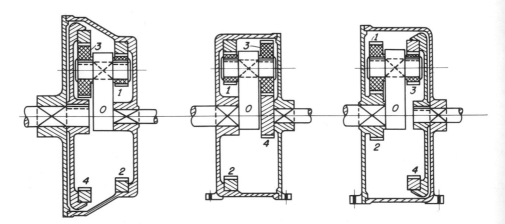

Figure 10-41 Problem 31.   **Figure 10-42** Problem 32.   **Figure 10-43** Problem 33.

ratio of shafts 4 and 0. A negative result means that gear 4 is rotating oppositely to arm 0.

$$Ans. \quad \frac{\text{rpm}_4}{\text{rpm}_0} = -\frac{N_2 N_3}{N_1 N_4} + 1.$$

32. Work Problem 31 with the epicyclic system shown in Fig. 10-42.

$$Ans. \quad \frac{\text{rpm}_4}{\text{rpm}_0} = 1 + \frac{N_2 N_3}{N_1 N_4}.$$

33. Work Problem 31 for Fig. 10-43.

$$Ans. \quad \frac{\text{rpm}_4}{\text{rpm}_0} = 1 + \frac{N_2 N_3}{N_1 N_4}.$$

34. Work Problem 31 for Fig. 10-44.

$$Ans. \quad \frac{\text{rpm}_4}{\text{rpm}_0} = -\frac{N_2 N_3}{N_1 N_4} + 1.$$

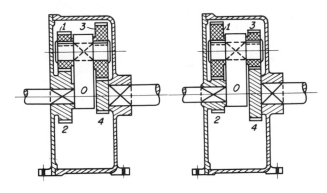

Figure 10-44 Problem 34.   **Figure 10-45** Problem 35.

35. Work Problem 31 for Fig. 10-45.

$$Ans. \quad \frac{\text{rpm}_4}{\text{rpm}_0} = 1 - \frac{N_2 N_3}{N_1 N_4}.$$

36. Two 20° full-depth gears are made of induction-hardened steel. Center distance is 9 in. Two-to-one reduction ratio. Pinion turns 1,725 rpm. AGMA Quality No. 8. If the gears are carrying 250 hp, find suitable values for P_d and b.

37. In Fig. 10-46, the gears are 20° full-depth of AGMA Quality No. 8.

(a) Find the horsepower transmitted by the 1–2 mesh.

(b) Find the required width b for the 3–4 mesh if they are carrying the same horsepower as the 1–2 mesh. *Ans.* (a) hp $= 22.9$; (b) $b = 2.29$ in.

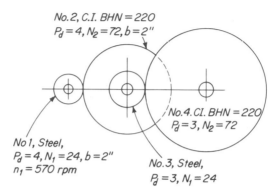

No.2, C.I. BHN $=220$
$P_d = 4, N_2 = 72, b = 2''$

No.4. CI. BHN $= 220$
$P_d = 3, N_2 = 72$

No 1, Steel,
$P_d = 4, N_1 = 24, b = 2''$
$n_1 = 570$ rpm

No. 3, Steel,
$P_d = 3, N_1 = 24$

Figure 10-46 Problem 37.

38. Same as Problem 37 except $P_d = 5$ and $b = 1.5$ in. for mesh 1–2, and $P_d = 4$ for mesh 3–4. *Ans.* (a) hp $= 11.6$; $b = 1.98$ in.

39. Two 20° full-depth gears have $P_d = 6$ and $b = 1.5$ in. AGMA Quality 9. Pinion turns 860 rpm. $N_1 = 30$ and $N_2 = 45$. Answer all questions asked in Problem 15.

40. Two 20° full-depth gears are made of flame-hardened steel and have 120 teeth each. $P_d = 5$ and $b = 2.5$ in. AGMA Quality No. 8. Rpm $= 150$.

(a) Find values for F_d and hp.

(b) Same, except the blanks are made with an inside radius of 10 in. Neglect inertia of spokes and hub. *Ans.* (a) $F_d = 3,900$ lb; hp $= 114.7$.
(b) $F_d = 2,810$ lb; hp $= 145.9$.

41. Two 20° full-depth gears are made of flame-hardened steel, and have 120 teeth each. $P_d = 6$. AGMA Quality No. 9. $b = 2$ in. and rpm $= 240$.

(a) Find the values for F_d and hp.

(b) Same except the blanks are made with an inside radius of 8.5 in. Neglect inertia of spokes and hub. *Ans.* (a) $F_d = 2,740$ lb; hp $= 96.8$.
(b) $F_d = 1,890$ lb; hp $= 128.9$.

42. Show that when a pair of gears are operating at their maximum horsepower capacity, the speed n_1 of the pinion must be such as to make F_d equal to one-half the smaller of either F_b or F_w.

Find the speed at which the gears of Problem 8 must be operated to obtain their maximum output, and find the value of this horsepower.

Ans. $n_1 = 896$ rpm; hp $= 469$.

11

Bevel, Worm, and Helical Gears

Bevel, worm, and helical gears are advanced forms of gearing capable of meeting special requirements of geometry or strength that cannot be obtained from spur gears. Bevel gears, with straight or spiral teeth cut on cones, can be used to connect intersecting shafts. A worm gear, consisting of a screw meshing with a gear, can be used to obtain a large speed reduction. Helical gears have teeth that lie in helical paths on the cylinders instead of teeth that are parallel to the shaft axis. The geometry of these different types of gearing is considerably more involved than for spur gears, and the problems of production and inspection are also more complicated.

b, face width of tooth
b', thickness of bevel gear parallel to axis
c, center distance
d, pitch diameter
d', pitch diameter, formative gear
e, error
F_b, bending capacity, pounds
F_d, dynamic tooth load, pounds

F_t, transmitted or horsepower load, pounds
F_w, capacity in wear, pounds
hp, horsepower
K, constant in wear equation
L, lead of worm or helical gear
n, speed of rotation, revolutions per minute
N, number of teeth in gear

N', number of teeth in formative gear
p, circular pitch in plane of rotation
p_n, circular pitch normal to teeth
p_o, circular pitch at outside radius of bevel gear
P_d, diametral pitch
P_{dn}, diametral pitch normal to teeth
Q, factor in wear equation, helical gears
Q', factor in wear equation, bevel gears
r, pitch radius
r', pitch radius, formative gear
r_o, outside radius of pitch cone of bevel gear
s, bending stress
T, torque
V, pitch-line velocity, feet per minute

V_s, velocity of sliding between teeth, feet per minute
y, Lewis or form factor
α, (alpha) pitch cone angle of bevel gear
β, (beta) ratio N_1/N_2
η, (eta) efficiency
λ, (lambda) helix angle of worm (between teeth and plane of rotation)
μ, (mu) coefficient of friction
φ (phi) pressure angle in plane of rotation
φ_n, (phi) pressure angle normal to teeth
ψ, (psi) helix angle of helical gear (between teeth and axis)
ω, (omega) speed of rotation, radians per second

1. Straight-Tooth Bevel Gears

When intersecting shafts must be connected by gearing, the pitch cylinders are replaced by pitch cones tangent to each other along an element, with their apexes at the intersection of the shafts. Teeth are placed on the cones, forming bevel gears, as illustrated in Fig. 11-1. The names of some of the principal parts of a bevel gear are given in Fig. 11-2. The back cone, as shown in the latter figure, has elements that are perpendicular to those of the pitch cone. The outlines of the teeth, as generated on the back cone, are shown in view A-A. Because of difficulties in securing uniform bearing along the tooth, the width of face b is generally not made longer than one-third of the pitch cone radius. Undercutting occurs in bevel gears having few teeth.

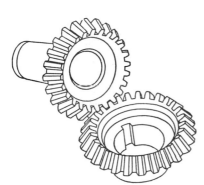

Figure 11-1 Straight-tooth bevel gears. (*From* Lubrication, *published by The Texas Co.*)

The size of bevel gear teeth is usually specified by a diametral pitch P_d or circular pitch p_o, which refers to the developed back cone of the gear as shown by view A-A in Fig. 11-2. Thus,

$$r_{o1} = \frac{N_1}{2P_d} \quad \text{and} \quad r_{o2} = \frac{N_2}{2P_d} \tag{1}$$

Here N_1 and N_2 represent the numbers of teeth in the gears.

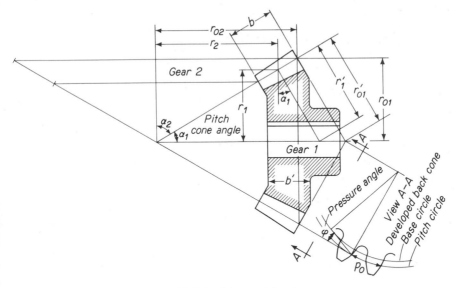

Figure 11-2 Straight-tooth bevel gear.

2. Beam Strength of Bevel Gears

The beam strength of bevel gear teeth is computed by the Lewis equation. It is assumed that the bevel gear tooth is equivalent to a spur tooth whose cross section is the same as the cross section of the bevel tooth at the midpoint of the face b.

Let r be the pitch radius at the midpoint of the face. The circular pitch p at the midpoint of the face, then, is

$$p = \frac{2\pi r}{N} \tag{2}$$

For the pinion or smaller gear, the radius r_1 to the midpoint of the face is found by

$$r_1 = r_{o1} - \tfrac{1}{2}b \sin \alpha_1 \tag{3}$$

where α_1 is the pitch cone angle for gear 1. The corresponding equation for the larger gear is

$$r_2 = r_{o2} - \tfrac{1}{2}b \sin \alpha_2 = r_{o2} - \tfrac{1}{2}b \cos \alpha_1 \tag{4}$$

where α_2 is the pitch cone angle for gear 2.

Although the number of teeth in a bevel gear is N_1, view A-A in Fig. 11-2 indicates that the shape of the teeth will be the same as those of a spur gear of radius r'_{o1}. The number of teeth in a fictitious gear of this radius is called the formative or virtual number of teeth N'_1. Radius r'_{o1} is equal to

$$r'_{o1} = \frac{r_{o1}}{\cos \alpha_1} \tag{5}$$

The circular pitch p_o at the outside radius has the value

$$p_o = \frac{2\pi r'_{o1}}{N'_1} = \frac{2\pi r_{o1}}{N_1} = \frac{2\pi r'_{o1} \cos \alpha_1}{N_1}$$

The last form on the right results from the substitution of Eq. (5). This equation yields

$$N'_1 = \frac{N_1}{\cos \alpha_1} \tag{6}$$

The y-factor, Table 10-1, Chapter 10, for use in the Lewis equation, is selected for a gear of N'_1 teeth. The Lewis equation, for the capacity of the tooth to carry a bending load, is

$$F_b = sbyp \tag{7}$$

3. Limit Load for Wear. Dynamic Load

The limit load for wear is determined by an equation similar to that used for spur gears but is based on the virtual number of teeth. However, because of the difficulty in securing bearing along the entire face width b, only about three-quarters of b is considered as effective.[1] Then

$$F_w = 0.75 d'_1 b Q' K \tag{8}$$

Here d'_1 is the diameter of the formative gear or $2r'_1$, and Q' is obtained by the use of the formative number of teeth. Hence by Eq. (6),

$$Q' = \frac{2N'_2}{N'_1 + N'_2} = \frac{2N_2/\sin \alpha_1}{(N_1/\cos \alpha_1) + (N_2/\sin \alpha_1)} = \frac{2N_2}{N_1 \tan \alpha_1 + N_2} \tag{9}$$

Factor K is obtained from Table 10-2 for spur gears.

The dynamic load can be approximated by the same equation used for spur gears if the thickness b' of the blanks in the axial direction is specified. When both gears are steel, the equation for dynamic load is

$$F_d = \frac{1.46 e n_1 N_1 b' r_1 r_2}{\sqrt{r_1^2 + r_2^2}} \tag{10}$$

Here e is the error as found in Table 10-3 for spur gears, and n_1 is the speed of the pinion in revolutions per minute.

As for spur gears, the transmitted force F_t for determining the horsepower capacity will be found from either F_b or F_w, whichever is the smaller. Then

$$F_t = F_b - F_d \tag{11}$$

or

$$F_t = F_w - F_d \tag{12}$$

[1] See p. 540, Reference 1, Chapter 10.

The value of the horsepower is given by

$$\text{hp} = \frac{F_t V}{33,000} \tag{13}$$

where

$$V = \frac{\pi d_1 n_1}{12} \text{ fpm} \tag{14}$$

Diameter d_1 in this equation is expressed in inches.

Example 1. Find the horsepower carried by a pair of 20° full-depth bevel gears of $N_1 = 40$ and $N_2 = 60$ teeth. Gears are steel and are hardened to BHN $= 350$. Diametral pitch at the outside diameter is equal to 4. Width of face is 2.5 in., but thickness b' in the axial direction is equal to 2 in. AGMA Quality is equal to No. 8. Pinion turns 600 rpm. See Fig. 11-3.

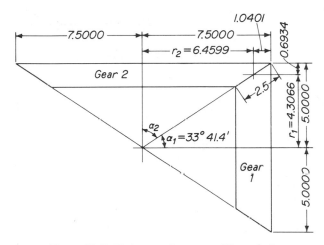

Figure 11-3 Pitch cones for gears of Example 1.

Solution. $\quad r_{10} = \dfrac{N_1}{2P_d} = \dfrac{40}{2 \times 4} = 5$ in., $\quad r_{20} = \dfrac{N_2}{2P_d} = \dfrac{60}{2 \times 4} = 7.5$ in.

$$\tan \alpha_1 = \frac{5}{7.5} = 0.66667$$

$$\alpha_1 = 33°41.4'$$

By Eq. (3): $\quad r_1 = r_{10} - \frac{1}{2}b \sin \alpha_1 = 5 - \frac{1}{2} \times 2.5 \times 0.55470$
  $= 4.3066$ in. at midpoint of face

By Eq. (4): $\quad r_2 = r_{20} - \frac{1}{2}b \cos \alpha_1 = 7.5 - \frac{1}{2} \times 2.5 \times 0.83205$
  $= 6.4599$ in. at midpoint of face

By Eq. (2): $\quad p = \dfrac{2\pi r_1}{N_1} = \dfrac{2\pi 4.3066}{40} = 0.6765$ in.,

at midpoint of face

By Eq. (5): $r'_1 = \dfrac{r_1}{\cos \alpha_1} = \dfrac{4.3066}{0.83205} = 5.176$ in.

By Eq. (6): $N'_1 = \dfrac{N_1}{\cos \alpha_1} = \dfrac{40}{0.83205} = 48.1$

By Table 10-1: $y = 0.129$

By Table 10-2: $s = 15,000$ psi

By Eq. (7): $F_b = sbyp = 15,000 \times 2.5 \times 0.129 \times 0.6765 = 3,270$ lb

By Eq. (9): $Q' = \dfrac{2N_2}{N_1 \tan \alpha_1 + N_2} = \dfrac{2 \times 60}{40 \times 0.66667 + 60} = 1.385$

By Eq. (8): $F_w = 0.75d'_1 bQ'K = 0.75 \times 2 \times 5.176 \times 2.5 \times 1.385 \times 163$
              $= 4,380$ lb

By Table 10-3: $e = 0.0035 + 0.0036 = 0.0071$ in.

By Eq. (10): $F_d = \dfrac{1.46 \times 0.0071 \times 600 \times 40 \times 2 \times 4.3066 \times 6,4599}{\sqrt{4.3066^2 + 6.4599^2}}$

              $= 1,780$ lb

The horsepower capacity will depend on F_b since it is smaller than F_w.

By Eq. (11): $F_t = F_b - F_d = 3,270 - 1,780 = 1,490$ lb

By Eq. (14): $V = \dfrac{\pi d_1 n_1}{12} = \dfrac{\pi 2 \times 4.3066 \times 600}{12} = 1,353$ fpm

By Eq. (13), $\text{hp} = \dfrac{F_t V}{33,000} = \dfrac{1,490 \times 1,353}{33,000} = 61.1$

Stress concentration, if present, may cause a reduction in the bending capacity of these gears.

4. Tooth Loads of Bevel Gears

Figure 11-4 indicates that there are three mutually perpendicular components of force acting at the midpoint of the tooth of a bevel gear. Transmitted force F_t acts tangentially to the pitch point. The force F_n is normal to the tooth surface and has the value $F_t \sec \varphi$. This latter force has the projection $F_t \tan \varphi$ in the axial plane, which is then divided into the radial and axial components $F_t \tan \varphi \cos \alpha$ and $F_t \tan \varphi \sin \alpha$, respectively.

These forces must be used in determining the bearing reactions, thrust, and torque on the shaft of the gear as shown by the following example.

Example 2. Consider the bevel gears of Example 1. Let the base of the pitch cone of the larger gear be located at the midpoint of a 24-in. simply supported shaft. Make a free-body diagram of the gear and shaft for the transmitted force and show all the reactions. Let the thrust and torque be resisted at the right-hand end of the shaft as in Fig. 11-5.

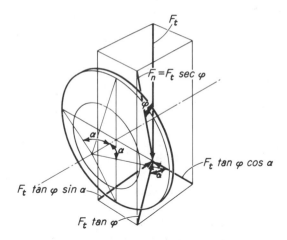

Figure 11-4 Forces at midpoint of bevel gear tooth.

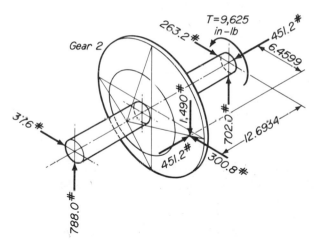

Figure 11-5 Tooth loads and shaft reactions on gear for Example 2.

Solution. From Example 1, $F_t = 1,490$ lb. As before, radius to midpoint of face, $r_2 = 6.4599$ in.

Axial force: $\quad F_t \tan \varphi \sin \alpha_2 = 1,490 \times 0.36397 \times 0.83205 = 451.2$ lb

Radial force: $\quad F_t \tan \varphi \cos \alpha_2 = 1,490 \times 0.36397 \times 0.55470 = 300.8$ lb

Torque: $\qquad\qquad\qquad T = r_2 F_t = 6.4599 \times 1,490 = 9,625$ in. lb

The forces above are shown on the gear in Fig. 11-5. The reactions at the bearings are found by simple statics.

The efficiency of bevel gearing is very high, providing the gears are properly mounted and adjusted. Not more than 1 or 2 per cent of the power should be lost by accurately manufactured gears that are properly mounted.

5. Spiral Bevel Gears

Bevel gears can also be made with teeth lying in spiral paths on the pitch cones. More teeth are then in contact simultaneously, and smoother and quieter action can be obtained from such gears. Accuracy of adjustment between a pair of bevel gears can be more readily secured with spiral teeth than with straight teeth.

The Gleason Works has developed a system of bevel gearing, both straight-tooth and spiral, which is in wide use. The addenda are arranged on the long and short plan to avoid undercutting and to give greater strength to pinions having few teeth. The addendum of the pinion is made longer than the usual value, and the addendum of the gear is shortened a corresponding amount. A variety of pressure angles are used in order to obtain the best operating conditions for each velocity ratio.

Hypoid gears have spiral teeth and shafts that do not intersect. It is thus possible to connect continuous nonintersecting shafts by such gears, as shown in Fig. 11-6. Hypoid gears have been used in automobiles so that the drive shaft can be placed beneath the level of the floor.

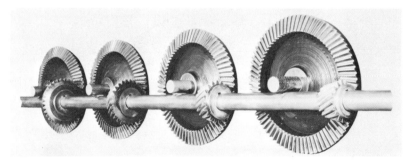

Figure 11-6 Hypoid gears on non-intersecting shafts. (*Courtesy Gleason Works.*)

In the rear axles of automobiles, the bevel gear stresses are very low under normal conditions of service. Such gears are therefore designed on the assumption that operation in low gear at full engine torque for a relatively short period, perhaps 0.1 per cent of the life of the car, would normally be sufficient to cause fatigue failure. The requirements are somewhat higher for trucks and busses. Gear steels for automotive service are selected mainly on

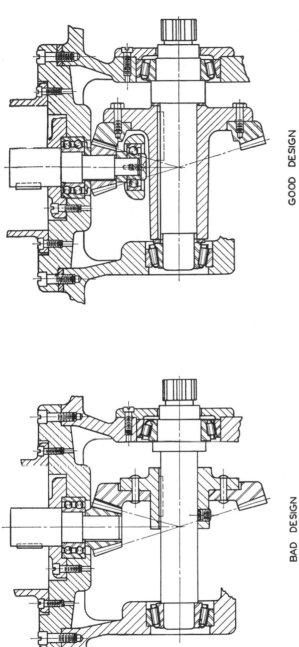

GOOD DESIGN

BAD DESIGN

Face width too great, more than one third of cone distance.
Metal at small end of pinion between teeth and bore too thin for proper strength. Cutter would interfere seriously with arbor in cutting operation.
Webbed section of steel ring gear adds to cost of material and machining.
Use of rivets to hold gear to hub introduces danger of runout.
A setscrew is inadequate to hold gear in correct axial position, and tends to "cock" gear on shaft.
Pinion held on shaft only by fit of bore.
An overhung pinion cannot be held in line by one double-row bearing.
No means of adjustment for definitely placing gears in correct mounting position.

Face width reasonable, less than one third of cone distance.
Sufficient metal at small end of pinion to provide strength and avoid interference of cutter with arbor.
Section of ring gear simpler and more direct in design. Gear is supported directly on back and is centered on large bore.
Screws to hold gear to hub are preferable.
Gear is positively held in position on shaft.
Pinion locked in position by washer and screw.
Pinion rigidly supported by addition of inboard bearing.
Adjusting washers provided, to be ground to thickness required to obtain correct position of pinion and gear.

Pinion and bearings can be assembled as a complete unit.

Figure 11-7 Design and arrangement of spiral bevel gears. Comparison of good and bad practice.

435

the basis of machining characteristics, cost, and resistance to stress concentration and warping in heat treatment.[2] A comparison between good and bad practice[3] in the mounting of bevel gears is shown in Fig. 11-7.

6. Worm Gears

Figure 11-8 Worm and wheel. (*Courtesy D. O. James Gear Mfg. Co.*)

Worm gears, as illustrated by Fig. 11-8, have spiral teeth and shafts at right angles. Worms can be made with single, double, triple, or more threads. The teeth of the wheel or gear envelop the worm threads and give line contact between the mating parts. The wheel is hobbed, and the worm is made by grinding or by milling with a disk cutter. Care must be exercised that the teeth of worm and gear are properly shaped to give conjugate surfaces. The geometry of worm gearing is very complicated, and reference should be made to the literature for complete information.

It is possible to secure a large speed reduction or a high increase of torque by means of worm gears. The velocity ratio does not depend upon diameters, but upon the numbers of teeth. The pressure angle should not be less than 20° for single- and double-thread worms, and 25° for triple- and multithread worms.

Case-hardened alloy steel is recommended for the worm. The worm wheel should be of bronze of approved composition. The wheel should usually contain not less than 29 teeth. The wheel should be provided with axial adjustment to obtain correct contact with the worm.

7. Geometric Relationships of Worm Gears

Figure 11-9 shows a worm of helix angle λ. Let p_1 and p_2 be the circular pitches in the planes of rotation for worm and gear, respectively. Let p_n be the circular pitch normal to the direction of the teeth. Then

$$p_n = p_1 \sin \lambda = p_2 \cos \lambda \tag{15}$$

[2]See Almen, J. A., and A. L. Boegehold, "Rear Axle Gears," *Proc. ASTM*, **35**, Part 2, 1935, p. 99.

[3]See Candee, A. H., "Industrial Applications of Spiral Bevel Gears and Hypoid Gears," *Trans. ASME*, **60**, 1938, p. 549. Other types of gears for right-angle drives are available. See, for example, Nelson, W. D., "Spiroid Gearing," *Machine Design*, **33**, 1961, Feb. 16, p. 136; Mar. 2, p. 93; and Mar. 16, p. 165.

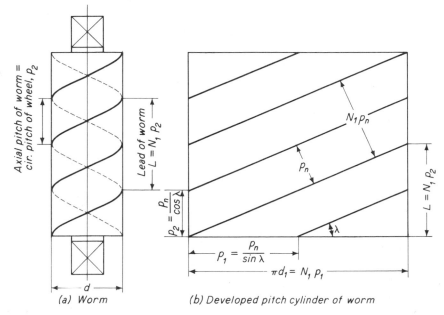

Figure 11-9 Geometry of the worm.

Equation (15) should be divided through by π, and π/p_n replaced by P_{dn}, the normal diametral pitch, to give

$$P_{dn} = \frac{\pi}{p_1 \sin \lambda} = \frac{\pi}{p_2 \cos \lambda}$$

Let N_1 be the number of starts or threads for the worm, and let N_2 be the number of teeth in the gear. Let d_1 and d_2 be the pitch diameters of worm and gear, respectively. Now $\pi d_1 = N_1 p_1$. The value of p_1 from Eq. (15) can be substituted to give

$$d_1 = \frac{N_1 p_1}{\pi} = \frac{N_1 p_n}{\pi \sin \lambda} = \frac{N_1}{P_{dn} \sin \lambda} \tag{16}$$

Similarly, $\pi d_2 = N_2 p_2$. Substitution for p_2 from Eq. (15) gives

$$d_2 = \frac{N_2 p_2}{\pi} = \frac{N_2 p_n}{\pi \cos \lambda} = \frac{N_2}{P_{dn} \cos \lambda} \tag{17}$$

Equations (16) and (17) can be combined to give

$$\tan \lambda = \frac{N_1 d_2}{N_2 d_1} \tag{18}$$

The values given above for d_1 and d_2 can be substituted into the equation for the center distance c to give

$$c = \frac{1}{2}(d_1 + d_2) = \frac{1}{2P_{dn}}\left(\frac{N_1}{\sin \lambda} + \frac{N_2}{\cos \lambda}\right) \tag{19}$$

or

$$\frac{2P_{dn}c}{N_2} = \frac{\beta}{\sin \lambda} + \frac{1}{\cos \lambda} \tag{20}$$

where

$$\beta = \frac{N_1}{N_2} \tag{21}$$

Equation (20) has been plotted in Fig. 11-10 and is helpful in the solution of problems.

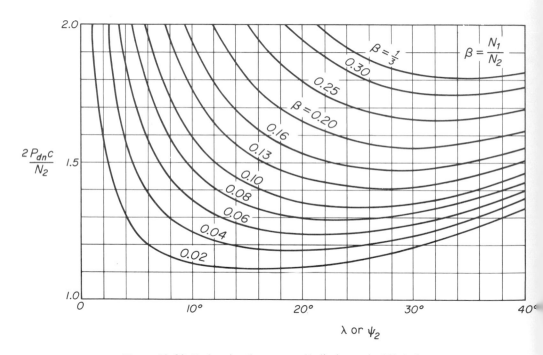

Figure 11-10 Design chart for worm and helical gears for 90° shafts.

Let L be the lead of the worm, or the distance the pitch circle of the wheel is advanced for each revolution of the worm. Then

$$L = N_1 p_2 \tag{22}$$

The pitch-line velocities V_1 and V_2 for the worm and wheel are for worm:

$$V_1 = \frac{\pi d_1 n_1}{12} = \frac{n_1 N_1 p_1}{12} \text{ fpm} \tag{23}$$

for wheel:

$$V_2 = \frac{\pi d_2 n_2}{12} = \frac{n_1 N_1 p_2}{12} \text{ fpm} \tag{24}$$

where n_1 and n_2 are the speeds, rpm, for the worm and wheel, respectively. The last form of Eq. (24) is obtained by observing in Fig. 11-9 that the pitch-line velocity $\pi d_2 n_2$ of the wheel is equal to the speed $n_1 L$ or $n_1 N_1 p_2$ along the worm axis.

From Fig. 11-9 and Eqs. (23) and (24),

$$\tan \lambda = \frac{L}{\pi d_1} = \frac{p_2}{p_1} = \frac{V_2}{V_1} \tag{25}$$

Example 3. A triple-thread worm has a pitch diameter of 4.7856 in. The hob for cutting the worm wheel has a diametral pitch P_{dn} equal to 2. Find the pitch diameter d_2 if the reduction is 12:1.

Solution. $N_1 = 3;$ $N_2 = 3 \times 12 = 36$

By Eq. (16): $\sin \lambda = \dfrac{N_1}{P_{dn} d_1} = \dfrac{3}{2 \times 4.7856} = 0.31344$

$$\lambda = 18°16'$$

By Eq. (17): $d_2 = \dfrac{N_2}{P_{dn} \cos \lambda} = \dfrac{36}{2 \times 0.94961} = 18.9552 \text{ in.}$

Example 4. A double-thread worm has a lead L of 2.18 in. The gear has 30 teeth and is cut with a hob of $P_{dn} = 3$. Find pitch diameters of the worm and gear, and center distance of the shafts.

Solution. $p_2 = \dfrac{L}{N_1} = \dfrac{2.18}{2} = 1.09 \text{ in.}$

$$\cos \lambda = \frac{p_n}{p_2} = \frac{1.0472}{1.09} = 0.96073$$

$$\lambda = 16°6.6'$$

$$d_1 = \frac{N_1}{P_{dn} \sin \lambda} = \frac{2}{3 \times 0.27747} = 2.4026 \text{ in.}$$

$$d_2 = \frac{N_2}{P_{dn} \cos \lambda} = \frac{30}{3 \times 0.96073} = 10.4087 \text{ in.}$$

Center distance: $c = \frac{1}{2}(d_1 + d_2) = \frac{1}{2}(2.4026 + 10.4087)$

$$= 6.4056 \text{ in.}$$

Example 5. A worm gear drive has a center distance of 8 in., and a P_{dn} equal to 4 in. The worm is quadruple and the wheel has 48 teeth. Find the pitch diameters of the worm and wheel and the value of the helix angle.

Solution. This type of problem must be solved by trial and error using Eq. (20). Approximate first trial values can be obtained from Fig. 11-10 using,

By Eq. (20):
$$\frac{2P_{dn}c}{N_2} = \frac{2 \times 4 \times 8}{48} = 1.3333$$

By Eq. (21):
$$\beta = \frac{N_1}{N_2} = \frac{4}{48} = 0.0833$$

The final values, after trial and error, are

$$\lambda = 16°45.7'; \qquad d_1 = 3.4676 \text{ in.}; \qquad d_2 = 12.5324 \text{ in.}$$

In general such problems have two solutions. The other one is

$$\lambda = 31° 37.6'; \qquad d_1 = 1.9070 \text{ in.}; \qquad d_2 = 14.0930 \text{ in.}$$

8. Horsepower Capacity of Worm Gears

The recommended value of the face width b_2 of the worm wheel can be determined by the following equation.

$$b_2 = \tfrac{1}{3}c^{0.875} \tag{26}$$

The curve of Fig. 11-11 can be used if desired.

The velocity of sliding V_s between the teeth is indicated by Fig. 11-12. Since the worm thread and gear tooth remain in contact, the normal component V_n of the worm and gear velocities V_1 and V_2 is the same for both members. Components OA and OB then represent the velocities along the tooth, and their sum is V_s, the total velocity of sliding between the teeth. When component OA is transplanted to BC, length OC can also be considered V_s since it has the same magnitude as AB. Velocity of sliding V_s is then given by

$$V_s = \frac{V_1}{\cos \lambda} \tag{27}$$

The horsepower capacity of a worm gear reduction can be found by the following equation.[4]

$$\text{hp} = \frac{n_1 c b_2 d_2 \beta (1 - \beta)^2 180}{330(180 + V_s^{0.85})} \tag{28}$$

where

$$n_1 = \text{speed of worm, rpm}$$

$$\beta = \frac{N_1}{N_2}, \qquad \text{speed reduction, Eq. (21)}$$

Values of $V_s^{0.85}$ can be had by calculation or from Fig. 11-13.

[4]From American Gear Manufacturers Association Standard 340.01, Sept. 1947, *Design and Rating of Worm Gears for Power Transmission.* A more recent Standard 440.03, Sept. 1959, will permit somewhat greater loadings, but the process of calculation is too involved and lengthy to be presented here.

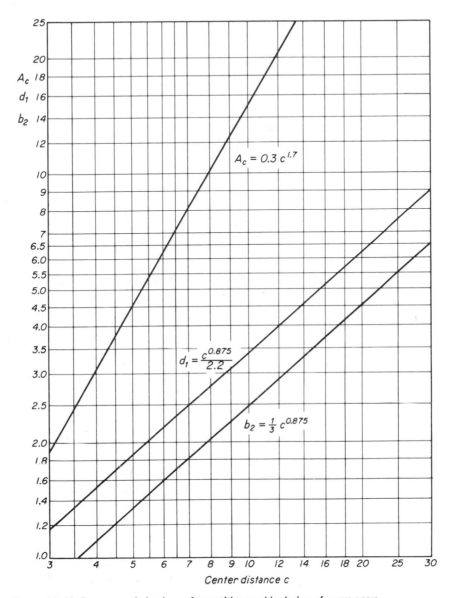

Figure 11-11 Recommended values of quantities used in design of worm gears.

Example 6. A 10:1 worm gear reduction has a quadruple-thread worm and a center distance of 8.9333 in. Worm turns 600 rpm. $P_{dn} = 3$. Find the horsepower.

Solution. $N_2 = 4 \times 10 = 40$

By Eq. (20): $$\frac{2P_{dn}c}{N_2} = \frac{2 \times 3 \times 8.9333}{40} = 1.3400$$

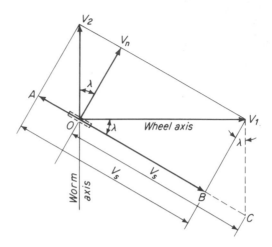

Figure 11-12 Relationship between V_1 and V_s.

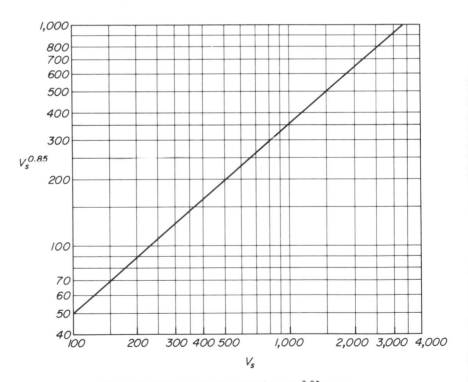

Figure 11-13 Curve for determination of $V_s^{0.85}$ from V_s.

By Eq. (21): $\beta = \frac{4}{40} = 0.1$

A trial value for λ should now be determined from Fig. 11-10. This is brought to the final form by trial and error, which gives

$$\lambda = 24°47.0'$$

By Eq. (16): $d_1 = \dfrac{N_1}{P_{dn} \sin \lambda} = \dfrac{4}{3 \times 0.41919} = 3.1807$ in.

By Eq. (17), $d_2 = \dfrac{N_2}{P_{dn} \cos \lambda} = \dfrac{40}{3 \times 0.90790} = 14.6859$ in.

By Eq. (26) or Fig. 11-11:

$$b_2 = 2.265 \text{ in.}$$

By Eq. (23): $V_1 = \dfrac{\pi d_1 n_1}{12} = \dfrac{\pi 3.1807 \times 600}{12} = 500$ fpm

By Eq. (27): $V_s = \dfrac{V_1}{\cos \lambda} = \dfrac{500}{0.90790} = 550$ fpm

By calculation or Fig. 11-13:

$$V_s^{0.85} = 213.6$$

By Eq. (28):   hp $= \dfrac{600 \times 8.9333 \times 2.265 \times 14.6859 \times 0.1 \times 0.9^2 \times 180}{330(180 + 213.6)}$

$$= 20.0$$

The result given by Eq. (28) is usually called the mechanical horsepower of the reduction.

The worm diameter at the base of the threads must be large enough to provide sufficient rigidity. When threads are integral with the shaft, it is recommended that the pitch diameter d_1 be not less than the value given by

$$d_1 = \frac{c^{0.875}}{2.2} \tag{29}$$

This equation should be used as a check and not for making design calculations.

9. Thermal Capacity of Worm Gear Reductions

The horsepower capacity of a worm gear reduction in continuous operation is usually limited by the heat-dissipating capacity of the case or container. The cooling rate for rectangular or box-type casings can be estimated by the curves[5] in Fig. 11-14. The cooling rate is greater at high velocities of the worm shaft, which causes a better circulation of the oil within the box. The cooling rate can be increased by a fan that directs a stream of air on the sides of the box.

[5]See Walker, H., "Thermal Rating of Worm Gear Boxes," *Proc. Inst. Mech. Engrs*, **151**, 1944, p. 326.

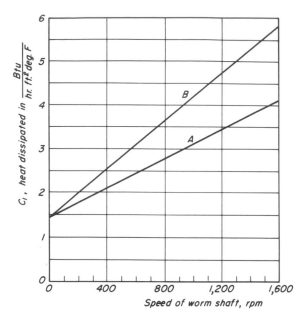

Figure 11-14 Heat dissipating capacity of worm gear box. *A*, without fan;
B, with fan on worm shaft.

Cooling rate C_1 is expressed in Btu/hr/ft²/°F. Values from the curve can be divided by 60 to obtain the heat lost per minute. The total heat lost depends on the area of the housing A_c in square feet, and the temperature difference ΔT between the housing surface and the surrounding air. The energy lost in Btu can be multiplied by 778 to give the equivalent energy in foot-pounds. The lost energy, in terms of horsepower, is then

$$\mathrm{hp}_c = \frac{778 C_1 A_c \Delta T}{60 \times 33{,}000} \qquad (30)$$

The housing should have a liberal clearance to avoid churning the oil and to provide for the dissipation of the heat. The cooling area can be estimated by the curve in Fig. 11-11 or by the following equation.

$$A_c = 0.3 c^{1.7} \qquad (31)$$

The lost hp_c is equal to the difference between the input and output horsepower. The output horsepower hp_o is equal to the input horsepower hp_i multiplied by the efficiency η. Hence

$$\mathrm{hp}_c = \mathrm{hp}_i - \mathrm{hp}_o = \mathrm{hp}_i - \mathrm{hp}_i \times \eta = \mathrm{hp}_i(1 - \eta)$$

or

$$\mathrm{hp}_i = \frac{\mathrm{hp}_c}{1 - \eta} \qquad (32)$$

The oil temperature should not exceed 180°F.

The efficiency equation for a worm gear reduction is the same as that used for a screw and nut. In the notation of this chapter, Eq. (16) of Chapter 5 becomes

$$\eta = \frac{\cos \varphi_n - \mu \tan \lambda}{\cos \varphi_n + \mu \cot \lambda} \tag{33}$$

Here φ_n is the pressure angle normal to the teeth, and μ is the coefficient of friction. This equation assumes that the axial thrust of the worm is taken by a rolling element thrust bearing. Values for μ, which depends on the velocity of sliding V_s, are given in Table 11-1.

Table 11-1 VALUES OF COEFFICIENT OF FRICTION μ FOR DIFFERENT VALUES OF SLIDING V_s

V_s, fpm	μ	V_s, fpm	μ	V_s, fpm	μ	V_s, fpm	μ
0	0.150	90	0.0560	500	0.0295	2,000	0.0160
5	0.099	100	0.0540	600	0.0274	2,200	0.0154
10	0.090	120	0.0519	700	0.0255	2,400	0.0149
20	0.080	140	0.0498	800	0.0240	2,600	0.0146
30	0.073	160	0.0477	900	0.0227	2,800	0.0143
40	0.0691	180	0.0456	1,000	0.0217	3,000	0.0140
50	0.0654	200	0.0435	1,200	0.0200	4,000	0.0131
60	0.0620	250	0.0400	1,400	0.0186	5,000	0.0126
70	0.0600	300	0.0365	1,600	0.0175	6,000	0.0122
80	0.0580	400	0.0327	1,800	0.0167		

From *American Gear Manufacturers Association Standard 440.03*, Sept. 1959.

Example 7. Find the input and output horsepower for the worm reduction in Example 6 for continuous operation based on the cooling of the housing in still air. The housing temperature rise is to be 90 deg. F. The pressure angle the normal section is $\varphi_n = 20°$.

Solution. By Fig. 11-11 or Eq. (31):

$$A_c = 12.4 \text{ ft}^2$$

By Fig. 11-14: $C_1 = 2.43$ Btu/hr/ft^2/°F

By Eq. (30): $\mathrm{hp}_c = \dfrac{778 C_1 A_c \Delta T}{60 \times 33,000} = \dfrac{778 \times 2.43 \times 12.4 \times 90}{60 \times 33,000} = 1.07$

By Eq. (27): $V_s = \dfrac{V_1}{\cos \lambda} = \dfrac{499.6}{0.90790} = 550$ fpm

By Table 11-1: $\mu = 0.0286$

By Eq. (33): $\eta = \dfrac{0.93969 - 0.0286 \times 0.46171}{0.93969 + 0.0286 \times 2.1659} = 0.925$

By Eq. (32): $\mathrm{hp}_i = \dfrac{\mathrm{hp}_c}{1 - \eta} = \dfrac{1.067}{1 - 0.925} = 14.21$ input hp

$$\mathrm{hp}_o = 14.21 - 1.07 = 13.14 \text{ output hp}$$

The output horsepower based on thermal capacity for this reduction is thus considerably less than the mechanical horsepower of Example 6.

A considerably greater output can be obtained in intermittent service if short periods of operation are followed by sufficiently long cooling periods. Artificial cooling also permits operation at larger loads.

The mounting of the worm and wheel must provide for the thrust loads that occur during operation. The axial position of the worm is not critical, but the wheel must be correctly located axially to secure proper contact between the teeth.

Rubbing speeds between the teeth for case-hardened worms and chill-cast phosphor bronze gears cut with good accuracy are limited to about 6,000 fpm. Lower speeds should be used if the materials and accuracy are of lower quality. On the other hand higher speeds can be permitted for superior materials and greater accuracy. Reduction in loading should be made for shock and impact conditions or for operation at elevated temperatures.

The helix angle should be limited to about 6° per thread. For example, if λ is 30°, then the worm should have at least five threads. For lower values of λ, the worm is self-locking; that is, a torque applied to the wheel axis will not rotate the worm.

When designing a worm gear reduction, it is a good plan, before deciding on the diameter of the worm, to ascertain whether a hob of the proper size is available for cutting the wheel. The purchase of a special tool may be avoided by changing the diameter of the worm so that the wheel can be cut by a hob that is already on hand. Gear-cutting firms sometimes catalog their worm gear hobs to enable their customers to take advantage of available tools.

10. Helical Gears

If a number of spur gears formed from thin plates are assembled with a small angular displacement between the teeth, the stepped gear of Fig. 11-15 results. A helical gear can be considered the limiting case of a stepped gear as the thickness of the plates and the angular displacements are made progressively smaller. A typical example of a pair of helical gears is given by Fig. 11-16. Because of the helix angle there will be end thrust on the shafts of helical gears, which must be provided for in the design of the bearings. End thrust can be eliminated by cutting a right-hand spiral over one-half the face, and using a left-hand spiral for the other half; herringbone gears are thus formed, as illustrated in Fig. 11-16. A central groove must usually be

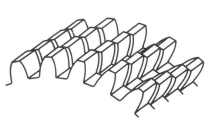

Figure 11-15 Stepped gears. Helical gears result if the plates are made progressively thinner.

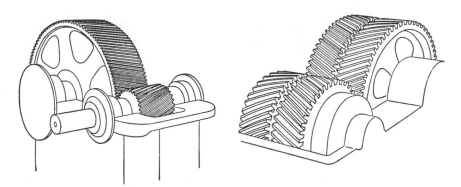

Figure 11-16 Types of spiral gears : left, helical gears ; right herringbone gears. (*From* Lubrication, *published by The Texas Co.*)

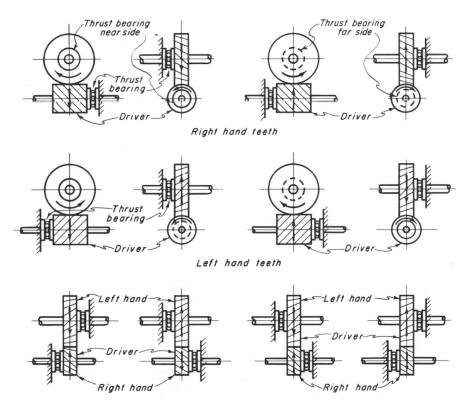

Figure 11-17 Directions of rotation and thrust for helical gears.

provided around the gear for clearance for hob or cutter. Herringbone gears made on the Sykes type of generating machine do not require the central clearance groove.

The helix angle ψ, illustrated in Fig. 11-19, usually varies from 15° to 30° for helical gears, and from 23° to 30°, or even 45°, for herringbone gears.

Helical gears can be cut by the same hobs used for making spur gears. When the gear shaper is used, however, a special cutter is required. The relationship between directions of rotation and thrust for helical gears is shown in Fig. 11-17.

Gears with helical teeth possess certain inherent advantages. More teeth are in contact simultaneously, and the load is transferred gradually and uniformly as successive teeth come into engagement. Helical gears thus operate more smoothly and carry larger loads at higher speeds than spur gears. The line of contact extends diagonally across the face of mating teeth. Since more teeth are in contact, undercutting causes little trouble in helical gearing.

Helical gears can also be used for transmitting power between nonparallel shafts, as shown in Fig. 11-18. When used in this way, the teeth have only point contact, which does not shift axially along the teeth during operation. Such gears are usually used only for the transmission of relatively small loads.

Figure 11-18   Helical gears on non-parallel shafts. (*Courtesy Socony-Vacuum Oil Co.*)

11. Pitch Diameter of Helical Gear

Let the pitch diameter of a helical gear be called d, and let the helix angle ψ be measured between an element of the tooth at the pitch cylinder and the center line of the shaft, as shown in Fig. 11-19. Let the symbol p represent the circular pitch or distance from tooth to tooth measured on the pitch cylinder in the plane of rotation. The normal circular pitch p_n is measured normal to the tooth elements. It should be noted that

$$p_n = p \cos \psi \tag{34}$$

The pitch diameter of a helical gear is then equal to

$$d = \frac{Np}{\pi} = \frac{Np_n}{\pi \cos \psi} = \frac{N}{P_{dn} \cos \psi} \tag{35}$$

where N is the number of teeth in the gear, and π/p_n is replaced by P_{dn}. The pitch diameter of a helical gear thus depends upon the helix angle ψ as well as upon the normal circular pitch and the number of teeth.

When a helical gear is cut with a hob, the circular pitch of the tool is

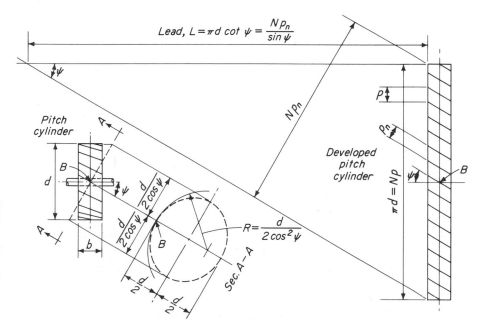

Figure 11-19 Geometry of helical gear.

equal to p_n of Fig. 11-19. However, when the gear shaper is used, the circular pitch of the cutter and p of Fig. 11-19 coincide. Angle ψ is usually made large enough to give an overlapping of tooth action; that is, $b \tan \psi$ is greater than p, where b is the width of the face.

12. Formative Number of Teeth

A plane normal to the element of the tooth at point B in Fig. 11-19 intersects the pitch cylinder (extended) in the ellipse shown by section A-A. The shape of the tooth at B would be that generated on a cylindrical surface having the same radius of curvature as the ellipse at B. From analytic geometry, the radius of curvature R at the end of a semiminor axis of an ellipse is known to be

$$R = \frac{d}{2 \cos^2 \psi} \qquad (a)$$

The formative number of teeth N' is defined as the number of teeth in a gear of radius R.

$$N' = \frac{2\pi R}{p_n} = \frac{\pi d}{p_n \cos^2 \psi} \qquad (b)$$

Substitution of the value for $\pi d/p_n$ from Eq. (35) gives the following equation for N'.

$$N' = \frac{N}{\cos^3 \psi} \tag{36}$$

When the beam strength of helical teeth is computed, the form factor y for N' teeth should be used.

13. Center Distance of Mating Gears

Figure 11-20 shows a pair of mating helical gears having pitch diameters d_1 and d_2 and helix angles ψ_1 and ψ_2, respectively. From Eq. (35), these dia-

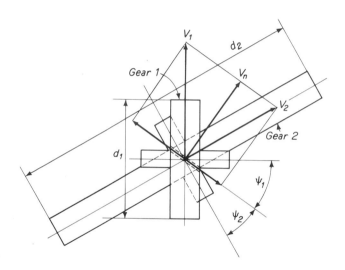

Figure 11-20 Components of velocity for mating helical gears.

meters can be written as

$$d_1 = \frac{N_1 p}{\pi} = \frac{N_1 p_n}{\pi \cos \psi_1} = \frac{N_1}{P_{dn} \cos \psi_1} \tag{37}$$

$$d_2 = \frac{N_2 p}{\pi} = \frac{N_2 p_n}{\pi \cos \psi_2} = \frac{N_2}{P_{dn} \cos \psi_2} \tag{38}$$

When meshed together, the center distance c for these gears is

$$c = \frac{d_1 + d_2}{2} = \frac{1}{2P_{dn}}\left[\frac{N_1}{\cos \psi_1} + \frac{N_2}{\cos \psi_2}\right] \tag{a}$$

Let the pitch-line velocities of the gears be V_1 and V_2, respectively. As illustrated by Fig. 11-20, the common velocity V_n normal to the tooth surfaces of both gears is

$$V_n = V_1 \cos \psi_1 = V_2 \cos \psi_2$$

or

$$V_2 = \frac{\cos \psi_1}{\cos \psi_2} V_1 \tag{b}$$

The angular velocities of the two gears are

$$\omega_1 = \frac{2V_1}{d_1}$$

and

$$\omega_2 = \frac{2V_2}{d_2} = \frac{2 \cos \psi_1}{d_2 \cos \psi_2} V_1 \tag{c}$$

The velocity ratio ω_1/ω_2 becomes

$$\frac{\omega_1}{\omega_2} = \frac{d_2 \cos \psi_2}{d_1 \cos \psi_1} \tag{d}$$

Substitution from Eqs. (37) and (38) gives

$$\frac{\omega_1}{\omega_2} = \frac{N_2}{N_1} \tag{39}$$

The velocity ratio of the gears thus depends on the number of teeth, and is independent of the helix angles. Substitution of Eq. (21) for velocity ratio $\beta = N_1/N_2$ into Eq. (a) gives

$$\frac{\beta}{\cos \psi_1} + \frac{1}{\cos \psi_2} = \frac{2P_{dn}c}{N_2} \tag{40}$$

Sometimes the center distance c, the angle between the shafts $\psi_1 + \psi_2$, and the velocity ratio β are fixed. Suitable values for N_2 and P_{dn} are chosen for the conditions of the problem, and the values of ψ_1 and ψ_2 are found by trial and error in Eq. (40).

Example 8. Two helical gears with 90° shafts have a center distance of 12 in. The gears have 21 and 77 teeth. The normal diametral pitch is 6. Find the required values for the helix angles. Find the corresponding pitch diameters of the two gears.

Solution. By Eq. (20): $\dfrac{2P_{dn}c}{N_2} = \dfrac{2 \times 6 \times 12}{77} = 1.87013$

By Eq. (21): $\qquad\qquad \beta = \dfrac{N_1}{N_2} = \dfrac{21}{77} = 0.27273$

Trial values are selected from Fig. 11-10 that are refined by trial and error to give

$$\psi_2 = 19°44'$$

$$\psi_1 = 70°16'$$

By Eq. (37): $\quad d_1 = \dfrac{N_1}{P_{dn} \cos \psi_1} = \dfrac{21}{6 \times 0.33764} = 10.3660 \text{ in.}$

By Eq. (38): $d_2 = \dfrac{N_2}{P_{dn}\cos\psi_2} = \dfrac{77}{6\times0.94127} = 13.6340$ in.

Such problems usually have two solutions. The other one is

$$\psi_2 = 48°22'$$
$$\psi_1 = 41°38'$$

By Eq. (37): $d_1 = \dfrac{21}{6\times0.74741} = 4.6829$ in.

By Eq. (38): $d_2 = \dfrac{77}{6\times0.66436} = 19.3171$ in.

14. Tooth Loads of Helical Gears

Figure 11-21 shows the forces acting on the tooth when the axes of the mating gears are parallel to each other. Force F_t is the transmitted or horsepower load found from the usual equation

$$F_t = \frac{33,000\ \text{hp}}{V} \tag{41}$$

where V is the pitch-line velocity in feet per minute.

Force F_n is normal to the tooth surface. It is inclined to the plane tangent to the pitch cylinder at the normal pressure angle φ_n. The projection of F_n in the tangent plane is inclined at helix angle ψ to the plane of rotation. The projection of F_n on the plane of rotation is inclined at angle φ to force F_t. The relationship between φ and φ_n is given by

$$\tan\varphi_n = \tan\varphi\cos\psi \tag{42}$$

The values of the various components of the tooth forces are shown in Fig. 11-21.

Helical gears on parallel shafts operate at high values for the efficiency.

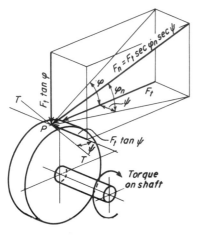

Figure 11-21 Forces acting on tooth of helical gear when axes are parallel.

15. Beam Strength, Dynamic Load, and Wear of Helical Gears with Parallel Axes

The equations for beam strength, dynamic load, and wear are similar to those for spur gears. However, adjustments must be made to take care of the effects of the helix angle ψ.

(a) *Beam Strength.* The same type equation used for estimating spur gears is used for the beam strength of helical gears.[6]

[6] See p. 491, Reference 1, Chapter 10. See also *AGMA 221.02, Tentative AGMA Standard for Rating the Strength of Helical and Herringbone Gear Teeth.*

$$F_b = sbyp_n \qquad (43)$$

The value of y should be taken for the formative number of teeth. Stress concentration may reduce the value of F_b given by this equation.

(b) *Dynamic Load.* The dynamic load for helical gears with parallel shafts can be estimated by the same equation used for spur gears.

$$F_d = \frac{1.46en_1N_1br_1r_2}{\sqrt{r_1^2 + r_2^2}}, \qquad \text{both gears steel} \qquad (44)$$

(c) *Limit Load for Wear.* The limit load for wear for external gears on parallel shafts can be taken as[7]

$$F_w = \frac{d_1bK}{\cos^2 \psi} Q \qquad (45)$$

where

$$Q = \frac{2N_2}{N_1 + N_2} \qquad (46)$$

and K can be taken from Table 10-2 for spur gears.

Example 9. Two helical gears have parallel shafts, and $P_{dn} = 4$. Normal pressure angle φ_n is 20°, and face width $b = 2.5$ in. Numbers of teeth are 32 and 44. Center distance c is 10 in. Both gears are steel, BHN $= 350$. AGMA Quality No. 8. Pinion turns 860 rpm. Find the value of the helix angle, and the horsepower the gears are carrying.

Solution. $d_1 = \frac{32}{76} \times 20 = 8.4211$ in.

$d_2 = \frac{44}{76} \times 20 = 11.5789$ in.

By Eq. (37): $\cos \psi_1 = \dfrac{N_1}{P_{dn}d_1} = \dfrac{32}{4 \times 8.4211} = 0.95000$

$\psi_1 = \psi_2 = 18°11.7'$

By Eq. (36): $N' = \dfrac{N}{\cos^3 \psi} = \dfrac{32}{0.95000^3} = 37.3$

By Table 10-1: $y = 0.121$

By Table 10-2: $s = 32,000$ psi,   in bending

By Eq. (43): $F_b = sbyp_n = 32,000 \times 2.5 \times 0.121 \times \dfrac{\pi}{4} = 7,600$ lb

By Eq. (46): $Q = \dfrac{2N_2}{N_1 + N_2} = \dfrac{2 \times 44}{76} = 1.158$

By Table 10-2: $K = 163$

By Eq. (45): $F_w = \dfrac{d_1bQK}{\cos^2 \psi} = \dfrac{8.4211 \times 2.5 \times 1.158 \times 163}{0.95000^2} = 4,400$ lb

The horsepower capacity will be based on F_w since it is smaller than F_b.

By Table 10-3, $e = 0.0035 + 0.0036 = 0.0071$ in.

[7]See p. 532, Reference 1, Chapter 10.

By Eq. (44), $F_d = \dfrac{1.46 \times 0.0071 \times 860 \times 32 \times 2.5 \times 4.211 \times 5.789}{\sqrt{4.211^2 + 5.789^2}}$

$\qquad\qquad\qquad = 2{,}430 \text{ lb}$

By Eq. (12): $F_t = F_w - F_d = 4{,}400 - 2{,}430 = 1{,}970 \text{ lb}$

By Eq. (14): $V = \dfrac{\pi d_1 n_1}{12} = \dfrac{\pi 8.421 \times 860}{12} = 1{,}896 \text{ fpm,}$

$\qquad\qquad\qquad$ pitch-line velocity

By Eq. (41): $\text{hp} = \dfrac{F_t V}{33{,}000} = \dfrac{1{,}970 \times 1{,}896}{33{,}000} = 113.4$

16. Load Capacity of Helical Gears with Crossed Axes

When helical gears with crossed axes are first put into service, only point contact between the teeth occurs. Contact usually increases to a short length of line after some wear has taken place. The load capacity is quite small when the gears are new, but the loading can be increased appreciably after careful wearing in.[8] The teeth should be of a size that causes two or more pairs always to be in contact. Nominal variations in center distance or axial locations of the gears have no effect on the action, and such gears are usually considered to be easy to mount.

REFERENCES

Standard reference works on gear theory are given at the end of Chapter 10.

PROBLEMS

1. Two 20° full-depth flame-hardened steel bevel gears have a diametral pitch P_d of 3 at the outside radius. Numbers of teeth are 24 and 60. Gears have AGMA Quality No. 8. Width of face b is 3 in., but blank thickness b' parallel to the shaft is 2.5 in. Pinion turns 570 rpm. Find the horsepower capacity of the pair.

Ans.   hp = 96.5.

2. Two 20° full-depth steel bevel gears have a diametral pitch P_d of 4 at the outside radius, and a BHN equal to 350. Numbers of teeth are 24 and 48. Gears have AGMA Quality No. 8. Width of face b is 2 in., but thickness of blank b' parallel to the shaft is 1.75 in. Pinion turns 1,140 rpm.

(a) Find the horsepower the gears are carrying.

(b) Let the base of the pitch cone for the pinion be mounted at the midpoint of a simply supported shaft 16 in. long. Thrust and torque are resisted as in Fig.

[8]See p. 136, Reference 4, Chapter 10, and p. 536, Reference 1, Chapter 10.

11-5. Make a free-body diagram for F_t acting at the midpoint of the tooth. Show numerical values for all forces and moments.   *Ans.* hp $= 51.9$.

3. Two 20° full-depth steel bevel gears have a diametral pitch P_d of 4 at the outside radius and a BHN $= 350$. Numbers of teeth are 30 and 45. Gears have AGMA Quality No. 8. Width of face b is 2.5 in., but thickness of blank b' parallel to the shaft is 2 in. If the gears are carrying 60 hp, find the speed at which the pinion is turning.   *Ans.* $n_1 = 1,208$ or 666 rpm.

4. A triple-thread worm has a lead angle of 17° and a pitch diameter of 2.2802 in. Find the center distance when the worm is mated with a wheel of 48 teeth.
     Ans. $c = 6.7171$ in.

5. A double-thread worm has a pitch diameter of 3 in. The wheel has 20 teeth and a pitch diameter of 5 in. Find the value of the helix angle.   *Ans.* $\lambda = 9°27.7'$.

6. A worm gear set gives a reduction of 18:1, and has a helix angle of 12°. The center distance is 4.6424 in. and the normal diametral pitch P_{dn} is 5. Find suitable values for N_1 and N_2.

7. A worm gear set has a helix angle of 19° and a center distance of 9.084 in. $P_{dn} = 4$. Find suitable values for N_1 and N_2.

8. A double-thread worm meshes with a 30-tooth wheel. Center distance is 8 in. $P_{dn} = 2.5$. Find the value of d_1 and d_2.

9. A double-thread worm meshes with a 50-tooth wheel. Center distance is 8 in. $P_{dn} = 4$. Worm turns 1,750 rpm. Find values for d_1, d_2, and the mechanical horsepower.   *Ans.* hp $= 10.5$.

10. A quadruple-thread worm meshes with a 40-tooth wheel. Helix angle is 23°. $P_{dn} = 3$. $\varphi_n = 20°$. Worm turns 1,050 rpm. Find the center distance, the mechanical horsepower, and the horsepower based on cooling for a temperature rise of 90 deg F.   *Ans.* $c = 8.9486$ in.; hp$_{mech} = 25.2$; hp$_o = 21.7$.

11. A triple-thread worm meshes with a 60-tooth wheel. $P_{dn} = 4$; $\varphi_n = 20°$. Pitch diameter of worm is 3.1838 in. Worm turns 1,150 rpm. Find the mechanical horsepower and the horsepower based on cooling for a temperature rise of 90 deg. F.   *Ans.* hp$_{mech} = 18.0$; hp$_o = 15.4$.

12. A triple-thread worm meshes with a 36-tooth wheel. $P_{dn} = 2$; $c = 12$ in.; $\varphi_n = 20°$. Pinion turns 860 rpm. Find values for d_1, d_2, the mechanical horsepower, and the horsepower based on cooling for a rise in temperature of 90 deg. F.
     Ans. $d_1 = 5.2013$ in.; hp$_{mech} = 36.1$; hp$_o = 26.9$.

13. Two helical gears have parallel shafts and are cut with a hob of p_n equal to 0.5236 in. Speed ratio is 2:1 and the center distance is equal to 9 in. Find the required value of the helix angle if the smaller gear has 35 teeth.
     Ans. $\psi_1 = 13°32'$.

14. Two helical gears have shafts at 90° and helix angles of 45°. Speed ratio is 3 : 1; p_n is equal to 0.7854 in. Find the center distance if the smaller gear has 20 teeth   *Ans.* $c = 14.1422$ in.

15. Work Problem 14 but with the helix angle ψ_1 for the smaller gear equal to $37\frac{1}{2}°$.   *Ans.* $c = 15.4713$ in.

16. Work Problem 14 but with the helix angle ψ_1 for the smaller gear equal to $52\frac{1}{2}°$.   *Ans.* $c = 13.5603$ in.

17. Two meshing helical gears have parallel shafts and a center distance of 9 in. Number of teeth are 35 and 70. $\varphi_n = 20°$; $P_{dn} = 6$; $b = 2$ in. Both gears are steel hardened to 350 BHN. AGMA Quality No. 8. Pinion turns 860 rpm. Find the horsepower capacity.   *Ans.* $hp = 55.5$.

18. Two helical gears have shafts at $90°$ and a center distance of 14 in. The speed ratio is 3:1. The smaller gear has 20 teeth; p_n is 0.7854 in. Find suitable values for the helix angles and the corresponding pitch diameters of the gears.

Ans. $\psi_1 = 63°38'$, $\psi_2 = 26°22'$;
or $\psi_1 = 46°13.8'$, $\psi_2 = 43°46.2'$.

19. Two helical gears have shafts at a $60°$ angle. Numbers of teeth are 35 and 105. Center distance is 8 in.; p_{dn} equals $\pi/10$. Find suitable values for angles ψ_1 and ψ_2 and the pitch diameters of the gears.

20. A pair of $20°$ helical gears have parallel shafts and a center distance of 20 in. $P_{dn} = 4$. Numbers of teeth are 32 and 44. Make a perspective sketch and show the tooth forces at the center of the tooth when the gear is carrying a transmitted force of 1,970 lb.

21. A helical gear has 30 teeth and a pitch diameter of 11 in. Gear has $P_{dn} = 3$ and $\varphi_n = 20°$. The force normal to the tooth surface is 1,000 lb. Find the horsepower transmitted at 570 rpm.   *Ans.* $hp = 42.5$.

22. A worm gear reducer has an input horsepower of 15. The worm shaft turns at 1,000 rpm. What is the efficiency of the reducer based on the cooling capacity for a 90 deg. F rise in temperature for continuous operation? The center distance is 12 in.   *Ans.* $Eff = 0.85$.

23. A worm gear reducer has a center distance of 8 in. and a helix angle of $8°33.6'$. Worm turns 1,750 rpm and has a diameter of 3.3592 in. $\varphi_n = 20°$. Find the horsepower based on cooling for a temperature rise of 90 deg. F.   *Ans.* $hp_o = 12.3$.

24. A single-thread worm meshes with a 50-tooth wheel. $P_{dn} = 4$; $\varphi_n = 20°$; $c = 8$ in.; $n_1 = 1,725$ rpm. Find d_1, d_2, hp_{mech}, and the horsepower based on a 90 deg. F rise in temperature.   *Ans.* $hp_{mech} = 5.32$; $hp_o = 2.19$.

25. A worm gear reduction has a center distance of 16 in. Worm turns 720 rpm. Worm has eight starts and wheel has 40 teeth. $P_{dn} = 2$; $\varphi_n = 25°$. Find d_1, d_2, hp_{mech}, and the horsepower based on a 90 deg. F rise in temperature.
Ans. $hp_{mech} = 111.9$; $hp_o = 76.0$.

26. A worm gear reduction has a 10-in. center distance. Worm has six starts and wheel has 40 teeth. $P_{dn} = 3$; $\varphi_n = 25°$. Worm turns 1,160 rpm. Find d_1, d_2, hp_{mech}, and the horsepower based on a temperature rise of 90 deg. F.
Ans. $hp_{mech} = 46.6$; $hp_o = 39.1$.

27. A triple-thread worm meshes with a 24-tooth wheel on a center distance of 6 in. $P_{dn} = 3$; $\varphi_n = 20°$. Worm turns 415 rpm. Find d_1, d_2, hp_{mech}, and the horsepower based on a temperature rise of 90 deg. F.   *Ans.* $hp_{mech} = 4.96$; $hp_o = 3.57$.

12

Miscellaneous Machine Elements

Design methods are given in this chapter for a miscellaneous group of machine elements that require only brief treatment.

1. Stresses in a Thick Cylinder

The thick-walled cylinder shown in Fig. 12-1 is subjected to uniform pressures p_i and p_o on the internal and external lateral surfaces, respectively. The top and bottom surfaces are assumed to be free from load. Since the body and the loading are symmetrical about the axis, shear stresses in the tangential and radial directions are not present, and only normal stresses s_t and s_r act on the element, as illustrated in Fig. 12-1(a).

When the wall of the cylinder is thin, the tangential or hoop stress s_t can be assumed to be uniform throughout the wall thickness, and its value can be found by the elementary equations of Chapter 7. This assumption cannot be made for a thick-walled cylinder, and the equations of the following derivation should be used for finding the stresses.[1]

[1]See Marin, Joseph, "Designing Shrink-Fit Assemblies," *Machine Design*, **14**, 1942, June, p. 68; July, p. 72, and Aug., p. 78. See also the following:

Horger, O. J., and C. W. Nelson, "Design of Press- and Shrink-Fitted Assemblies," *Trans. ASME*, **59**, 1937, p. A-183, and **60**, 1938, p. A-32;

Peterson, R. E., and A. M. Wahl, "Fatigue of Shafts at Fitted Members with a Related Photoelastic Analysis," *Trans. ASME*, **57**, 1935, p. A-1 and A-69, and **58**, 1936, p. A-74;

Horger, O. J., and J. L. Maulbetsch, "Increasing the Fatigue Strength of Press-Fitted Axle Assemblies by Surface Rolling," *Trans. ASME*, **58**, 1936, p. A-91, and **59**, 1937, p. A-37.

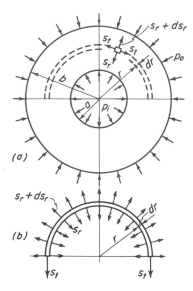

Figure 12-1 Thick-walled cylinder subjected to internal and external pressures.

Consider the stresses acting on the semicircular element of Fig. 12-1(b). The thickness in the direction perpendicular to the paper has been taken equal to unity. The vertical component of the inward radial stresses across the diameter of the element is equal to $2s_r r$, and for the outward component of the stresses, $2(s_r + ds_r)(r + dr)$. The equilibrium equation for this element is then

$$2s_r r + 2s_t dr = 2(s_r + ds_r)(r + dr) \qquad (a)$$

The right-hand side should be expanded and terms of higher order should be neglected. The equation can then be rewritten as

$$r\frac{ds_r}{dr} + s_r = s_t \qquad (1)$$

The strain or unit deformation ϵ_o in the direction perpendicular to the paper can be found from the stresses s_t and s_r as follows:

$$\epsilon_o = -\frac{\mu s_t}{E} - \frac{\mu s_r}{E} \qquad (b)$$

where μ is Poisson's ratio and E is the modulus of elasticity. This equation may be rearranged as

$$s_t + s_r = -\frac{\epsilon_o E}{\mu} \qquad (c)$$

The right-hand side is a constant. For convenience it will hereafter be called $2C_1$. Substitution into Eq. (1) gives

$$r\frac{ds_r}{dr} + 2s_r = 2C_1 \qquad (d)$$

or

$$r^2\frac{ds_r}{dr} + 2rs_r = 2rC_1 \qquad (e)$$

or

$$\frac{d}{dr}(r^2 s_r) = 2rC_1$$

Integration gives

$$r^2 s_r = C_1 r^2 + C_2 \qquad (f)$$

where C_2 is the constant of integration. Equation (f) may be written as follows:

$$s_r = C_1 + \frac{C_2}{r^2} \qquad (g)$$

This value for s_r should be substituted into Eq. (c), to give

$$s_t = C_1 - \frac{C_2}{r^2} \tag{h}$$

At the inner boundary, $r = a$, the radial stress s_r becomes equal to $-p_i$, and Eq. (g) is written

$$-p_i = C_1 + \frac{C_2}{a^2} \tag{i}$$

At the outer boundary, $r = b$, radial stress s_r is equal to $-p_o$, and Eq. (g) becomes

$$-p_o = C_1 + \frac{C_2}{b^2} \tag{j}$$

Constants C_1 and C_2 can now be found by solving Eqs. (i) and (j) simultaneously:

$$C_1 = \frac{a^2 p_i - b^2 p_o}{b^2 - a^2} \tag{k}$$

$$C_2 = -\frac{a^2 b^2 (p_i - p_o)}{b^2 - a^2} \tag{l}$$

Substitution into Eqs. (g) and (h) gives[2] the values of the stresses s_r and s_t.

$$s_r = \frac{a^2 p_i - b^2 p_o}{b^2 - a^2} - \frac{a^2 b^2 (p_i - p_o)}{r^2 (b^2 - a^2)} \tag{2}$$

$$s_t = \frac{a^2 p_i - b^2 p_o}{b^2 - a^2} + \frac{a^2 b^2 (p_i - p_o)}{r^2 (b^2 - a^2)} \tag{3}$$

For many applications, the outer pressure p_o is equal to zero. Equations (2) and (3) then reduce to the following forms,

$$s_r = \frac{a^2 p}{b^2 - a^2} \left(1 - \frac{b^2}{r^2} \right) \tag{4}$$

$$s_t = \frac{a^2 p}{b^2 - a^2} \left(1 + \frac{b^2}{r^2} \right) \tag{5}$$

where p_i is now called p. The stresses are a maximum at the inner edge, where $r = a$, and the value of s_r is $-p$. The tangential stress for this point is seen to be

$$s_t = p \left[\frac{1 + (a/b)^2}{1 - (a/b)^2} \right] \tag{6}$$

At the inner boundary, $r = a$, the tangential elongation ϵ_t is equal to

$$\epsilon_t = \frac{1}{E_h} (s_t - \mu s_r) \tag{m}$$

where E_h is the modulus of elasticity for the material. The total increase in length of the inner boundary is equal to $2\pi a \epsilon_t$, giving an increase u_h in the radius of the hole equal to $2\pi a \epsilon_t / 2\pi$, or $a\epsilon_t$. Hence,

[2]These are known as the Lamé equations. See p. 239 of Reference 5, end of chapter.

$$u_h = \frac{a}{E_h}(s_t - \mu s_r) = \frac{ap}{E_h}\left[\frac{1 + (a/b)^2}{1 - (a/b)^2} + \mu\right] \tag{7}$$

Radial displacement u_h is, of course, outward.

2. Shrink and Press Fit Stresses

In a shrink or press fit, pressure p is caused by the interference of metal between the shaft and the hub. Because of this pressure, the radius of the hole is increased, and the radius of the shaft is decreased. The decrease u_s in the radius of the shaft is equal to

$$u_s = a\epsilon_t = \frac{a}{E_s}(s_t' - \mu s_r') \tag{a}$$

where E_s is the modulus of elasticity for the material composing the shaft. For a solid shaft, stresses s_r' and s_t' are both equal to $-p$, and the radial deformation becomes

$$u_s = -\frac{ap}{E_s}(1 - \mu) \tag{8}$$

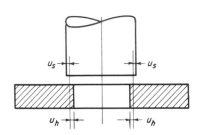

This displacement is inward, as indicated by the minus sign. The sum of the absolute values for u_h and u_s is the radial interference of metal for the shrink or press fit as shown by Fig. 12-2.

When hub and shaft are composed of the same material, the modulus of elasticity for both parts is the same and can be represented by E. The double sum of the absolute values of u_h and u_s is the diametral interference of metal Δ. By Eqs. (7) and (8), $\Delta = 2(|u_h| + |u_s|)$.

Figure 12-2 Interference of metal in press fit.

$$\Delta = \frac{4ap}{E}\left[\frac{1}{1 - (a/b)^2}\right] \tag{9}$$

If Eq. (9) is solved for p, the value of the radial stress corresponding to a given value for the diametral interference is obtained.

$$s_r = -p = -\frac{E\Delta}{4a}\left[1 - \frac{a^2}{b^2}\right] \tag{10}$$

When this value for p is substituted into Eq. (6), the equation for s_t for the hub is obtained.

$$s_t = \frac{E\Delta}{4a}\left[1 + \frac{a^2}{b^2}\right] \tag{11}$$

Example 1. An 8 in.-diam steel shaft is to have a press fit in a 20 in.-diam cast iron disk. The maximum tangential stress in the disk is to be 5,000 psi. The modulus of elasticity for steel is 30,000,000 psi, and for cast iron, 15,000,000 psi. Poisson's ratio is equal to 0.3.

(a) Find the required diametral interference of metal.

(b) Plot values of s_t along a radius of the disk.

(c) If the disk is 10 in. thick in the axial direction, find the force required to press the parts together if the coefficient of friction is equal to 0.12. Also find the torque which the joint could carry because of the shrink fit pressure.

Solution. (a) $a = 4$ in.; $b = 10$ in.; $b/a = 2.5$; $a/b = 0.4$.

From Eq. (6): $p = \dfrac{s_t[1 - (a/b)^2]}{1 + (a/b)^2} = \dfrac{5,000(1 - 0.16)}{1 + 0.16} = 3,620$ psi

By Eq. (7): $u_h = \dfrac{4 \times 3,620}{15,000,000}\left[\dfrac{1.16}{0.84} + 0.3\right] = 0.00162$ in.

increase in hole radius

By Eq. (8): $u_s = -\dfrac{4 \times 3,620}{30,000,000}(1 - 0.3) = -0.00034$ in.

decrease in shaft radius

$$\Delta = 2(0.00162 + 0.00034) = 0.0039 \text{ in.}$$

diametral interference

(b) The application of Eq. (5) for various values of r gives the curve of Fig. 12-3 for stresses s_t.

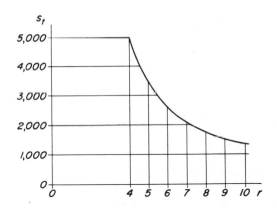

Figure 12-3 Values of tangential stress s_t for Example 1.

(c) Force required for assembly,

$$F = 8\pi \times 10 \times 3,620 \times 0.12 = 109,200 \text{ lb}$$

Torque carried by press fit,

$$T = 109,200 \times 4 = 436,800 \text{ in. lb}$$

3. Stress Concentration Caused by Press Fit

A shaft with bending stress has a stress concentration at the edge of a press fitted member as at A in Fig. 12-4(a). Some improvement in fatigue

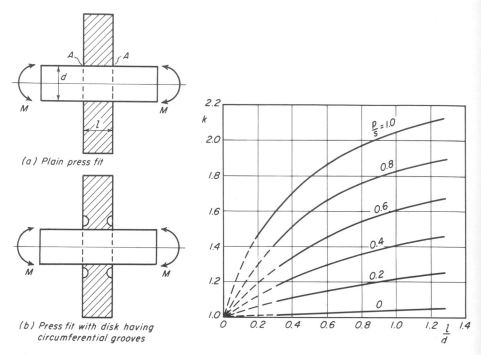

Figure 12-4 Stress concentration for press fit.

Figure 12-5 Stress concentration factors for press fit.

strength can be obtained by grooving the disk as shown in Fig. 12-4(b). Experiments have shown that rolling the shaft between small diameter rollers under pressure high enough to deform the surface has also resulted in improved fatigue strength. Stress concentration factors K for plain press fitted assemblies are given by Fig. 12-5.

Example 2. Find the stress concentration factor for the shaft of Example 1 if the shaft carries an alternating bending moment of 350,000 in. lb.

Solution.

$$I = \frac{\pi d^4}{64} = \frac{\pi 8^4}{64} = 201.06 \text{ in.}^4$$

$$s = \frac{Mc}{I} = \frac{350,000 \times 4}{201.06} = 6,963 \text{ psi}$$

$$\frac{p}{s} = \frac{3,620}{6,963} = 0.52$$

$$\frac{l}{d} = \frac{10}{8} = 1.25$$

By Fig. 12-5: $K = 1.56$

Max. bending stress: $s_{max} = 1.56 \times 6,963 = 10,860 \text{ psi}$

When subjected to reversed bending, press fitted joints are subject to a mode of failure known as *fretting corrosion*.[3] Surface pitting is accompanied by an oxide debris which is caused either by a grinding action or by the alternate welding and tearing away of the high spots. The removed particles become oxidized and form an abrasive powder which continues the destructive process.

There is no completely satisfactory method for stopping fretting unless all relative motion can be prevented. When relative motion cannot be eliminated, solid lubricants such as MoS may maintain a lubricating film for a long period of time. An increase in the wear resistance of the surfaces so as to reduce surface welding may also be helpful.

4. Stresses in Disk Flywheel

Flame-cut disk flywheels are widely used because of the uniformity and high strength of steel plate.

Let Fig. 12-1(a) be a view of such a flywheel. Pressures p_o and p_i are not present. In addition to stresses s_r and s_t, the element is loaded by an outwardly directed centrifugal force because of the rotation at ω rad/sec. Let the material have a weight of γ lb/in.[3] Let the flywheel have unit thickness. Consider an element at radius r of length dr in the radial direction and of unit length in the tangential direction. The weight of the element is $\gamma\, dr$. It is subjected to an acceleration of $r\omega^2$. The centrifugal force on the element is $\gamma r\omega^2\, dr/g$. This force has a total upward component on the semicircular element shown in Fig. 12-1 (b) of $2\gamma r^2\omega^2\, dr/g$. This term should be included with the others in Eq. (a) of Section 1. When the resulting equation is reduced to lowest terms, the result is

$$s_t - s_r - r\frac{ds_t}{dr} - \frac{\gamma r^2\omega^2}{g} = 0 \tag{12}$$

The solution to this equation is well known[4] and will not be given here. In most cases stress s_t controls the design. It is a maximum at the inner boundary and is equal to

$$s_{tmax} = \frac{\gamma\omega^2}{4g}\left[(3 + \mu)b^2 + (1 - \mu)a^2\right] \tag{13}$$

where μ is Poisson's ratio.

Example 3. A circular steel plate flywheel is 24 in. in outside diameter and has a 6 in. hole. The flywheel rotates at 3,000 rpm. Find the value of the maximum tangential stress. $\mu = 0.3$. $\gamma = 0.283$ lb/in.[3]

Solution. $g = 12 \times 32.174 = 386$ in./sec[2]

[3]See Uhlig, H. H., "Mechanism of Fretting Corrosion," *Trans. ASME*, **76**, 1954, p. 401.
[4]See p. 245 of reference 5, end of chapter.

$$\omega = \frac{2\pi n}{60} = \frac{2\pi 3,000}{60} = 314.16 \text{ rad/sec}$$

In Eq. (13), $\quad s_{tmax} = \dfrac{0.283 \times 314.16^2}{4 \times 386} [(3 + 0.3) \times 12^2 + (1 - 0.3) \times 3^2]$

$$= 18.09(475.2 + 6.3) = 8,710 \text{ psi}$$

If the plate is drilled near the inner boundary for bolting to a flange, stress concentration occurs at the bolt holes.

The polar moment of inertia J about the axis of a disk flywheel with a hole is given by

$$J = \frac{1}{2} \frac{W_o}{g} b^2 - \frac{1}{2} \frac{W_i}{g} a^2$$

where W_o represents the weight of a disk of radius b, and W_i is the weight of a disk of radius a.

$$W_o = \gamma \pi l b^2 \quad \text{and} \quad W_i = \gamma \pi l a^2$$

where l represents the thickness of the plate in the axial direction. Substitution gives

$$J = \frac{\gamma \pi l}{2g}(b^4 - a^4) \tag{14}$$

The energy ΔKE delivered by a change in speed from ω_{max} to ω_{min} is given by

$$\Delta KE = \tfrac{1}{2} J(\omega_{max}^2 - \omega_{min}^2)$$

5. Flywheel with Spokes and Rim

Figure 12-6 shows a cast flywheel with spokes and rim. When the wheel rotates, a uniformly distributed force F_c acts outwardly on the rim and produces the tensile force F. The spokes exercise a restraining influence on

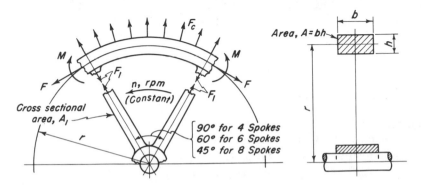

Figure 12-6 Forces and moments on segment of fly wheel.

the expansion of the rim, and cause moments M to act as shown. Equations for F and M, as well as for the tensile force F_1 in the spoke,[5] are given in Table 12-1. Here w is the weight of the rim per inch of circumference, and symbols r, n, h, A, and A_1 have the meanings indicated in Fig. 12-6. In these equations, the depth of section h is assumed to be small as compared to radius r.

Table 12-1 EQUATIONS FOR FORCES AND MOMENTS IN FLYWHEEL SPOKES AND RIM

Spokes	H	F_1, lb	F, lb	M, in. lb
4	$\dfrac{2/3}{(0.0730r^2/h^2) + 0.643 + (A/A_1)}$	$\dfrac{wr^2n^2}{35,200} \times H$	$\dfrac{wr^2n^2}{35,200}(1 - 0.500H)$	$0.1366F_1r$
6	$\dfrac{2/3}{(0.0203r^2/h^2) + 0.957 + (A/A_1)}$	$\dfrac{wr^2n^2}{35,200} \times H$	$\dfrac{wr^2n^2}{35,200}(1 - 0.866H)$	$0.0889F_1r$
8	$\dfrac{2/3}{(0.0091r^2/h^2) + 1.274 + (A/A_1)}$	$\dfrac{wr^2n^2}{35,200} \times H$	$\dfrac{wr^2n^2}{35,200}(1 - 1.207H)$	$0.0662F_1r$

The tensile stress in the spoke is equal to F_1/A_1. The tensile stress in the rim is equal to F/A plus the effect of the bending moment $6M/bh^2$. Stress concentration effects are present at the junction between the spoke and rim. Sudden changes in the velocity of rotation, as well as belt tensions, cause additional bending stresses in spokes and rim.

Care must be exercised to assure that dimensions and speed of rotation for cast iron flywheels have such values that the resulting tensile stresses do not exceed safe working values for this material. Tensile stress values for cast iron vary over a considerable range, depending on the composition and quality of the material.

Example 4. Find the stresses in a cast iron flywheel with 6 spokes rotating uniformly at 600 rpm; radius $r = 36$ in. to center of rim; width of rim $b = 12$ in.; depth of rim $h = 4$ in.; and area of spokes $A_1 = 10$ in.[2] Cast iron weighs 0.256 lb/in.[3]

Solution. $r/h = 9$, and $A/A_1 = 4.8$

By Table 12-1: $H = \dfrac{2/3}{0.0203 \times 81 + 0.957 + 4.8} = 0.0901$

$w = 0.256 \times 12 \times 4 = 12.29$ lb per in. of rim

$F_1 = \dfrac{12.29 \times 36^2 \times 600^2}{35,200} \times 0.0901$

$= 162,900 \times 0.0901 = 14,680$ lb

[5]For derivation, see p. 98 Reference 5, end of chapter.

Tensile stress in spoke: $s = \dfrac{14{,}680}{10} = 1{,}470$ psi

$$F = 162{,}900(1 - 0.866 \times 0.0901) = 150{,}200 \text{ lb}$$

Tensile stress in rim: $s = \dfrac{150{,}200}{12 \times 4} = 3{,}130$ psi

$$M = 0.0889 \times 14{,}680 \times 36 = 46{,}980 \text{ in. lb}$$

Bending stress in rim: $s = \dfrac{6M}{bh^2} = \dfrac{6 \times 46{,}980}{12 \times 16} = 1{,}470$ psi

Resultant tensile stress in rim:

$$s = 3{,}130 + 1{,}470 = 4{,}600 \text{ psi}$$

Suppose the spokes in the foregoing example are not present and the rim freely expands under the influence of centrifugal force. The weight per lineal inch of rim is w, and the acceleration is $r\omega^2$. The centrifugal force F_c per inch of rim is

$$F_c = \frac{w}{g} r\omega^2 = \frac{wr}{g}\left(\frac{2\pi n}{60}\right)^2 = \frac{wrn^2}{35{,}200} \tag{15}$$

Force F in the rim is equal to $F_c r$.

Example 5. Find the stresses for the data of Example 4 if the spokes are not present and the flywheel consists of a freely rotating ring.

Solution. By Eq. (15): $F_c = \dfrac{12.29 \times 36 \times 600^2}{35{,}200} = 4{,}520$ lb/in.

Force in rim: $F = F_c r = 4{,}520 \times 36 = 162{,}900 \text{ lb}$

Stress: $s = \dfrac{F}{A} = \dfrac{162{,}900}{48} = 3{,}390$ psi

It can be seen from this example that the elementary method which considers the flywheel as a rotating ring free to expand gives stresses on the unsafe side. Also, the elementary method gives no indication of the value of the stress in the spokes.

6. Flywheel Requirements

The shafts in many kinds of machinery are subjected to torque loading that varies throughout the work cycle. A smaller driving motor can be used, and smoother operation obtained, if the shaft is provided with a flywheel. During the portion of the work cycle that requires a high torque, energy from the flywheel assists the driving motor, but at a reduction in speed. This energy, as well as the speed, are returned to their original values during those portions of the cycle where the torque requirements are low.

Such a shaft, and a typical loading diagram, are shown in Figs. 12-7(a)

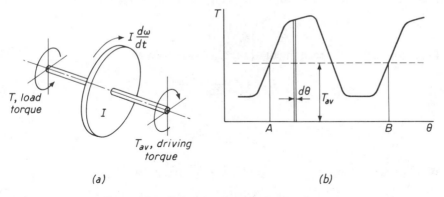

Figure 12-7 Flywheel and torque loading diagram.

and (b). It is assumed that the shaft is very stiff, and the operating speed sufficiently low so that any effects from torsional vibration are negligible. The shaft is in equilibrium from the driving torque T_{av}, the load torque T, and the inertia torque $I d\omega /dt$. Driving torque T_{av} is assumed to be constant. Then

$$I\frac{d\omega}{dt} = T - T_{av} \qquad\qquad (a)$$

but

$$\omega \, dt = d\theta \qquad\qquad (b)$$

These equations should be multiplied together:

$$I\omega \, d\omega = (T - T_{av}) \, d\theta$$

The equation can be integrated between any two points A and B.

$$I\int_A^B \omega \, d\omega = \int_A^B (T - T_{av}) d\theta \qquad\qquad (c)$$

The right side represents the energy supplied to the shaft by the flywheel between points A and B. It can be represented by U_{AB}. The left side should be integrated, and the limits substituted.

$$\frac{1}{2}I(\omega_B^2 - \omega_A^2) = U_{AB} \qquad\qquad (16)$$

On the diagram for torque loading, Fig. 12-7(b), let points A and B be selected so that the energy supplied by the flywheel is a maximum for the chosen interval along the θ-axis. Equation (16) can then be written

$$\frac{1}{2}I(\omega_{max}^2 - \omega_{min}^2) = U_{max} \qquad\qquad (17)$$

Equation (17) represents the maximum change in kinetic energy of the system because of variation in speed. Equation (17) can be written

$$I\left(\frac{\omega_{max} + \omega_{min}}{2}\right)(\omega_{max} - \omega_{min}) = U_{max} \tag{18}$$

Let a coefficient of speed fluctuation, C_f, be defined as

$$C_f = \frac{\omega_{max} - \omega_{min}}{\omega_{av}} \tag{19}$$

where

$$\omega_{av} = \frac{\omega_{max} + \omega_{min}}{2} \tag{20}$$

When these are substituted,[6] Eq. (18) can be arranged as

$$I = \frac{U_{max}}{C_f \omega_{av}^2} \tag{21}$$

Equations (19) and (20) can be combined to give

$$\omega_{max} = \omega_{av}\left(1 + \frac{1}{2}C_f\right) \tag{22}$$

$$\omega_{min} = \omega_{av}\left(1 - \frac{1}{2}C_f\right) \tag{23}$$

Typical values for the coefficient of fluctuation, C_f, for different types of equipment are shown in Table 12-2.

Table 12-2   COEFFICIENT OF FLUCTUATION*

Type of Equipment	C_f
Crushing machinery	0.200
Electrical machinery	0.003
Electrical machinery, direct-driven	0.002
Engines with belt transmission	0.030
Flour-milling machinery	0.020
Gear-wheel transmission	0.020
Hammering machinery	0.200
Machine tools	0.030
Paper-making machinery	0.025
Pumping machinery	0.030–0.050
Shearing machinery	0.030–0.050
Spinning machinery	0.010–0.020
Textile machinery	0.025

*Kent's Mechanical Engineers' Handbook, 12th ed., "Design and Production," p. 7-40.

[6]See Biezeno, C. B., and R. Grammel, *Engineering Dynamics*, Vol. 4, London: Blackie & Son, 1954, p. 107. See also Timoshenko, S., and D. H. Young, *Advanced Dynamics*, New York: McGraw-Hill Book Company, 1948, p. 185, and Den Hartog, J. P., *Mechanics*, New York: McGraw-Hill Book Company, 1948, p. 268.

Example 6. The variation in torque loading for a work cycle of one revolution for a shaft is shown in Fig. 12-8. Input torque T_{av} is assumed to be constant.

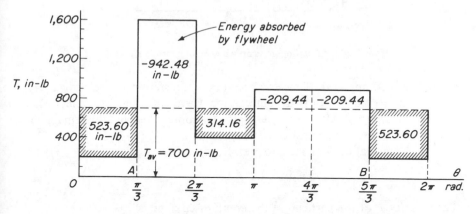

Figure 12-8 Torque loading diagram for Example 6.

Average speed is 860 rpm. Coefficient of fluctuation C_f is 0.10. Find the diameter of the required flywheel if it is cut from a 1-in. steel plate.

Solution. $2\pi T_{av} = \dfrac{\pi}{3}(200 + 1{,}600 + 400 + 900 + 900 + 200)$

$$T_{av} = 700 \text{ in. lb}$$

The areas of the torque diagram above and below the value of T_{av} are shown in the figure. The maximum value of U occurs between points A and B. It has the value

$$U_{max} = 942.48 - 314.16 + 209.44 + 209.44$$
$$= 1{,}047.20 \text{ in. lb}$$

$$\omega_{av} = \frac{2\pi n}{60} = \frac{2\pi 860}{60} = 90.06 \text{ rad/sec}$$

By Eq. (21), $I = \dfrac{U_{max}}{C_f \omega_{av}^2} = \dfrac{1{,}047.20}{0.10 \times 90.06^2}$

$$= 1.2911 \text{ in. lb sec}^2$$

By Eq. (14) for $a = 0$, $b^4 = \dfrac{2gI}{\gamma \pi l} = \dfrac{2 \times 386 \times 1.2911}{0.283\pi \times 1} = 1{,}121.1$

$$b^2 = 33.48, \qquad b = 5.78 \text{ in.}$$

$$\text{Diam} = 11.57 \text{ in.}$$

7. Impact of Elastic Bodies

Let a uniform stress s_0 be suddenly applied to the end of a long bar as shown in Fig. 12-9. At the first instant, an infinitely thin layer of material

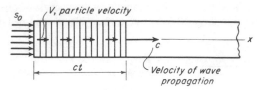

Figure 12-9 Compression wave in long elastic bar.

in the bar is compressed.[7] This compression is transferred to the next layer, and so on. In other words a compression wave, moving with velocity c, travels along the bar. After time t, a length ct is compressed and the remainder of the bar is at rest in the unstressed condition.

The deformation δ of the compressed zone is

$$\delta = \frac{s_0 ct}{E} \tag{24}$$

The velocity V of the particles in the compressed zone is

$$V = \frac{\delta}{t} = \frac{s_0 c}{E} \qquad \text{or} \qquad c = \frac{VE}{s_0} \tag{25}$$

The velocity V of the particles should be distinguished from the velocity c of the wave front.

Let the material in the bar have a weight of γ lb/in.³ The density ρ, mass/in.³, is equal to γ/g, where g is the gravitational constant, 386 in./sec². The mass of the moving particles in the wave is equal to ρAct, where A is the cross-sectional area of the bar. The force on the end of the bar is equal to As_0. The momentum and impulse equation states that mass times velocity is equal to force times time. Hence

$$\rho ActV = As_0 t \qquad \text{or} \qquad s_0 = c\rho V$$

Substitution of Eq. (25) gives

$$s_0 = V\sqrt{E\rho} \tag{26}$$

The force k required to compress the bar a unit distance is equal to AE/l, where l is the length of the bar; hence $E = kl/A$. The weight W_b of the bar is equal to $Al\gamma$ or $Alg\rho$; hence $\rho = W_b/Alg$. Substitution in Eq. (26) gives

$$F = s_0 A = V\sqrt{\frac{kW_b}{g}} \tag{27}$$

where F is the force due to impact. Equations (25) and (26) can be combined to give

$$c = \sqrt{\frac{E}{\rho}} \tag{28}$$

[7]See Donnell, L. H., "Longitudinal Wave Transmission and Impact," *Trans. ASME*, **52**(1), 1930, p. APM-52-14-153. See also Johnson, R. C., "Impact Forces in Mechanisms," *Machine Design*, **30**, June 12, 1958, p. 138, and Arnold, R. N., "Impact Stresses in Freely Supported Beams," *Proc. Inst. Mech Engrs.*, **137**, 1937, p. 217.

Example 7. Find the velocity of propagation for a compression wave in steel. $\gamma = 0.283$ lb/in.^{3}

Solution. In Eq. (28): $c = \sqrt{\dfrac{E}{\rho}} = \sqrt{\dfrac{Eg}{\gamma}} = \sqrt{\dfrac{30,000,000 \times 386}{0.283}}$

$$= 202,280 \text{ in./sec} = 3.19 \text{ miles/sec}$$

Impact stresses can be caused by a moving weight striking the end of the bar as shown in Fig. 12-10. A compression wave travels along the bar and is reflected at the wall; it returns to weight W and is again reflected; and so on. The maximum stress may thus be several times as great as that given by Eq. (26). The exact analysis of this problem is very involved. However an approximate solution, based on energy, can be obtained if simplifying assumptions are made.

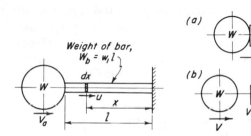

Figure 12-10 Free end of horizontal bar struck by moving weight.

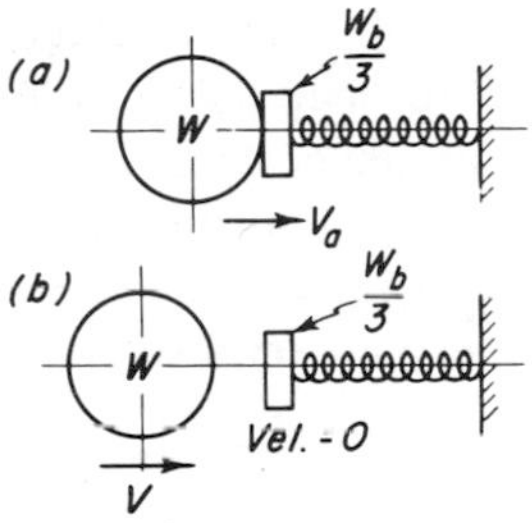

Figure 12-11 Impact when bar of Fig 12-10 is replaced by weight at end of spring.

Weight W has velocity V before striking, but after impact the velocity is reduced to V_a. If it is assumed that the weight and bar remain in contact after striking, the end of the bar also has the velocity V_a. The solution to the problem by the energy method is based on the assumption that all elements in the bar upon impact are instantly given velocities which are proportional to their distances from the wall. Thus, if u is the velocity for the element at distance x from the wall, the following proportion will hold:

$$\frac{u}{x} = \frac{V_a}{l} \quad \text{or} \quad u = \frac{V_a}{l}x \tag{a}$$

The kinetic energy of the system after impact is given by

$$KE = \frac{W}{2g}V_a^2 + \int_0^l \frac{u^2}{2g}w_1 dx$$

where w_1 is the weight per lineal inch of the bar.

Substitution of the value of u from Eq. (a) and integration gives

$$KE = \frac{WV_a^2}{2g}\left[1 + \frac{W_b}{3W}\right] \tag{29}$$

Thus, by Eq. (29), the kinetic energy after impact would be unaffected if the bar were replaced by a weight of magnitude $W_b/3$ at the end of a weightless spring having the same elastic stiffness as the bar. Such a system is shown in Fig. 12-11(a). The situation before impact has occurred, when W has its original velocity V and when $W_b/3$ is not moving, is illustrated in Fig. 12-11(b). The total momentum before and after impact remains unchanged. Hence,

$$WV = \left[W + \frac{W_b}{3}\right]V_a \quad \text{or} \quad V_a = \frac{1}{1 + W_b/3W}V \qquad \text{(b)}$$

Substituion in Eq. (29) gives the value of the kinetic energy of the system after impact.

$$KE = \frac{WV^2}{2g}\left[\frac{1}{1 + W_b/3W}\right] \qquad (30)$$

Upon impact the force in the bar increases uniformly from zero to a maximum value of F. The average value is $\frac{1}{2}F$. If δ represents the maximum shortening of the bar under impact, the strain energy of the bar at the instant of maximum compression is $\frac{1}{2}F\delta$. If k represents the force required to compress the bar a unit distance, then $\delta = F/k$ and the energy is equal to $\frac{1}{2}F^2/k$. This is equal to the KE in Eq. (30). When these values are equated, the following expression results.

$$F = V\sqrt{\frac{kW}{g}\left[\frac{1}{1 + (W_b/3W)}\right]} \qquad (31)$$

The stress in the bar is found by dividing both sides of this equation by the cross-sectional area A. The equation can then be rearranged as follows.

$$s = \frac{V}{A}\sqrt{\frac{kW_b}{g}}\sqrt{\frac{W}{W_b}\left[\frac{1}{1 + (W_b/3W)}\right]}$$

or

$$\frac{s}{s_0} = \sqrt{\frac{W}{W_b}\left[\frac{1}{1 + (W_b/3W)}\right]} \qquad (32)$$

where s_0 is given by Eq. (27). Equation (32), derived from the energy assumption, is plotted in Fig. 12-12.

The exact solution to this problem is obtained by more involved methods. The results can be closely represented by the following empirical equation, which is also plotted in Fig. 12-12.

$$\frac{s}{s_0} = \sqrt{\frac{W}{W_b} + \frac{2}{3}} + 1 \qquad (33)$$

Examination of the curves indicates that large discrepancies are present when the energy method is used and when the ratio of striking weight to weight of bar, W/W_b, is small.

Example 8. A 1 in.-diam round steel bar 20 in. long is struck by a 40 lb weight moving at a velocity of 30 in./sec. Find the value of the stress due to impact by the energy method and compare with the results of the exact solution.

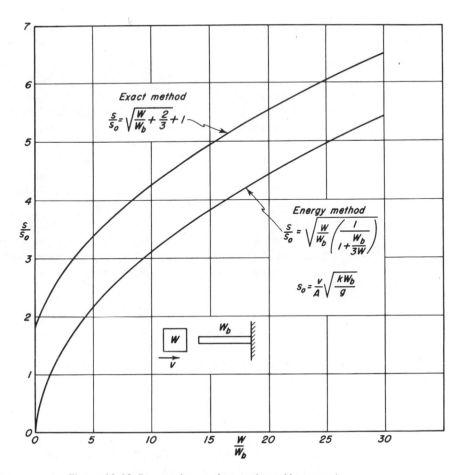

Figure 12-12 Exact and approximate values of impact stresses.

Solution. $$W_b = \frac{\gamma \pi d^2 l}{4} = \frac{0.283\pi 1^2 \times 20}{4} = 4.445 \text{ lb}$$

$$\frac{W}{W_b} = \frac{40}{4.445} = 9$$

$$k = \frac{AE}{l} = \frac{0.7854 \times 30,000,000}{20} = 1,178,000 \text{ lb/in.}$$

By Eq. (27): $$s_0 = \frac{30}{0.7854} \sqrt{\frac{1,178,000 \times 4.445}{386}} = 4,445 \text{ psi}$$

By Eq. (32), energy: $$s = 4,445 \sqrt{9 \left[\frac{1}{1 + 1/(3 \times 9)} \right]} = 4,445 \times 2.946$$

$$= 13,080 \text{ psi}$$

By Eq. (33), exact: $$s = 4,445 \left[\sqrt{9 + \frac{2}{3}} + 1 \right] = 4,445 \times 4.108$$

$$= 17,820 \text{ psi}$$

8. Force Produced by Falling Weight

The solution will be obtained by the energy method. Let the weight W in Fig. 12-13 fall through the height h, and let W and the bolt head remain in contact after striking. Let δ be the maximum value for the stretch of the bolt, and let F be the corresponding impact force. The stiffness of the bolt is such that k lb are required for a stretch of 1 in.

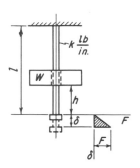

After impact, and at the lowest point of travel, the elastic energy gained by the bolt is equal to $F\delta/2$, as illustrated by the force-deformation graph in Fig. 12-13. This energy comes from the kinetic energy of the weight, as given by Eq. (30), together with the potential energy $W\delta$ given up by the additional lowering of the weight. Hence,

Figure 12-13 Force and deformation caused by falling weight.

$$\frac{F\delta}{2} = \frac{WV^2}{2g} \times \frac{1}{1 + (W_b/3W)} + W\delta \qquad (a)$$

When the substitutions for the impact deflection, $\delta = F/k$, and for the statical deflection, $\delta_{st} = W/k$, are made in the expression above, a quadratic equation results, which, upon solution for F/W, gives the following result.

$$\frac{F}{W} = 1 + \sqrt{1 + \frac{2h}{\delta_{st}} \times \frac{1}{1 + (W_b/3W)}} \qquad (34)$$

The value of the impact force F can be obtained by substitution in this equation. It can be written in terms of velocity V for the falling weight, if desired, by use of the relationship $V^2 = 2gh$, where g is the gravitational constant. Equation (34) indicates that if the weight is held just above the bolt head, $h = 0$, and then released, $F = 2W$.

The value of k for a cylindrical bar is equal to AE/l, where A is the cross-sectional area, and E is the modulus of elasticity. This value is found by taking the equation for deformation, $\delta = Fl/AE$, and letting δ be equal to unity when the load F becomes equal to k. In all the foregoing, it is assumed, of course, that the elastic limit for the bolt material is not exceeded.

Example 9. In Fig. 12-13 let the bar be round, with a diameter of $\frac{1}{2}$ in. and length of 12 in. If weight W is 10 lb, find the maximum stress in the bolt if height h is 4 in. Material weighs 0.283 lb/in.³; $E = 30,000,000$ psi.

Solution. $A = \dfrac{\pi d^2}{4} = 0.1964$ in.², area of bolt

$$\delta_{st} = \frac{Wl}{AE} = \frac{10 \times 12}{0.1964 \times 30,000,000} = 0.00002037 \text{ in.}$$

$$W_b = 12 \times 0.1964 \times 0.283 = 0.6668 \text{ lb}$$

In Eq. (34), $\quad \dfrac{F}{10} = 1 + \sqrt{1 + \dfrac{2 \times 4}{0.00002037} \times \left(\dfrac{1}{1 + 0.6668/(3 \times 10)}\right)}$

$\qquad\qquad F = 6{,}200 \text{ lb}$

$\qquad\qquad s = \dfrac{6{,}200}{0.1964} = 31{,}570 \text{ psi}$

This problem indicates that high stresses are readily produced by impact loads. It may also be noted that the weight of the bar W_b, when small, has but little effect on the force caused by impact.

If the yield point stress of the material is exceeded, a portion of the energy of the falling weight is used in giving a permanent stretch to the bar.

If the lower end of the bolt in the foregoing problem were threaded, and if the weight were stopped in its fall by the nut, the entire force of 6,200 lb would pass through the net section at the threads, and the stress would thereby be considerably increased. Thus, for $\frac{1}{2}$ in.-13UNC with an area at the root of thread of 0.1416 in.^{2}, the stress would be 43,790 psi. Furthermore, there would be stress concentration effects at the root of the threads.

The impact stress can be reduced by decreasing the stiffness of the stressed member. Thus, suppose it were permissible to reduce the bolt, for the entire 12 in. length above the nut, to the stress area of the thread. A further substitution in Eq. (34) shows that force F is reduced to 5,290 lb. The stress becomes 37,360 psi, as compared with 43,790 psi when the full $\frac{1}{2}$-in. body was used.

Hence, decreasing the cross section of the main body of the bolt, which must absorb the energy of the impact, to that of the minimum size through which the force must pass, has brought about a substantial reduction in the magnitude of the force. A further reduction in size, however, would cause the stress in the bolt to increase.

9. Impact of Weight on Beam

Impact between a weight and a simply supported beam is illustrated in Fig. 12-14. The energy method can also be applied to this problem; it gives the following equation for the ratio between the impact force F and the weight W.

$$\frac{F}{W} = 1 + \sqrt{1 + \frac{2h}{\delta_{st}} \times \frac{1}{1 + (17W_b/35W)}} \qquad (35)$$

where W_b is the weight of the beam, and h and δ_{st} have the same meanings as in the preceding section. Equation (35) is known as the Cox equation. It has been in use for almost a century. Like all energy solutions, it is of somewhat limited usefulness.

For most metals, the yield point and ultimate strength are raised when the load is very quickly applied and is of short duration. Because of the many uncertainties present, designers are rarely

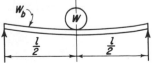

Figure 12-14 Impact of weight on beam.

inclined to take advantage of these raised properties when deciding upon suitable values for the working stresses.

10. Gaskets and Seals

Gaskets and seals are divided into two main classes: static and dynamic. The static gasket is used to prevent the loss of fluid in a pressure vessel. A dynamic seal is used to prevent loss of fluid in a sliding or rotating joint. Numerous types of materials are used such as organic and mineral fibers, leather, natural and synthetic rubber, cork, paper, and soft metals.

For static seals, a flat rigid surface and a thin gasket are preferable, and a minimum amount of packing surface should be exposed to the fluid.

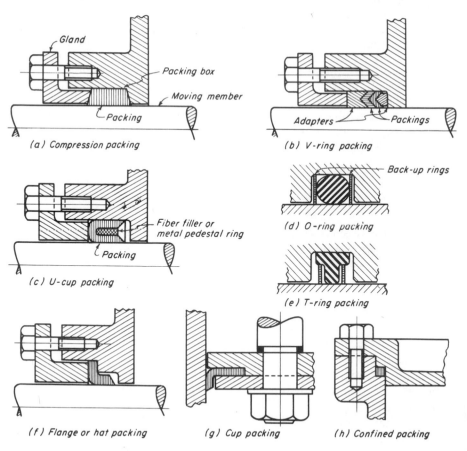

Figure 12-15 Various types of gaskets and seals.

Sometimes the sealing is created by friction. Sometimes the gasket is confined to a chamber and distorted to effect a seal by jamming the packing across the leakage path. Allowance must sometimes be made for a change in volume if the gasket material swells upon contact with the fluid. For a dynamic seal it is most important that the sliding member have a very smooth hard surface.

Various shapes for a number of different gaskets and seals are shown in Fig. 12-15. The stuffing box shown in (a) has the disadvantage of high friction when the pressure of the gasket is high enough to prevent loss of fluid. The V-ring or chevron packing shown in (b) has less friction and tends to seal more tightly with increase of pressure. The molded U-cup packing is widely used for low and medium pressures. The O-ring shown in (d) has the advantage of low friction and simplicity of design. The ring should be slightly compressed after assembly. This seal is popular in aeroplane hydraulic mechanisms. The plastic back-up rings prevent extrusion of the soft rubber into the clearance space under very high pressures. The flange or hat packing shown in (f) is used in a number of different forms. Sometimes a circumferential spring is used to press the lip of the packing against the shaft.

The mechanical seal[8] of Fig. 12-16 is very effective for sealing a rotating shaft against gas or liquid under pressure. The sealing member may be

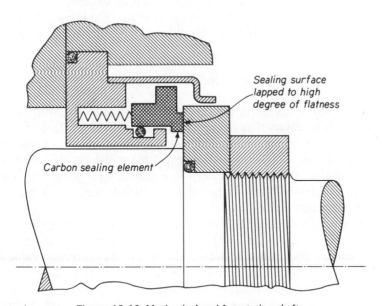

Sealing surface lapped to high degree of flatness

Carbon sealing element

Figure 12-16 Mechanical seal for rotating shaft.

[8]See Alexander, T. E., "Mechanical Seals— Uses and Abuses," *Mech. Eng.*, **84**, Dec. 1962, p. 51. See also Mayer, E., "Leakage and Wear in Mechanical Seals," *Machine Design*, **32**, Mar. 3, 1960, p. 106, and Brkich, A., "Mechanical Seals," *Prod. Eng.* **21**, Apr. 1950, p. 85.

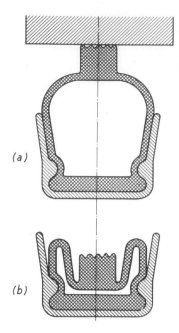

(a)

(b)

Figure 12-17 Inflatable seal.

either stationary or rotating and is made from refined carbon. It is spring loaded to compensate for wear. The mating surface is made of a hard material such as Stellite and must be lapped to a high degree of flatness. The edge of the carbon member should be sharp and square to prevent the entrance of foreign matter which may cause a rapid increase in the rate of wear.

The inflatable seal, Fig. 12-17, is convenient for sealing large openings with a relatively large amount of clearance.

Fluid-tight joints can also be formed by coating the surfaces with a thermosetting liquid plastic[9] which cures in the absence of air after the connection has been tightened. The material will flow into surface irregularities and machining costs can sometimes be reduced.

In piping service, rubber and nonmetallic materials, as well as lowmelting metals, are generally limited to temperatures below 250°F. Materials such as asbestos, with or without metal jacketing, may be used up to 850°F. For temperatures higher than 850°F, all-metal gaskets must be used.

11. Design of Gaskets for Static Loads

Various kinds of gaskets are illustrated in Table A. To maintain a fluid-tight joint it is necessary that the parts be tightly bolted together. The initial or bolting-up pressure must be enough to cause local yielding of the gasket where it is in contact with the asperities of the metal flange surfaces. This minimum contact pressure, necessary to secure a tight joint even for low values of the internal pressure, is called the "yield" value y of the gasket. Values are given in Table A.

The internal fluid pressure in the pipe or vessel reduces the gasket contact pressure. Experience has shown that the ratio between the resultant contact pressure and the fluid pressure should not be less than a certain value if the joint is to remain tight. This ratio is called the gasket factor m. As indicated by Table A, it varies for different kinds of gaskets.

The bolting-up and fluid pressures cause a distortion in the flanges and adjacent parts of the joint. The chief effect is a rotation that tends to concentrate the entire gasket force on the outer edge of the gasket. Gasket pres-

[9]See Whittemann, R. F., and M. B. Pearce, "Anerobics—A New Approach to Gaskets," *Mech. Eng.*, **91**, Aug. 1969, p. 26.

Table A

TABLE A: GASKET FACTORS m AND YIELD VALUES y		GASKET FACTOR m	YIELD VALUE y	SKETCH & NOTES	REFER TO TABLE B: FACING LIMITATION	USE COL.
GASKET MATERIAL						
GUM RUBBER SHEET		.50	0			
HARD RUBBER SHEET		1.00	180		USE ①④⑥ ONLY	
CLOTH INSERTED SOFT RUBBER		.75	50			
CLOTH INSERTED HARD RUBBER		1.25	400			
VEGETABLE FIBRE SHEET (HEMP OR JUTE)		1.75	1120			
RUBBERIZED WOVEN WIRE INSERTED ASBESTOS	3-PLY	2.25	2200			
	2-PLY	2.50	2880			
	1-PLY	2.75	3650			
ASBESTOS COMPOSITION OR COMPRESSED ASBESTOS	⅛ THICK	2.00	1620		USE ①④⑥ ONLY	
	1/16″ THICK	2.75	3650			
	1/32″ THICK	3.50	6480			
SPIRAL-WOUND METAL, ASBESTOS FILLED	CARBON STEEL	2.50	2880			II
	KA25 OR TYPE 316	3.00	4500			
SERRATED STEEL, ASBESTOS FILLED		2.75	3650			
CORRUGATED METAL, ASBESTOS INSERTED OR CORRUGATED METAL JACKET, ASBESTOS FILLED	SOFT ALUMINUM	2.50	2880		USE ① ONLY	
	SOFT COPPER OR BRASS	2.75	3650			
	IRON OR SOFT STEEL	3.00	4500			
	MONEL OR 4-6% CHROME	3.25	5450			
	11-13% CHROME, KA25 OR TYPE 316	3.50	6480			
CORRUGATED METAL	SOFT ALUMINUM	2.75	3650			
	SOFT COPPER OR BRASS	3.00	4500			
	IRON OR SOFT STEEL	3.25	5450			
	MONEL OR 4-6% CHROME	3.50	6480			
	11-13% CHROME, KA25 OR TYPE 316	3.75	7600			
FLAT METAL JACKET, ASBESTOS FILLED	SOFT ALUMINUM	3.25	5450			
	SOFT COPPER OR BRASS	3.50	6480			
	IRON OR SOFT STEEL	3.75	7600			
	MONEL OR 4-6% CHROME	4.00	8820			
	11-13% CHROME, KA25 OR TYPE 316	4.25	10 120			
GROOVED IRON OR SOFT STEEL CORE, METAL JACKETED	SOFT ALUMINUM	3.25	5450		USE ①④⑥ ONLY	
	SOFT COPPER OR BRASS	3.50	6480			
	IRON OR SOFT STEEL	3.75	7600			
	MONEL OR 4-6% CHROME	4.00	8820			
	11-13% CHROME, KA25 OR TYPE 316	4.25	10 120			
SOLID METAL, ⅛ THICKNESS OR MORE (FOR THINNER GASKETS SEE NOTE)	SOFT ALUMINUM	4.00	8820	NOTE: FOR EACH 1/32″ REDUCTION IN THICKNESS BELOW ⅛″ INCREASE m BY .25, COMPUTE $y = 180 (2m-1)^2$		I
	SOFT COPPER OR BRASS	4.75	13 000			
	IRON OR SOFT STEEL	5.50	18 000			
	MONEL OR 4-6% CHROME	6.00	21 780			
	11-13% CHROME, KA25 OR TYPE 316	6.50	25 920			

(*Reproduced from* Mechanical Engineering, *reference 7, bibliography.*)

sures are accordingly not computed for the entire width, but only for a narrow band on the outer edge, called the effective gasket yielding width *b*. Equations for determining this width are found in Table B. Although these equations are empirical, they are based on long experience in the design of fluid-tight joints.[10]

The area subjected to fluid pressure is assumed to extend to the edge of the effective gasket area.

[10]For Tables A and B see Rossheim, D. B., and A. R. C. Markl, "Gasket Loading Constants," *Mech. Eng.* **65,**, 1943, p. 647, and **66**, 1944, p. 72. See also the following:
Roberts, Irving, "Gaskets and Bolted Joints," *Trans. ASME*, **72**, 1950, 169;
Smoley, E. M., "Sealing with Gaskets," *Machine Design*, **38**, Oct. 27, 1966, p. 172;
Linderoth, L. S., "Selecting Hydraulic Seals," *Machine Design*, **16**, Sept. 1944, p. 119.

Table B

TABLE B: EFFECTIVE GASKET WIDTH b		
FACING SKETCH	BASIC GASKET YIELDING WIDTH b_o	
	COL. I	COL. II
(1a)	$\dfrac{n}{2}$	$\dfrac{n}{2}$
(1b)	$\dfrac{w+t}{2}\left(\dfrac{w+n}{4}max\right)$	$\dfrac{w+t}{2}\left(\dfrac{w+n}{4}max\right)$
(2)	$\dfrac{w+n}{4}$	$\dfrac{w+3n}{8}$
(3)	$\dfrac{w}{2}:\left(\dfrac{n}{4}\ min\right)$	$\dfrac{w+n}{4}:\left(\dfrac{3n}{8}\ min\right)$
(4)	$\dfrac{3n}{8}$	$\dfrac{7n}{16}$
(5)		
(6)	$\dfrac{n}{4}$	$\dfrac{3n}{8}$
(7)		
(8)	$\dfrac{w}{8}$	

EFFECTIVE GASKET YIELDING WIDTH b
$b = b_o$, when $b_o \leqq \frac{1}{4}"$
$b = \frac{\sqrt{b_o}}{2}$, when $b_o > \frac{1}{4}"$

LOCATION OF GASKET LOAD ATTACK
O.D. CONTACT FACE / ₵ NUBBIN / I.D. BOLTS / ₵ BETWEEN O.D. BOLTS & O.D. OF FLANGE
INSIDE GASKET / WITH NUBBIN / FULL FACE GASKET

(*Reproduced from* Mechanical Engineering, *reference 7, bibliography.*)

Figure 12-18 shows the forces acting in a gasketed joint. Before the fluid pressure is acting, the force F_g in the bolt and on the gasket is

$$F_g = A_g q \qquad (36)$$

where A_g is the effective area of the gasket, and q is pressure on the gasket caused by tightening the bolts.

After the fluid pressure p is acting, a force $A_i p$ is produced, where A_i is the area subjected to internal pressure. The force on the gasket required to prevent loss of pressure is $A_g mp$. The bolt force F_b then is

$$F_b = p(A_i + A_g m) \qquad (37)$$

Forces F_g and F_b can be taken as substantially equal to each other. Then

$$A_g q = p(A_i + A_g m) \qquad (38)$$

Example 10. A cross section through a gasketed manhole joint is shown in Fig. 12-19. The internal fluid pressure is 500 psi gage. If 24 bolts are used for holding the cover in place, find the tensile stress in the net area of the threads.

Solution. From Table B:

$$b_o = \frac{n}{2} = \frac{2}{2} = 1 \text{ in.}$$

$$b = \frac{\sqrt{b_o}}{2} = \frac{\sqrt{1}}{2} = 0.5 \text{ in.,}$$

$$\text{effective gasket width}$$

Effective gasket area:

$$A_g = 27.5\pi \times 0.5 = 43.2 \text{ in.}^2$$

Area subjected to internal pressure:

$$A_i = \frac{\pi}{4} \times 27^2 = 572.6 \text{ in.}^2$$

By Table A: $m = 3.5$

In Eq. (38): $q = \dfrac{500(572.6 + 43.2 \times 3.5)}{43.2}$

$$= 8,380 \text{ psi}$$

This value for the gasket pressure is satisfactory, since Table A indicates that a compression of 6,480 psi is sufficient to seat the gasket properly. For $1\frac{1}{2}$ in.-12 UNF: Stress area = 1.58 in.²

By Eq. (37): $F_b = 500(572.6 + 43.2 \times 3.5) = 361,900 \text{ lb}$

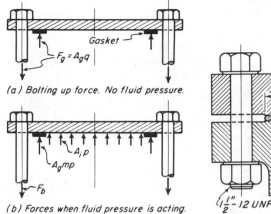

(a) Bolting up force. No fluid pressure.

(b) Forces when fluid pressure is acting.

Figure 12-18 Equilibrium of gasketed joint.

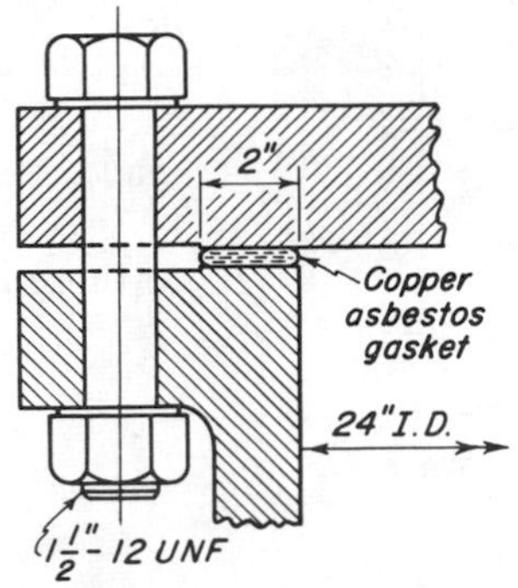

Figure 12-19 Gasketed man-hole cover. Example 10.

Tensile stress: $$s = \frac{361,900}{24 \times 1.58} = 9{,}540 \text{ psi}$$

For small diameters or low pressures, the initial gasket contact pressure, when computed as in the foregoing example, may be smaller than the y-value given in Table A. Additional bolting-up force should be applied until the contact pressure is equal to, or slightly greater than, the tabulated value. However, for large diameters or high pressures, care must be exercised to assure that the bolting-up pressure on the effective area does not exceed the true yield point value in compression for the gasket material. The gasket must retain sufficient elasticity to maintain a tight joint after the initial compression has been reduced by the fluid pressure. Published information is lacking at the present time on the maximum permissible contact pressure for the various types of gaskets. For solid metal gaskets, the yield point value in compression would be used.

Considerable experience is required to design high-pressure joints properly. The maximum internal pressure that a particular type of gasket can retain depends on a number of factors, such as flange design, type of facing, bolt location and arrangement, and relaxation of bolts at elevated temperature. Stresses and deformations at the joint arising from the thermal expansion of the connected parts must also be considered. For severe service conditions, the ring-type joint illustrated in Fig. 12-20 has proved very successful. It is easily made tight and the gasket can be used over and over again. The cross section is either oval or octagonal, and the temper should be dead soft. This type of gasket will not blow out if a leak develops in service.

It is sometimes difficult, in high-pressure joints, to secure sufficient initial bolt force. Bolts with larger diameters will carry more force, but it is

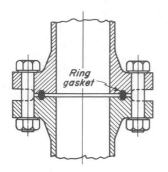

Figure 12-20 Ring-type flange joint for high pressure service.

difficult to tighten larger bolts to the same stress that is easily secured in bolts of smaller diameter. A torque wrench must, of course, be used if a predetermined stress is to be secured in the bolt. Experience has shown that installation stresses due to torque loads for various sizes of alloy-steel studs made in accordance with the 8-pitch thread series in sizes larger than 1 in. are approximately equal to those given by the following empirical equation.

$$s = \frac{45,000}{\sqrt{d}} \tag{39}$$

where d is the nominal diameter of bolt stud.

12. Wire Rope

Wire rope is used in conveying and hoisting equipment, and also in such stationary applications as guy wires and stays.[11] A number of wires, such as 7, 19, or 37, are first twisted into a strand. A number of strands, usually 6 or 8, are then twisted about a core or center to form the rope. Cores may be of hemp or of wire. Several typical cross sections for wire rope are given in Fig. 12-21. The wires are made of various grades of steel, although wrought

Figure 12-21 Cross sections of typical wire rope.

iron is also used. For long life the core, as well as the entire rope, must be continuously saturated with lubricant.

Flexibility in wire ropes is secured by using a large number of small-diameter wires. Ropes of few wires of relatively large size, such as the 6 × 7, are used for guy wires, but are too stiff for hoisting service unless the sheaves have very large diameters. The 6 × 19 and 6 × 37 types are the most widely

[11]See Drucker, D. C., and H. Tachau, "A New Design Criterion for Wire Rope," *Trans. ASME*, **67**, 1945, p. A-33, Discussion **68**, 1946, p. A-75. See also Starkey, W. L., and H. A. Cress, "An Analysis of Critical Stresses and Mode of Failure of a Wire Rope," *Trans. ASME*, **81**, 1959, *Jour. of Engrg. for Industry*, p. 307.

used ropes for hoisting service. The radius of the groove in the sheave should be just large enough to provide clearance for the rope without pinching. The circular cross section of the rope is preserved and the stresses are more evenly distributed. The condition and proper alignment of the sheaves are also important factors affecting the useful life. When wire rope is wound on drums in two or more layers, abrasion and severe crushing stresses are induced, particularly where the rope must cross over the depressions made by the preceding layers.

Wire ropes are subjected to various kinds of stresses. There is the direct tension stress T/A, where T is the force in the rope and A is the cross-sectional area. Bending stresses also occur in the wires when the rope is passing over a sheave. From elementary mechanics, it is known that

$$\frac{M}{EI} = \frac{1}{r} = \frac{2}{d_s}$$

$$s = \frac{Mc}{I} = \frac{Md_w}{2I}$$

(a)

where d_w is the diameter of the wire, d_s is the diameter of the sheave, I is the moment of inertia of a single wire, and E is the modulus of elasticity for the material. Elimination of M/I between these equations gives the following expression for the bending stress.

$$s = \frac{d_w E}{d_s}$$

(40)

An increase in wire size is thus seen to give an increase in stress. A decrease in sheave size also gives an increase in bending stress, and care must be exercised that wire rope is not operated on excessively small sheaves.

The pressure p between rope and sheave can be determined by consideration of Fig. 12-22. This pressure is assumed to be uniformly distributed over the curved surface of the groove which receives the rope and which runs circumferentially around the sheave. The value of the tensile force T

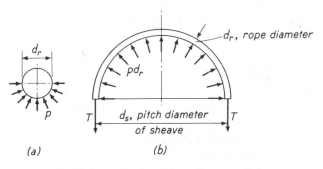

Figure 12-22 Forces in wire rope passing around sheave.

in the rope is then equal to $\frac{1}{2}pd_rd_s$, where d_r is the diameter of the rope. Hence

$$p = \frac{2T}{d_rd_s} \tag{41}$$

Because of the small areas in actual contact between the wires, or between wires and groove, merely nominal values for p cause very high compressive stresses in the materials.

Failure of a wire rope occurs as a result of fatigue and wear in passing over the sheaves. Recent investigations have shown that the ratio of pressure p to the ultimate strength of the material is a significant variable for determining the fatigue life. Figure 12-23 shows the results of plotting p/s_{ult} vs.

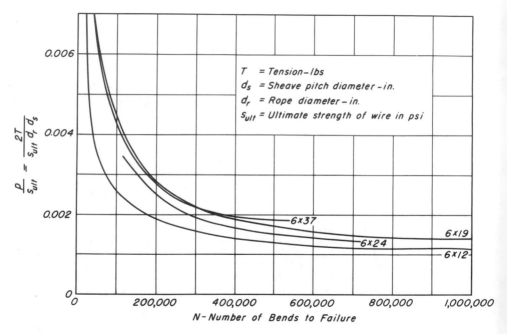

Figure 12-23 Fatigue life of wire rope as determined by experiments.

bends to failure for various kinds of wire rope. A flexing and unflexing of the rope in passing over a sheave counts as a single bend. This figure indicates, for example, that if a 6 × 19 rope has a p/s_{ult} value of 0.0014, a very long life can be expected so far as fatigue failure is concerned. The product d_rd_s in the denominator of the variable indicates that a decrease in the diameter of the sheave permits an increase in the size of the rope; however, a practical lower limit exists for sheave size beyond which it is not safe to go because of high bending stresses and abrasion of the wires.

Dimensional data for ropes and information on the physical properties of the materials are given in Table 12-3. The minimum diameters for sheaves for iron rope are 50 per cent greater than the values for steel rope. Values for the minimum factor of safety for various kinds of applications are shown in Table 12-4.

Table 12-3   WIRE ROPE DATA

Metallic Cross-Sectional Area ($d_r = $ diam rope)		Diameter of Outer Wire, d_w	Minimum Diameter d_s for Sheaves for Steel Rope	Modulus of Elasticity of Steel Rope	Breaking Stress	
Type	in.^{2}	in.	in.	psi	Type of Material	psi
6 × 7	$0.380d_r^2$	$1/9d_r$	$42d_r$	14,000,000	Improved plow steel	200,000
6 × 19	$0.404d_r^2$	$1/16d_r$,*	$24d_r$	12,000,000	Plow steel	175,000
6 × 37	$0.404d_r^2$	$1/22d_r$	$18d_r$	11,000,000	Extra-strong cast steel	160,000
8 × 19	$0.352d_r^2$	$1/19d_r$	$20d_r$	10,000,000	Cast steel	140,000
					Iron	65,000

*Filler wire type. For Warrington type large outer wires, $d_w = {}^1/_{14}$.

Table 12-4   MINIMUM FACTORS OF SAFETY

Type of Service	FS	Type of Service			FS
		Hot-ladle cranes			8
Track cables	3.2	Slings			8
Guys	3.5				
Mine shafts, 500-ft depth	8	Elevators—Carspeed, fpm	Passenger	Freight	Dumbwaiters
1,000–2,000-ft depth	6				
3,000 ft-depth and more	4	50	7.50	6.67	5.33
Miscellaneous hoisting equipment	5	150	8.20	7.32	5.98
Haulage ropes	6	300	9.17	8.20	6.88
Overhead and gantry cranes	6	500	10.25	9.14	8.00
Jib and pillar cranes	6	800	11.25	10.02	
Derricks	6	1,100	11.67	10.43	
Small electric and air hoists	7	1,500	11.87	10.61	

Example 11. (a) What diameter of 6 × 19 wire rope of improved plow steel would be required for a 5-ton load supported on 8 lines? Compute for long life and continuous operation.

   (b) What is the permissible static load for the rope of part (a)? Use a factor of safety of 6.

(c) If the rope is always loaded to the full static value, find the number of bends, using minimum size pulleys, which the rope can sustain before fatigue failure may be expected. On the basis of 100 bends per working day, what would be the expected life of the cable?

(d) What minimum size sheave would be required for the rope to be free from fatigue if operation takes place at the static load value?

Solution. (a) By Table 12-3: $\qquad s_{ult} = 200{,}000$ psi, and $d_s = 24d_r$.

$$T = \frac{5 \times 2{,}000}{8} = 1{,}250 \text{ lb}$$

From Fig. 12-23: $\qquad \dfrac{2T}{s_{ult}d_r d_s} = 0.0014$

Hence: $\qquad d_r d_s = 24d_r^2 = \dfrac{2T}{0.0014 s_{ult}}$

$$= \frac{2 \times 1{,}250}{0.0014 \times 200{,}000} = 8.93$$

Solving: $\qquad d_r = 0.61$ in. Use $\frac{5}{8}$-in. diam rope

$$d_s = 24 \times \tfrac{5}{8} = 15 \text{ in. recommended}$$
$$\text{minimum diameter of sheave}$$

(b) By Table 12-3: $\qquad A = 0.404 d_r^2 = 0.404 \times 0.625^2$

$$= 0.158 \text{ in.}^2 \text{ metallic cross-sectional area}$$

$$\text{Max. static load} = \frac{0.158 \times 200{,}000}{6} = 5{,}200 \text{ lb}$$

The maximum permissible static load is over 4 times as great as the working load when fatigue is considered.

(c) $\qquad \dfrac{2T}{s_{ult}d_r d_s} = \dfrac{2 \times 5{,}200}{200{,}000 \times 0.625 \times 15} = 0.00555$

From Figure 12-23:

No. of bends to failure $= 65{,}000$

$$\text{Expected life} = \frac{65{,}000}{100} = 650 \text{ working days}$$

(d) $\qquad d_s = \dfrac{2T}{s_{ult}d_r \times 0.0014} = \dfrac{2 \times 5{,}200}{200{,}000 \times 0.625 \times 0.0014} = 59.5 \text{ in.}$

This result is the diameter of the sheave for operation at the permissible static load and for freedom from fatigue. It is obviously an impractical result. Therefore, a reduced load must be used, as was done in part (a), if sheaves of reasonable size are to be used, and if freedom from fatigue is to be achieved as well.

Catalogs of wire rope manufacturers contain a large amount of valuable information on the selection and maintenance of rope. Their knowledge of field conditions, collected over many years of experience, should be utilized by the purchaser.

13. Curved Beams

Curved beams in the form of hooks and brackets are frequently used as machine elements.[12] When such bodies are subjected to bending moments, the stress distribution is not linear on either side of the neutral axis, but increases more rapidly on the inner side. For curved beams, the assumption can usually be made with sufficient accuracy that cross sections normal to the curved elements of the unloaded beam remain plane and perpendicular to the elements after the bending moments are acting. The deformations thus vary directly with the distance from the neutral axis.

Consider the elements of the curved beam of Fig. 12-24 lying between

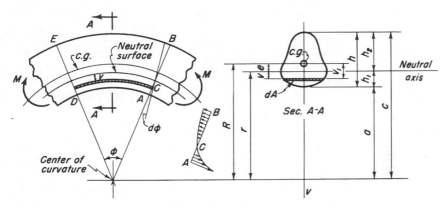

Figure 12-24 Curved beam in pure bending.

two axial planes AB and DE separated by the angle φ. Let the total angular deformation of plane AB with respect to DE be represented by $d\varphi$. A shorter element on the inner side thus deforms the same amount as a longer element, symmetrically located with respect to the neutral surface, deforms on the outer side. The inner element, however, has the higher stress, since its deformation has taken place over a shorter length element.

The cross sections of the beam of Fig. 12-24 are assumed to be symmetrical with respect to the v-axis, as is indicated by view A-A. Because of the higher stresses on the inner elements, the neutral surface for a curved beam no longer passes through the center of gravity of the cross section, but shifts inward a small amount, designated e in the figure. The total deformation for an element located a distance v below the neutral surface is $v\,d\varphi$. This deformation takes place in an element whose length is $(r - v)\varphi$. The elongation or unit deformation ϵ then is

[12]See p. 65, Reference 5, end of chapter. See also, Wahl, A. M., "Calculation of Stress in Crane Hooks," *Trans. ASME*, **68**, 1946, p. A-239.

$$\epsilon = \frac{v \, d\varphi}{(r - v)\varphi} \tag{a}$$

Multiplication by the modulus of elasticity E gives the stress at this point. Thus

$$s = \frac{Ev \, d\varphi}{(r - v)\varphi} = \frac{E \, d\varphi}{\varphi}\left[\frac{r}{r - v} - 1\right] \tag{b}$$

This equation indicates that the distribution of stress over the cross section is hyperbolic.

From the condition of static equilibrium, the sum of the stresses over a cross section must add to zero, and the moment made by the stresses must be equal to the applied moment M. Hence

$$\int s \, dA = \frac{E \, d\varphi}{\varphi} \int \frac{v \, dA}{r - v} = 0 \tag{c}$$

$$\int sv \, dA = \frac{E \, d\varphi}{\varphi} \int \frac{v^2 \, dA}{r - v} = M \tag{d}$$

The integrations in the equations above are to extend over the entire cross section. By division of numerator by denominator, Eq. (d) becomes

$$M = -\frac{E \, d\varphi}{\varphi} \int v \, dA + \frac{Er \, d\varphi}{\varphi} \int \frac{v \, dA}{r - v} \tag{e}$$

By Eq. (c), the second integral of Eq. (e) is equal to zero. The first integral of Eq. (e) represents the moment of area of the cross section about the neutral axis, that is, $-eA$. The minus sign is required because the center of gravity lies on the negative side of the neutral axis. Therefore

$$M = \frac{E \, d\varphi}{\varphi} eA \tag{f}$$

Substitution in Eq. (b) yields

$$s = \frac{M}{eA}\frac{v}{r - v} \tag{g}$$

Substitution of the limits for v gives the following equations.

$$s_{max} = \frac{Mh_1}{Aea} \qquad \text{when} \qquad v = h_1 \tag{42}$$

$$s_{min} = -\frac{Mh_2}{Aec} \qquad \text{when} \qquad v = -h_2 \tag{43}$$

Since these equations are sensitive to small variations in the value of e, considerable care must be exercised in determining this quantity accurately.

Additional useful equations can be obtained by letting $v = v_1 - e$ in Eq. (c).

$$\int \frac{v}{r - v} \, dA = \int \frac{v_1 - e}{r - v_1 + e} \, dA = \int \frac{v_1 - e}{R - v_1} \, dA = 0 \tag{h}$$

The integral on the right may be written

$$\int \frac{v_1\, dA}{R - v_1} = e \int \frac{dA}{R - v_1} \tag{i}$$

$$= \frac{e}{R} \int \frac{R - v_1 + v_1}{R - v_1}\, dA \tag{j}$$

$$= \frac{e}{R} \left[\int dA + \int \frac{v_1\, dA}{R - v_1} \right] \tag{k}$$

Let

$$\int \frac{v_1\, dA}{R - v_1} = m_1 A \tag{44}$$

where A is the area of the cross section, and m_1 is another constant. Substitution in Eq. (k) will give the following equation for e.

$$e = \frac{m_1 R}{1 + m_1} \tag{45}$$

Substitution in Eq. (i) gives

$$\int \frac{dA}{R - v_1} = \frac{A}{R}(1 + m_1) \tag{46}$$

The integral of Eq. (46) can be evaluated for simple types of cross sections and the value of m_1 can then be found. Eccentricity e can then be found by use of Eq. (45), and the stress by use of Eqs. (42) and (43).

14. Curved Beam of Rectangular Cross Section

The foregoing general equations for the stresses in a curved beam are now adapted to the rectangular cross section, Fig. 12-25. Equation (46) can be written

$$A(1 + m_1) = \int_{-h/2}^{+h/2} \frac{dA}{1 - (v_1/R)} \tag{a}$$

When the numerator is divided by the denominator, the following result is obtained.

$$A(1 + m_1) = \int_{-h/2}^{+h/2} \left[1 + \frac{v_1}{R} + \frac{v_1^2}{R^2} + \frac{v_1^3}{R^3} \right. $$
$$\left. + \frac{v_1^4}{R^4} + \cdots \right] b\, dv_1$$

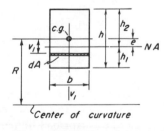

Figure 12-25 Curved beam of rectangular cross section.

This expression should now be integrated term by term and the limits substituted.

$$A(1 + m_1) = b\left[h + \frac{h^3}{12R^2} + \frac{h^5}{80R^4} \right] \tag{b}$$

or

$$m_1 = \frac{h^2}{12R^2} + \frac{h^4}{80R^4} \tag{c}$$

Let $c_1 = 2R/h$, where c_1 is called the index of curvature. When this substitution is made, Eq. (c) becomes

$$m_1 = \frac{1}{3c_1^2} + \frac{1}{5c_1^4} = \frac{1}{3c_1^2}\left[1 + \frac{0.6}{c_1^2}\right]$$

$$= \frac{1}{3c_1^2}\left[\frac{c_1^2 + 0.6}{c_1^2}\right] \approx \frac{1}{3c_1^2}\left[\frac{c_1^2}{c_1^2 - 0.6}\right] = \frac{1}{3c_1^2 - 1.8} \tag{d}$$

The eccentricity e is found by substitution into Eq. (45).

$$e = \frac{hc_1}{2(3c_1^2 - 0.8)} \tag{47}$$

Substitution for e, h_1, and a in Eq. (42) gives the maximum stress.

$$s_{max} = \frac{6M}{bh^2} \times \frac{3c_1^2 - c_1 - 0.8}{3c_1(c_1 - 1)} = K_c \frac{6M}{bh^2} \tag{48}$$

where

$$K_c = \frac{3c_1^2 - c_1 - 0.8}{3c_1(c_1 - 1)} \tag{49}$$

Equation (49) gives the value of the stress concentration factor for curvature for a rectangular cross section.

Example 12. Let the width b of the rectangle be 4 in. and the depth of section h be 6 in. The radius of curvature R to the center of gravity is also equal to 6 in. Find the value of the eccentricity e, the stress concentration factor K_c, and the maximum stress.

Solution. $$c_1 = \frac{2R}{h} = 2$$

In Eq. (47): $$e = \frac{6 \times 2}{2(3 \times 2^2 - 0.8)} = 0.536 \text{ in.}$$

In Eq. (49): $$K_c = \frac{3 \times 2^2 - 2 - 0.8}{3 \times 2(2 - 1)} = 1.533$$

In Eq. (48): $$s_{max} = 1.533\frac{6M}{4 \times 6^2} = 0.0639M$$

Sometimes the cross section of a curved beam is so shaped that no analytic expression can be derived for the eccentricity e. The integration of Eq. (44) can be performed numerically if the equation is written

$$m_1 A = \sum \frac{v_1 \Delta A}{R - v_1} \tag{50}$$

Accurate results can be obtained by this method if the cross section is divided into a sufficient number of elemental strips ΔA.

Example 13. Find the value of the maximum stress for the beam of Example 12 by use of Eq. (50) and compare with the results obtained by use of Eq. (48).

Solution. Divide the cross section into 12 strips, as shown by Fig. 12-26. The computations of Eq. (50) have been carried out in Table 12-5. Substitution of the summation of the right-hand column in Eq. (50) gives

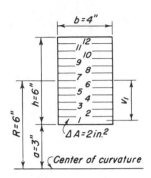

$$m_1 A = 2.3422$$

Solving: $$m_1 = \frac{2.3422}{24} = 0.0976$$

In Eq. (45): $$e = \frac{0.0976 \times 6}{1 + 0.0976} = 0.534 \text{ in.}$$

Hence: $$h_1 = 2.466 \text{ in.}$$

$$h_2 = 3.534 \text{ in.}$$

Figure 12-26 Cross section of Example 13 divided into elementary strips.

In Eq. (42): $$s_{max} = \frac{2.466M}{24 \times 0.534 \times 3} = 0.0642M$$

Numerical integration thus gives accurate results, and the method can therefore be used for odd-shaped cross sections.

Table 12-5

No.	v_1	$R - v_1$	$\dfrac{v_1 \, \Delta A}{R - v_1}$
1	2.75	3.25	1.6923
2	2.25	3.75	1.2000
3	1.75	4.25	0.8235
4	1.25	4.75	0.5263
5	0.75	5.25	0.2857
6	0.25	5.75	0.0870
7	−0.25	6.25	−0.0800
8	−0.75	6.75	−0.2222
9	−1.25	7.25	−0.3448
10	−1.75	7.75	−0.4516
11	−2.25	8.25	−0.5454
12	−2.75	8.75	−0.6286

$$\sum \frac{v_1 \, \Delta A}{R - v_1} = 2.3422$$

15. Curved Beam of Circular Cross Section

By a similar process, the equation for the maximum stress in a curved beam of round cross section can be shown to be

$$s_{max} = K_c \frac{32M}{\pi d^3} \tag{51}$$

where $$K_c = \frac{4c_1^2 - c_1 - 1}{4c_1(c_1 - 1)} \tag{52}$$

The index of curvature c_1 is equal to $2R/d$.

16. Angular Deflection of Curved Bar

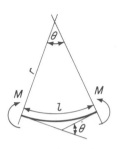

Unless the index of curvature, $2R/d$ or $2R/h$, is very small, the angular change between two cross sections of a curved beam caused by the moments M is approximately the same as would take place if the beam were straight. For a straight beam, the relationship between bending moment and radius of curvature is given by the equation $1/r = M/EI$. Since the moment M is the same at all points, the radius r is also constant, and the elastic line is a circle, as shown by Fig. 12-27. Since $r\theta = l$, substitution for r gives the value of θ as

Figure 12-27 Angular deformation caused by moments.

$$\theta = \frac{Ml}{EI}$$ (53)

17. Cams

A cam provides a convenient means for transforming rotary motion into reciprocating motion. Since the cam outline can be given a wide variety of shapes, many different types of motion can be secured. The plate cam, illustrated in Fig. 12-28, moves the follower in a direction at right angles to the cam axis. For the cylindrical cam of Fig. 12-29, the follower moves in a direction parallel to the cam axis. The cam must have the proper shape if the desired type of motion is to be imparted to the follower.

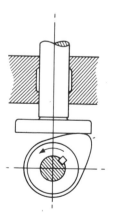

Figure 12-28 Plate cam with mushroom follower.

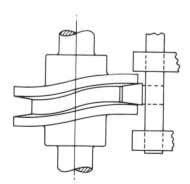

Figure 12-29 Cylindrical cam.

For many commonly used cams, the rise of the follower is parabolic, harmonic, or cycloidal with respect to the rotation of the cam. Equations for the displacement, velocity, and acceleration of the follower of such cams are given in Table 12-6.

Table 12-6 EQUATIONS FOR DISPLACEMENT, VELOCITY, AND ACCELERATION FOR THE PITCH CURVE OF VARIOUS TYPES OF CAMS

Type of Cam	Displacement	Velocity	Acceleration
Parabolic or constant acceleration	For $\frac{\theta}{\beta} \leq 0.5$, $y = 2h\frac{\theta^2}{\beta^2}$ For $\frac{\theta}{\beta} \geq 0.5$, $y = h\left[1 - 2\left(1 - \frac{\theta}{\beta}\right)^2\right]$	$\frac{dy}{dt} = \frac{4h\omega\theta}{\beta^2}$ $\frac{dy}{dt} = \frac{4h\omega}{\beta}\left(1 - \frac{\theta}{\beta}\right)$	$\frac{d^2y}{dt^2} = \frac{4h\omega^2}{\beta^2}$ $\frac{d^2y}{dt^2} = -\frac{4h\omega^2}{\beta^2}$
Harmonic	$y = \frac{h}{2}\left(1 - \cos\frac{\pi\theta}{\beta}\right)$	$\frac{dy}{dt} = \frac{\pi h\omega}{2\beta}\sin\frac{\pi\theta}{\beta}$	$\frac{d^2y}{dt^2} = \frac{\pi^2 h\omega^2}{2\beta^2}\cos\frac{\pi\theta}{\beta}$
Cycloidal	$y = h\left(\frac{\theta}{\beta} - \frac{1}{2\pi}\sin\frac{2\pi\theta}{\beta}\right)$	$\frac{dy}{dt} = \frac{h\omega}{\beta}\left(1 - \cos\frac{2\pi\theta}{\beta}\right)$	$\frac{d^2y}{dt^2} = \frac{2\pi h\omega^2}{\beta^2}\sin\frac{2\pi\theta}{\beta}$

Lift h of the cam takes place during a rotation β. The rise y for a rotation θ can be found by substitution in the equations of the table.

Curves showing the rise y for the three types of cams, plotted for the nondimensional coordinate θ/β, are shown in Fig. 12-30. The velocity dy/dt is shown in Fig. 12-31, and the acceleration d^2y/dt^2 is shown in Fig. 12-32.

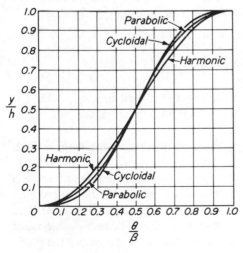

Figure 12-30 Displacement of follower for various types of cams.

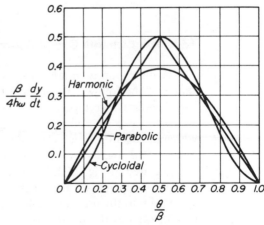

Figure 12-31 Velocity of follower for various types of cams.

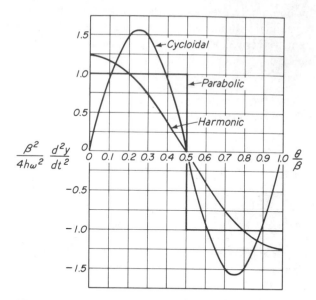

Figure 12-32 Acceleration of follower for various types of cams.

The inertia forces in a cam-driven system are proportional to the acceleration of the follower. The parabolic cam has the smallest maximum acceleration, but has the undesirable property of having sudden changes in value at the start, the midheight, and the end of the rise. The harmonic cam has a gradual change in the value of the acceleration except at the start and end of cam action. The cycloidal cam has a higher peak value for the acceleration than the others, but its value does not change as abruptly.

18. Plate Cam with Central Roller Follower

A plate cam with a central roller follower is shown in Fig. 12-33. The center of the follower roller has the path of the pitch curve, and the actual cam is everywhere tangent to the roller.

The force exerted by the cam is directed normal to the cam surface, which in general, for a roller follower, is not in the direction of the follower motion. The angle between these directions is known as the pressure angle α as illustrated in Fig. 12-33(b). The sidewise force exerted by the follower on its guides depends on the magnitude of the pressure angle. It is sometimes specified that the pressure angle should not exceed 30°. The permissible value is influenced by the speed of operation and the weight of the connected parts.

Reference to Fig. 12-33(b) indicates that the pressure angle α can be obtained by the following equation.

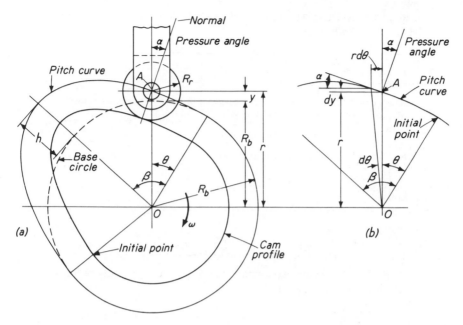

Figure 12-33 Radial cam with translating roller follower.

$$\tan \alpha = \frac{1}{r}\frac{dy}{d\theta} \tag{54}$$

where r is the distance between the centers for cam and roller.

Example 14. A cycloidal cam with a central roller follower has a rise h of 1 in. in an angle β of 70°. Base circle radius R_b is 4 in. and the follower roller radius R_r is 0.8 in.

Calculate the values of the pressure angle α at intervals of θ/β of 0.1. Find the maximum value of the acceleration of the follower. Speed of rotation is 600 rpm.

Solution. By Table 12-6, the equation for the rise y of the cycloidal cam is

$$y = h\left(\frac{\theta}{\beta} - \frac{1}{2\pi}\sin\frac{2\pi\theta}{\beta}\right)$$

To find the pressure angle α, the derivative $dy/d\theta$ is required for use in Eq. (54). Thus

$$\frac{dy}{d\theta} = \frac{h}{\beta}\left(1 - \cos\frac{2\pi\theta}{\beta}\right)$$

$$\beta = 70° = 1.22173 \text{ rad}$$

$$\frac{h}{\beta} = 0.81851$$

The calculations can be conveniently carried out in tabular form as shown in Table 12-7.

Table 12-7 CALCULATIONS FOR PRESSURE ANGLE α FOR CYCLOIDAL CAM OF EXAMPLE 14

$\dfrac{\theta}{\beta}$	$\dfrac{2\pi\theta}{\beta}$	$\sin\dfrac{2\pi\theta}{\beta}$	$\cos\dfrac{2\pi\theta}{\beta}$	$\dfrac{\sin(2\pi\theta/\beta)}{2\pi}$	y	$r = R_b + y$	$\dfrac{dy}{d\theta}$	$\tan\alpha = \dfrac{1}{r}\dfrac{dy}{d\theta}$	α
0	0	0	1	0	0	4	0	0	0
0.1	36°	0.58779	0.80902	0.09355	0.00645	4.00645	0.15632	0.03902	2° 14.1′
0.2	72°	0.95106	0.30902	0.15137	0.04863	4.04863	0.56558	0.13970	7° 57.2′
0.3	108°	0.95106	−0.30902	0.15137	0.14863	4.14863	1.07145	0.25827	14° 28.9′
0.4	144°	0.58779	−0.80902	0.09355	0.30645	4.30645	1.48070	0.34383	18° 58.5′
0.5	180°	0	−1	0	0.5	4.5	1.63702	0.36378	19° 59.4′
0.6	216°	−0.58779	−0.80902	−0.09355	0.69355	4.69355	1.48070	0.31548	17° 30.6′
0.7	252°	−0.95106	−0.30902	−0.15137	0.85137	4.85137	1.07145	0.22086	12° 27.2′
0.8	288°	−0.95106	0.30902	−0.15137	0.95137	4.95137	0.56558	0.11423	6° 31.0′
0.9	324°	−0.58779	0.80902	−0.09355	0.99355	4.99355	0.15632	0.03130	1° 47.6′
1.0	360°	0	1	0	1	5	0	0	0

$$\omega = \frac{2\pi n}{60} = \frac{2\pi 600}{60} = 20\pi \text{ rad/sec}$$

By Fig. 12-32, the maximum value of the follower acceleration will occur for θ/β equal to 0.25. Then by Table 12-6,

$$\left(\frac{d^2y}{dt^2}\right)_{max} = \frac{2\pi h\omega^2}{\beta^2} = \frac{2\pi \times 1 \times 400\pi^2}{1.22173^2} = 16{,}620 \text{ in./sec}^2$$

To obtain a smooth surface, the calculations for an actual cam would be made for much finer intervals—perhaps for a degree or less apart.

19. Manufacture of Cam with Central Roller Follower

One method for making a plate cam with a central roller follower is to mill or grind the surface on a milling machine. The radius R_g of the cutter or grinding wheel, in general, will not be the same as the radius R_r of the follower roller.

The situation will be as in Fig. 12-34 where the system of Fig. 12-33 is rotated to make the cam center O, and the cutter center B on the same vertical line. To find the distance r_g between the axes of cam and cutter, let the cosine theorem be applied to triangle OAB.

$$r_g^2 = r^2 + (R_g - R_r)^2 + 2r(R_g - R_r)\cos\alpha \tag{55}$$

When the cam is inclined at angle $\theta - \eta$ from the initial point, then the cutter should be brought into the cam until the distance between the axes at O and B is equal to r_g.

Angle η can be found by application of the sine theorem to triangle OAB. Thus

$$\sin\eta = \frac{R_g - R_r}{r_g}\sin\alpha \tag{56}$$

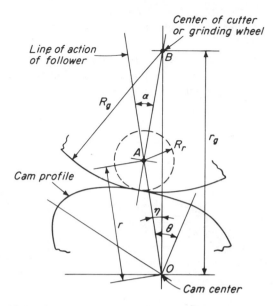

Figure 12-34 Geometry for locating center of cutter or grinding wheel for cam of Fig. 12-27.

Example 15. Calculate the values of r_g and η for the plate cam of Example 14 for the same θ/β values. Radius R_g of the cutter is equal to 2.5 in.

Solution. $R_g - R_r = 2.5 - 0.8 = 1.7$ in.

The calculations can be conveniently carried out in tabular form and are shown in Table 12-8.

Table 12-8 CALCULATIONS FOR LOCATION OF CUTTER CENTER FOR CYCLOIDAL CAM OF EXAMPLE 15

$\dfrac{\theta}{\beta}$	$\sin\alpha$	$\cos\alpha$	r^2	$2r(R_g - R_r) \times \cos\alpha$	r_g^2	r_g	$\sin\eta$	η
0	0	1	16	13.60000	32.49000	5.70000	0	0
0.1	0.03899	0.99924	16.05164	13.61158	32.55322	5.70554	0.01162	0°40.0′
0.2	0.13836	0.99038	16.39140	13.63292	32.91432	5.73710	0.04100	2°21.0′
0.3	0.25006	0.96823	17.21113	13.65722	33.75835	5.81019	0.07316	4°11.8′
0.4	0.32515	0.94566	18.54551	13.84629	35.28180	5.93985	0.09306	5°20.4′
0.5	0.34186	0.93975	20.25	14.37818	37.51818	6.12521	0.09488	5°26.6′
0.6	0.30086	0.95367	22.02941	15.21873	40.13814	6.33547	0.08073	4°37.8′
0.7	0.21566	0.97647	23.53579	16.10654	42.53233	6.52168	0.05622	3°13.4′
0.8	0.11350	0.99354	24.51606	16.72591	44.13197	6.64319	0.02904	1°39.8′
0.9	0.03129	0.99951	24.93554	16.96975	44.79529	6.69293	0.00795	0°27.3′
1.0	0	1	25	17.00000	44.89000	6.70000	0	0

20. Plate Cam with Flat-faced Follower

A plate cam with a flat-faced follower is shown in Fig. 12-35. The follower is located at B at the rise y given by the cam equation for the pitch curve. The actual cam outline contacts the follower at point C located a distance

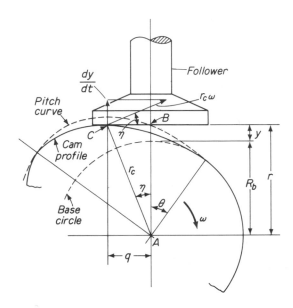

Figure 12-35 Plate cam with flat-faced follower.

q to the left of the center line. Point C is at radius r_c from center of rotation A, and at angle η from the center line. The upward velocity of the follower is represented by dy/dt. Then, by similar triangles,

$$\frac{dy/dt}{r_c\omega} = \frac{q}{r_c}$$

or

$$q = \frac{1}{\omega}\frac{dy}{dt} = \frac{dy}{d\theta} \tag{57}$$

Also,

$$r_c = (r^2 + q^2)^{1/2} \tag{58}$$

and

$$\tan \eta = \frac{q}{r} \tag{59}$$

Example 16. A cycloidal cam with a flat-faced follower has a rise h of 1 in. in an angle β of 70°. Radius R_b of the base circle is equal to 4 in. Calculate the pitch curve and coordinates r_c and η for the cam for values of θ/β of 0, 0.2, 0.4, 0.6, 0.8, and 1.0.

Solution. By Table 12-6 $y = h\left(\dfrac{\theta}{\beta} - \dfrac{1}{2\pi}\sin\dfrac{2\pi\theta}{\beta}\right)$

$$q = \frac{dy}{d\theta} = \frac{h}{\beta}\left(1 - \cos\frac{2\pi\theta}{\beta}\right) = \frac{1}{1.22173}\left(1 - \cos\frac{2\pi\theta}{\beta}\right)$$

The calculations are carried out in Table 12-9 below. The reader should plot the data for both the pitch curve and the cam outline.

Table 12-9 CALCULATIONS FOR PITCH CURVE AND CAM OUTLINE OF CYCLOIDAL CAM WITH FLAT-FACED FOLLOWER OF EXAMPLE 16

$\frac{\theta}{\beta}$	$2\pi\frac{\theta}{\beta}$	$\sin\frac{2\pi\theta}{\beta}$	$\cos\frac{2\pi\theta}{\beta}$	y	r^2	q	r_c	$\tan\eta = \frac{q}{r}$	η
0	0	0	1	0	16	0	4	0	0
0.2	72°	0.95106	0.30902	0.04863	16.39140	0.56558	4.0879	0.13970	7° 57.2′
0.4	144°	0.58779	−0.80902	0.30645	18.54551	1.48070	4.5539	0.34383	18° 58.5′
0.6	216°	−0.58779	−0.80902	0.69355	22.02941	1.48070	4.9216	0.31548	17° 30.5′
0.8	288°	−0.95106	0.30902	0.95137	24.51606	0.56558	4.9836	0.11423	6° 31.0′
1.0	360°	0	1	1	25	0	5	0	0

21. The Polydyne Cam

Many other types of curves are used for cams. The rise y, for example, can be expressed as a polynomial[13] in various powers of x or θ/β. The general equation for the so-called 3-4-5 cam would then be written

$$y = h(C_0 + C_1 x + C_2 x^2 + C_3 x^3 + C_4 x^4 + C_5 x^5) \qquad (60)$$

The various constants are evaluated from the desired boundary conditions for the follower. For example, suppose it is desired that displacement y, velocity dy/dx, and acceleration d^2y/dx^2 be zero for $x = 0$. Also let it be desired that rise y be equal to h, and dy/dx and d^2y/dx^2 be zero when $x = 1$. These initial conditions should be substituted into the above equation, and the equations obtained by differentiation, and the resulting equations solved for the C's. The result is

$$y = h(10x^3 - 15x^4 + 6x^5) \qquad (61)$$

The equation can be differentiated to obtain expressions for velocity dy/dt and acceleration d^2y/dt^2. When the results are plotted, the resulting curves are very close to those of the cycloidal cam.

22. Remarks on Cam Design

Cam design is a broad and involved subject, and only a few of the more obvious points have been touched upon in this chapter.[14] The object of the

[13] See Dudley, W. M., "New Methods in Cam Design," *SAE Quart. Trans.*, **2**, 1948, p. 19. See also Stoddard, D. A., "Polydyne Cam Design," *Machine Design*, **25**, Jan., 1953, p. 121; Feb., p. 146, and Mar., p. 149.

[14] For a complete and detailed treatment of the subject, see Rothbart, H. A., *Cams*, New York: John Wiley & Sons, 1956. This book contains extensive bibliographies. See also Mischke, C. R., "Optimal Offset on Translating Follower Plate Cams," *Trans. ASME* **92**, 1970, *Jour. of Engrg. for Industry*, p. 172.

cam is to have the follower deliver a prescribed motion in a smooth and vibrationless manner. Vibration may be instigated by a sudden break in the acceleration curve. It is beneficial if the acceleration curve be continuous and have the smallest maximum values. The rise should be kept as small as possible and achieved in the longest possible time. Acceleration varies as the square of the velocity.

The accuracy with which the cam is cut will influence the vibration effects. Play or backlash should be kept to a minimum. Vibration, together with noise and wear, will occur in the dynamical system if it is operated at the speed of its natural frequency. The forces and stresses from vibration are usually superposed on those resulting from normal operation. The cycloidal form is perhaps the best cam so far as vibration is concerned.

The operation is affected by the flexibility or elastic deformation of the parts of the system. The parts of the system act as springs of various stiffnesses. The moving parts should be both as rigid and as light as possible. Faulty operation can occur from the difference between the movement at the end of the mechanical chain and the initial movement imposed by the cam. The polydyne cam permits the design of the cam shape to be such that the desired motion will occur at the end of the follower chain.

An increase in the pressure angle usually means an increase in the forces which must be provided for. The guide for a translating follower should have adequate bearing length. Offsetting the follower may reduce the pressure angle during that portion of the cycle where the forces are the greatest. The mathematical calculations, however, are more involved. A pivoted oscillating follower will reduce the side thrust and permit a smaller cam to be used.

The spring of a compression system must be strong enough to keep the follower in contact with the cam at all times. This is important in flexible high-speed systems. The follower must not be allowed to jump or leave the cam surface. Sometimes a ramp or small precam is located at the start of the main part of the cam to remove the play and elastic deformations from the system so that the end of the chain will start to move as soon as the main part of the cam begins to act.

For light loads and slow speeds, a cam can be composed of sections of circular arcs. Such cams are relatively easy to manufacture and check dimensionally. The acceleration curve, however, has abrupt changes in value.

23. Snap Rings

Snap rings made of hardened steel have many useful applications in holding cylindrical parts in position. The external ring of Fig. 12-36(a) and the internal ring of sketch (b) have ears for expanding or contracting the ring during assembly. Special pliers are required for this operation. Sometimes the ring is merely made of hardened steel wire of rectangular or circular

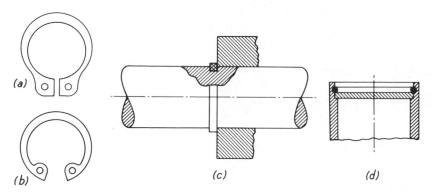

Figure 12-36 Snap rings and applications.

cross section. Sketch (c) shows how a ring can be used for holding a hub in the desired location along a shaft. In (d) the ring is used for retaining the closure at the end of a tube.

24. Flat Plates

Flat plates frequently occur in machines and structures, and the designer should be able to calculate the stresses and deflections produced by the loading. Unfortunately, the derivation of the necessary equations is very complex and beyond the scope of this book. Some results, for simple cases, will be set down here for ready reference.

(a) *Circular Plate, Uniform Load, Simply Supported Edge*
Deflection at center,

$$w = \frac{(5 + \mu)qa^4}{64(1 + \mu)D} \tag{62}$$

Moment at center,

$$M = \frac{(3 + \mu)qa^2}{16} \tag{63}$$

Maximum stress (at center),

$$s = \frac{3(3 + \mu)qa^2}{8h^2} \tag{64}$$

where a = radius of circular plate

$$D = \frac{Eh^3}{12(1 - \mu^2)}, \text{ flexural rigidity of plate} \tag{65}$$

E = modulus of elasticity, psi
h = thickness of plate, in.
M = bending moment, in. lb/in.

q = uniform load, psi

μ = (mu) Poisson's ratio (usually 0.3 for metals)

(b) *Circular Plate, Uniform Load, Clamped Edge*

Deflection at center, $w = \dfrac{qa^4}{64D}$ (66)

Moment at center, $M = \dfrac{(1 + \mu)qa^2}{16}$ (67)

Maximum stress (at edge), $s = \dfrac{3qa^2}{4h^2}$ (68)

(c) *Circular Plate, Concentrated Load at Center, Simply Supported Edge.* Deflection at center,

$$w = \frac{(3 + \mu)Pa^2}{16\pi(1 + \mu)D}$$ (69)

Stress at center, on lower surface,

$$s = \frac{P}{h^2}\left[(1 + \mu)\left(0.485 \ln \frac{a}{h} + 0.52\right) + 0.48\right]$$ (70)

where P = concentrated load

ln = logarithm to base e

(d) *Circular Plate, Concentrated Load at Center, Clamped Edge*

Deflection at center, $w = \dfrac{Pa^2}{16\pi D}$ (71)

Stress at center, on lower surface,

$$s = \frac{P}{h^2}(1 + \mu)\left(0.485 \ln \frac{a}{h} + 0.52\right)$$ (72)

Equations are available for numerous other types of loadings and edge conditions.[15]

(e) *Rectangular Plate, Uniform Load, Simply Supported Edges.* The corners of a uniformly loaded rectangular plate tend to rise if the plate is merely supported at the edges. For a simply supported edge, vertical movement must be prevented, but the edge must be free to rotate. An edge constructed as in Fig. 12-37 fulfills these conditions. The coordinate system for this plate is shown by Fig. 12-38. A good approximation for the stress and deflection can be had by the following empirical equations.

Maximum stress,

$$s_x = \frac{0.75qa^2}{h^2[1.61(a/b)^3 + 1]} \quad \text{at} \quad x = \frac{a}{2}, y = 0$$ (73)

[15]See References 2, 4, 5, and 6, end of chapter. See also, Wahl, A. M., and G. Lobo,. "Stresses and Deflections in Flat Circular Plates with Central Holes," *Trans ASME*, **52**(1), 1930, *J. Appl. Mech.*, p. 29, and Georgian, J. C., "Uniformly Loaded Circular Plates with a Central Hole and Both Edges Supported," *Trans. ASME*, **79**, 1957, *J. Appl. Mech.*, p. 306.

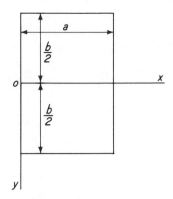

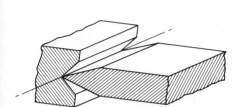

Figure 12-37 Simple support for plate.

Figure 12-38 Simply supported rectangular plate.

Deflection at center,

$$w = \frac{0.142qa^4}{Eh^3[2.21(a/b)^3 + 1]} \tag{74}$$

(f) *Rectangular Plate, Uniform Load, Clamped Edges.* The coordinate system for this plate is shown by Fig. 12-39.

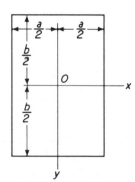

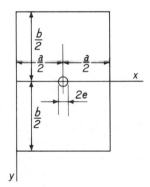

Figure 12-39 Rectangular plate-clamped edges.

Figure 12-40 Rectangular plate-simply-supported edges concentrated load at center.

Maximum stress,

$$s_x = \frac{qa^2}{2h^2[0.623(a/b)^6 + 1]} \qquad \text{at} \qquad x = \frac{a}{2}, y = 0 \tag{75}$$

Deflection at center,

$$w = \frac{0.0284qa^4}{Eh^3[1.056(a/b)^5 + 1]} \tag{76}$$

These two equations are also empirical, but give reasonably good results.

(g) *Rectangular Plate, Concentrated Load at Center, Simply Supported Edges.* Instead of a point load, the load P is assumed to be uniformly distributed over a small circle of radius e, as shown in Fig. 12-40.

Deflection at center, $w = \alpha \dfrac{Pa^2}{Eh^2}$ $\qquad\qquad$ (77)

Maximum stress, $s_x = \dfrac{1.5P}{\pi h^2}\left[(1 + \mu)\ln\dfrac{2a}{\pi e} + 1 - \gamma_1\right]$ $\qquad$ (78)

Constants α and γ_1 are given below for different values of b/a.

b/a	1.0	1.1	1.2	1.4	1.6	1.8	2.0	3.0	∞
α	0.1265	0.1381	0.1478	0.1621	0.1714	0.1769	0.1803	0.1846	0.1849
γ_1	0.564		0.349	0.211	0.124	0.072	0.041		0

(h) *Rectangular Plate, Concentrated Load at Center, Clamped Edges.* The coordinate system for this plate is given by Fig. 12-39.

Deflection at center, $w = \alpha \dfrac{Pa^2}{Eh^3}$ $\qquad\qquad$ (79)

Maximum stress, $s_x = \beta \dfrac{P}{h^2}$ at $x = \dfrac{a}{2}, y = 0$ $\qquad$ (80)

Constants α and β are given below for different values of b/a.

b/a	1.0	1.2	1.4	1.6	1.8	2.0
α	0.0611	0.0706	0.0755	0.0777	0.0786	0.0788
β	0.754	0.894	0.962	0.991	1.000	1.004

In general, the foregoing plate equations are valid only for small deflections—deflections which are only a fraction of the plate thickness. In cases where the deflection is not small compared to the thickness, but is small compared to the other dimensions, the strain of the middle surface of the plate must be considered and the theory becomes more involved. For still larger deflections, the bending effects are of less significance. The plate becomes a membrane and the load is carried largely by the tension stresses resulting from the stretching of the plate.

REFERENCES

1. Faupel, J. H., *Engineering Design*, New York: John Wiley & Sons, 1964.

2. Roark, R. J., *Formulas for Stress and Strain*, 3d ed., New York: McGraw-Hill Book Company, 1954.

3. Rothbart, H. A., ed., *Mechanical Design and Systems Handbook*, New York: McGraw-Hill Book Company, 1964.

4. Spotts, M. F., *Mechanical Design Analysis*, Englewood Cliffs, N.J.: Prentice-Hall, Inc., 1964.

5. Timoshenko, S., *Strength of Materials*, Vol. II, 2d ed., New York: Van Nostrand Reinhold Co., 1941.

6. Timoshenko, S., and Woinowsky-Krieger, S., *Theory of Plates and Shells*, 2d ed., New York: McGraw-Hill Book Company, 1959.

7. Rossheim, D. B., and Marke, A. R. C., "Gasket Loading Constants," *Mechanical Engineering*, **65**, 1943, p. 647; **66**, 1944, p 72.

PROBLEMS

1. (a) A steel disk is shrunk on a steel shaft. The radius of the disk is very large as compared to the radius of the shaft. The diametral interference is equal to 0.001 times the shaft diameter. Find the values of tangential and normal stress. $E = 30,000,000$ psi.

(b) Work part (a) but with $b = 2a$.

(c) Work part (a) but with the disk made of cast iron, $E = 15,000,000$ psi. The shaft is steel. $\mu = \frac{1}{3}$.

(d) Repeat part (c) except that $b = 2a$.

    Ans. (a) $s_t = s_r = 15,000$ psi: (b) $s_t = 18,750$ psi, $s_r = 11,250$ psi;

       (c) $s_t = s_r = $ 9,000 psi; (d) $s_t = 10,710$ psi, $s_r = 6,430$ psi.

2. A steel shaft is pressed on a steel disk twice the shaft diameter, and with a thickness $1\frac{1}{2}$ times the shaft diameter. The shearing stress in the shaft caused by the torque which the pressed fit can carry is equal to 12,000 psi. Find the value of the diametral interference. $E = 30,000,000$ psi. The coefficient of friction is 0.12.

    Ans. $\Delta = 0.00148a$.

3. A 2-in. shaft has a torsional shear stress of 12,000 psi. It has a shrink fit in an 8-in. steel disk 3 in. thick. Find the tangential stress and diametral interference if the entire torque of the shaft is to be transmitted through the shrink-fit friction. $E = 30,000,000$ psi. The coefficient of friction is 0.12.   *Ans.* $\Delta = 0.00118$ in.

4. A disk whose diameter is 3 times that of the shaft, and whose thickness is twice the shaft diameter, has a diametral interference of 0.0005 times the shaft diameter. For the shaft, $E = 30,000,000$ psi. For the disk, $E = 15,000,000$ psi. Poisson's ratio $= 0.25$. If the torsional stress in the shaft is equal to 10,000 psi, find the coefficient of friction required if the entire torque is to be carried by the shrink-fit friction.   *Ans.* Coefficient of friction $= 0.156$.

5. A disk of Class 30 cast iron is to be shrunk on a 6-in. steel shaft. The tangential stress in the disk is not to exceed 8,000 psi. Diametral interference of the metal is 0.004 in. $\mu = 0.3$. Find the shrink-fit pressure between disk and shaft. Find the minimum permissible outside diameter of the disk. Find the factor of safety.

    Ans. $p = 2,610$ psi; $b = 4.21$ in.; $FS = 3.44$.

6. A 10-in. diam steel shaft is to be pressed into a steel disk 18 in. in diameter and 8 in. thick. The interference of the metal must be such that the force required to press the parts together must lie between 144 and 200 tons. If the coefficient of friction is 0.15, find the maximum and minimum values for the diametral interference of metal.   *Ans.*   0.0074–0.0102 in.

7. A bronze bushing 2 in. OD and $1\frac{3}{8}$ in. ID is to be pressed into a hollow cylinder of 4 in. OD. The diametral interference of the metal is 0.002 in. Find the tangential and normal stresses for the steel and bronze at the boundary between the two parts.

Ans. $p = 4{,}320$ psi; bronze, $s_t = -12{,}050$ psi; steel, $s_t = 7{,}190$ psi.

8. A 4-in. diam steel shaft is pressed into a steel disk 12 in. in diameter and 5 in. thick. The shaft has a shearing stress equal to 8,000 psi with the torque resisted by the shrink fit. The coefficient of friction is 0.15. Find the diameter of the hole in the disk before assembly.   *Ans.*   3.9984 in.

9. A 3-in. diam steel shaft is to be press fitted into a Class 30 cast iron hub of 7 in. OD. Find the diametral interference of the metal if the *FS* against breakage of the hub from the press fit is to be 4.   *Ans.* $\Delta = 0.0019$ in.

10. A disk flywheel is to be cut from 3-in. rolled steel plate. The bore is 0.2 of the outside diameter. Maximum speed is 3,000 rpm with a maximum stress from rotation equal to 12,000 psi. Find the diameter of the flywheel and the amount of energy delivered for a 10 per cent drop in speed.

Ans.   Diam $= 28.24$ in.; $\Delta KE = 107{,}080$ ft lb.

11. A disk flywheel is cut from $2\frac{1}{2}$-in. steel plate. Outside diameter is 36 in., bore is 6 in. Rpm $= 3{,}000$. Find the value of maximum stress in the material. Find *KE* delivered for a 5 per cent drop in speed.

Ans. $s_t = 19{,}460$ psi; $\Delta KE = 121{,}100$ ft lb.

12. A rolled steel disk flywheel is 24 in. in outside diameter and 3 in. thick. It rotates on a 2.5-in. diam shaft. Normal speed is 2,400 rpm with a 20 per cent drop during the working cycle. Find the maximum stress in the material and the energy delivered per cycle.   *Ans.* $s_t = 5{,}510$ psi; $\Delta KE = 67{,}890$ ft lb.

13. A 42-in. outside diameter cast iron flywheel has 4 spokes and rotates 800 rpm. The rim is 4×4 in. The spoke is 2×3 in. Find the stress in rim and spokes.

Ans.   Rim, $s = 2{,}450$ psi; spoke, $s = 600$ psi.

14. The load cycle for each revolution of a shaft is shown in Fig. 12-41. The shaft is driven by a belt and can be assumed to receive a constant torque from the driving motor. Find the energy delivered per cycle for a flywheel mounted on the shaft for a coefficient of fluctuation of 0.05. The speed is 600 rpm. Find the outside diameter of the flywheel if it is solid and cut from a 1-in. steel plate.

Ans. $\Delta KE = 108$ ft lb; diam $= 17.38$ in.

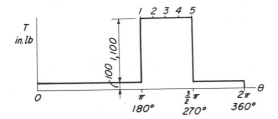

Figure 12-41 Problem 14.

15. The load cycle for a shaft is shown in Fig. 12-42. The shaft is driven by a belt and can be assumed to receive a constant driving torque. Find the energy delivered per cycle for a flywheel mounted on the shaft for a coefficient of fluctuation of 0.10. Speed is 150 rpm. Find the outside diameter if the flywheel is solid and cut from $2\frac{1}{2}$-in. steel plate.   *Ans.* $\Delta KE = 1,100$ ft lb; diam $= 41.5$ in.

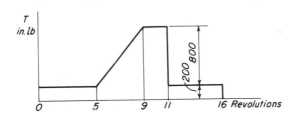

Figure 12-42 Problem 15.

16. Let the bolt in Fig. 12-13 be $\frac{3}{4}$ in. in diameter, and let length l be equal to 15 in. The threads at the bottom are 10 per inch, stopping immediately above the nut. Ignore the effect of the weight of the bolt. The bolt is made of steel.

(a) If a weight W of 70 lb drops 0.1 in. to the nut, find the stress in the gross and net sections of the bolt.

(b) Find the stress if the entire body of the bolt above the nut is reduced to the stress area of the threads.

(c) Find the stress if one-half the length of the bolt is reduced to the stress area.

(d) If bolt is made of phosphor bronze, and $\gamma = 0.32$ lb/in.^{3}, find the stress in the net section of the thread.

Ans.   (a) $s = 8,120$ psi; $s = 10,740$ psi; (b) $s = 9,370$ psi; (c) $s = 9,980$ psi; (d) $s = 7,660$ psi.

17. If the bolt in Fig. 12-13 is $\frac{1}{2} \times 18$ in., and weight W is 10 lb, find the height h which will cause an impact stress of 40,000 psi in the gross area of the bolt. $E = 30,000,000$ psi; $\gamma = 0.283$ lb/in.^{3} *Ans.* $h = 9.7$ in.

18. A 4-in. × 4-in. × 4-ft wooden post stands on a rigid foundation. From what height must a 500 lb weight be dropped on the upper end of the post to give an impact stress of 4,000 psi? Ignore the effect of the weight of the post. $E = 1,500,000$ psi.   *Ans.* $h = 8.06$ in.

19. In Fig. 12-43, the crane had been lowering the load at the uniform rate of 2 fps when the brakes on the drum were suddenly applied. Assuming that the upper end of the cable had been stopped instantaneously, find the additional stretch in the cable due to stopping. Let E for the twisted wire cable be 16,000,000 psi. $A = 0.5$ in.^{2} *Ans.* $y = 0.473$ in.

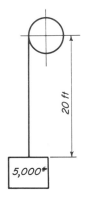

Figure 12-43 Problem 19.

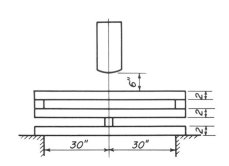

Figure 12-44 Problem 20.

20. The steel beams in Fig. 12-44 are 3 in. wide. Find the maximum stress due to impact. Weight = 1,000 lb.   *Ans.* $s = 62,780$ psi

21. What must be the rate of the spring, in pounds per inch, if the force caused by the falling weight in Fig. 12-45 is equal to 1,000 lb? Rod and spring are of brass.   *Ans.* $k_s = 557$ lb/in.

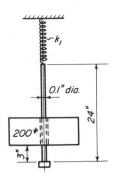

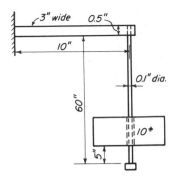

Figure 12-45 Problem 21.

Figure 12-46 Problem 22.

22. In Fig. 12-46 all parts are of steel. Find the stress in the bolt and in the beam due to impact.   *Ans.*   Bolt, $s = 51,560$ psi; beam, $s = 32,400$ psi.

23. The gasket in Fig. 12-47 is of cloth-inserted hard rubber. If the bolts are only tightened sufficiently to seat the gasket, find the permissible value of the internal pressure.   *Ans.* $p = 16.2$ psi.

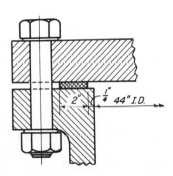

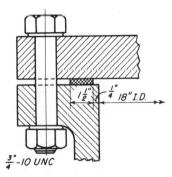

Figure 12-47 Problem 23.

Figure 12-48 Problem 24.

24. The asbestos composition gasket shown in Fig. 12-48 is $\frac{1}{8}$-in. thick. The cover is held in place by 24 bolts. If the bolts are tightened only sufficiently to seat the gasket, find the permissible value of the internal pressure. By use of Eq. (8), Chapter 5, find the force at the end of a 2-ft wrench handle required to tighten the bolts.   *Ans.* $p = 118.5$ psi; $F = 12$ lb.

25. The gasket shown in Fig. 12-49 is of soft copper $\frac{3}{16}$ in. thick, held in place by 24 bolts. If the bolts are only tightened sufficiently to seat the gasket, find the permissible value of the internal pressure. Find the torque required to tighten the bolts.   *Ans.* $p = 580$ psi; $T = 1,920$ in. lb.

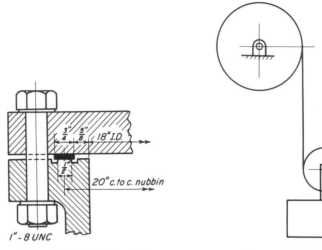

Figure 12-49 Problem 25.   **Figure 12-50** Problem 26.

26. The cable in Fig. 12-50 is 6×37 of plow steel wire. Its diameter is $1\frac{1}{4}$ in. The load is 20,000 lb and the sheave has a 24-in. pitch diameter.

    (a) Find expected life on the basis of 300 bends per week.

    (b) Find the *FS* on the basis of direct stress in the cable.

    (c) Find the resultant stress when the cable is passing around the sheave.

    (d) Find the minimum pitch diameter of the sheave and drum if fatigue effects are to be avoided.

                 Ans.   (a) 7.7 years; (b) $FS = 11$; (c) $s = 86,870$ psi, $d_s = 51$ in.

27. The cable in Fig. 12-50 is 6×19 and is 1 in. in diameter. The sheave is 24 in. in diameter. Resulting bending and direct stress in the cable when passing around the sheave is equal to 100,000 psi. Find the value of *W*.

                 Ans. $W = 17,680$ lb.

28. A 1-in. diam 6×19 wire rope is made of improved plow steel wire. The resultant bending and direct stress is 100,000 psi when passing around the sheaves. Expected life is 8 years at 400 bends per week and 50 weeks per year. Find the force in the rope and compute the minimum permissible sheave diameter.

                 Ans. $T = 7,630$ lb, $d_s = 23.1$ in.

29. Find the value of *m* in Fig. 12-51 that gives a resultant stress value of 10,000 psi on cross section *AB*.   *Ans.* $m = 11.8$ in.

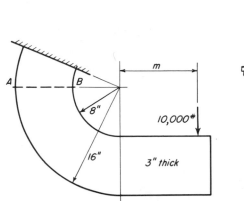

Figure 12-51 Problem 29.

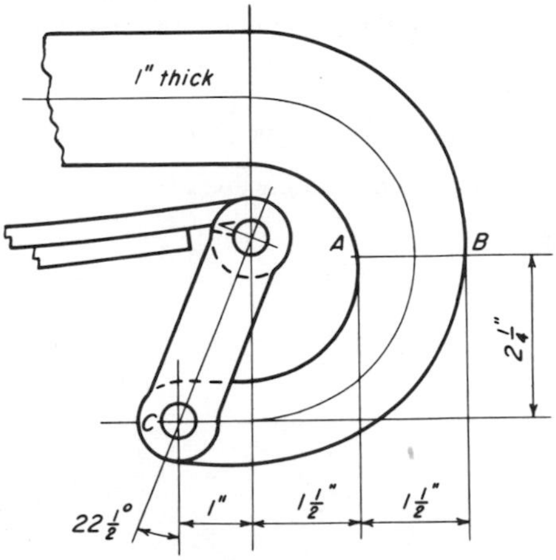

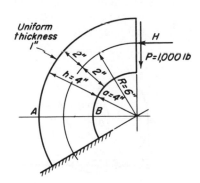

Figure 12-52 Problem 30.

30. Figure 12-52 shows the shackle at one end of a symmetrically arranged leaf spring. The spring supports a load of 4,000 lb at its center. Find the maximum normal stress on cross section AB. *Ans.* $s = -17,260$ psi.

31. Find the value of length m in Fig. 12-53 that causes the tensile stress at A to be numerically equal to the compressive stress at B. *Ans.* $m = 7.24$ in.

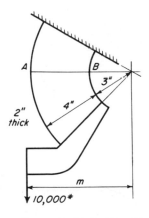

Figure 12-53 Problem 31.

Figure 12-54 Problem 32.

32. Find the value of force H if the resultant stresses normal to the horizontal cross section in Fig. 12–54 are to be numerically equal at points A and B. *Ans.* $H = 1,465$ lb.

33. (a) Starting with Eq. (48) for the maximum bending stress in a curved beam of rectangular cross section, add the effect of the direct stress, and arrive at an expression for the resultant normal stress for point B in Fig. 12-55 in terms of load P, area of cross section A, and ratio c_1.

(b) If $s = 7{,}500$ psi at B, $P = 5{,}000$ lb, $h = 4$ in., and thickness equals 2 in., the corresponding value of a.

$Ans.$ (a) $s = \dfrac{P}{A}\left[\dfrac{3c_1^2 - 1.8}{c_1 - 1}\right].$

(b) $a = 3.55$ in.

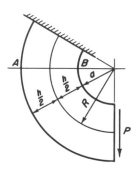

Figure 12-55 Problem 33.

34. What value of c_1 will make the bending stress at the inner edge of a curved beam of rectangular cross section loaded in pure bending 5 per cent higher than that in a straight beam of the same cross section? 10 per cent higher?

$Ans.$ $c_1 = 13.95$; $c_1 = 7.30$.

35. A crane hook has the loading and dimensions given in Fig. 12-56.

(a) Draw cross section AB to full-size scale and divide into 18 strips $\frac{1}{4}$ in. wide running at right angles to line AB.

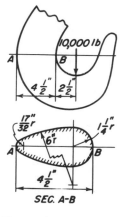

Figure 12-56 Problem 35.

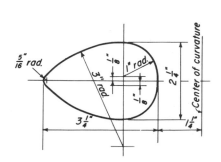

Figure 12-57 Problem 36.

(b) Locate the center of gravity by use of the integral $A\bar{x} = \int x\, dA$, where A is the area of the cross section, x is the distance from the center of gravity to one end, say point B, x is the distance from point B to the center of gravity of each strip, and dA is the area of the strip. This operation should be carried out in tabular form.

(c) Find the value of m_1 by use of Eq. (50), and the value of the maximum and minimum stresses.   *Ans.* $\bar{x} = 2.06$ in.; $s_{max} = 14{,}600$; $s_{min} = -6{,}200$ psi.

36. Perform the operations of Problem 35 for the cross section of Fig. 12-57. Divide into 13 quarter-inch strips. Load is 5,000 lb.
$$Ans. \quad \bar{x} = 1.46 \text{ in.}, s_{max} = 11{,}000 \text{ psi}, s_{min} = -3{,}750 \text{ psi.}$$

37. A harmonic cam has a central roller follower. $\beta = 120°$, $h = 1$ in., $R_b = 4$ in., $R_r = 0.625$ in., and $R_g = 2.5$ in. Calculate, in tabular form, values of α, r_g, and η for θ/β values of 0, 0.2, 0.4, 0.6, 0.8, and 1.0.
$$Ans. \quad \text{For } \theta/\beta = 0.4: \alpha = 9°19.3', r_g = 6.20316 \text{ in.}, \eta = 2°48.4'.$$

38. A cycloidal cam has a central roller follower. $\beta = 120°$, $h = 1$ in., $R_b = 4$ in., $R_r = 0.625$ in., and $R_g = 2.5$ in. Calculate, in tabular form, values of α, r_g, and η for θ/β values of 0, 0.2, 0.4, 0.6, 0.8, and 1.0.
$$Ans. \quad \text{For } \theta/\beta = 0.4: \alpha = 11°20.5', r_g = 6.1559 \text{ in.}, \eta = 3°26.0'.$$

39. A harmonic cam has a flat-faced follower. $\beta = 150°$, $h = 1$ in., $R_b = 1.5$ in. Calculate, in tabular form, values of r_c, q, and η for θ/β of 0, 0.2, 0.4, 0.6, 0.8, and 1.0.   *Ans.*   For $\theta/\beta = 0.6$: $r_c = 2.2288$ in., $q = 0.5706$ in., $\eta = 14°50.0'$.

40. A harmonic cam has a flat-faced follower. $\beta = 105°$, $h = 1$ in., $R_b = 1.5$ in. Calculate, in tabular form, values for r_c, q, and η at 15-degree intervals.
$$Ans. \quad \text{For } \theta = 60°: r_c = 2.2706 \text{ in.}, q = 0.8356 \text{ in.}, \eta = 21°35.6'.$$

41. A harmonic cam has a flat-faced follower. $\beta = 90°$, $h = 0.8$ in., $R_b = 1.6$ in. Calculate, in tabular form, values of r_c, q, and η for θ/β of 0, 0.2, 0.4, 0.6, 0.8, and 1.0.   *Ans.*   For $\theta/\beta = 0.4$: $r_c = 2.0248$ in., $q = 0.7608$ in., $\eta = 22°4.3'$.

42. Calculate and plot the values for y/h, $(dy/dx)/h$, and $(d^2y/dx^2)/h$ for the 3-4-5 ploynomial cam. Make calculations, in tabular form, for $x = \theta/\beta$ of 0, 0.2, 0.4, 0.6, 0.8, and 1.0.
$$Ans. \quad \text{For } \theta/\beta = 0.4: y/h = 0.3174, (dy/dx)/h = 1.728, (d^2y/dx^2)/h = 2.88.$$

43. Derive the constants in the displacement equation for the 3-4-5 polynomial cam for the initial conditions stated in Section 21.

44. A 2-in. diam steel shaft is to be press-fitted into a 7-in. diam $\times$ 3-in. thick steel disk. Diametral interference of the metal is to be 0.0018 in. Find the force required to assemble them if the coefficient of friction is 0.15.

45. A 2-in. diam steel shaft is to be press-fitted into a 6-in. diam $\times$ $2\frac{1}{2}$-in. thick steel disk, with a Class 7 fit. Find the maximum required assembly force for a coefficient of friction of 0.15. Assume the tightest permissible fit.

46. A 4-in. diam steel shaft is to be press-fitted into an 8-in. diam Class 50 cast iron hub. *FS* for hub is to be 5. Find the value of the radial and tangential stresses for the hub at the hole.

47. A steel shaft 3.5 in. in diameter is to be press-fitted into a Class 50 cast iron hub 8 in. in diameter. Find the required diametral interference of the metal if the hub has a *FS* of 4 from the press-fit stresses.

48. A hardened steel beam is 25 in. long with simple supports. An 80 lb weight drops 4 in. and strikes the beam at the center. Find the value of the bending stress for the following conditions. Ignore W_b in the calculations.

(a) The beam is $\frac{1}{2} \times 2$ in. set on edge.

(b) The beam is 1×1 in.

(c) The beam is $\frac{1}{2} \times 2$ in. laid on the flat.

49. Derive the equation for displacement for the 2-3 polynomial cam for the initial conditions: when $x = 0$: $y = 0$, and $dy/dx = 0$; and when $x = 1$: $y = h$, and $dy/dx = 0$.

13 Dimensioning and Details

The design of a machine includes many factors other than those of determining the loads and stresses and selecting the proper materials. Before construction or manufacture can begin, it is necessary to have complete assembly and detail drawings to convey all necessary information to the shop men. The designer usually does not make such drawings, but he frequently is called upon to supervise the draftsmen who do this detailing or to check the drawings before they are sent to the shop. Much experience and familiarity with manufacturing processes are needed before one can become conversant with all phases of production drawings.

Drawings should be carefully checked to see that the dimensioning is done in a manner that will be most convenient and understandable to the production departments. It is obvious that a drawing should be made in such a way that it has one and only one interpretation. In particular, shop personnel should not be required to make trigonometric or other involved calculations before the production machines can be set up.

Dimensioning is an involved subject and long experience is required for its mastery.

The connection between the English and metric systems is made by taking the inch as being exactly 2.54 cm.

515

1. Dimensioning

Tolerances must be placed on the dimensions of a drawing to limit the permissible variations in size because it is impossible to manufacture a part exactly to a given dimension. Although small tolerances give higher quality work and a better operating mechanism, the cost of manufacture increases rapidly as the tolerances are reduced, as indicated by the typical curve of Fig. 13-1. It is therefore important that the tolerances be specified at the largest values that the operating or functional considerations permit.

The development of production processes for large-volume manufacture at low cost has been largely dependent upon interchangeability of component parts. Thus the designer must determine both the proper tolerances for the individual parts, and the correct amount of clearance or interference to permit assembly with the connecting parts. The manner of placing tolerances on drawings depends somewhat on the kind of product or type of manufacturing process. If the tolerance on a dimension is not specifically stated, the drawing should contain a blanket note which gives the value of the tolerance for such dimensions. However, some concerns do not use blanket notes, on the supposition that if each dimension is considered individually, wider tolerances than those called for in the note could probably be specified. In any event it is very important that a drawing be free from ambiguities and be subject only to a single interpretation.

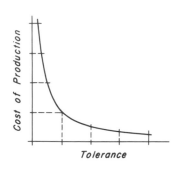

Figure 13-1 Relationship between size of tolerance and cost of production.

2. Redundant Dimensioning

In a given direction, a point should be located by one and only one dimension. Much confusion and expense arise from violation of this rule.[1]

For example, consider the horizontal dimension of the part shown in Fig. 13-2(a). For a part made as in sketch (b), lengths AB and AC are in accord with the drawing, but BC is not. Perhaps length BC is the important one for proper functioning, but a production man could argue that technically he had followed the drawing by making AB and AC correctly. Similarly, in sketch (c), lengths BC and AC are in accord with the drawing, but AB is not.

The difficulty can be corrected merely by omitting one of the dimensions in Fig. 13-2(a). The two dimensions that should be retained are determined by the functional requirements of the design.

[1]See p. 48 of Reference 2, end of chapter.

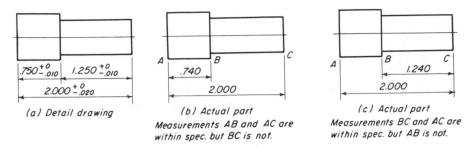

(a) Detail drawing

(b) Actual part
Measurements AB and AC are
within spec. but BC is not.

(c) Actual part
Measurements BC and AC are
within spec. but AB is not.

Figure 13-2 Example of incorrect production parts caused by redundant dimension.

3. Dimensioning of Clearance Fit—Maximum Material

The dimensions for a typical clearance fit for a hole and shaft are shown in Fig. 13-3. The tolerances shown here have been determined by the designer from consideration of the functioning of the assembly and the permissible cost of production. The bar diagram for this fit is shown in Fig. 13-4.

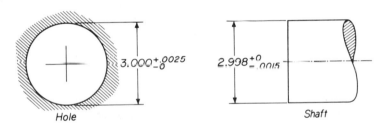

Figure 13-3 Maximum material tolerancing of clearance fit.

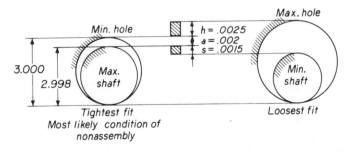

Figure 13-4 Exaggerated view of conditions of clearance fit of Fig. 13-3.

The hole dimension of 3.000 in. and the shaft dimension of 2.998 in. represent parts at the maximum material condition, *MMC*. An assembly of two parts with these dimensions will give the tightest fit, or the fit with the least possibility of assembly. In Fig. 13-3 one tolerance is zero, and the other tolerance gives parts which recede from the tightest assembly or the most dangerous condition. Clearance fits are usually dimensioned on the basis of *MMC*.

This type of dimensioning has another advantage. Should the workman aim at the principal dimensions, but through error produce an oversized hole or an undersized shaft, the parts might still be acceptable providing the dimensions do not exceed the limits specified by the drawings.

Figure 13-3 represents what would nominally be called a "3-in. shaft." It should be noted, however, that the 3 in. dimension is placed, not on the shaft, but on the dimension for the hole. When this is done, the dimensioning is said to be on the *basic hole system*. This method is advantageous because the manufacturing department can use standard-sized reamers for the holes, and then machine the shafts to fit.

On the other hand, the 3-in. dimension could just as well have been placed on the shaft, and the other dimensions adjusted to fit. When this is done, the dimensioning is known as the *basic shaft system*.

4. Dimensioning of Interference Fit—Minimum Material

Many joints are made as shrink or press fits, in which there is an interference of metal caused by the hole being smaller than the shaft. Here the most dangerous condition arises when the interference of metal is the least, giving a joint with the smallest holding ability. Such fits are frequently dimensioned on the basis of the least material condition, *LMC*, as shown by Fig. 13-5. Dimensions of 2.000 in. and 2.001 in. represent parts with the minimum

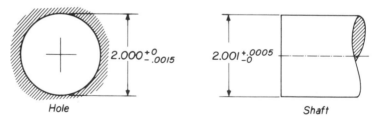

Hole   Shaft

Figure 13-5 Minimum material tolerancing of interference fit

amount of material. As indicated by Fig. 13-6, each dimension has one tolerance equal to zero with the other tolerance extending in the direction that will give tighter fit.

In minimum material dimensioning, should the workman aim at dimen-

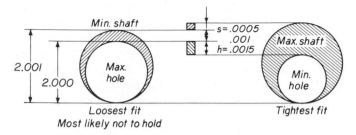

Figure 13-6 Exaggerated view of condition of interference fit of Fig. 13-5.

sions of 2.000 in. and 2.001 in., but by error produce an oversized hole or an undersized shaft, the parts will then not be in accord with the drawings.

5. Unilateral and Bilateral Tolerances

The tolerances in Fig. 13-3 and 13-5 are known as unilateral tolerances. One tolerance is zero, and all the variations of the dimensions are given by the other tolerance. This method has the advantage that a tolerance revision can be made with the least disturbance to the other dimensions. Suppose in Fig. 13-7 that a fit was originally dimensioned with the hole and shaft tolerances h_1 and s_1 and with an allowance or minimum clearance of a_1. Suppose experience with this fit indicates that larger tolerances can be accepted. These can be easily changed on the drawings to h_2 and s_2 without disturbing other dimensions already present.

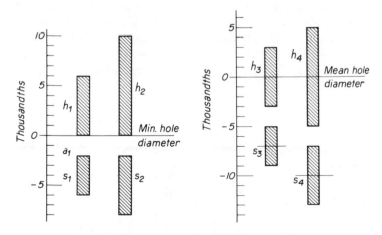

Figure 13-7 Unilateral toler-   **Figure 13-8** Bilateral tolerances
ances.

In bilateral dimensioning, a mean dimension is used which extends to the midpoint of the tolerance zone with equal plus and minus variations extending each way from this dimension, as in Fig. 13-8. It should be noted that a revision in the tolerance involves a change in at least one of the mean dimensions. Drawings made with unilateral tolerances are usually easier to check than those made with bilateral tolerances.

In some situations, the bilateral method of tolerancing is very appropriate. The locations of hole centers, for example, when the variation from the basic dimension is equally critical in both directions, are usually so dimensioned. For practical manufacturing purposes, welded assemblies may be dimensioned with bilateral tolerances. The same applies to loosely toleranced dimensions. If the tolerances are large, it is sometimes more convenient to give the mean dimension, and the variation each way as plus and minus values.

Sometimes, instead of showing a dimension with tolerances, the two extreme limits of the dimension are given on the drawing.

6. Selective Assembly

No difficulty arises in maintaining suitable tolerances if the fit is very loose. The same is true for very tight fits or fits having considerable interference. However, when the difference between allowance and maximum clearance must be small, the tolerances for hole and shaft may become excessively small, with consequent increase in the cost of production.

These costs may be minimized by selective assembly, which consists of first making each of the mating parts with large tolerances and then sorting them into groups having small tolerances. Suppose, for example, in Fig. 13-9(a), it is necessary that the clearance between a hole and shaft should vary from 0.0004 in. to 0.0010 in. The tolerances for each part, if divided equally, would thus be only 0.0003 in., and the cost of manufacture would be high. If selective assembly is used, the tolerances can be increased to larger values, as shown in Fig. 13-9(b). The bar diagram for this fit, given in Fig.

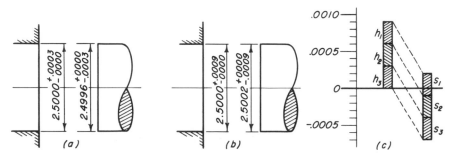

Figure 13-9 Dimensioning for selective assembly.

13-9(c), indicates that the parts are to be sorted into three groups, each with a variation in size of 0.0003 in., as follows:

	Holes			*Shafts*	
Group h_1	2.5006	$^{+0.0003}_{-0.0000}$	Group s_1	2.5002	$^{+0.0000}_{-0.0003}$
h_2	2.5003	$^{+0.0003}_{-0.0000}$	s_2	2.4999	$^{+0.0000}_{-0.0003}$
h_3	2.5000	$^{+0.0003}_{-0.0000}$	s_3	2.4996	$^{+0.0000}_{-0.0003}$

If the group of parts with the h_1 holes is now assembled with the s_1 shafts, the resulting fit will vary only by the desired amount of 0.0004 to 0.0010 in. The h_2 holes are assembled with the s_2 shafts, and so on. Selective assembly is thus a systematic process and is not made through mere trial and error.

Selective assembly gives a closely controlled fit, even with relatively large shop tolerances. The method, however, has a number of disadvantages. The sorting operation and the requirements for additional gages increase production costs. Universal interchangeability is no longer maintained, servicing is more complicated, and a whole subassembly may have to be replaced rather than a single part. After sorting, unless the parts are made in large quantities, the corresponding groups of components, which must be assembled together, may not contain equal numbers of parts. However, selective assembly is advantageous for interference fits where small changes in size may either cause the fit to become too loose to hold, or so tight that the press-fit stresses are excessive.

7. Standardized Cylindrical Fits

Fits between cylindrical bodies and holes have been standardized into eight classes by the American Standards Association. When the class number is specified, the hole tolerance, shaft tolerance, and allowance or average interference can be determined from the diameter by means of the formulas given in Fig. 13-10. Classes 1 to 4 are clearance fits, and Classes 7 and 8 are interference fits. However, Classes 5 and 6 may be either clearance or interference fits, depending on the random assembly of the parts. They are accordingly called transitional fits.

As shown by the bar diagrams, Class 1 is the loosest fit and has the largest tolerances on the parts. It is used for fits where accuracy is not essential. The allowance and tolerances for Class 2 are also liberal. Class 3 is used for sliding fits and for the more accurate machine tool and automotive parts. Considerable precision is required in making parts for Class 4 fits. Selective assembly is usually practiced with Class 5 fts. Class 6 is suitable for drive fits for gears, pulleys, rocker arms, and so forth. It is also used for drive fits in thin sections or for extremely long fits in other sections. It is suitable for automotive, ordnance, and general machine manufacturing. Considerable pressure is

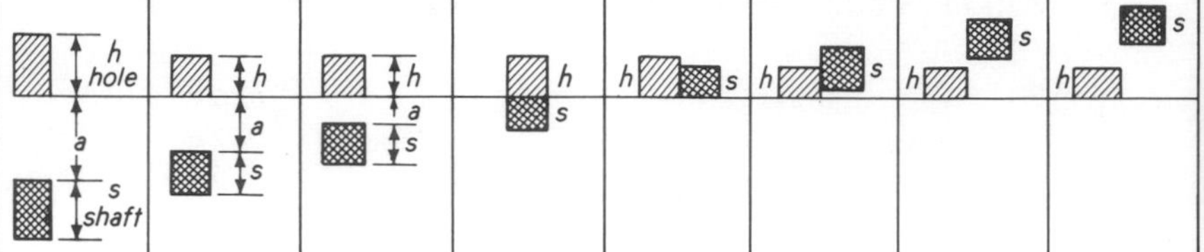

Class of fit	1	2	3	4	5	6	7	8
Bar diagram (basic hole) system	Loose fit	Free fit	Medium fit	Snug fit	Wringing fit	Tight fit	Medium force fit	Heavy force and shrink fit
Hole tolerance, h	$.0025\sqrt[3]{d}$	$.0013\sqrt[3]{d}$	$.0008\sqrt[3]{d}$	$.0006\sqrt[3]{d}$	$.0006\sqrt[3]{d}$	$.0006\sqrt[3]{d}$	$.0006\sqrt[3]{d}$	$.0006\sqrt[3]{d}$
Shaft tolerance, s	$.0025\sqrt[3]{d}$	$.0013\sqrt[3]{d}$	$.0008\sqrt[3]{d}$	$.0004\sqrt[3]{d}$	$.0004\sqrt[3]{d}$	$.0006\sqrt[3]{d}$	$.0006\sqrt[3]{d}$	$.0006\sqrt[3]{d}$
Allowance, a	$.0025\sqrt[3]{d^2}$	$.0014\sqrt[3]{d^2}$	$.0009\sqrt[3]{d^2}$	0				
Av. interference					0	$.00025\,d$	$.0005\,d$	$.001\,d$

Figure 13-10 ASA B4a-1925 classification of cylindrical fits. d = diameter.

required for the assembly of parts for Class 7. This fit is used for fastening the shafts to locomotive and railroad car wheels. It is also used for crank disks and for the armatures of motors and generators. It is the tightest fit that is recommended when the part with the hole is of cast iron. Class 8 is suitable for heavy force or shrink fits when the part with the hole is made of steel.

The fits shown in Fig. 13-10 have long since been superceded by larger and more flexible systems which permit a far wider variety of fits to be specified. This old standard, however, does serve a useful purpose in helping form a mental picture as to how a cylindrical fit can vary from a very loose clearance fit to one where heavy pressure is required for assembly.

Table 13-1 PREFERRED BASIC SIZES, INCHES

	0.0100		0.0500
	0.0125	1/16,	0.0625
1/64,	0.015625		0.0800
	0.0200	3/32,	0.09375
	0.0250		0.1000
1/32,	0.03125	1/8,	0.1250
	0.0400	5/32,	0.15625

3/16 to 3/4 by 1/16ths
7/8 to 3 by 1/8ths
3-1/4 to 4 by 1/4ths

Table 13-2 RECOMMENDED TOLERANCES AND ALLOWANCES, INCHES

0.0001	0.0012	0.0100*
0.00015	0.0015	0.0120
0.0002*	0.0020*	0.0150
0.00025	0.0025	0.0200*
0.0003	0.0030	0.0250
0.0004	0.0040	0.0300*
0.0005*	0.0050*	(Preferred
0.0006	0.0060	values are
0.0008	0.0080	indicated
0.0010*		by asterisks.)

In order to reduce the expense of tools and gages, cylindrical sizes called for on drawings should be kept as few as possible. Preferred basic sizes are given in Table 13-1. The number of tolerances that may be used with each diameter should also be kept to a minimum to save the expense of an excessive number of gages. Recommended values for tolerances and allowances are given in Table 13-2.

8. Eccentricity and Out-of-Roundness

Coaxial cylindrical parts must have a tolerance on the permissible deviations of the axes of the cylinders from each other. Inspection of such parts is frequently done by rotating the part in a V-block with an indicator resting on the other cylinder. As shown in Fig. 13-11, the total indicator reading, *TIR*, is double the displacement of the axes of the cylinders from each other.

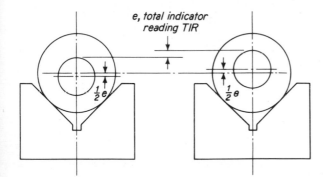

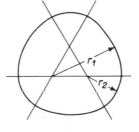

Figure 13-11 Total indicator reading TIR is double the diametrical eccentricity $\frac{1}{2}e$.

Figure 13-12 Lobed outline has same diametrical reading for all directions.

It is customary to check the roundness of a cylinder by measuring the diameter in several angular positions. It is possible, however, to obtain identical diametral readings and yet for the body to be out-of-round. This is illustrated by the part in Fig. 13-12. A micrometer will give the same reading in all directions and yet the part is not round. Shafting produced by centerless grinding machines sometimes has a shape like this due to a vibration in the machine. The error is uncovered by rotating the part in a V-block with an indicator.

9. Cumulative and Noncumulative Tolerances

Unnecessarily small tolerances sometimes result from the use of cumulative or compound tolerances. For example, the functional requirements of a part may be such as to permit a tolerance of 0.004 in. from the datum plane for each of the vertical surfaces in Fig. 13-13(b). When each surface is dimensioned individually from the datum, full advantage is taken of the permissible variation in size. But if the dimensioning is done chain fashion from surface to surface, as shown in Fig. 13-13(a), the variation of 0.004 in. in the over-all length must be divided among the four dimensions, and the permissible tolerance is reduced to 0.001 in., with consequent increase in manufacturing difficulties. Thus, in general, cumulative or compound tolerances should be avoided.

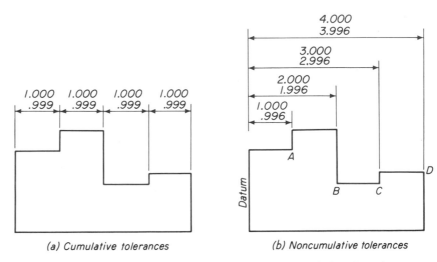

(a) Cumulative tolerances   (b) Noncumulative tolerances

Figure 13-13 Increase in tolerances obtained by noncumulative tolerancing.

However, suppose that the relationship between certain features must be held within small limits between themselves, but that a relatively large variation in their location from the datum plane is permissible. In such cases, the introduction of a compound tolerance may be preferable to noncumulative dimensioning.

10. Dimensioning of Hole Centers

The cylindrical surface of a hole can be located by means of dimensions for the center and for the hole diameter. When these dimensions have tolerances, a cumulative or compound tolerance results.

When the hole center is located by means of right-angle dimensions with tolerances, as shown in Fig. 13-14(a), the tolerance zone for the hole center consists of a small square, which, for this example, is 0.002 in. on the side. Note that the tolerance zones for the centers in Figs. 13-14(b) and (c) are not completely specified, although shop workmen would probably interpret the

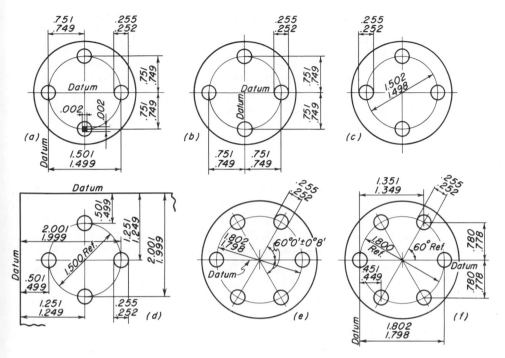

Figure 13-14 Various methods of dimensioning hole centers.

drawing as implying that the same tolerance is intended in both directions. When holes are to be made on a jig borer, it is customary to select data on the left and top sides of the part, as shown in Fig. 13-14(d), and run all dimensions to the right and downward. Tolerance zones may also be completely defined by tolerances on the diameter of the hole circle and on the angular dimensions, as shown in Fig. 13-14(e). For sheet-metal work or once-only job shops, this method is very convenient. In many cases, the horizontal and vertical coordinates are used for locating the hole centers, as illustrated by Fig. 13-14(f).

11. True Position Dimensioning

Other methods for the dimensioning of holes are employed which avoid the use of compound tolerances. These usually use untoleranced dimensions to locate the true geometrical center, and then permit the center to lie anywhere within a circular tolerance zone of specified diameter. An example is shown in Fig. 13-15. The word "BASIC" or "BSC" is added to the untoleranced dimensions which locate the hole centers. Sometimes a basic dimension is indicated on a drawing by enclosing it in a rectangular box.

A basic dimension represents a theoretical value used to describe the

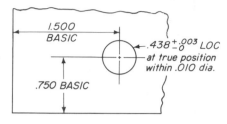

Figure 13-15 True position method for locating hole center.

exact size, shape, or location of a feature. It is used as a basis from which variations, usually by notes, are given of the permissible tolerances. This method permits the use of chain dimensioning without involving the accumulation of tolerances.

A reference dimension, marked "REF.," is an untoleranced dimension used on a drawing for informational purposes only, and does not govern machining or inspection operations.

12. Datum and Functional Surfaces

The surfaces of a part perform various duties in fulfilling the objective of the design. The surfaces to which the various features of a part are located by means of dimensions are called the datum planes. The functional surfaces are those operating or positioning surfaces that control the operation or location of the mechanism. Clearance surfaces provide continuity in the part, but have no functioning characteristics. Atmospheric surfaces have no contact relationship to the other surfaces of the part. Examples would be the outside of castings and exposed bolt heads.

The datum surfaces for a part, in general, consist of three mutually perpendicular planes, as shown in Fig. 13-16. For castings or other parts with

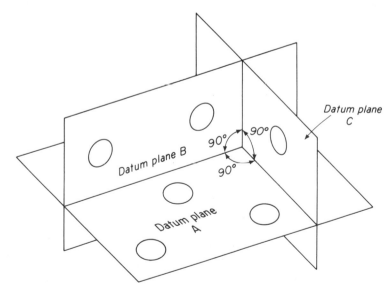

Figure 13-16 Datum planes and locating points required for supporting a body in three dimensions.

irregular surfaces, tooling and measuring (as well as dimensioning,) should be done from small locating surfaces. As indicated in Fig. 13-16, six such points are required for locating a body in space. Three points are required for locating with respect to the first plane, two points for locating with respect to the second plane, and one point for the third plane. In order to facilitate manufacture and inspection, it is important that the datum surfaces pass through actual physical features on the part and not be merely imaginary planes in space.

For example, it is both natural and convenient in the T-slide of Fig. 13-17,

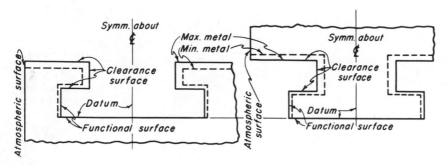

Figure 13-17 Tolerance zones for *T*-slide.

to run the horizontal dimensions from the vertical center line. Manufacture and inspection, however, may be difficult if this imaginary plane is devoid of all physical features.

One of the vertical functional surfaces could of course be used for the datum instead of the imaginary plane through the center. In general, however, it is found that once a datum is chosen and the tolerance zones established in relationship to it, it is not possible to shift the dimensioning to a new datum and preserve the same tolerance zones as originally planned.

13. Geometric and Positional Tolerances

In addition to size, detail drawings must provide for such qualities as flatness, squareness, concentricity, parallelism, and so forth. The control of such properties presents the most troublesome phase of proper dimensioning procedure.[2] As a rule, such information cannot be conveyed by dimensions,

[2]See Spalding, F. L., "How and When to Specify Tolerances of Form," *Machine Design*, **30**, May 15, 1958, p. 104, and Pohs, H. A., "Geometric and Positional Tolerance Symbol System," *Machine Design*, **36**, Jan. 30, 1964, p. 102; also July 2, 1964, p. 124. See also Reference 5, end of chapter.

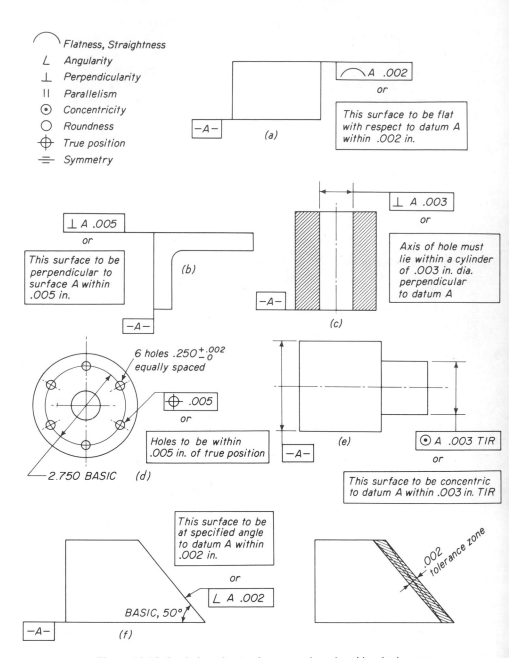

Figure 13-18 Symbols and notes for geometric and positional tolerances.

but must be handled by standardized symbols or suitably worded notes. Care and attention must be exercised that everyone who will use the drawing will arrive at the same interpretation of it.

Standardized symbols and examples of their use, as well as equivalent notes, are shown in Fig. 13-18. Although the notes are more bulky, the meaning sometimes can be made clearer than when symbols are used.

Dimensions with tolerance for size are also placed on the drawing. The tolerance provides a zone within which the finished part must lie. The combination of positional and size tolerances must frequently be interpreted. For example, in Fig. 13-18(a), let the vertical dimension for the distance between the upper and lower planes have a tolerance of 0.010 in. According to the drawing, the irregularities of the top surface must be confined between two planes a distance 0.002 in. apart. As drawings are usually interpreted, this 0.002 in. region can be permitted to lie anywhere within the 0.010 in. tolerance zone, and can be either parallel or not parallel to the datum if nothing is said about parallelism.

If the size tolerance is given and nothing is said on the drawing about form tolerances, it is customary to interpret the drawing as permitting parts of any shape to be acceptable so long as they lie completely within the given tolerance zone.

Datum surfaces are not perfect, but an effort should be made to make the irregularities of the datum smaller in value than those of the surfaces dimensioned from them.

14. Dimensioning of Tapers

Tapers or conical surfaces must be carefully dimensioned to avoid ambiguities and conflicts of dimensions that might permit the drawing to be interpreted in more than one way. In general, a taper can be completely specified by means of three dimensions, covering location, diameter, and amount of taper. Several examples are illustrated in Fig. 13-19. Strictly speaking, no tolerance on the taper or angle in sketches (a) and (b) would be

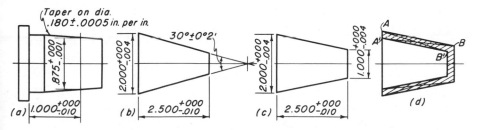

Figure 13-19 Various methods of dimensioning tapers.

required in laying out either the maximum metal condition or the minimum metal condition. A compound tolerance is introduced by use of a tolerance on these taper and angle dimensions.

In sketch (c) dimensions for both large and small diameters and the length are given. The tolerance zone specified by these three dimensions is represented by Fig. 13-19(d). Since no tolerance for the taper is given in sketch (c), the interpretation can be placed on the drawing that any part falling within the tolerance zone will be acceptable. The angle for the taper could then vary from that represented by line AB' to line $A'B$. Thus the tolerances on the diameters can cause a considerable variation in the angle of the taper. This variation must be checked carefully by the designer to make certain that it will be acceptable.

When dimensions are made to corners, as in Figs. 13-19(b) and (c), it is understood that the theoretically sharp corner is referred to. Actual measurements on a part must be made by means of rolls or by a comparator, since a theoretically sharp corner does not exist. When a fourth dimension is placed on a drawing for a taper, ambiguities immediately arise. For example, suppose an additional dimension for the small diameter were given in Fig. 13-19(b). Examination indicates that a variety of interpretations could then be placed on the drawing and much confusion arise.

15. Manufacturing and Gage Tolerances

The accuracy of the different machining operations varies considerably. In addition, accuracy is usually more difficult to achieve on large parts than on small parts. Figure 13-20 may be taken as a suitable guide of expected accuracies if actual experience on the part in question is lacking.[3] These curves are plotted as multiples of i, where

$$i = \sqrt{20d} + 1 \tag{1}$$

and where d is the diameter or length.

Tolerances are also necessary for the manufacture of gages, and may be from 5 per cent to 10 per cent of the tolerances on the work. Thus, if unnecessarily small work tolerances are specified, gage costs will be very high. Figure 13-21 indicates tolerances for gages of different qualities or accuracies. The designer tries to specify tolerances on the work which will permit manufacture and inspection by gages of "good" quality rather than by the more expensive "high" quality gages. For example, if a 2-in. hole has a tolerance of 0.003 in. and the gage tolerance is 10 per cent of that of the hole, Fig. 13-21 indicates that a "good" quality gage is sufficient. Should the hole tolerance be 0.0015 in., a "high" quality gage must be used.

[3]See Gladman, C. A., "Drawing Office Practice in Relation to Interchangeable Components," *Proc. Inst. Mech. Engrs.* (London), **152**, 1945, p. 388.

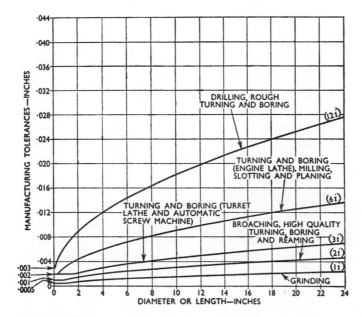

Figure 13-20 Average relationship between manufacturing tolerance and size of part for various kinds of machining operations.

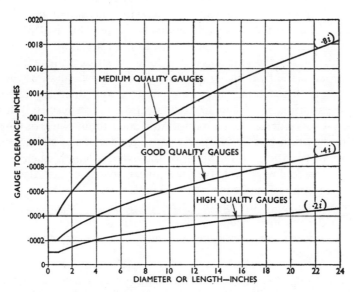

Figure 13-21 Average relationship between gage tolerance and size of part for gages of different qualities.

16. Production Process in Statistical Control

A variation in length or other quality can occur from two general types of causes. A variation can occur as a result of a large number of chance causes. These occur at random and are inherent in the entire production process and method of measurement.

The other type of variation does not occur at random, and is due to an assignable cause. When all the variations due to assignable causes are located one by one and corrected, the desired state of control is approached, and the system is then said to be in statistical control, or is called a constant cause system.

When a process is in statistical control, and measurements are made on the product, it is found that the number of parts with small variations are more numerous than the parts with large variations. When a plot is made of the frequency of occurrence vs. variation or error for a large number of parts, a smooth curve, known as the normal frequency curve, is obtained.[4]

Although unilateral tolerances have advantages for production drawings, it is usually more convenient to rewrite the dimensions with bilateral tolerances for consideration from the standpoint of statistics. The mean or average dimension is usually designated by placing a bar on top of the symbol.

Application of the normal frequency curve to the variations in the length of a part is illustrated by Fig. 13-22. The area under the curve represents the total production. The curve is asymptotic to the plus and minus horizontal or error axis. The so-called "natural tolerance of a process" is that spread of the curve that includes 99.73 per cent of the total production.

The drawing for the part specifies a total variation in length of $2u$. If the part can be made by a highly accurate production process, as shown in Fig. 13-22(d), with a natural tolerance equal to about three-quarters of the permissible variation, ideal and trouble-free production results. When the process is not so accurate, with a natural tolerance about equal to the specifications as shown in Fig. 13-22(e), the entire production is still acceptable. However small variations in the value of the average length, such as might be occasioned by tool wear, will give parts longer or shorter than the required value. Sketch (f) shows the frequency curve for a process whose natural tolerance is greater than specification $2u$. Some few parts are longer and some are shorter than the specified values, as shown by the areas in the tails of the curve. One hundred per cent inspection may be necessary to locate such unacceptable parts.

[4]See Rice, W. B., "Setting Tolerances Scientifically," *Mech Eng.*, **66**, 1944, p. 801; Burr, I. W., "Some Theoretical and Practical Aspects of Tolerances for Mating Parts," *Ind. Quality Control*, **15**, Sept. 1958, p. 18; Acton, F. S., and E. G. Olds, "Tolerances— Additive or Pythagorean," *Ind. Quality Control*, **5**, Nov. 1948, p. 6.

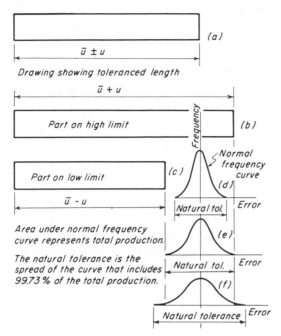

Figure 13-22 Distribution of lengths when process is in statistical control.

17. Dimensioning of Assemblies

The assembly of a number of parts into a final product is shown schematically in Fig. 13-23. If parts 1, 2, 3, and 4 have the dimensions shown, the mean value $\bar{w}$ of the resultant dimension is $-\bar{a} + \bar{b} + \bar{c} - \bar{d}$. The maximum value of the opening occurs when 1 and 4 are at the low limit, and 2 and 3 are at the high limit. Hence

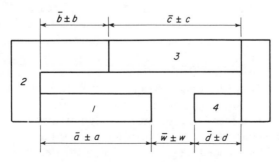

Figure 13-23 Tolerances for assembly.

$$\bar{w} + w = -(\bar{a} - a) + (\bar{b} + b) + (\bar{c} + c) - (\bar{d} - d) \qquad \text{(a)}$$

The minimum value occurs when 1 and 4 are at the high limit and 2 and 3 are at the low limit. Hence

$$\bar{w} - w = -(\bar{a} + a) + (\bar{b} - b) + (\bar{c} - c) - (\bar{d} + d) \qquad \text{(b)}$$

From the two equations above, tolerance w is found to be

$$w = a + b + c + d \qquad \text{(2)}$$

This equation assumes that the tolerances for an assembly are arithmetic or additive.

The simultaneous occurrence of all parts at the extreme limits assumed by Eqs. (a) and (b) is exceedingly rare. Suppose the dimensions of the different members of an assembly are in no way dependent on each other. In addition, let the lengths of the production parts be distributed normally, with the mean at the center of the tolerance zone. Also, let it be assumed that the ratio of the natural tolerances of the production process to the tolerances of the drawing be constant for all members. Then the following equation may give a more realistic value for the tolerance w of the assembly.

$$w = \sqrt{a^2 + b^2 + c^2 + d^2 + } \cdots \qquad \text{(3)}$$

Example 1. In Fig. 13-23 let $a = b = c = d = 0.002$ in. Find variation w of the assembly by the additive and Pythagorean rules.

Solution. By additive rule: $w = 4 \times 0.002 = 0.008$ in.

By Pythagorean rule: $w = \sqrt{4 \times 0.002^2} = 0.004$ in.

When the above conditions are not wholly fulfilled, the resulting tolerance w will probably lie intermediate to the values given by Eqs. (2) and (3).

18. Preferred Numbers

The factor of size or capacity must be considered in all manufactured articles or items of commerce. The term "size" may refer to linear dimensions, volume, weight, speed, power, or other features of the product. When an item is produced in a number of sizes, it is logical that the progression of sizes should be geometric rather than arithmetic. Each size is then larger than the preceding by a fixed percentage. Small sizes will then differ from each other by small amounts, and large sizes by larger amounts.

It is important from an economic standpoint that the number of sizes in a series be standardized at the smallest number that permit the product to take care of its intended range. The value of the step from one size to the next will of course depend upon the product under consideration. The system of *preferred numbers* has been devised in order that the various sizes of a series can be determined in an orderly fashion.[5]

In the range 10–100, the 5-series has five numbers, each of which is approximately 60 per cent greater than the preceding. The 10-series gives 10 numbers approximately 25 per cent apart, and the 20-series gives 20 numbers approximately 12 per cent apart. These numbers are shown in Table 13-3.

Table 13-3 BASIC PREFERRED NUMBERS, DECIMAL SYSTEM, 10–100

5-Series, 60 Per Cent Steps	10-Series, 25 Per Cent Steps	20-Series, 12 Per Cent Steps
10	10	10
		11.2
	12.5	12.5
		14
16	16	16
		18
	20	20
		22.4
25	25	25
		28
	31.5	31.5
		35.5
40	40	40
		45
	50	50
		56
63	63	63
		71
	80	80
		90

Preferred numbers below 10 are formed by dividing the numbers between 10 and 100 by 10. If still smaller numbers are desired, division is by 100. Preferred numbers above 100 are formed by multiplying the numbers between 10 and 100 by 10 or by 100. Tables for 40-series and 80-series numbers are available when the steps must be closer together. Tables using fractional numbers rather than decimal have also been constructed.

[5] See "Preferred Numbers," ASA Z 17.1-1936. New York: American Standards Association.

19. Surface Roughness

The properties and performance of machine parts may be affected in many ways by the degree of roughness of the various surfaces.[6] Among these effects, the following may be mentioned.

(1) The friction and wear between unlubricated surfaces are increased with increase of roughness.

(2) Friction and danger of seizure in lubricated surfaces are reduced for smooth surfaces.

(3) Interference fits made on rough surfaces may have a reduced area of contact and a consequent reduction in the holding ability of the joint.

(4) Fatigue strength of parts is increased as the surfaces become more smooth.

(5) Fluid flow in small passageways is reduced as the roughness of the walls is increased.

A demand has gradually grown up for a more exact method of specifying roughness than simply by marking the drawing "finish," "grind," or "lap." The roughness of a surface can be specified by calling for the height of the irregularities above and below the mean plane.

A true profile section, magnified equally in the horizontal and vertical directions, is shown in Fig. 13-24(a). The condition of the surface can be studied in more detail by exaggerating the roughness effect. Figure 13-24(b) shows the same profile when an additional magnification of 25 times has been given to the vertical direction with no change in the horizontal direction. Waviness also occurs in surfaces, as shown in Fig. 13-24(c). No exact differentiating point exists between roughness and waviness, although the pitch of the undulations must be about 0.040 in. or more before they are spoken of as waves. Specimens can be prepared for viewing under the microscope by a process called taper sectioning, in which the surface is ground at a small angle, thus accentuating the heights of the irregularities. An angle of 2°17' gives a magnification of 25. It is necessary to support the surface with a coating of nickel plating to prevent smearing of the contour during the sectioning process. A surface that has been prepared in this way[7] is illustrated in Fig. 13-25.

Heights above and below the mean surface are sometimes expressed in microinches or millionths of an inch, 0.000001 in., in terms of a special kind of average distance called the root-mean-square height, h_{rms}. This height is found by taking the ordinates at equidistant points on the profile, squaring, adding, dividing by the number of terms, and then taking the square root of

[6]See Broadston, J. A., "Standards for Surface Quality," *Prod. Eng.*, **15**, 1944, pp. 622, 704, 756, and 806; and Norton, M. R., "Development of Standards for Army Ordnance Finishes," *Mech. Eng.*, **64**. 1942, p. 703.

[7]See Tarasov, L. P., "Relation of Surface Roughness Readings to Actual Surface Profile," *Trans. ASME*, **67**, 1945, p. 189.

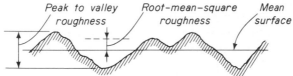

(a) *True profile section magnified equally in horizontal and vertical directions*

Peak to valley　　Root-mean-square　　Mean
/ roughness　　　/ roughness　　　　/ surface

(b) *Exaggerated appearance of surface (a) by further magnification of 25 times in vertical direction only, no change in horizontal magnification*

(c) *Roughness consisting of fine irregularities superposed on larger or wavelike variations*

Figure 13-24 Cross section of surface showing irregularities of roughness and waviness.

Figure 13-25 Taper section of cylindrical surface finished by grinding.

this average. This method of calculating gives more weight to the higher peaks of the surface, since it is felt that high, narrow peaks may have considerable influence on the surface qualities, but would have negligible effect in shifting the position of the mean line. The root mean square height is about 10 per cent larger than the arithmetic average of the ordinates.

A symbol for surface finish in wide use is the check mark with the bar on the top shown in Fig. 13-26. The sides of the mark are inclined at 60° to the surface, and the roughness height in microinches is represented as shown. The drawing should contain a note stating whether root mean square, arithmetic average, or peak-to-valley height is referred

Roughness height microinches → **63**

Surface of part

Figure 13-26 Symbol and marking for surface finish.

to. Sometimes the roughness width and height of wave for a wavy surface are given by means of additional numbers on the symbol. The lay of the surface, or the direction of the predominant surface pattern as formed by the machining or grinding operation, can also be placed on the symbol if desired.

Instruments are available for measuring surface roughness. A diamond tracing point is slowly drawn over the surface, and the roughness is indicated by a trace on a moving paper tape or by a meter. Some instruments give results in terms of the arithmetic average from the mean surface or in terms of rms roughness.

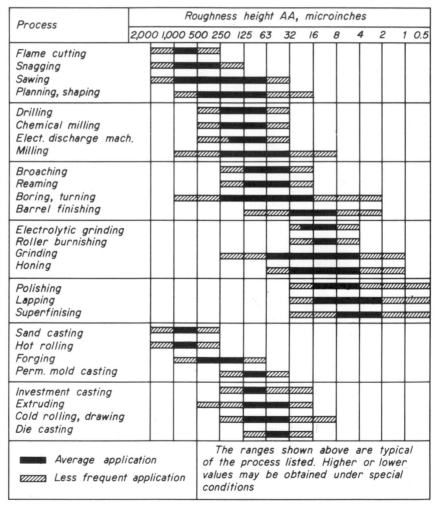

Figure 13-27 Surface roughness available by common production methods. Arithmetic average of deviations from mean surface.

Sometimes roughness can be gaged with sufficient accuracy by comparing the surface with a calibrated specimen which has been prepared by the same machining operation. This comparison can be made by rubbing the fingernails across the work piece and then over the sample. Samples can be obtained for this purpose in a variety of roughnesses, such as produced by turning, milling, grinding, lapping, and so forth.

Different types of machining operations give surfaces of various degrees of roughness with a considerable range of variation for any particular operation. The approximate range of such variations[8] is shown in Fig. 13-27. Among the numbers that are recommended for specifying the average height of irregularities, the following may be mentioned: 1, 2, 4, 8, 16, 32, 63, and 125 microinches.

The inverse problem is also of interest to designers. In other words, when the rms height of a surface has been specified, how far below the peaks are the deepest valleys? It is easy to visualize a surface in terms of peak-to-valley distance, but it is difficult to determine what this range would probably be when only the arithmetic average or rms height is known. Experiments have shown that the average peak-to-valley roughness can be obtained from the rms value by multiplying by $4\frac{1}{2}$ for cylindrical ground surfaces, by 6 or 7 for other types of fixed-abrasive finishes, and by 10 for loose-abrasive lapped surfaces. The factors quoted give average values for "predominant peak" roughness which occurs more or less uniformly over the entire surface. Another type of irregularity that occurs at intervals that are large as compared to the width of the irregularities themselves is termed "deepest maximum" roughness. The foregoing factors should be doubled to obtain deepest maximum values.

20. Dimensional Requirements for Assembly

When two or more parts with toleranced dimensions must be assembled, the tolerances on the individual parts must be so related to each other that the entire production is amenable to assembly.

Consider, for example, simple parts made from the detail drawing of Fig. 13-28(a) which are to be assembled with loose bolts of maximum diameter B. The hole of minimum diameter H has a mean dimension from the edge equal to M. The tolerance m_1 on M, and the allowance between the bolt and hole, must be related to each other in a manner that will permit assembly of the entire production.

Sketch (b) shows a maximum diameter bolt central in a minimum diameter hole which is at the mean dimension from the edge. It is obvious that assembly will just be possible for a plate with a hole center which exceeds M by $\frac{1}{2}(H - B)$ with another plate whose center is less than M by

[8]See *SAE Handbook*, New York: Society of Automotive Engineers.

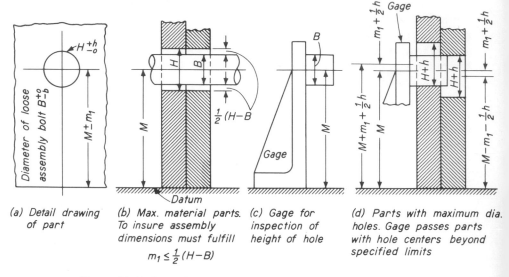

(a) Detail drawing of part

(b) Max. material parts. To insure assembly dimensions must fulfill

$$m_1 \leq \tfrac{1}{2}(H-B)$$

(c) Gage for inspection of height of hole

(d) Parts with maximum dia. holes. Gage passes parts with hole centers beyond specified limits

Figure 13-28 Determination of tolerance $m_1 = \tfrac{1}{2}(H - B)$ and results of using fixed limit functional gage.

$\tfrac{1}{2}(H - B)$. Tolerance m_1 on M thus cannot exceed $\pm\tfrac{1}{2}(H - B)$. If m_1 had a value greater or less than this amount, the space B required for the assembly bolt would be invaded and the bolt could not be inserted. The relationship between m_1, H, and B is thus governed by the equation $m_1 \leq \tfrac{1}{2}(H - B)$.

For quick and easy inspection, sketch (b) will serve as the basis for the design of the gage of sketch (c). Sketch (d) shows the results of using this gage on parts not at the maximum material condition. It is seen that the location of the centers for these larger holes does not conform to sketch (a). Assembly is in no way impaired, but actually the parts do not agree with the dimensions of the print. If desired, a note "at MMC," maximum material condition, may be included in the detail drawing to indicate that inspection with a gage such as that shown in (c) will be satisfactory.

The detail drawing, sketch (a), will ordinarily not specify the bolt diameter. The gage pin diameter can nevertheless be determined from the print as $B = H - 2m_1$. It is understood that a go-and-no go plug gage will be used to check the diameter of the holes.

A similar example is shown in Fig. 13-29(a), which shows the details for parts with two holes to be assembled with loose bolts or rivets. Hole diameters will be maintained by a go-and-no go plug gage. Sketch (b) shows two maximum material parts with the holes at the mean distance apart and with maximum diameter bolts central in the holes. The clearance on either side of a bolt is equal to one-half the allowance between bolt and hole or $\tfrac{1}{2}(H - B)$. This represents the permissible variation in the hole center in one direction, or $\tfrac{1}{2}m_2 \leq \tfrac{1}{2}(H - B)$ or $m_2 \leq (H - B)$.

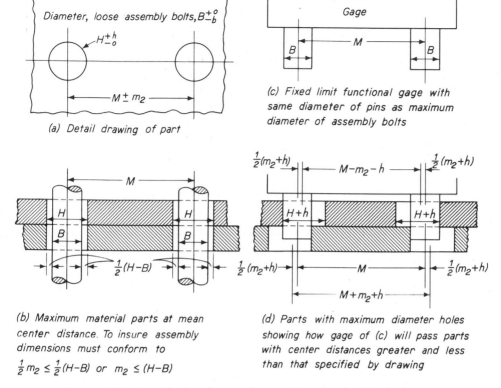

(a) Detail drawing of part

(c) Fixed limit functional gage with same diameter of pins as maximum diameter of assembly bolts

(b) Maximum material parts at mean center distance. To insure assembly dimensions must conform to

$$\tfrac{1}{2}m_2 \le \tfrac{1}{2}(H-B) \text{ or } m_2 \le (H-B)$$

(d) Parts with maximum diameter holes showing how gage of (c) will pass parts with center distances greater and less than that specified by drawing

Figure 13-29 Determination of center distance tolerance and results of using fixed limit functional gage.

When the fixed position gage of sketch (c) is used to check the location of the hole centers, sketch (d) shows that parts in violation of the drawing will be passed. However, assembly will be in no way impaired. If for any reason the dimension $M \pm m_2$ must be maintained, then a suitable note "regardless of feature size" or "RFS" must be placed on the drawing to so indicate.

Gage pin diameter $B = H - m_2$ can be derived from the print when all reference to the assembly bolt diameter has been omitted.

21. Dimensioning of Parts with Lugs and Notches

The detail drawings for two parts which must afterwards be assembled are shown in Fig. 13-30(a). Diameter of studs and holes is to be maintained by suitable go-and-no go plug gages. Sketch (b) shows two maximum material parts with mean center distances M. The clearance on either side between a stud and hole is equal to one-half the allowance between these features, or

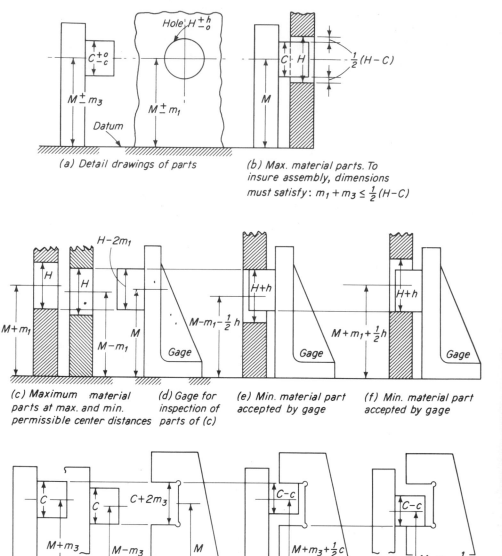

(a) Detail drawings of parts

(b) Max. material parts. To insure assembly, dimensions must satisfy: $m_1 + m_3 \leq \frac{1}{2}(H-C)$

(c) Maximum material parts at max. and min. permissible center distances

(d) Gage for inspection of parts of (c)

(e) Min. material part accepted by gage

(f) Min. material part accepted by gage

(g) Maximum material parts at max. and min. permissible center distances

(h) Gage for inspection of parts of (a)

(i) Min. material part accepted by gage

(j) Min. material part accepted by gage

Figure 13-30 Fixed limit functional gages for inspecting parts with holes and studs.

$\frac{1}{2}(H - C)$. If the total production is to be assembled, the drawing shows that $m_1 + m_3$ must be equal to or less than $\frac{1}{2}(H - C)$. Hence, for assembly the following equation must be fulfilled: $(m_1 + m_3) \leq \frac{1}{2}(H - C)$.

Sketch (c) of Fig. 13-30 shows two maximum material parts with holes at the maximum permissible variation in their centers. The pin for the accompanying inspection gage will have a diameter of $H - 2m_1$. When this gage is used to inspect parts not at MMC, sketches (e) and (f) show that such parts will have hole centers located at dimensions in excess of those permitted by sketch (a).

Sketch (g) of Fig. 13-30 shows two maximum material parts with studs at the maximum and minimum permissible center distances. The accompanying gage will have a gap or hole equal to $C + 2m_3$, sketch (h). Such a gage

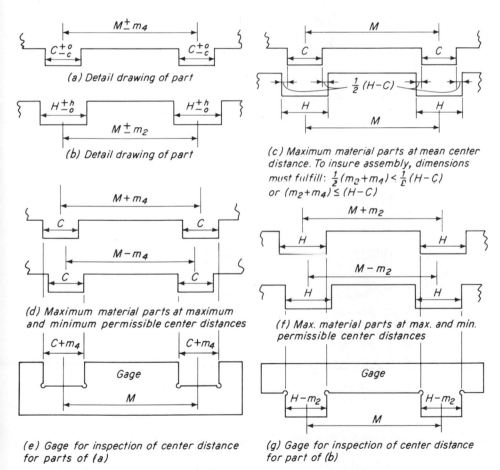

(a) Detail drawing of part

(b) Detail drawing of part

(c) Maximum material parts at mean center distance. To insure assembly, dimensions must fulfill: $\frac{1}{2}(m_2 + m_4) < \frac{1}{2}(H - C)$ or $(m_2 + m_4) \leq (H - C)$

(d) Maximum material parts at maximum and minimum permissible center distances

(f) Max. material parts at max. and min. permissible center distances

(e) Gage for inspection of center distance for parts of (a)

(g) Gage for inspection of center distance for part of (b)

Figure 13-31 Fixed limit functional gage accepts parts with center distances beyond those shown on drawing.

will pass parts in violation of sketch (a) when the stud is not at the maximum material conditions, sketches (i) and (j).

Figures 13-31(a) and (b) show parts with two lugs and two notches which must afterwards be assembled. Sketch (c) indicates that the dimensions must fulfill the equation $(m_2 + m_4) \le (H - C)$ if the entire production is to be assembled. Sketches (e) and (g) give the widths of the notches and lugs of the corresponding inspection gages. As before, for parts not at the maximum material condition, such gages will pass parts with center distances greater or less than the limits imposed by sketches (a) and (b). This figure, and the accompanying discussion, will apply equally well to the assembly of a part with two holes with another part with two fixed studs.

In general, the details can be considerably simplified by using true position dimensioning for the location of hole centers. The situation for the assembly of parts with loose bolts, or the assembly of a part with a hole to a part with a fixed stud, is summarized in Fig. 13-32.

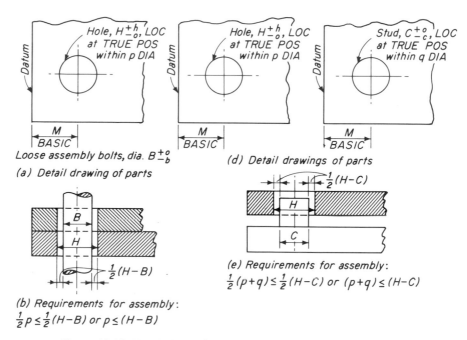

Figure 13-32 Fixed limit functional gage accepts parts with center distances beyond those shown on drawing.

22. Dimensioning for Assembly of Concentric Cylinders

The cylindrical parts of Fig. 13-33(a) and (b) are later to be assembled. Each diameter has a tolerance with the permissible eccentricities restricted by the notes giving the total indicator runout, *TIR*. Sketch (c) shows the

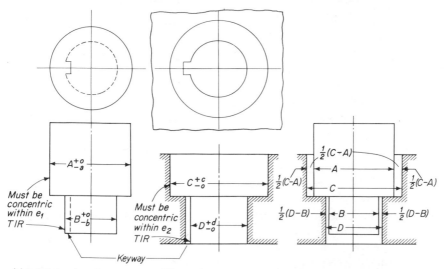

(a) Detail drawing of part   (b) Detail drawing of part

Keyways to be aligned after assembly

(c) Max. metal parts concentrically
located. To insure assembly
dimensions must satisfy

$$\tfrac{1}{2}(e_1 + e_2) \le \tfrac{1}{2}\left[(C-A) + (D-B)\right]$$
$$\text{or } (e_1 + e_2) \le \left[(C-A) + (D-B)\right]$$

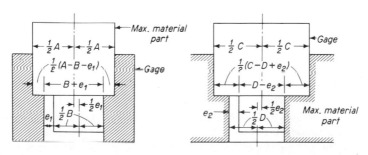

(d) Gage for checking eccentricity   (f) Gage for checking eccentricity

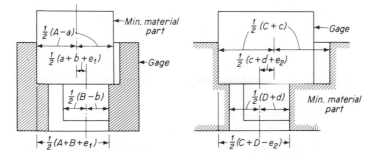

(e) Eccentricity of minimum material
part passed by gage

(g) Eccentricity of minimum material
part passed by gage

Figure 13-33 Dimensional requirements for assembly. Results when
inspection is made by fixed-limit functional gages.

clearances along the sides of two maximum material parts concentrically located with respect to each other. It is obvious that if all parts are to be amenable to assembly then the following equation must be fulfilled: $(e_1 + e_2) \leq [(C - A) + (D - B)]$.

A gage suitable for inspecting a stud with maximum material is shown in sketch (d). However, if this gage is used on a part with minimum material it is possible for a part with an eccentricity of $\frac{1}{2}(a + b + e_1)$ to be passed by the gage ss shown in sketch (e).

A gage suitable for inspecting the body part of sketch (b) is shown in sketch (f). If this gage is used on a part with minimum material, however, it is possible for a part with an eccentricity of $\frac{1}{2}(c + d + e_2)$ to be accepted by the gage as shown in sketch (g).

If such increases in eccentricity are acceptable to the designer, he can so inform the user of the print by adding "at *MMC*" to the eccentricity note on each of Figs. 13-33(a) and (b).

No allowance on the foregoing figures and equations has been made for any clearances or gage maker's tolerances that may be desired. Such adjustments can be easily included when the final dimensions of the design are being determined.

23. Detailing

When the design has been completed, detail or shop drawings must be made before actual manufacture or construction can begin. The detail draftsmen perform a very important function in industry, since the success of a project depends on correct details and shopwork as well as on design calculations. Skill and cleverness in arranging the mechanical details can be acquired only after considerable time has been spent in actual practice in this line of work. A few of the more obvious examples of good and bad detailing are illustrated in Figs. 13-34 to 13-53 inclusive.[9]

REFERENCES

1. *Dimensioning and Tolerancing for Engineering Drawings*, USAS Y14.5-1966, New York: American Society of Mechanical Engineers.

2. Buckingham, Earle, *Principles of Interchangeable Manufacture*, 2d ed., New York: The Industrial Press, 1941.

3. Hume, K. J., *Engineering Metrology*, 2d ed., London: Macdonald & Co., 1963.

4. Kissam, O., *Optical Tooling*, New York: McGraw-Hill Book Company, 1962.

[9]See Geiger, Ernest, "The Rights and Wrongs of Details," *Prod. Eng.*, **12**, 1941, pp. 72, 122, and 248.

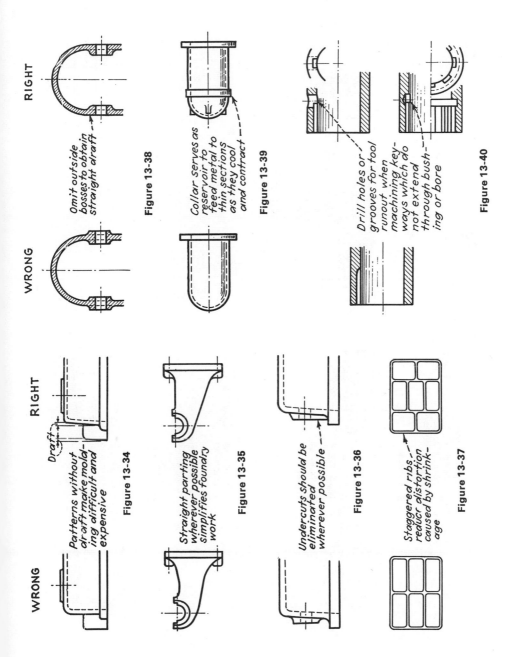

WRONG   **RIGHT**

Patterns without draft make molding difficult and expensive

Figure 13-34

Straight parting wherever possible simplifies foundry work

Figure 13-35

Undercuts should be eliminated wherever possible

Figure 13-36

Staggered ribs reduce distortion caused by shrinkage

Figure 13-37

WRONG   **RIGHT**

Omit outside bosses to obtain straight draft

Figure 13-38

Collar serves as reservoir to feed metal to thin sections as they cool and contract

Figure 13-39

Drill holes or grooves for tool runout when machining keyways which do not extend through bushing or bore

Figure 13-40

547

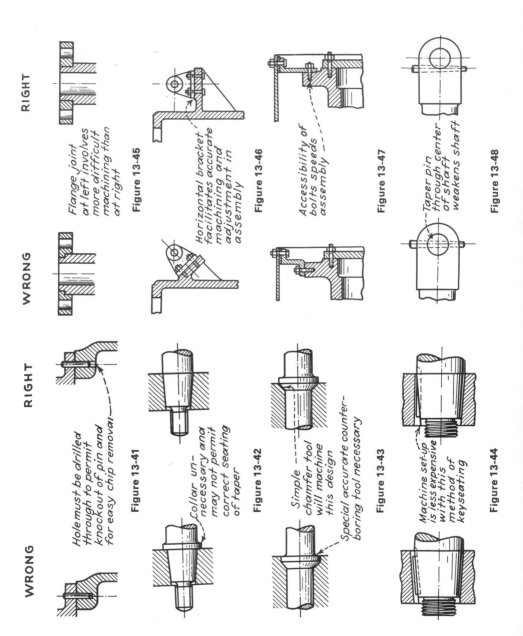

WRONG   RIGHT   WRONG   RIGHT

Hole must be drilled through to permit knockout of pin and for easy chip removal

Figure 13-41

Collar unnecessary and may not permit correct seating of taper

Figure 13-42

Simple chamfer tool will machine this design

Special accurate counter-boring tool necessary

Figure 13-43

Machine set-up is less expensive with this method of keyseating

Figure 13-44

Flange joint at left involves more difficult machining than at right

Figure 13-45

Horizontal bracket facilitates accurate machining and adjustment in assembly

Figure 13-46

Accessibility of bolts speeds assembly

Figure 13-47

Taper pin through center of shaft weakens shaft

Figure 13-48

548

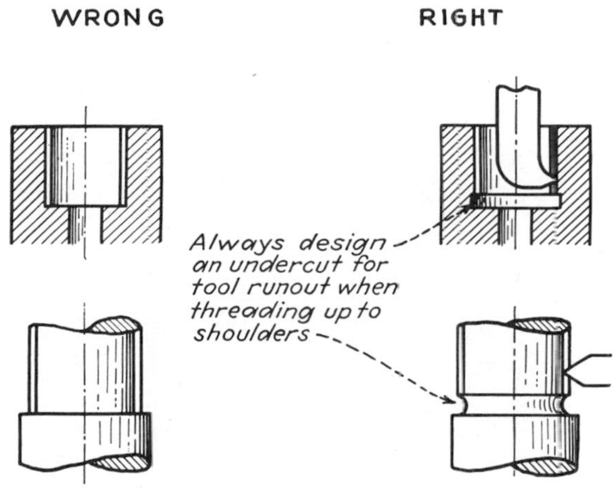

WRONG

RIGHT

WRONG

RIGHT

Always design an undercut for tool runout when threading up to shoulders

Figure 13-49

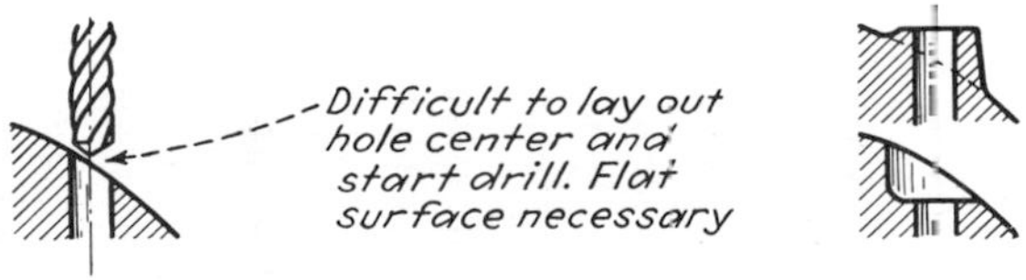

Clearance grooves or undercuts avoid impracticability of machining or grinding into sharp corners

Figure 13-52

Difficult to lay out hole center and start drill. Flat surface necessary

Figure 13-50

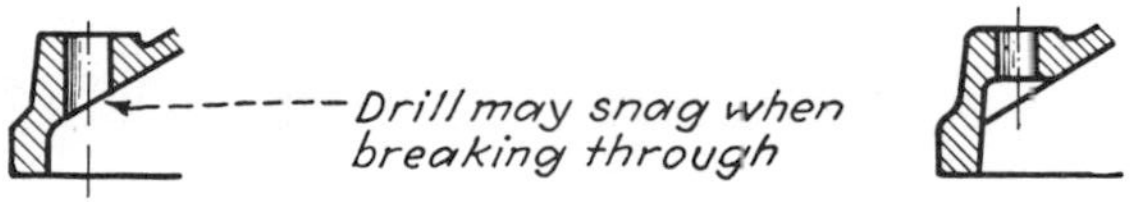

Clearance for drills and taps should be checked

Figure 13-53

Drill may snag when breaking through

Figure 13-51

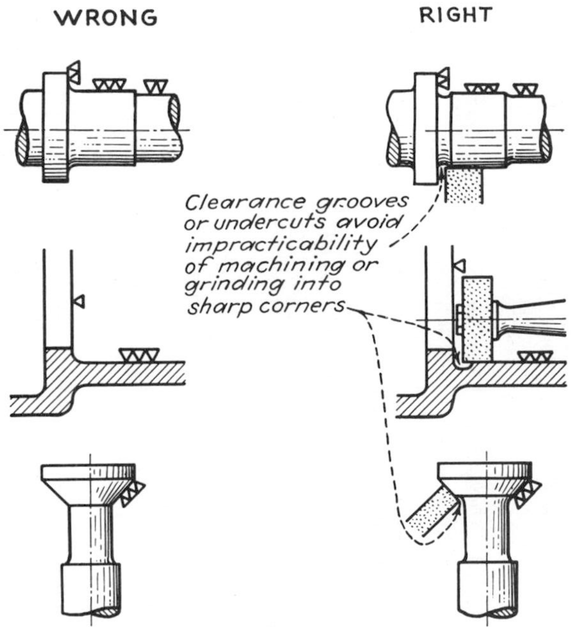

549

5. *Military Standard, Dimensioning and Tolerancing*, MIL-STD-8C, 16 Oct. 1963. Washington, D.C.: U.S. Government Printing Office.

6. Miller, L., *Engineering Dimensional Metrology*, London: Edward Arnold, 1962.

7. Rolt, F. H., *Gauges and Fine Measurements*, Vols. I and II, London: Macmillan & Co., 1929.

8. Rothbart, H. A., ed., *Mechanical Design and Systems Handbook*, New York: McGraw-Hill Book Company 1964, Section 19, "Fabrication Principles."

9. Whitehead, T. N., *Design and Use of Instruments and Accurate Mechanisms*, New York: Dover Publications, 1933.

10. Wilson, F. W., ed., *Tool Engineers Handbook*, 2d ed., New York: McGraw-Hill Book Company, 1959, Section 86, "Fits Tolerances and Allowances," and Section 87, "Inspection Equipment and Methods."

PROBLEMS

1. Draw and dimension the bar diagram for a $3\frac{3}{8}$ in.-shaft, Class 3 fit. Draw and dimension a typical hole and shaft for this fit. Use unilateral tolerances and the basic hole system for maximum metal. Use dimensions to nearest 0.0001 in.

2. Draw and dimension the bar diagram for a $3\frac{3}{4}$ in.-shaft, Class 2 fit. Draw and dimension a typical hole and shaft for this fit. Use unilateral tolerances and the basic shaft system for minimum metal. Use dimensions to nearest 0.0001 in.

3. Draw a typical detail for a $3\frac{7}{16}$ in.-shaft and bearing with an allowance of 0.0025 in. Shaft tolerance is 0.0015 in.; bearing tolerance is 0.0020 in. Use unilateral tolerances.

    (a) Dimension for basic shaft system and maximum metal.
    (b) Dimension for basic shaft system and minimum metal.
    (c) Dimension for basic hole system and maximum metal.
    (d) Dimension for basic hole system and minimum metal.

4. Draw and dimension the bar diagram for an 8 in.-shaft, Class 8 fit. Draw and dimension a typical hole and shaft for this fit. Use unilateral tolerances and the basic hole system for maximum metal. Use dimensions to nearest 0.0001 in.

5. Draw and dimension the bar diagram for a 4 in.-shaft, Class 3 fit. Compute the value of the ratio of the diametral clearance to the shaft diameter for the fit at the tightest condition. Do the same for the loosest permissible fit. Draw and dimension the bar diagram for a 4 in.-bearing for a minimum value for the c/r ratio of 0.0015 and for a maximum value of c/r of 0.0020. Compare the tolerance conditions of the bearing with those of a Class 3 fit.

6. A part having a hole of 1.000 in. diam is mated with a shaft of diameter 0.995 in. What class or classes of standardized cylindrical fits can the resulting assembly be designated? Answer the question for shaft diameters of 0.997, 0.9975, 0.9988, and 0.9992 in.

7. (a) Draw and dimension the bar diagram for a 6 in.-shaft, Class 7 fit. Draw and dimension a typical hole and shaft for this fit. Use unilateral tolerances and basic shaft system for minimum metal. Use dimensions to nearest 0.0001 in.

(b) Suppose the shaft is of steel, and the disk is of cast iron 12 in. in diameter. What is the value of the maximum tangential stress in case the maximum size shaft is assembled with a minimum diameter hole? What is the value of the stress if the minimum diameter shaft is assembled with the maximum size hole? $\mu = 0.3$. For cast iron, $E = 15,000,000$ psi. *Ans.* (b) Tightest fit, $s_t = 7,370$ psi. Loosest fit, $s_t = 3,420$ psi.

8. Two parts, each with two holes 1.001 in., plus 0.002 in., minus 0, are to be assembled with loose bolts and inspected with a gage as in Fig. 13-29(c). The total variation in the center distance for the holes in the parts passed by the gage is to be limited to 0.006 in.

(a) Find the value for m_2 that should be placed on the drawing.

(b) Find the maximum permissible diameter of the assembly bolt.

9. A part has two fixed studs with centers dimensioned 4.000 plus and minus 0.002 in. Maximum stud diameter is 0.750 in. This part is to be assembled with another with two holes of diameter 0.753 in., plus 0.002 in., minus 0.

(a) Find the tolerance m_4 for the center distance of the holes.

(b) Find the total variation in center distance for the holes if inspection is made by the gage of Fig. 13-31(g).

10. Find the value of e_1 and B' in Fig. 13-54 if it is desired that the stem of the stud have a clearance of 0.001 in. along each side when two maximum material parts are assembled with the maximum permissible eccentricities facing each other. Make a drawing showing this situation and dimension completely.

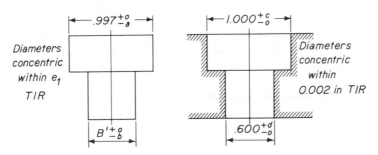

Figure 13-54 Problem 10.

11. In the assembly of Fig. 13-55, find the maximum and minimum values of dimension d if the tolerances are additive. Find the values by the Pythagorean rule.

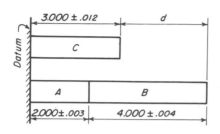 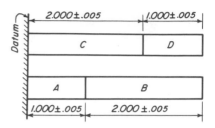

Figure 13-55 Problem 11.   **Figure 13-56** Problem 12.

12. In the assembly of Fig. 13-56, find the maximum displacement of the right ends with respect to each other if the tolerances are additive. Find the displacement if the tolerances are Pythagorean.

13. In the assembly of Fig. 13-57, find the maximum variation of the right end of D with respect to the datum if the tolerances are additive. Find the variation if the tolerances are Pythagorean.

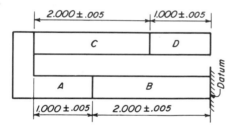 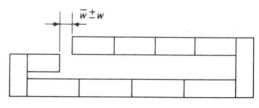

Figure 13-57 Problem 13.   **Figure 13-58** Problem 14.

14. The tolerance $\pm w$ on the gap $\bar{w}$ in Fig. 13-58 must be equal to ± 0.009 in.

(a) If the tolerances on the separate parts are all equal, what tolerance must be placed on each part if it is assumed that the additive rule applies?

(b) Suppose the production lengths fulfill all the necessary statistical assumptions. Find the tolerance that can now be placed on each part.   *Ans.* ± 0.003 in.

15. Bolts of maximum diameter 0.500 in. are to be used for the assembly of parts with two holes of minimum diameter 0.505 in. What should be the tolerance for the center distance for the holes?

16. Bolts of maximum diameter 0.625 in. are to be used for the assembly of parts with two holes of minimum diameter 0.629 in. What is the tolerance on the center distance for the holes?

17. Parts have two holes with center distance dimensioned 2.875 ± 0.003 in. Holes are dimensioned 0.937 in. plus 0.002 in. minus 0. What is the maximum permissible diameter of the assembly bolt?

18. Two bolts of diameter 0.999 ± 0.001 are used for the assembly of two parts. Tolerance for the holes in the parts is ±0.0015 in. The mean center distance for the holes is 2.5 in. Make a detail drawing for the bolt and part on the basic shaft system for maximum metal. Allowance between bolt and hole is 0.002 in.

19. Two bolts of nominal diameter $1\frac{1}{4}$ in. are used to assemble two loose parts. The total hole tolerance is 0.002 in. Total tolerance on the bolt diameter is 0.004 in. Mean center distance for the holes is 4 in. Determine that tolerance on the center distance to be placed on the drawing if the maximum variation on center distance passed by the gage is to be ±0.005 in. Make detail drawings and show all necessary dimensions. Use the basic shaft system for maximum metal. Use unilateral tolerances on diameters and bilateral tolerance on center distance.

20. Two parts with two holes each are to be assembled with $1\frac{9}{16}$ in. bolts. The mean center distance is 5 in. A Class 2 fit is desired.

(a) Draw and dimension the bar diagram for this fit using dimensions to the nearest 0.0001 in.

(b) Make a detail drawing for the part and give complete dimensions. Use the basic shaft system for maximum metal.

(c) Make a drawing for the bolt giving correct dimensioning for the diameter.

21. A hole is located by the dimensions given in Fig. 13-59. Find the x- and y-coordinates of the hole.   *Ans.* $x = 0.45$ in., $y = 0.60$ in.

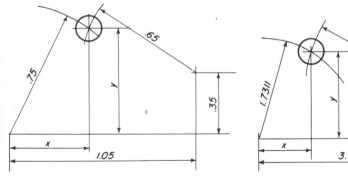

Figure 13-59 Problem 21.

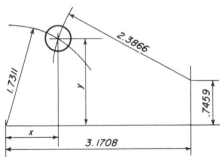

Figure 13-60 Problem 22.

22. A hole is located by the dimensions given in Fig. 13-60. Find the x- and y-coordinates correct to the nearest 0.0001 in.

Ans. $x = 0.8996$ in., $y = 1.4790$ in.

23. (a) Find the value of B in Fig. 13-61 to ensure assembly of total production.

(b) If two maximum material parts are assembled with permissible eccentricities facing each other, find the clearance along the left side of the stem of the stud. Make a drawing and dimension completely.

(c) Suppose two minimum material parts are passed by the gages of Fig. 13-33 assembled with the eccentricities facing each other. Find the clearance at the head of the stud and along the stem. Make a drawing and dimension completely.

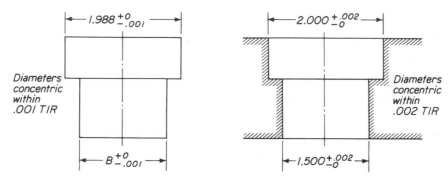

Figure 13-61 Problem 23.

14 Engineering Materials

The composition and physical properties of the principal materials used in machine construction will be discussed in this chapter. It is fully as important that a machine element be made of a material that has properties suitable for the conditions of service as it is for the loads and stresses to be accurately determined. Frequently, the limitations imposed by the material are the controlling factors in a design. The designer must be acquainted with the effects that the methods of manufacture and heat treatment have on the properties of engineering materials. The manufacturing process used for fabrication of the part will also influence the type of material that can be used. Sometimes the help of a professional metallurgist is needed to insure the best possible choice of material and heat treatment.

1. The Tension Test

The tension test is widely used for determining yield point stress, ultimate or tensile stress, breaking or rupture stress, elongation, and reduction of area for various materials. Mild- or low-carbon steel is unique among the metals in having a pronounced elongation at the yield point. The phenomenon of the upper and lower yield points, shown in Fig. 14-1(b), is sometimes explained by defining the upper yield point as the stress which will start the flow process, and the lower yield point as the lowest stress at which the extension proceeds still further. The shape of the stress-strain curve is influenced to a large extent by the type of machine on which the tests are made. The constant-strain-rate type of machine, shown in Fig. 14-1(a), stretches the specimen at a uniform rate by means of power-driven screws. As soon as yielding starts, the poise must be reverted to the left to keep the beam in balance between the stops at

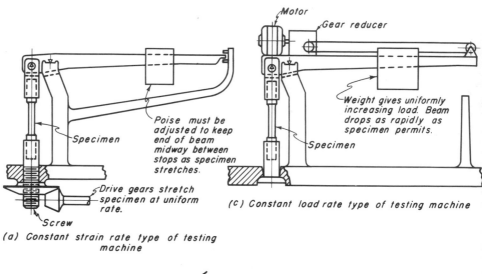

(a) Constant strain rate type of testing machine

(c) Constant load rate type of testing machine

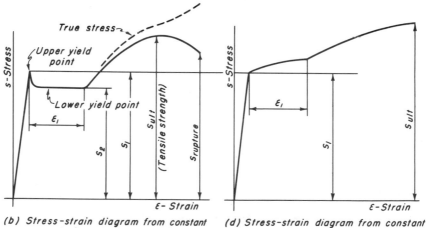

(b) Stress-strain diagram from constant strain rate type machine

(d) Stress-strain diagram from constant load rate type machine

Figure 14-1 Schematic drawings of testing machines and corresponding stress-strain diagrams for low-carbon steel.

the right end. The material continues to yield until it is strengthened by strain or work hardening. A higher stress then is necessary for additional stretching.

In the second type of machine, illustrated in Fig. 14-1(c), the beam is allowed to drop as rapidly as the specimen stretches from the effect of a constantly increasing load.[1] No reduction in stress can occur with this kind of loading. The stress-strain diagram which appears looks like that shown in (d). The diagram from the constant-load-rate type of machine depends on the rapidity with which the load is applied. Experiments have shown that the yield point stress s_1 and yield point elongation ϵ_1 are increased as the speed of loading increases. Some diagrams obtained on $\frac{1}{2}$ by $\frac{3}{16}$ in. mild-steel strips are shown in Fig. 14-2.

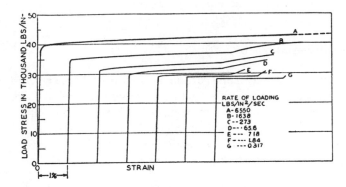

Figure 14-2 Stress-strain diagrams for constant load rate tests for mild steel.

The conditions under which constant-strain-rate tests are made are somewhat analogous to drawing or rolling operations in metal forming. However, machine parts are employed in such a manner that if an overload causes the yield point of the material to be exceeded, the deformations can take place as rapidly as the stretch of the material permits.

Alloy and heat-treated steels, steels which have been cold worked, cast iron, and nonferrous metals do not exhibit the well-defined yield point of mild steel. The yield strength for such materials is generally taken as the stress corresponding to some arbitrarily selected value of permanent defor-

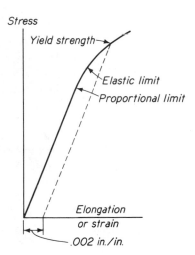

Figure 14-3 Yield strength for material not having a well-defined yield point.

[1]See Davis, E. A., "The Effect of Speed of Stretching and the Rate of Loading on the Yielding of Mild Steel," *Trans. ASME*, **60**, 1938, p. A-137, and **61**, 1939, p. A-89.

mation which the material can have without undergoing appreciable structural damage. This elongation is often taken as 0.2 per cent, as shown in Fig. 14-3, although smaller values are sometimes used.

Stretching of a soft steel specimen is accompanied by a decrease in diameter. After the ultimate stress is reached, a pronounced reduction of area, called necking down, occurs at some point in the stressed region. For the constant-strain-rate type of machine, the total load which the specimen can then support with the beam in balance is reduced. It is standard practice to compute the stresses for the stress-strain diagram using the original cross-sectional area. There is thus an apparent decrease of stress in the right-hand portion of the diagram. This supposed reduction does not occur if the actual values of the cross sections are determined and used for computing the stress as shown by the dashed curve in Fig. 14-1(b).

2. Physical Constitution of Steel

Steel is an alloy of iron and carbon, or of iron, carbon, and other alloying elements. Carbon must be present to the extent of about 0.05 per cent by weight in order for the material to be known as steel rather than commerical iron. The quantity of carbon is much higher in cast iron, about 2 to 4 per cent. Steel, like other metals, has a crystalline structure when viewed under the microscope, with grains varying in size from about 0.001 in. to 0.010 in. in diameter.

A number of different metallurgical terms are used in describing the structure of steel. Pure iron, for example, is known as ferrite. Ferrite is soft and

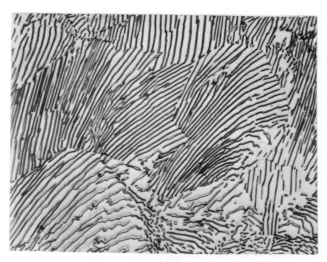

Figure 14-4 Pearlite, 0-9% carbon steel, magnification 1,000 diameters. (*Courtesy United States Steel Corp, Research Laboratory.*)

ductile and has a tensile strength of about 40,000 psi. Cementite is the chemical compound of iron and carbon, Fe_3C. It is nonductile and very hard. Another constituent of steel is pearlite. As shown in Fig. 14-4, pearlite grains have a laminar structure of alternate layers of ferrite and cementite. The carbon content of pearlite is 0.83 per cent by weight. When slowly cooled from above the upper critical temperature, a steel of less than 0.83 per cent carbon is composed of a mixture of grains of ferrite and pearlite. For carbon percentages higher than 0.83, the steel is a mixture of grains of pearlite and cementite upon slow cooling. When the percentage of carbon is exactly 0.83, the so-called "eutectoid structure" results, which is wholly composed of grains of pearlite.

When steels are heated, transformations in the crystal structure take place at certain temperatures. The magnified grains of a 0.30 per cent carbon steel are shown schematically in Fig. 14-5. When the steel is heated to the lower critical temperature of 1330°F, a new constituent, called austenite, appears. Austenite is formed from pearlite by the carbon of the cementite going into solid solution in the ferrite. The situation is analogous to the solution of a salt in water, except that the solvent, in the case of metals, is in the solid condition. As the temperature continues to rise, the austenite dissolves the free ferrite until at the upper critical temperature the steel consists wholly of austenite.

The upper critical temperature depends on the carbon content, as indicated by a portion of the iron-carbon equilibrium diagram shown in Fig. 14-6. The upper critical temperature, for any percentage of carbon, is given by line *ACE*, and the lower critical temperature is represented by line *BCD*. Above the lower critical temperature, the pearlite, for all values of the carbon content, has become austenite. In region *ACB*, the austenite dissolves the free ferrite as the temperature rises, and in region *ECD*, the austenite dissolves the free cementite. Above the upper critical temperature, the steel is composed wholly of austenite. A reversal of these changes takes place if the steel is slowly cooled from above the upper critical temperature. Austenite rarely exists at room temperature except in certain highly alloyed steels.

The grain structure at room temperature, for different amounts of carbon, is shown by the three circles in the lower part of the figure.

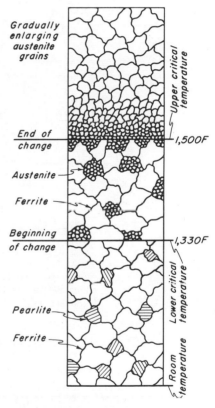

Figure 14-5 Effect of heating on the microstructure of 0.30% C steel.

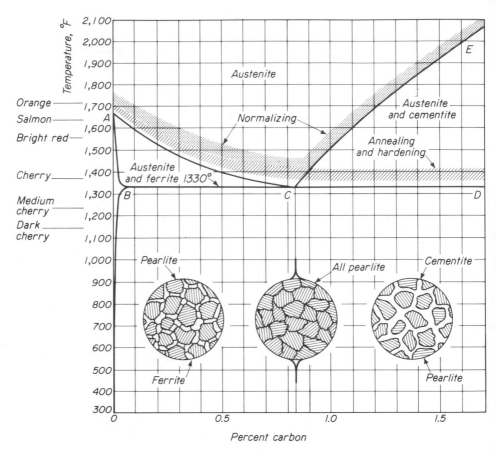

Figure 14-6 Portion of iron-cementite diagram showing compositions, critical temperatures, and temperature ranges for normalizing, annealing, and hardening.

3. Types of Steel Used in Machine Construction

The selection of a suitable steel for a machine part depends on a full understanding of the distribution and fluctuation of the stress. Knowledge is also required of any unusual conditions that may be present, such as stress concentration, impact, corrosion, abrasion, and high or low temperatures. The type of steel selected depends on whether or not the part is to be heat treated. The following summary gives a general classification of steels as used in machine construction:

Not Heat Treated:
(a) Plain, low-carbon steel, hot rolled or cold drawn.

   (b) Free cutting steels or steels resulphurized for ease of machining.

   (c) Low-carbon sheets for forming, drawing, or spinning.

   (d) High-strength low-alloy steels, HSLA.

Heat Treated:

   (a) Carburized. These are low-carbon or low-alloy steels.

   (b) Oil or water quenched and then tempered. These steels contain considerable amounts of carbon or alloying elements.

4. Numbering Systems for Carbon and Alloy Steels

Many different carbon and alloy steels are used for the construction of machinery. The numbering system of the Society of Automotive Engineers and the American Iron and Steel Institute is based on the chemical composition, and provides a simple means whereby any particular steel can be specified. In general, this catalog system uses a number composed of four digits. The first two digits indicate the type or alloy classification; the last two (and in some cases, three) digits give the carbon content. Carbon has such a pronounced effect on the strength and hardness of steel that the content is thus incorporated in the catalog number. Plain carbon steel, for example, is denoted by the basic numeral 10. Thus, steel 1045 indicates a plain carbon steel containing 0.45 per cent carbon. The basic numerals for the various types

Table 14-1   BASIC NUMERALS FOR SAE AND AISI STEELS

Carbon steels	1xxx	Molybdenum steels	4xxx
Plain carbon	10xx	Carbon-molybdenum	40xx
Free-cutting screw stock	11xx	Chromium-molybdenum	41xx
		Chromium-nickel-molybdenum	43xx
Manganese steels	13xx	Nickel-molybdenum; 1.75% Ni	46xx
		Nickel-molybdenum; 3.50% Ni	48xx
Nickel steels	2xxx		
3.50% Ni	23xx	Chromium steels	5xxx
5.00% Ni	25xx	Low chromium	51xx
		Medium chromium	52xxx
Nickel-chromium steels	3xxx	Corrosion and heat resisting	51xxx
1.25% Ni, 0.60% Cr	31xx		
1.75% Ni, 1.00% Cr	32xx	Chromium-vanadium steels	6xxx
3.50% Ni, 1.50% Cr	33xx	1.00% Cr	61xx
Corrosion and heat resisting	30xxx	Silicon-manganese steels	9xxx
		2.00% Si	92xx
			AISI
		Chromium-nickel-molybdenum steels	86xx
		Chromium-nickel-molybdenum steels	87xx

of steels are given in Table 14-1. The arrangement of Table 14-2 may serve as an aid in committing some of the series numbers to memory.

Table 14-2   BASIC NUMERALS FOR SEVERAL WIDELY USED STEELS

Nickel	23xx, 25xx
Nickel-chromium	31xx, 32xx, 33xx
Chromium- molybdenum	41xx
Chromium-nickel-molybdenum	43xx
Nickel-molybdenum	46xx, 48xx

Table 14-3   CHEMICAL COMPOSITION OF SELECTED CARBON AND ALLOY AISI AND SAE STEELS (PREFIXES A, B, or C OMITTED IN THE SAE SYSTEM)

AISI	Carbon	Manganese	Sulphur	AISI	Carbon	Manganese	Nickel	Chromium	Molybdenum
Plain Carbon Steels				*Nickel-Chromium Steels*					
C1010	0.08–0.13	0.30–0.60	0.05 max	A3140	0.38–0.43	0.70–0.90	1.10–1.40	0.55–0.75	
C1020	0.18–0.23	0.30–0.60	0.05 max						
C1035	0.32–0.38	0.60–0.90	0.05 max	*Chromium-Molybdenum Steels*					
C1045	0.43–0.50	0.60–0.90	0.05 max						
C1060	0.55–0.65	0.60–0.90	0.05 max	A4140	0.38–0.43	0.75–1.00		0.80–1.10	0.15–0.25
C1095	0.90–1.05	0.30–0.50	0.05 max	*Nickel-Chromium-Molybdenum Steels*					
Free-Cutting Steels				A4320	0.17–0.22	0.45–0.65	1.65–2.00	0.40–0.60	0.20–0.30
				A4340	0.38–0.43	0.60–0.80	1.65–2.00	0.70–0.90	0.20–0.30
B1112	0.08–0.13	0.70–1.00	0.16–0.23	*Nickel-Molybdenum Steels*					
B1113	0.08–0.13	0.70–1.00	0.24–0.33						
C1117	0.14–0.20	1.00–1.30	0.08–0.13	A4620	0.17–0.20	0.45–0.65	1.65–2.00		0.20–0.30
C1137	0.32–0.39	1.35–1.65	0.08–0.13	A4640	0.38–0.43	0.60–0.80	1.65–2.00		0.20–0.30
C1141	0.37–0.45	1.35–1.65	0.08–0.13	A4815	0.13–0.18	0.40–0.60	3.25–3.75		0.20–0.30
Nickel Steels				*Plain Chromium Steels*					
			Nickel	E52100	0.98–1.10	0.25–0.45		1.30–1.60	
A2317	0.15–0.20	0.40–0.60	3.25–3.75	*Chromium-Vanadium Steels*					
High Strength Low-Alloy Steel*				A6150	0.48–0.53	0.70–0.90		0.80–1.10	0.15 min. V
Cor-Ten	0.12 max	0.20–0.50	0.25–0.75 Si						
	0.25–0.55 Cu	0.30–1.25 Cr	0.65 max Ni	*Nickel-Chromium-Molybdenum Steels*					
T₁ type A	0.12–0.21	0.70–1.00	0.20–0.35 Si	A8620	0.18–0.23	0.70–0.90	0.40–0.70	0.40–0.60	0.15–0.25
	0.40–0.65 Cr	0.15–0.25 Mo	0.03–0.08 V	A8640	0.38–0.43	0.75–1.00	0.40–0.70	0.40–0.60	0.15–0.52
T₁ type B	0.12–0.21	0.95–1.30	0.20–0.35 Si	*Silicon-Manganese Steels*					
	0.30–0.70 Ni	0.40–0.65 Cr	0.20–0.30 Mo	A9260	0.56–0.64	0.70–1.00			1.8–2.2 Si
	0.03–0.08 V								

*Proprietary steels, USS Corp. Similar products of other makers are marketed under different trade names.

The AISI number for a steel is similar to the SAE number, but capital letter prefixes are included to indicate the process of manufacture as follows:

A—Basic open-hearth alloy steel
B—Acid Bessemer carbon steel
C—Basic open hearth carbon steel
D—Acid open hearth carbon steel
E—Electric furnace alloy steel

The chemical compositions for a selected number of steels used for machinery are given in Table 14-3. The mechanical properties for these

Table 14-4 MECHANICAL PROPERTIES OF TYPICAL LOW-CARBON AND CASE-HARDENING STEELS*

Steel	Condition	Tensile Strength, psi	Yield Strength, psi	% Elong. in 2 in.	% Red. in area	BHN	Rock-well	Machin-ability, %
1010	Hot-rolled	51,000	29,000	38	70	101		40
	Cold-drawn	56,000	33,000	35	65	113		45
1020	Hot-rolled	65,000	43,000	36	59	143	B79	50
	Cold-drawn	78,000	66,000	20	55	156	B83	60
1112	Hot-rolled	67,000	40,000	27	47	140		
	Cold-drawn	82,500	71,000	15	43	170	B87	100
1113	Cold-drawn	83,500	72,000	14	40	170	B87	130
1117	Hot-rolled	71,000	45,000	33	64	143	B78	90
	Cold-drawn	86,000	75,000	22	52	170	B87	95
2317†	Hot-rolled	85,000	56,000	29	60	163		50
	Cold-drawn	95,000	75,000	25	58	197	C12	
4320†	Hot-rolled	87,000	59,000	30	60	179	C10	55
	Cold-drawn	99,000	65,000	23	54	207	C16	
4620	Hot-rolled	85,000	63,000	28	64	183	B90	58
	Cold-drawn	101,000	85,000	22	60	207	C15	64
4815†	Hot-rolled	105,000	73,000	24	58	212	C15	55
	Cold-drawn	110,000	78,000	23	55	217	C17	
8620	Hot-rolled	89,000	65,000	25	63	192	B90	60
	Annealed	102,000	85,000	22	58	212	C16	63
HSLA‡								
	Cor Ten	70,000	50,000	24				
	T₁ type A	115–135,000	100,000	18	40			
	T₁ type B	115–135,000	100,000	18	40			

*Taken mainly from *Ryerson Data Book*, Joseph T. Ryerson & Son, Inc., and Hoyt, S. L., *Metals and Alloys Data Book*, New York: Van Nostrand Reinhold Company, 1943, Table 35. Table indicates probable median expectation of the steels listed based on a 1-in. round section. In individual tests, these steels may develop results differing considerably from those shown, particularly for cold drawn materials.

†Steels 2317, 4320, and 4815 usually not stocked by steel warehouses.

‡High Strength Low Alloy. These are proprietary steels, USS Corp. Similar products of other makers are marketed under different trade names.

materials are given in Tables 14-4 and 14-5. The SAE and AISI lists of steels are constantly being revised, and it is necessary to consult an up-to-date list if one is in doubt as to whether or not a particular steel is in regular production.[2]

Table 14-5   MECHANICAL PROPERTIES OF TYPICAL MEDIUM-CARBON OR DIRECT-HARDENING STEELS*

Steel	Condition	Tensile Strength, psi	Yield Strength, psi	% Elong. in 2 in.	% Red. in area	BHN	Rock-well	Machin-ability, %
1035	Hot-rolled	85,000	54,000	30	53	183	B90	65
	Cold-drawn	92,000	79,000	25	50	201	B94	67
	Water Q1550°F							
	Drawn 1000°F	103,000	72,000	23	59	201	B94	
1045	Hot-rolled	98,000	59,000	24	45	212	C16	56
	Cold-drawn	103,000	90,000	14	40	217	C18	60
1060†	Hot-rolled	95,000	59,000	25	52	197	C14	53
	Oil Q1550°F							
	Drawn 1000°F	122,000	90,000	19	53	255	C25	
1095	Hot-rolled	142,000	83,000	18	38	293	C28	
	Water Q1450°F							
	Drawn 800°F	200,000	138,000	12	37	388	C42	
1137	Hot-rolled	92,000	57,000	27	61	192	B92	70
	Cold-drawn	105,000	90,000	15	38	207	C15	75
	Oil Q1550°F							
	Drawn 1000°F	112,000	88,000	21	56	255	C25	
3140†	Hot-rolled	96,000	64,000	26	56	195	C12	57
	Cold-drawn	115,000	98,000	17	45	248	C24	
4140	Hot-rolled	89,000	62,000	26	58	187	B91	57
	Cold-drawn	102,000	90,000	18	50	223	C19	66
4340	Hot-rolled	101,000	69,000	21	45	207	C15	45
	Cold-drawn	111,000	99,000	16	42	223	C19	55
	Oil Q1550°F							
	Drawn 1000°F	182,000	162,000	15	40	363	C39	
4640†	Hot-rolled	100,000	87,000	21	50	201	C12	60
	Cold-drawn	126,000	97,000	14	39	269	C27	
52100†	Hot-rolled	109,000	80,000	25	57	235	C22	45
	Oil Q1550°F							
	Drawn 1000°F	185,000	170,000	9	34	415	C43	
6150	Hot-rolled	103,000	70,000	27	51	217	C18	50
	Cold-drawn	118,000	94,000	20	43	255	C25	
8740	Hot-rolled	95,000	64,000	25	55	190	B92	56
	Cold-drawn	107,000	96,000	17	48	223	C19	66
9260†	Hot-rolled	142,000	92,000	18	38	302	C31	

*See notes at bottom of Table 14-4.

†Steels 1060, 3140, 4640, 52100, and 9260 usually not stocked by steel warehouses.

[2]"Hot-rolled" and "as-rolled" are usually considered synonymous terms. Similarly "cold-drawn", "cold-rolled", and "CRS" (if steel) are used interchangeably.

Tests have shown that the tensile strength and Brinell Hardness Number have a direct relationship. For most steels, the tensile strength is found to be approximately 500 times the BHN. Tests show that the ductility becomes less as the hardness increases.

5. Plain Carbon Steel

When the only alloying element in steel is carbon, it is known as plain carbon steel. Carbon is a powerful alloying agent, and wide changes in strength and hardness can be secured by varying the amount of this element. The range of desired properties can be further enhanced by heat treatment. Some of the many uses for plain carbon steels are given in Table 14-6. This

Table 14-6   DIFFERENT USES OF PLAIN CARBON STEELS

Carbon Range, %	Uses of Carbon Steel
0.05–0.10	Stampings, sheets, wire, rivets, welding stock, cold-drawn parts.
0.10–0.20	Structural shapes, machine parts, carburized parts, screws.
0.20–0.30	Gears, shafting, levers, welded tubing, carburized parts.
0.30–0.40	Seamless tubing, shafts, connecting rods, crane hooks, axles. Can be heat treated.
0.40–0.50	Forgings, shafts, gears, studs
0.60–0.70	Drop hammer dies, set screws, locomotive tires, lock washers, hard-drawn spring wire.
0.70–0.80	Plow beams, cultivator disks, anvil faces, band saws, hammers, wrenches.
0.80–0.90	Plow shares, shovels, harrow blades, punches, rock drills, cold chisels, hand tools, music wire, leaf springs.
0.90–1.00	Springs, knives, axes, dies, hay rake teeth, harrow blades.
1.00–1.10	Drills, taps, milling cutters, knives.
1.10–1.20	Drills, lathe tools.
1.20–1.30	Files, reamers, knives, metal cutting tools.
1.25–1.40	Razors, saws, wire-drawing dies, metal-cutting saws.

type of steel is usually easier to anneal and machine than is alloy steel. Plain carbon steel is the cheapest, and the designer will normally use it except where severity of service conditions or difficulty of heat treatment necessitates the use of a more expensive alloy steel. Carbon steels are manufactured in larger quantities then any of the others.

The thickness of the standard gages of carbon steel sheets are shown in Table 14-7.

Carbon steels can be hardened and therefore strengthened by heat treatment when the carbon content is about 0.30 per cent or more. Heat treatment

Table 14-7 THICKNESS OF STANDARD GAGES FOR CARBON STEEL SHEETS, INCHES

Gage	Thickness	Gage	Thickness	Gage	Thickness	Gage	Thickness
7	0.1793	13	0.0897	19	0.0418	25	0.0209
8	0.1644	14	0.0747	20	0.0359	26	0.0179
9	0.1494	15	0.0673	21	0.0329	27	0.0164
10	0.1345	16	0.0598	22	0.0299	28	0.0149
11	0.1196	17	0.0538	23	0.0269	29	0.0135
12	0.1046	18	0.0478	24	0.0239	30	0.0120

generally consists of heating the part above the upper critical temperature, as shown by the zones in Fig. 14-6, and then quenching by submerging in a cooling medium such as water. After quenching, the steel will be very strong, but it may not be suitable for use because of lack of ductility and the presence of residual stresses. Quenching stresses are removed and ductility is restored, but tensile strength is reduced, by a subsequent operation called tempering or drawing. This operation consists of reheating the part and then quenching or cooling it in air. The reduction in the tensile strength and the increase in the ductility depend on the temperature used for tempering.

During quenching, the heat can be dissipated only through the surface where the part is in contact with the cooling medium. For parts of large cross section and large amounts of heat, the cooling rate for the interior may be so low that the metallurgical changes required for hardening cannot be secured. Under such conditions, only the material near the surface will be hardened and therefore strengthened; the interior of the body will largely be unaffected. The situation is generally referred to as mass effect or the effect of section size.

Curves for the mechanical properties of plain carbon steel 1045 for various tempering temperatures and various section sizes are given in Fig. 14-7. The curves show average values only. Fluctuations are to be expected; they are due principally to the permissible variations in composition, as indicated by Table 14-3.

The principal shortcoming of plain carbon steel lies in the shallow penetration of the hardening. The cooling rate must be very rapid, and high residual stresses, distortion, lack of ductility, and even quenching cracks will occur. The use of plain carbon steel in applications requiring only a hardened surface is sometimes prevented by the foregoing disadvantages.

Wrought iron, although its method of manufacture is different, is essentially a very low carbon steel with slag inclusions. Due to rolling, the slag inclusions are elongated into long stringers and do not greatly reduce the strength in the longitudinal direction. Wrought iron is tough and welds easily. It is widely used for making iron pipe.

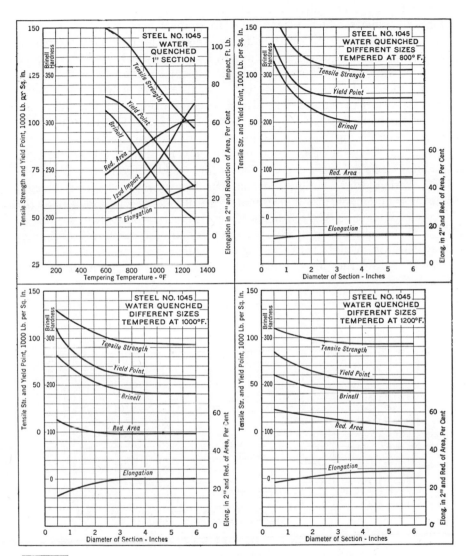

In section ½″ to 2″ incl., quenched from 1475/1525° F.; over 2″ to 4″ incl., from 1500/1550° F.; over 4″, from 1525/1575° F.

Figure 14-7 Properties of water quenched and tempered carbon steel 1045 in different sizes. (*Courtesy International Nickel Co.*)

6. Alloy Steels

When a suitable alloy is present in addition to the carbon, the metallurgical changes which take place during quenching occur at a faster rate, the cooling effects penetrate deeper, and a larger portion of the part is strengthened. The quenching can usually be done in oil. The heat is removed less drastically, and the residual stresses are smaller than for carbon steel. Hence there is less danger of distortion or formation of hardening cracks. A carbon steel part that might distort or crack from water quenching can sometimes be made of an alloy steel and quenched in oil. This ability to confer depth of hardening, or hardenability, is one of the principal reasons for the use of alloys. Of the various alloying elements, manganese is used in almost all alloy steels to confer depth of hardening. Nickel, chromium, and molybdenum, singly or in various combinations, are widely used. The maximum hardness obtainable in a thin section of heat-treated steel depends upon the carbon content and is largely unaffected by the lack or presence of any alloys.

The mechanical properties for 1 in.-round test pieces of chromium-molybdenum steel 4140 at various tempering temperatures are shown in Fig. 14-8. The properties for this material in different-sized sections are given in Table 14-8.

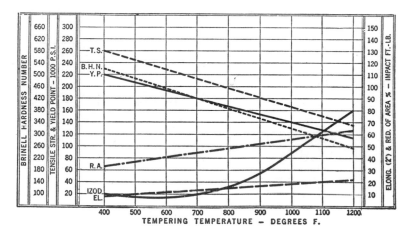

Figure 14-8 Properties of normalized and tempered chromium-molybdenum steel 4140. (*Courtesy Climax Molybdenum Co.*)

Curves showing average mechanical properties for steels 4340 and 8742 are given in Figs. 14-9 and 14-10, respectively. The properties of steel A8640 are similar to those of 8742 for small sections, but 8742 is somewhat deeper hardening in the larger sizes.

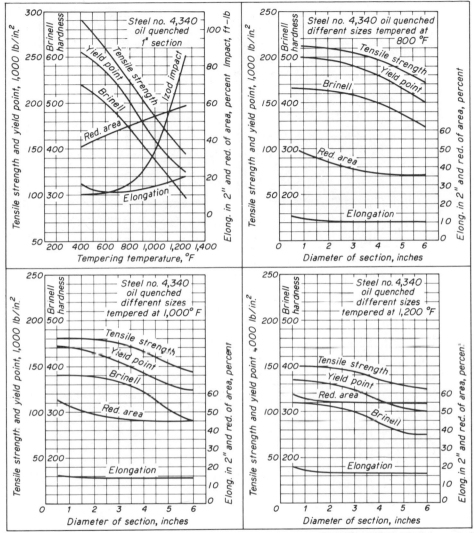

In sections 1/2" to 2" incl., quenched from 1,500 to 1,550 °F; over 2" to 4" incl., from 1,525 to 1,575 °F; over 4", from 1,550 to 1,600° F.

Figure 14-9 Properties of oil quenched and tempered nickel-chromium-molybdenum steel 4340 in different sizes. (*Courtesy International Nickel Co.*)

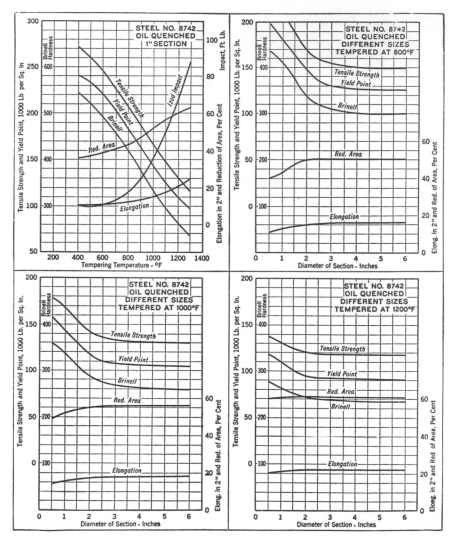

In sections ½″ to 2″ incl., quenched from 1500/1550° F . over 2″ to 4″ incl., from 1525/1575° F .
over 4″, from 1550/1600° F.

Figure 14-10 Properties of oil quenched and tempered nickel-chromium-molybdenum steel 8742 in different sizes. (*Courtesy International Nickel Co.*)

Table 14-8   EFFECT OF MASS ON THE PHYSICAL PROPERTIES OF
CHROMIUM-MOLYBDENUM STEEL 4140*
(Normalized and tempered at 1000°F)

Size (in.)	Outside Edge of Section						Center of Section					
	Tensile Strength	Yield Point	% Elong. in 2 in.	% Red. Area	BHN	Izod, ft lb	Tensile Strength	Yield Point	% Elong. in 2 in.	% Red. Area	BHN	Izod, ft lb
2	122,650	93,775	18.8	49.1	252	25.4	117,425	87,450	18.7	48.9	241	31.8
3	124,200	90,850	18.5	50.0	255	24.9	118,525	75,625	19.9	51.1	238	30.8
6	108,950	62,150	24.0	56.8	217	44.3	96,265	53,020	25.0	50.6	192	38.3
9	107,290	58,310	21.0	47.8	210	32.2	90,145	48,730	16.9	52.6	175	36.4
12	104,050	58,205	21.8	50.8	207	25.8	93,060	50,215	20.3	37.3	185	27.4

*Data from Climax Molybdenum Co.

The larger steel warehouses and the various alloy producers will supply expert metallurgical advice upon request.

7. High-strength Low-alloy Steels, HSLA

These steels have superior strength qualities, and, because of the reduced dead weight, are extensively used for mobile equipment. They exhibit good cold-forming and fabricating properties, and are readily welded with no tendency to harden on rapid cooling. In general, preheating and stress relieving are not required. Corrosion resistance is several times as great as for ordinary structural steel. Resistance to abrasion, fatigue, and impact is also good. The over-all cost is somewhat more than for ordinary structural steel.

8. Cost of Steel

The cost of materials is usually of prime importance in machine construction. In general, alloy steels are not employed if a satisfactory design can be secured by the use of plain carbon steel. If an alloy steel must be used, the lower-priced alloys should be given first consideration. When steel is ordered in large quantities, it can be purchased at the mill at what is known as the mill base price. Such prices are published in the metal trade journals for various products such as bars, shapes, plates, and sheets. The price of an alloy steel is determined by an "alloy extra" that is added to the base price. Table 14-9 gives the relative costs of certain classes of steel based on a price of 100 for hot-rolled plain carbon bars.

When steel is not purchased at the mill, but from a steel warehouse, the warehouse base price for such materials applies. This price is higher than

Table 14-9   APPROXIMATE WAREHOUSE METAL PRICES COMPARED
TO HOT-ROLLED CARBON STEEL BARS

Hot-rolled carbon steel bars	100
Hot-rolled alloy steel bars	170
Cold-finished carbon steel bars	130
Cold-finished alloy steel bars	210
High-strength, low-alloy, steel plate	140
304 Stainless steel bars	720
Aluminum bars	580

the mill base because of extra handling charges, transportation, and ware-houseman's profit.

Various other factors add to the price of steel. One of these is the "size extra." The base price applies only to sections of nominal size, and very small or very large sections cost an additional amount. A "quantity extra" is added to the price when the amount ordered is less than a specified mini-mum amount. A "cutting extra" is also added when stock lengths are not ordered. Other extras, such as annealing, machine straightening, or truck delivery, may further increase the pound price.

For many products where the cost of labor is relatively small, and where the cost of the materials is important, a small increase in the mill price of one steel is all that is needed to cause a widespread change to another type of steel which also gives satisfactory results. When labor costs are relatively large, as, for example, in aircraft parts, a change in the price of steel may have practically no effect on the composition used.

9. Heat Treatment of Steel

The microstructure that is formed when steel is quenched at a sufficiently rapid cooling rate from above the upper critical temperature is called *marten-site*. Steel in the martensitic condition is nonductile and is very hard and strong. Needle-like crystals in angular arrangement appear when prepared specimens are viewed under a microscope. Martensite is unstable and is generally regarded as being a supersaturated solid solution of carbon in iron. Fully hardened steel having the martensitic structure is not only too brittle to use, but contains quenching stresses. As was previously mentioned, a second operation of heating, called *tempering*, is required to relieve the internal stress and restore a suitable amount of ductility. The products formed when martensite is tempered have the structure of very finely divided pearlite. Tempering is usually carried out at temperatures between 800°F and 1,200°F. Ductility is increased and tensile strength reduced for an increase of tempering temperature. The tempering temperature must therefore be properly chosen so that the physical properties of the steel after heat treatment will be suitable for the working stress and design of the part. Steel must be heated to a temperature high enough to change the structure into austenite if hardening is to take place upon quenching. Quenching has no effect upon ferrite and pearlite.

When the cooling rate is not sufficiently rapid to form martensite, a finely divided pearlitic structure results from quenching. The ductility is greater and the strength less than for martensite. A further decrease of the cooling rate causes the thickness of the lamellae to increase until the coarse pearlitic structure of annealed or furnace cooled steel is secured.

Prolonged heating of steel within the range 1,250–1,350°F causes the cementite to coalesce into spherical particles. The resulting structure is called *spheroidized cementite*. Steel in this condition is soft and ductile and is readily machined. Heat treatment after machining may be required to secure the requisite degree of strength and hardness.

Steel, as received from the mill, is frequently given a treatment called *normalizing* to secure a uniform grain structure before it is machined. Normalizing consists of heating the steel above the upper critical temperature, removing it from the furnace, and cooling it in air. Sometimes the mechanical properties developed from normalizing are very satisfactory, and no other heat treatment need be given. The results obtained from annealing and hardening are usually improved if the steel is first normalized. Normalizing is frequently applied to forgings and castings to insure a uniform grain structure.

When the greatest softness and ductility are desired, the steel should be annealed. The heating is done slowly in the furnace and the maximum temperature, as shown by Fig. 14-6, must be maintained long enough to refine the grain structure. Cooling is done very slowly in the furnace or in an insulated container.

By a recent discovery in heat treating, it has been found possible to produce steels of the desired structure and mechanical properties by direct transformation of austenite. The method is patented and is called *austempering*. It consists of giving the heated steel a hot quench in a bath of molten lead or salt maintained at suitable temperatures. The tempering or drawing operations are eliminated when this treatment is used.

The heated parts must be protected from the oxygen of the atmosphere during the heat-treating process to prevent decarburization or loss of carbon by the surface material. If the part in service is subjected to fluctuating tension stress which exceeds the endurance limit, the weak surface layer of low-carbon material will be vulnerable to the formation of a fatigue crack. This crack will propagate inward and cause ultimate failure, even though the interior of the body has been strengthened by the heat treatment. Heat-treating furnaces can be constructed which surround the parts at the quenching temperature with an atmosphere of inert gases. Resistance to decarburization is a valuable property for a heat-treating steel. For good success in heat treatment, a part should be free from reentrant angles, sudden transitions from thin to thick sections, and thin projections attached to heavy masses.

Permanent deformation of metals is nearly always a shearing phenomenon. The atoms in the crystals are arranged in definite and repeating patterns. This uniformity of arrangement produces planes of weakness to shearing stress. The portions of the crystal, although adhering tightly together, will glide or slip on each other along these planes whenever the loading becomes sufficiently great.

The reasons for the increased strength and hardness resulting from quenching have been of great interest to investigators. The slip-interference theory is perhaps the best known. It is believed that during quenching exceedingly small cementite particles are thrown out of solution and dispersed throughout the crystal structure. These particles serve as mechanical keys which resist any motion along the slip planes and thereby give a large increase to the elastic strength of the material.

Hardenability, or depth of hardening, in steels is a question of great importance to the designer. Curves such as those shown in Figs. 14-7 to 14-10 show the average heat treating properties for these steels. It is frequently necessary to have precise information on the hardening ability for each particular lot of incoming steel. The Jominy test has been devised for this purpose.

A 1 in.-diam specimen is made from the steel whose hardness after quenching is to be determined. The specimen is heated and then quenched by hanging it vertically over a jet of water that impinges gently on the bottom surface. Flats are ground on the sides of the specimen, Rockwell C hardness readings are taken at $\frac{1}{16}$-in. intervals from the quenched end, and the results are plotted. The resulting curve shows the manner in which the hardness decreases with distance from the quenched end. When a round bar of any size, or a machine part, is quenched in the conventional manner, the surface and interior points have cooling rates equal to those of different points of the Jominy specimen. Cooling rate curves and tables are available[3] which show the correlation that exists between distances from the quenched end of the Jominy specimen and the diameters of bars of various sizes for cooling of the surface, the center, and intermediate points. The hardness, as given by the Jominy curve of the specimen at the proper distance from the quenched end, will be the hardness of the chosen point of the quenched object. The strength of the particular lot of steel from which the specimen was taken can thus be estimated.

As a typical example, Jominy curves for steel 8642 are shown in Fig. 14-11. The effect of tempering temperatures on the hardness is illustrated by the various curves. The mechanical properties for this steel as interpreted from the hardness results are given in Table 14-10.

[3]See "Selection of Construction Steels," *Metal Progress*, **66**, July 15, 1954, p.l.

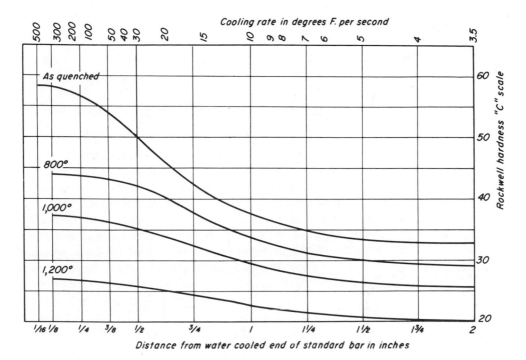

Figure 14-11 Typical end quench hardenability curves of AISI Steel 8642.
(*Courtesy Joseph T. Ryerson & Sons, Inc.*)

Table 14-10 MECHANICAL PROPERTIES INTERPRETED FROM
HARDENABILITY RESULTS OF FIG. 14-11 FOR STEEL 8642.
[*Quenched in oil from 1550°F and tempered (drawn as shown)*]

Size of Round	Tempering Temperature	Tensile Strength, psi	Yield Point, psi	% Elong. in 2 in.	% Red. Area	Brinell Hardness	Rockwell C Hardness
1 in.- round center	800	198,000	189,000	13	46	401	43
	1000	170,000	159,000	16	52	341	37
	1200	131,000	115,000	20	61	262	27
2 in.- round 1/2 radius	800	185,000	174,000	14	51	375	40
	1000	158,000	146,000	17	56	321	34
	1200	127,000	110,000	21	62	255	25
3 in.- round 1/2 radius	800	161,000	150,000	17	55	331	35
	1000	145,000	133,000	19	58	293	31
	1200	120,000	102,000	22	63	241	23
4 in.- round 1/2 radius	800	145,000	133,000	19	58	293	31
	1000	131,000	115,000	20	61	262	27
	1200	114,000	96,000	22	64	229	21

10. Residual Stresses from Heat Treatment

The increased strength and hardness resulting from the heat treatment of machine parts is usually accompanied by residual stresses. Since these may be either favorable or unfavorable to the engineering functioning of the part, the subject should be given careful consideration by the designer. Such stresses arise from plastic deformation which takes place while the material is at elevated temperature.

Consider first the purely thermal effects caused by quenching a through-heated steel from a temperature somewhat below the lower critical temperature of 1330°F. Upon immersion in the cooling medium, the surface cools at a very rapid rate. Contraction of the outer layers is largely prevented by the rigid core at the higher temperature. Tensile stresses are thus set up on the surface which are balanced by compressive stresses in the interior. With the decrease in temperature of the core, and the accompanying thermal contraction, the tension on the surface may be reduced to zero and then turned into compression. There will then be a balancing tensile stress in the interior. The situation is as shown in Fig. 14-12, which represents the final stress distribution across the diameter of a long cylindrical body.

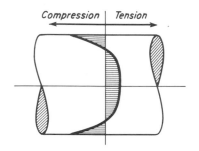

Compression | Tension

Figure 14-12 Distribution of residual stress along diameter of a long cylindrical body.

Now suppose the material is through-heated above the critical temperature into the austenitic range. Let it also be assumed that the material contains a sufficient proportion of carbon or alloying elements so that martensite will be formed upon sudden cooling. The thermal tension on the surface forms as before. However the phase change to martensite throughout the body involves an increase in volume which tends to increase the tension on the surface and the compression in the interior. The continued thermal shrinkage, however, reduces the tension on the surface or even turns it into compression.

These thermal and transformation effects occur simultaneously, but the exact manner in which they combine is highly complex and has never been explained to the extent that easily followed rules have been made available. The final quenching stresses on the surface may therefore be either tension or compression, depending on the quenching temperature, cooling rate, size and shape of the part, carbon and alloy content, and the temperature at which the transformation to martensite occurs. The stress distribution may also be more complex than that shown in Fig. 14-12. Quenching must be very carefully carried out. A large surface tensile stress may cause warping and distortion or even cracking of the part. The tendency to crack is augmented by the presence of stress raisers such as notches, grooves, holes, reentrant corners, and so on. Oil quenching is less severe than water quenching. The heat is removed less quickly, and the difference in temperature between the surface and the center is less than for water quenching. The tendency for distortion and cracking is therefore less with oil quenching.

A residual surface compressive stress is beneficial for parts with repeated or fatigue loading, and such a stress can effect a worthwhile increase in the factor of safety. It is known, however, that surface residual stresses tend to lessen or fade gradually with repeated stressing. Tempering to restore a portion of the ductility causes a loss of the surface compression. In fact, a tempering temperature of 1,000°F may reduce the compression to zero or turn it into tension.

A surface with residual tension is vulnerable to the formation of fatigue cracks. Such conditions have undoubtedly been the cause of many failures. Residual tension can be removed and the surface left in compression by mechanical operations such as rolling between rollers at high pressure, shot peening, tumbling, or peen hammering.[4]

11. Carburizing and Nitriding

Sometimes a machine part requires a hard surface to resist wear, abrasion, or local deformation from impact loads, combined with a tough, strong core for strength. The case hardening or carburizing process can be used for obtaining such a structure. The parts are packed in a compound rich in carbon and held for a number of hours in the furnace at a temperature within the austenitic range. Carbon is absorbed by the surface layer. The same effect is secured by heating the steel in the presence of gases rich in carbon. The high-carbon surface material is transformed to martensite by quenching, followed by tempering at 300–350°F for relief of quenching stresses. The period of heating may extend from 3 to 24 hr. A depth of case of about $\frac{1}{16}$ in. can be secured after about 8 hr in the furnace. Freedom from warping during the extended heating period is a very desirable property for a carburizing steel. Carburizing steels in general are low in carbon and have good machinability ratings, but all machine work must be completed before the part is carburized. The core properties for a number of steels used for carburizing are given in Table 14-11.

Nitriding produces a very hard surface layer by the absorption of nitrogen during heating that lasts for an extended period at 900–1,000°F in the presence of ammonia gas. Special nitriding steels must be used, but no heat treatment subsequent to the hardening is required. Both the carburizing and nitriding processes leave the part with a residual stress of compression

---

[4]For additional information on residual stress see Hetenyi, M., ed., *Handbook of Experimental Stress Analysis*, New York: McGraw-Hill Book Company, 1950, p. 495.

Table 14-11 CORE PROPERTIES OF VARIOUS CARBURIZING STEELS*

Steel	Tensile Strength	Yield Strength	% Elong. in 2 in.	% Red. Area	Izod, ft lb	Steel	Tensile Strength	Yield Strength	% Elong. in 2 in.	% Red. Area	Izod, ft lb
1020	96,000	60,000	19	41	48	4119	137,000	112,000	16	42	
1117	96,000	59,000	23	53	33	4615	110,000	75,000	23	66	67
2315	138,000	108,000	22	61	42	6115	104,000	62,000	28	58	
3115	125,000	85,000	22	54	36	A8620	135,000	108,000	17	48	

*One-inch rounds carburized 8 hr 1,650–1,700°F, cooled in box, reheated to 1,475°F. Steels 1020 and 1117 are water quenched and tempered at 350°F. Others are oil quenched and tempered at 300°F.

throughout the hardened zone. If the working load produces a fluctuating tensile stress, the presence of the residual surface compression gives a favorable stress condition for preventing the formation of a fatigue crack.

The cyaniding process forms a very hard surface layer 0.002 in.–0.010 in. thick through immersion for a short period, 15 min to 1 hr, in a bath of molten sodium cyanide, and then by quenching in water or oil. The hardness is derived from compounds of nitrogen and carbon absorbed by the surface layer.

12. Flame Hardening

Localized surface heating with a flame or induction coil, followed by rapid cooling, usually leaves the surface in tension. The cool interior of the body does not expand, and the attached heated layer must expand plastically outward. This enlarged material is unable to return to its original size when cooled, and residual tension results.

Grinding cracks can be accounted for by the process above. Brake drums, clutch plates, railroad wheels and rails, and other friction surfaces are affected in the same way. Flame-cut parts are usually left with a surface layer of tension. Contraction of weld metal deposited on relatively cold-base metal produces a residual stress of tension. Bent shafts can be straightened by heating them on the outside of the bend. If properly carried out, the residual tensile stress in the region that has been heated will cause the shaft to straighten.

Flame or induction hardening leaves the surface in a state of residual compression only when the steel has a sufficient carbon or alloy content to undergo the phase change to martensite upon quenching from the austenitic condition. The increase in volume from the martensite formation must be sufficient to overcome the thermal contraction of the heated layer.

At the edges of the heated region, the martensite gradually merges into the weaker surrounding material. When the part must be heated in sections, the junctions between the heated regions should be located at points where the loading stresses are low. For example, in hardening the teeth of a gear, the junctions should be placed at the tips of the teeth and not at the base where the maximum bending stress occurs.

13. Strain Hardening

Plastic deformation, resulting from drawing metals through a die, or rolling them between rollers, raises the values of the yield point and tensile strength of ductile materials such as mild steel, stainless steel, copper, brass, and aluminum. This increase in strength is accompanied by a loss of ductility and an increase of hardness. The effect of cold drawing on the properties of steel can be estimated by the curves in Fig. 14-13. The strengthening becomes

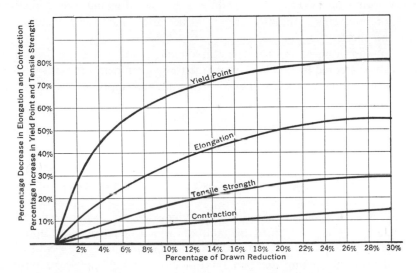

Figure 14-13 Effect of cold drawing on the tensile properties of steel. (*Courtesy Bethlehem Steel Co.*)

less at distances removed from the surface, and the properties of bars of large cross section undergo less change because of the cold work. The stress-strain diagram for low-carbon steel loses its characteristic shape if the material is cold worked. The large elongation at the yield point does not take place, but a gradual bending of the curve occurs after the proportional limit has been passed.

Materials that are too weak and soft for use in the annealed condition are sometimes strengthened by cold working. After such treatment, the mechanical properties may no longer be the same in all directions, and usually a definite loss of ductility results in the direction normal to the deformation. In addition, any increase of the yield point in tension produced by stretching is usually accompanied by a decrease of the yield point in compression. This fact must be kept in mind if the material in normal operation is subjected to reversing stresses.

In general, cold working leaves the part with residual stress on the surface. This is an undesirable situation, since parts made from such materials will warp if a portion of the surface is removed by machining. A tensile residual stress on the surface causes the part to be vulnerable to the formation of a fatigue crack should the stress be fluctuating.

The hardening and loss of ductility produced by plastic deformation disappears if the material is subjected to annealing temperatures. When the fabrication process involves successive steps of cold working, the material may become so brittle that rupture may occur. This danger can be avoided by restoring the ductility periodically by annealing the metal between the cold-working operations. Cold rolling and cold drawing are sometimes used for producing a smooth, bright surface and dimensional accuracy on bars and sheets.

An improvement in the strength, ductility, and impact resistance of steel can be secured by hot working at the forging temperature. Hot working gives a finer and more uniform grain structure, and improves the soundness of the material.

14. Hardness

The term hardness is used to designate certain mechanical properties of a material, such as resistance to penetration, scratching, abrasion, or cutting. Since a rigorous definition of hardness has not as yet been written, this quality must be specified by numbers which are dependent upon the method of making the test. Manufacturing operations such as cold work, quenching and tempering, and precipitation heat treating change the hardness of many engineering materials. Hardness testing is therefore one of the principal methods for determining the suitability of a material for its intended purpose. Hardness testing is a valuable inspection tool for maintaining uniformity of quality in heat-treated parts. A useful correlation exists between the hardness of a material and the tensile strength, endurance limit, and wear resistance.

The Brinell is one of the best known of the penetration methods for determining hardness. In the standard test, a hardened 10-mm diam steel ball is pressed into the specimen by a force of 3,000 kg. The diameter of the impression is determined by a hand microscope, and the area of the contact surface between ball and material is computed. The Brinell Hardness Number, BHN, is found by dividing the load in kilograms by the area in square millimeters. The harder the material, the smaller the indentation, and the higher the Brinell number. The tensile strength of steel in pounds per square inch is usually about 500 times the BHN.

The Rockwell machine for measuring hardness is also in wide use. It is a rapid method, since the hardness is read directly by a dial on the machine which eliminates the necessity of making a separate measurement of the impression. Two types of Rockwell readings are in general use: the B and the C. The B reading is made by use of a steel ball penetrator $\frac{1}{16}$ in. in diameter. The dial gives a reading which is dependent upon the penetration caused by a load of 90 kg, after an initial load of 10 kg has first been applied to seat the ball on the specimen. Rockwell B hardness tests are used for un-hardened steel and nonferrous metals which are not extremely hard. Both letter and number must be used in specifying Rockwell hardnesses.

For the Rockwell C reading, a 120° diamond cone called a brale is used as the penetrator. The initial load is again 10 kg, but the main load con-sists of 140 additional kg. For very thin specimens, or for parts covered with a shallow, hardened skin, the B or C Rockwell tests penetrate too deeply, and the superficial Rockwell test must be used. In the superficial test, both the initial and major loads are considerably reduced, depending on the thick-

Table 14-12   APPROXIMATE EQUIVALENT HARDNESS NUMBERS

Steel*								Nickel and High Nickel Alloys†				
R_c	BHN	R_c	BHN	R_b	R_c	BHN	R_b	BHN	R_b	R_c	BHN	R_b
65	739	45	421		25	253	101.5	298	106	32.0	118	66
64	722	44	409		24	247	101.0	275	104	28.5	114	64
63	705	43	400		23	243	100.0	258	102	25.5	111	62
62	688	42	390		22	237	99.0	241	100	22.5	108	60
61	670	41	381		21	231	98.5	228	98	20.0	106	58
60	654	40	371		20	226	97.8	215	96	17.0	103	56
59	634	39	362		18	219	96.7	204	94	14.5	100	54
58	615	38	353		16	212	95.5	194	92	12.0	98	52
57	595	37	344		14	203	93.9	184	90	9.0	95	50
56	577	36	336	109.0	12	194	92.3	176	88	6.5	93	48
55	560	35	327	108.5	10	187	90.7	168	86	4.0	91	46
54	543	34	319	108.0	8	179	89.5	161	84	2.0	89	44
53	525	33	311	107.5	6	171	87.1	155	82		87	42
52	512	32	301	107.0	4	165	85.5	149	80		85	40
51	496	31	294	106.0	2	158	83.5	144	78		83	38
50	481	30	286	105.5	0	152	81.7	139	76		81	36
49	469	29	279	104.5				134	74		79	34
48	455	28	271	104.0				129	72		78	32
47	443	27	264	103.0				125	70		77	30
46	432	26	258	102.5				121	68			

*See p. 1236, *Metals Handbook*, 8th ed., 1961. BHN values for 10 mm tungsten carbide ball, 3000 kg load.

†See ASTM Std. E 140–65, Vol. 30. BHN values for 10 mm standard ball, 3000 kg load.

ness of the specimen or of the hardened zone. The relationship between Brinell and Rockwell numbers is shown in Table 14-12.

Another instrument for hardness testing is the Shore Scleroscope, in which a small diamond-tipped weight is allowed to fall on the specimen. The height of the rebound is used as the measure of the hardness of the material. This instrument has the desirable feature of portability, and can be used on specimens too large to be placed in a regular hardness-testing machine.

The Vickers pyramid test uses a small diamond pyramid to indent the specimen. The hardness number, as for the Brinell test, is found by dividing the load by the area of the indentation. Since the loads are light, a high magnification is required to determine the width of the impression.

Scratch hardness tests are also in use. A diamond or sapphire point is drawn over a polished surface under a definite load and the width of the scratch is measured with a microscope. The hardness number is usually considered as varying inversely with the square of the width of the scratch.

15. Machinability

The relative ease with which a given material may be machined, or cut with sharp edged tools, is called its machinability. The ratings given in Tables 14-4 and 14-5 are based on Bessemer screw stock B 1112 as 100 per cent. These refer to steels "when turned with a suitable cutting fluid at 180 ft per min under normal cutting conditions."

When a part requires a large amount of machining, the designer tries to specify a composition and grain structure best adapted to the cutting operations. Sometimes certain elements, principally sulfur and manganese, are added to steel to improve the cutting properties. Steels with very low carbon content usually have poor machinability. For greatest cutting ease, the carbon content should be about 0.10 per cent for Bessemer steels and 0.20 per cent for open-hearth steels. Alloy steels, because of their higher strength and toughness, are usually harder to machine than plain carbon steels. In general, the machining properties of a steel can be materially improved by cold drawing. Annealing and normalizing heat treatments are beneficial for medium- and high-carbon and alloy steels. For best machining, a steel should be soft and brittle. This is not possible since steel is usually soft and tough or hard and brittle. For high surface speeds, medium-size pearlite grains, with each grain surrounded by an envelope of ferrite, are desirable. For low speeds and heavy feeds and depths of cut a spheroidal microstructure is best. In general, the machinability rating cannot be correlated with the Brinell Hardness Number. Steels of best machinability have BHN of 187 to 217.

16. Grain Size

A relationship exists between the mechanical properties of steel and the size of the grains or crystals. Coarse-grained steels, in general, harden deeper and are best for rough machining operations. They have good hot-forming characteristics and offer better resistance to high-temperature creep. Fine-grained steels, in contrast, are tougher and are superior for impact resistance, especially at low temperatures. They exhibit less distortion, internal stress, and cracking from heat treatment.

Grain sizes in steel are designated by ASTM grain size numbers 1 to 8. Number 1 is the largest, and has a calculated diameter of equivalent spherical grain of 0.0113 in. Number 8 is the smallest, with a spherical grain diameter of 0.0010 in. Steels of grain sizes 1 to 5, inclusive, are generally called coarse-grained, and fine-grained steels have grain sizes 6 to 8, inclusive.

The grains of austenite that are formed when steel is heated have the smallest size at the upper critical temperature, as indicated by Fig. 14-5. A fine-grained steel therefore results if the material is cooled from the fine-grained austenitic condition. If the steel is heated above the upper critical temperature, the grains of austenite increase in size and a coarse-grained steel results upon cooling. The grain size is therefore to a large extent under the control of the heat treater. Hence, if a fine-grained steel is desired in a hot-rolled or forged product, the hot working must be continued until the temperature drops almost to the upper critical point, or the grains will coarsen in cooling from the high temperature at which hot working was stopped.

Grain size in steel refers to the size of the previously existing austenite grains at the quenching temperature. Therefore, the sample must be quenched from the desired temperature and tempered lightly, and a specimen must then be prepared for examination. The martensite grains reflect the austenitic grain size. The behavior of carburizing steel is usually evaluated by means of a standardized method called the McQuaid–Ehn test.

An uncertainty sometimes exists in the purchase of steel as to whether lots from different sources, but of the same chemical composition, will have identical physical properties when given the same heat treatment. The physical properties, as mentioned in the foregoing, depend on the grain size, which in turn depends upon the grain-coarsening characteristics of the material. It is possible for one piece of steel to remain fine-grained for a considerable temperature range above the upper critical temperature, whereas another piece of steel of the same chemical composition, but from another lot, may coarsen when raised to this temperature. A knowledge of the inherent grain growth tendency is thus very important if consistent physical properties are to be expected after heat treatment. In fact, the term fine-grained is sometimes applied to a steel which does not begin to coarsen until the temperature rises a considerable amount above the upper critical. A coarse-grained steel, then, is one which coarsens rapidly as the temperature rises beyond the upper critical. Closer temperature control is required for the heat treatment of coarse-grained steel. Fine-grained steels have a wider range of safe hardening temperatures, and more predictable results can be obtained in the heat treatment process. The tendency for grain growth at elevated temperatures can be closely controlled by the addition during steel making of such inhibiting agents as minute quantities of finely divided aluminum.

17. Corrosion

Corrosion is the deterioration or destruction of a material because of reaction with its environment. All environments are corrosive to some degree: all atmospheres and most kinds of water, chemical and fuel gases, industrial and food acids. Contamination by the products of corrosion may affect the purity and market value of manufactured products. Failure of mechanical equipment due to corrosion may be hazardous to the operating personnel. Corrosion is electrochemical in nature and is frequently a very complex process.

Rusting is a term reserved for the oxidation of iron and steel. It has been estimated that the loss from the rusting of iron and steel products in the United States is $6 or $7 billion per year. An iron surface exposed to dry air forms a thin film of iron oxide. So long as the atmosphere remains dry, the reaction is seldom destructive as the film prevents the further penetration of oxygen. However, with increasing humidity, the amount of rust produced may be many times that formed in a dry atmosphere. The oxidation rate also increases with increase in temperature.

On the other hand, the oxidation rate is slow for a tightly adhering scale, such as occurs in aluminum. Aluminum oxidizes readily, but the scale adheres so tightly that it gives good protection to the parent metal.

Corrosion can take place uniformly over a surface, or may be highly localized in nature. The placing of two dissimilar metals in contact provides an electron flow between them which results in an increase in the corrosion rate. Corrosion is sometimes confined to crevices formed by holes, bolt and rivet heads, lap joints, and gasket surfaces. Here the crack is wide enough to admit the liquid, but small enough to provide a stagnant zone. Smooth surfaces will sometimes pit and the corrosion goes on at a increasingly rapid rate as the pits become deeper.

Intergranular corrosion, or the failure of grain boundaries, may occur especially if impurities are present. Selective leaching may remove one component of an alloy without affecting the others. Corrosion is sometimes hastened by relative movement between the metal and corrosive fluid which prevents the formation of a passive surface film on the metal. Impingement, when the fluid must change directions (as in an elbow), can cause failure in the metal.

Stress corrosion cracking may occur from the simultaneous presence of a tensile stress and a specific corrosive medium. The metal as a whole is largely unaffected, but a network of fine cracks spreads over its surface. This type of failure includes what is called "season cracking" of brass and "caustic embrittlement" of steel. The cracking starts at tensile stress smaller than the yield point value, but the process is not as yet fully understood. It may proceed along the grain boundaries or may occur across the crystals. Hydrogen embrittlement of steel can also be classified as a corrosion phenomenon.

A corrosive environment may reduce the fatigue strength of a metal to but a fraction of its normal value. The pits and cracks serve as stress raisers.

18. Prevention of Corrosion

There are a number of natural material combinations where corrosion can be reduced to a very low value. In general, such combinations will give the maximum amount of corrosion protection at the lowers cost. These[5] are listed in Table 14-13. Pure metals are generally more resistive to corrosion than those containing impurities or small amounts of other elements.

Table 14-13   NATURAL CORROSIVE-RESISTIVE COMBINATIONS

Corrosive Element	Resisting Medium
Concentrated sulfuric acid	Steel
Diluted sulfuric acid	Lead
Nitric acid	Stainless steels
Hot hydrochloric acid	Hastelloys, Chlorinets*
Hydrofluoric acid	Monel metal
Caustic	Nickel and nickel alloys
Distilled water	Tin
Hot strong oxidizing solutions	Titanium
Nonstaining atmospheric exposure	Aluminum
Ultimate resistance of all materials	Tantalum

*Nickel-base alloys with molybdenum.

Careful attention should be given to the details of construction for the containers of corrosive materials. Low stresses should be used, and stress concentrations and uneven heat distribution should be avoided. Vessels should be welded rather than riveted or bolted. Bottoms should be rounded or sloped with round corners for good drainage. Drain-out valves and plugs should be flush with the bottom. Baffles and stiffners should provide for free

[5]See p. 195, Reference 5, end of chapter. This book should be consulted for information on the details of application of these materials.

drainage. Whenever possible the same material should be used throughout. Means should be provided for easy cleaning and easy replacement of failed elements. Air should be excluded from the system whenever possible.

Many processes are in use for applying a corrosion-resistant coating to the surface of steel and iron products. Coatings of zinc, tin, lead, copper, nickel, chromium, and cadmium are in wide use. Application is made by hot dipping, electroplating, or by a molten metal spray, depending on the metals involved. The Parkerizing and Bonderizing processes utilize a coating of iron phosphate as the protective agent or as a base for the application of paint or enamel.

Resistance of steel to atmospheric corrosion can be increased by the addition of about 0.20 per cent of copper. The HSLA steels described in Section 7 are still more resistant to this destructive action.

Paint, varnish, lacquer, and other organic coatings will provide limited protection against certain kinds of corrosion. Success will depend not only on the selection of the proper top coat, but on satisfactory preparation of the surface, and on a suitable prime coat. Cathodic and anodic protection can reduce corrosion substantially by means of electric currents.[6]

19. Stainless Steel

The so-called stainless steels are in wide use for resisting corrosion. There are many alloys of this kind, and the proper alloy must be selected for its intended purpose. Chromium is the principal alloying element, and occurs in percentages from 11.5 to 30. In some alloys nickel is also present in amounts up to 20 per cent. Close control must be exercised on the carbon content of these materials. There are three general classifications of stainless steels, and Table 14-14 shows a few typical alloy numbers in each class. Because of the alloys, the cost is considerably more than for plain carbon steel.

Under corrosive environments the film that forms on the surface is very stable, and is self-healing. However some stainless steels are suseptible to localized corrosion in stagnant areas. These materials should not be used for chloride-containing media.

Considerable experience may be required for the successful fabrication of these materials. Unless a free-cutting type is specified, machining is difficult with a resulting poor surface.

(a) *Austenitic, 18 per cent Cr–8 per cent Ni (302, 304, 316, 347)*. About one-half of all stainless steel produced is austenitic, and of these, the greatest tonnage is number 304. The nickel is used principally to increase the ease of fabrication. When high strength is required it must be produced by cold work, but this reduces the corrosion resistance. It is possible by cold work to achieve strengths several times those shown in Table 14-14 for annealed materials.

[6]See p. 205, Reference 5, end of chapter.

Table **14-14**   COMPOSITION AND MECHANICAL PROPERTIES OF
SELECTED STAINLESS STEELS

	Austenitic (Cr 18%, Ni 8%) Hardenable by Cold Working; Nonmagnetic when Fully Annealed				Martensitic (Cr 12%) Hardenable by Heat Treatment		Ferritic (Cr 17%) Nonhardenable; Can Be Hot Worked; Magnetic	
Type No.	302	304	316	347	410	440A	430	446
Cr	17.0–19.0	18.0–20.0	16.0–18.0	17.0–19.0	11.5–13.5	16.0–18.0	14.0–18.0	23.0–27.0
Ni	8.0–10.0	8.0–12.0	10.0–14.0	9.0–12.0	0.5 max	0.5 max	0.5 max	0.5 max
C, max.	0.15	0.08	0.10	0.08	0.15	0.60–0.75	0.12	0.2
Mn, max.	2.0	2.0	2.0	2.0	1.0	1.0	1.0	1.5
Si, max.	1.0	1.0	1.0	1.0	1.0	1.0	1.0	1.0
Others			Mo 2.0–3.0	Cb stab.		Mo .75 max		N. 25 max
Yield str.*	30,000	30,000	30,000	30,000	32,000	55,000	35,000	45,000
Tensile str.	80,000	80,000	75,000	80,000	60,000	95,000	60,000	75,000
El. in 2 in. %	50	50	40	40	20	20	20	20
Red. Area	60	60	50	50	50	40	40	40
Brinell max.	180	180	200	200	200	240	200	200
Yield str.†					35–180,000	55–240,000		
Tensile str.					60–200,000	95–275,000		
El. in 2 in. %					25	20		
Brinell					120–400	200–555		
Rockwell					B70–C45	B95–C55		

*Strength values are for annealed or furnace-cooled material.
†Strength values are for heat-treated material.

As usual, this increase in strength is accompanied by a reduction in ductility
and an increase in hardness.

At temperatures 950–1,450°F, these steels are susceptible to intergranular
corrosion unless special alloys or treatments are provided. This temperature
range occurs in welding and may result in a decay zone on either side of the
weld and somewhat removed from it. The spot-welding characteristics,
however, are very good.

(b) *Martensitic (410, 440A)*. These steels can be hardened by quenching,
and then tempered to get the desired mechanical properties. For corrosion
resistance the surface must be smooth with all scale and foreign matter removed. A softening treatment must usually be given before machining,
although free-cutting grades are available. These steels are difficult to weld
because of the strong air-hardening tendency. Welding must usually be followed by immediate annealing. Stainless steel cutlery is usually of the martensitic type. When the carbon content is high, the properties resemble those
of tool steel. With lower carbon content, the steel is similar to ordinary
machine steel.

(c) *Ferritic (430, 446)*. The forming and drawing properties of these
materials resemble those of mild steel. Number 430 is used in architectural
work, automobile trim, hardware, and equipment in the chemical and food
industries. It has many applications in high-temperature service, and has good

scaling resistance up to 1,550°F. Welding may cause grain coarsening which reduces the toughness and impact resistance. Number 446 has good heat resistance and is used in furnaces and heat-treating equipment.

20. Wear

Wear and corrosion are the two most adverse factors for reducing the life of mechanical equipment.[7] The loss of a relatively small amount of material at certain critical locations may be the only difference between a new machine and one that is worn out. High-carbon martensite, or martensite tempered below 450°F has the best wear resistance of the steel structures.

(a) *Cutting Wear*. Cutting wear results when a hard rough surface is rubbed over a softer one with or without a lubricant. The projections on the harder metal act like cutting tools to remove material from the softer one. The action is more severe at low speeds. Unless the metals have widely different hardnesses, such wear soon ceases as the parts become worn in.

(b) *Abrasive Wear*. This is the most common type of wear, as there is often some grit such as metal particles, oxides, or dust carried between the rubbing surfaces. Rapid and continuing wear may result. Particles too small to be removed by a filter can produce wear. The abrasive can be either embedded or rolling. Abrasive embedded in the softer material can wear the harder member. Cast-iron plates are used for lapping hard steel since the abrasive is embedded in the graphite flakes of the iron. Although abrasive particles embedded in a soft bearing shell will wear the shaft, the effect is less than the scoring that might result from loose particles. A relatively thick oil film that will keep the parts separated is effective in preventing wear. Under conditions of boundary lubrication, the separation of the parts does not change appreciably with continued operation, and such wear will continue even after the initial surface roughness has been eliminated.

High-manganese or Hadfield steel possesses exceptional ability to resist abrasion. Its impact properties are also very good. The steel contains 10–14 per cent manganese and 1.0–1.4 per cent carbon. This material work hardens so rapidly, and is so tough, that machining is difficult and must frequently be done by grinding.

(c) *Adhesion or Galling*. This is a form of frictional failure that manifests itself as a welding of one surface to another. Portions of the surfaces stick together and when they are torn apart material is removed from one of them. A film of any type on the surfaces affords protection against galling. For galling to occur, the rubbing conditions must be such as to break down the film, at least locally, for the short time necessary to bring clean metal into

[7]See Dayton, R. W., "Mechanisms of Wear: Their Relation to Laboratory Testing and Service," in *Proceedings of Special Summer Conference on Friction and Surface Finish*, Cambridge: Massachusetts Institute of Technology, June 1940, p. 127. See also Kragelski, I. V., *Friction and Wear*, Washington: Butterworths, 1965, and Rabinowicz, E., *Friction and Wear of Materials*, New York: John Wiley & Sons, 1965.

contact with clean metal. Finely divided grit in the oil tends to clean the surfaces and promote galling. Like metals, when rubbed together, are more apt to gall than metals of different metallurgical structure. An exception to this rule is cast iron, which runs well with cast iron.

Any material which tends to form a tightly adhering film over the rubbing surfaces will decrease the tendency for galling. For example, extreme pressure, or EP, lubricants act in this way to prevent galling in hypoid gears. Coatings such as phosphate, oxide, tin, and cadmium are useful in preventing galling during the running-in periods.

(d) *Corrosive Wear.* Chemical attack will affect the wear rate—either making it less or making it greater. If the corrosion product is a tightly adhering film, the wear rate is reduced, as mentioned above for EP lubricants. When the resulting film is nonadhering, the wear rate is usually increased. Sometimes the film adheres but gradually builds up to such a thickness that it peels off, causing a sudden acceleration in the wear rate. Corrosion may be caused by the presence of water or certain small but active agents in the lubricant or atmosphere.

21. Short-term Effects of High Temperatures

A rise in temperature generally causes a change in the physical properties of a metal. Tests on medium-carbon rolled and cast steels indicate that the tensile stress increases until the temperature rises to the blue-heat point at about 500-550°F. As shown by Fig. 14-14, this increase of strength is accompanied by loss of ductility; it is termed "blue brittleness." A further increase in temperature causes a rapid decrease of strength.[8] Other tests have shown that the impact strength at about 800-900°F is considerably less, but at 1,300°F is somewhat greater, than at room temperature. A further rise in temperature, however, is accompanied by a rapid decrease in impact strength. The endurance limit for a change of temperature generally follows the same trend as the tensile strength. The modulus of elasticity drops for a rise in temperature, as is shown by Fig. 14-15.

Cast iron is but slightly affected by temperatures up to 800°F, but above 900°F a sharp decrease in strength occurs. Malleable iron, above 800°F, shows a decrease in strength and an increase in ductility with increasing temperature.

At subnormal temperatures, there will be an increase in the values for the tensile strength, yield strength, endurance limit, and hardness, and a decrease in the ductility of both ferrous and nonferrous metals.[9] At low temperatures, however, many steels have a marked reduction in impact strength.

[8]See "Cooperative Short-Time High Temperature Tension Tests of Carbon Steel K6," *Proc. ASTM*, **33**(1), 1933, p. 213.

[9]See "Impact Resistance and Tensile Properties of Metals at Subatmospheric Temperatures," ASME–ASTM Committee Report, Aug. 1941. See also, "Behavior of Ferritic Steels at Low Temperatures," ASME–ASTM Committee Report, Dec. 1945.

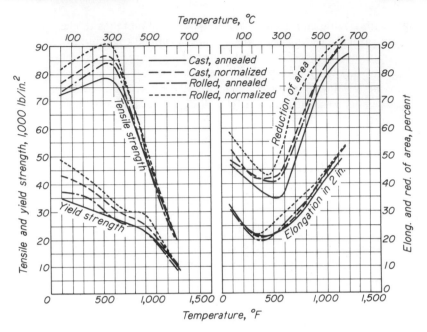

Figure 14-14 Tensile properties of cast, annealed or normalized, and of rolled, annealed or normalized, 0.28 per cent carbon steel at elevated temperatures.

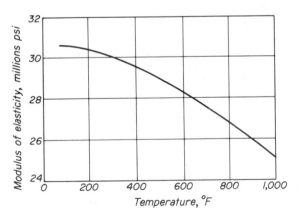

Figure 14-15 Reduction of modulus of elasticity with temperature for SAE 4130 annealed steel.

Consequently, the designer must use special care in avoiding or reducing stress concentrations. A fine-grained steel is superior to a coarse-grained steel for such service. Three and one-half per cent nickel steel, stainless steel of the 18:8 type, or nonferrous alloys are in widest use for impact loads at subnormal temperatures.

22. Creep of Steel at High Temperature

If a part is subjected to stress at high temperature for a long period of time, it will undergo a slow and permanent deformation called creep. When the part operates with close clearances, the designer must make a careful attempt to predict the amount of creep that will probably take place during the useful life of the machine. For applications where close clearances are not involved, the permissible deformations are limited only by the deformations that may cause embrittlement and failure.

The amount that a member deforms or creeps depends upon the stress, the temperature, and the kind of material. Materials which have the same properties at room temperature may differ greatly in creep characteristics at elevated temperature. Hence suitable values for working stresses at high temperatures can be determined only by experiments on each material. Reliable data can be secured only by conducting longtime creep tests which duplicate as nearly as possible the conditions under which the steel will be used. The testing equipment is expensive, and must embody great precision of temperature control and measurement of elongation.

The tensile test, in which the total elongation is plotted as a function of the time, is the most significant. A creep curve, as that shown by the typical examples of Fig. 14-16, consists of three parts. In the first stage, the material is becoming stronger because of the strain hardening, and the creep rate is therefore decreasing. In the second stage, the strengthening due to strain hardening is counterbalanced by the weakening due to annealing, and the creep rate is approximately constant. In the third or last stage, the annealing effect predominates, and the deformations take place at an accelerated rate until failure results.

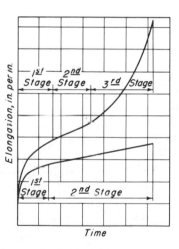

Figure 14-16 Creep of mild steel.

For sufficiently low stresses and temperatures, the third stage may not appear within the duration of the test. In fact, since it is usually impossible to conduct the test for as long a time as the service life of the part, designers usually extrapolate the straight portion of the curve beyond the duration of the test in order to predict the expected amount of creep at a future time. However, an element of danger lies in this practice, since the only way to be certain that the curve will remain straight for a service life of 10 or 20 years is to conduct the test for that length of time. Although experiments of such duration are manifestly impractical, it has been recommended that tests extending to 10 per cent of the expected service life be performed whenever feasible. The reliability of short-term creep tests for predicting long-term deformations is decidedly doubtful at the present time.

The selection of a suitable steel for continuous service at high temperatures is a very difficult task. The following steels have been recommended[10] for use in oil refinery equipment: up to 900°F, 0.08–0.18 C killed (or degasified) steel; 1,000°F, 0.08–0.18 C plus 0.50 Mo; 1,100°F, C-Mo plus 1.25 Cr or 1.50 Si; 1,200°F, 2.0 to 5.0 Cr plus 0.50 Mo; 1,300°F, 18 Cr, 8 Ni; above 1,300°F, 16 Cr, 13 Ni, 3 Mo.

23. Cast Iron

Cast iron or gray iron is characterized by high percentages of carbon and silicon. The average composition of cast iron is shown by Table 14-15.

Table 14-15   COMPOSITION OF CAST IRON

Carbon	2.00–4.00%
Silicon	0.50–3.00%
Manganese	0.20–1.00%
Phosphorus	0.05–0.80%
Sulfur	0.04–0.15%
Iron	Remainder

Gray iron is the most extensively used cast construction material. It is cheap, easily cast, and readily machined. Although strong in compression, cast iron has the disadvantage of being weak in tension. When a casting must sustain high tensile loads, some other material, such as alloy cast iron, malleable iron, or cast steel, must be used. Such materials, however, are more expensive than gray iron. Cast iron is notable for its lack of ductility; the elongation for a standard tensile specimen is usually less than 1 per cent. Cooling stresses can be relieved and machinability improved by annealing. For bodies of intricate shape, castings have inherent advantages over built-up or fabricated products.

It is general practice to classify cast irons with respect to tensile strength. Some properties[11] of cast irons are shown in Table 14-16. Suitable compositions to meet the strength requirements are chosen by the producing foundry unless the customer specifies otherwise. The cost of the iron increases with increase of strength.

Cast iron contains about 0.70 per cent carbon combined with iron in a pearlitic structure. The remaining carbon appears as more or less finely divided graphite flakes. The flake graphite is responsible for the relatively low

[10]See Fleishman, M., "Selection of Steel for High Temperatures," *Steel*, **102**, Jan. 17 1938, p. 34.
[11]See "Symposium on Cast Iron," *Proc. ASTM*, **33**(2), 1933, p. 115. See also ASTM Spec. A 48–64.

Table 14-16   STRENGTH OF GRAY CAST IRONS

Class No.	Tensile Strength Min., psi	Average Transverse Load,* lb	Compressive Strength† psi	Average Shear Strength, psi	Modulus of Elasticity, psi	BHN	Usual Min. Wall Thickness, in.
20	20,000	1,800	80,000	32,500	11,600,000	110	1/8
25	25,000	2,000	100,000	34,000	14,200,000	140	1/8
30	30,000	2,200	110,000	41,000	14,500,000	170	1/4
35	35,000	2,400	125,000	49,000	16,000,000	200	3/8
40	40,000	2,600	135,000	52,000	18,100,000	230	1/2
50	50,000	3,000	160,000	64,000	22,600,000	250	1/2
60	60,000	3,400	150,000	60,000	19,900,000	275	3/4

*Specimen 1.2-in. diameter, 18 in. supports, load at center.
†Subject to variations up to ±10 per cent.

hardness and strength of cast iron: the larger and coarser the flakes, the lower the strength. Dispersal of the carbon is aided by the presence of silicon, but is retarded by sulfur. However, sufficient manganese is usually present to form MnS, which alleviates the effect of the sulfur. Phosphorus forms the hard constituent, steatite, which has no effect on the graphite dispersal. A finer dispersal and a higher tensile strength are secured by an increase in the cooling rate. Since the cooling rate is affected by the section size, the interior of thick sections is weaker than the material near the surface or thin sections of the same composition.

The microstructure for different forms of cast iron is shown[12] in Fig. 14-17.

When higher strength and hardness are desired, alloying elements such as nickel, chromium, molybdenum, and copper are added to cast iron. In general, the alloying elements serve to improve the characteristics of the graphite. Compositions are available which will produce castings suitable for resisting impact, abrasion, corrosion, and high and low temperatures. Modern research has developed alloys of cast iron which are suitable for service as gears, crankshafts, and forging and drawing dies. Heat treatments of quenching and tempering are usually practiced with alloy cast iron to obtain full benefit of the expensive alloy additions. The designer should consult with an experienced foundryman on the composition and heat treatment for specialized applications. The engineering staffs of firms which market the various alloys also render expert metallurgical advice.

Cast iron is difficult to weld. Its tendency to crack can be reduced by careful preheating before welding, followed by slow cooling. Bronze welding

[12]See Moffatt, W. G., G. W. Pearsall, and J. Wulff, *The Structure and Properties of Materials, Vol. 1, Structures*, New York: John Wiley & Sons, 1964, p. 195.

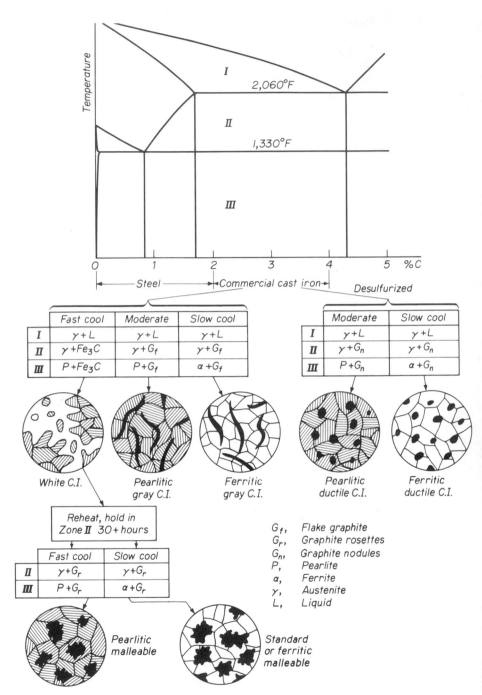

Figure 14-17 Cast iron microstructures and summary of the phases coexisting at various temperatures.

rods are extensively used on cast iron. Cast iron has excellent vibration-damping characteristics as well as good resistance to wear. The resistance to corrosion is only fair. It is difficult to secure satisfactory results with metal inserts cast in place.

White cast iron results if the cooling rate after pouring is very rapid. The carbon is retained in solution, and the metal is very hard and resistant to wear. White cast iron cannot be machined; if cast iron must be machined, care must be taken that thin sections of a casting are not chilled so rapidly that white iron forms. White cast iron has important applications. For example, the tread or outer surface of the rim of a freight-car wheel can be cooled so rapidly that white iron results while the rest of the casting remains soft and machinable. To do this, a portion of the mold, called a chill, is made of iron. The chill removes heat very rapidly from the molten metal. Other applications of white cast iron include rolling-mill rolls, railroad brake shoes, and plow shares.

Although the basic idea of forming castings is very simple, great skill and long experience are necessary to master the various techniques of producing quality castings of intricate form.[13]

24. Malleable Iron. Nodular Iron

As shown in Fig. 14-17, white cast iron, if completely free from graphite flakes can be converted into malleable iron by a prolonged annealing process. The long heating and slow cooling induces the carbon to form into clusters or rosettes called temper carbon. The resulting structure has better continuity and is much more ductile than gray iron. Mechanical properties are given in Table 14-17. The matrix surrounding the rosettes can be either pearlite or

Table 14-17 PROPERTIES OF MALLEABLE AND DUCTILE CAST IRON*

Material	ASTM Spec.	Grade or Type	Tensile Strength, psi	Yield Strength, psi	Elong in 2 in., %
Malleable	A47-61	32510	50,000	32,500	10
		35018	53,000	35,000	18
Nodular or ductile	A339-55	60-45-10	60,000	45,000	10
	A339-55	80-60-03	80,000	60,000	3
	A396-58	100-70-03	100,000	70,000	3
	A396-58	120-90-02	120,000	90,000	2

*For malleable iron, $E = 24,000,000$–$27,000,000$ psi; for ductile iron, $E = 25,000,000$ psi.

[13]See *Cast Metals Handbook*, 4th ed., Des Plaines, Ill.: American Foundrymen's Society, 1957. See also the author's *Design Engineering Projects*, Englewood Cliffs, N. J.: Prentice-Hall, Inc., 1968., Chapter 3.

ferrite, depending on the rate of cooling. Grade 32510 is in widest use for malleable parts.

Ductile or nodular iron, as shown by Fig. 14-17, is obtained directly from the cupola and dispenses with the prolonged heating necessary to produce malleable iron. The addition of a small amount of magnesium causes the carbon to appear in spherical or nodular form. Like malleable iron it is much more ductile than gray iron, as indicated by Table 14-17. The use of this material is increasing rapidly as it can be used in many places, such as crankshafts, where a more expensive forging was formerly required. Nodular iron shrinks more than gray iron, but its melting temperature is lower than for cast steel.

The machinability of ductile iron is similar to that of gray iron of equivalent hardness. Ductile iron can be brazed, but, because of the high carbon content, should be given careful handling in much the same way as for gray iron. The damping capacity is only slightly better than that of steel. Annealed nodular iron has good shock resistance. It is suitable for subzero temperature service, and can also be used in applications at elevated temperatures. It has a low creep rate at high temperatures, and can resist corrosion as well as gray iron.

Meehanite iron is made by a patented process in which a fine dispersal of graphite is secured by the addition of a calcium-silicon alloy.

25. Cast Steel

Cast steels have approximately the same chemical composition as the wrought steels, except that the percentages of silicon and manganese are larger. Steel is difficult to cast. It shrinks about $\frac{1}{4}$ in. per foot and makes a rough looking casting. The casting first undergoes a reduction in size, known as solidification shrinkage, which is followed by a considerable thermal contraction as the casting cools to room temperature. The casting should be designed in such a manner that the solidification will occur progressively, starting in the portions farthest removed from the molten metal. Uniformity of sections or a gradual transition from thin to thick sections is desirable, and stress concentrations should be reduced as much as possible. Projections should be avoided in long castings, or collapsible molds and cores should be used, to prevent hot tears caused by the large amount of cooling contraction.

Steel castings are in wide use in railway equipment, heavy machinery, oil refineries, and rolling mills. The design of the part must be suitable to the casting characteristics and physical properties of the material. A satisfactory microstructure depends on the conditions present during solidification, and thus on correct foundry practice. Heat treatment, such as annealing, normalizing, or quenching and drawing, gives a more uniform internal structure and improves the physical properties. Castings are usually

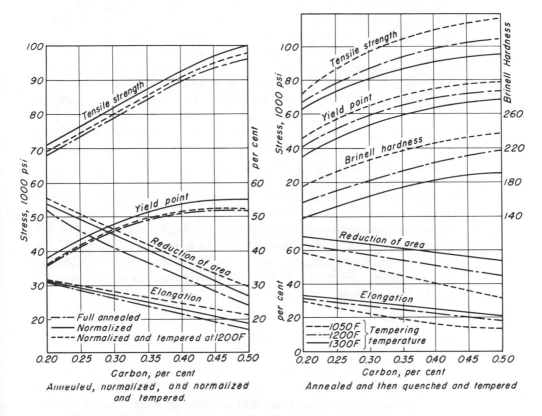

Figure 14-18 Mechanical properties of medium-carbon cast steel. (*Courtesy Steel Founders' Society of America.*)

inferior to corresponding wrought metals in impact resistance. The shock or impact qualities are improved by heat treatment, and also by the use of alloys.

The physical properties of cast steel vary with the carbon content, as shown in Fig. 14-18. Smaller values for the strength and ductility will be found in the interior of castings of large section size, as is shown in Fig. 14-19.

Low-carbon steel castings, containing 0.15 per cent to 0.30 per cent carbon, have desirable elevated temperature and carburizing properties. Parts made of this material frequently compete with castings made of gray or malleable iron. Cast steel can be readily welded. The microstructure of deposited weld metal is essentially the same as that of cast

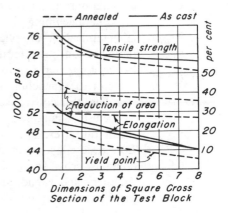

Figure 14-19 Effect of mass on the mechanical properties of cast carbon steel specimens taken at center of test block.

steel. If the carbon content is high enough to cause hardening and loss of ductility from cooling in air, it may be necessary to preheat the casting before welding, and then follow this by an annealing treatment afterwards.

Medium-carbon steel castings, in the range 0.30–0.40 per cent carbon, have greater strength but somewhat less ductility than low-carbon castings. The increase in the amount of carbon effects an improvement in the casting qualities of the metal. Heat treatment gives improved physical properties over the as-cast condition. High-carbon steel castings contain 0.40 per cent carbon and over. The greater hardness makes this material suitable for applications involving abrasion and impact. Machining must usually be preceded by an annealing heat treatment.

A wide variety of alloy steel castings is available for specific applications. Both low- and high-alloy castings are made. Heat treatment is usually practiced to secure the full benefit of the alloy additions. All the principal alloying elements used in wrought steel are employed in making steel castings. Certain alloys of high chromium and nickel content are suitable for high-temperature service or where corrosion conditions are present.

Steel casting design is highly involved, and in general should be conducted with the cooperation of the foundryman.[14]

26. Tool Steel

Materials used for tools must be very hard and resistant to deformation and wear. High-carbon and alloy steels of high purity meet this requirement. Plain carbon steel of about 0.65 per cent carbon is used for tools which must withstand considerable battering, such as hammers, heading dies, and hot forging dies. A carbon content of 1.10 per cent is used for general purpose cutting and hand tools. Razors, drawing dies, files, and other tools requiring extreme hardness have percentages of carbon up to 1.40 per cent.

The machining of such high-carbon steels must be preceded by normalizing, annealing, or spheroidizing heat treatments. The quenching and tempering of tool steels require expert handling. The tempering temperatures are low as compared to those of machine steels and are used only for relieving some of the quenching stresses.

The abrasion resistance and toughness of tool steel can be increased by the use of suitable alloys. Also, tool steel can be given a less rapid quenching, in oil for example, and residual stress and distortion will thereby be less. Low residual stress and low distortion are especially valuable if the body contains cross sections of both large and small diameters. Chromium, tungsten, and vanadium, either singly or in various combinations, are the principal alloys used. Compositions are available which hold their hardness

[14]See *Steel Castings Handbook*, Cleveland, Ohio: Steel Foundrymen's Association, 1944. See also Briggs, C. W., R. A. Gezelius, and A. A. Donaldson, "Steel Casting Design for Engineers and Foundrymen," *Trans. Am. Foundrymen's Ass.*, **46**, 1938, p. 605.

and cutting edges even when raised to a red heat from the friction of the cutting.

High-speed steel of the 18-4-1 type contains 18 per cent tungsten, 4 percent chromium, 1 per cent vanadium, and 0.70 per cent carbon, and has superior cutting qualities. Its manufacturing and heat-treating processes must be closely controlled. Stellite is a cobalt-chromium-tungsten-carbon alloy also used for cutting tools. Cemented carbide cutting tools are made by powdered metallurgy, and in general will outlast even high-speed steel.

27. Aluminum Alloys

Lightweight aluminum alloys have found wide applications in manufactured products. The high-strength alloys of this material have practically the same strength as mild steel. As is shown in Table 2-1, the weight per cubic inch is only about one-third as great as for steel. The stiffness of a part made of aluminum is less and the deflection greater than that of a corresponding design in steel. Aluminum is a very versatile metal. It can be extruded, forged, or rolled into sheets. It is suitable for making sand, permanent mold, or die castings, and is also used for automatic screw machine products. Depending on the type of alloy, the strength can be increased by cold working or heat treatment.

Aluminum objects secure considerable protection from atmospheric corrosion by a coating of aluminum oxide, which forms naturally in air. Sometimes this oxide layer is built up artificially by anodic oxidation to give better protection. High-strength aluminum alloys are not so resistant to corrosion as the pure metal. In Alclad sheets, protection from corrosion is secured by rolling a coating of pure aluminum on both sides of a high-strength alloy core.

Soft aluminum is somewhat difficult to machine. The machinability of the alloys and the full heat-treated grades is superior to the machinability of the pure metal. Considerable experience and skill are required for the satisfactory welding of aluminum. Torch, metallic and carbon arc, and electric resistance methods are in use. Unlike steel, aluminum exhibits no change in color to warn the operator that the welding temperature is being approached. The presence of the oxide film adds to the diffculties of securing sound welds. The high strength secured by cold working or heat treatment is usually lost in the heated region. Cold working of welds to restore the strength is sometimes possible. Brazing techniques suitable for certain compositions have also been developed.

28. Alloy and Temper Designations

The numbering system for wrought aluminum alloys uses a four-digit number, the first of which designates the alloy type. The second digit identifies the specific alloy modification, and the last two digits identify the specific aluminum alloy or indicate the aluminum purity. The principal alloying elements and their digits are indicated by Table 14-18.

Table 14-18   WROUGHT ALUMINUM ALLOY DESIGNATION SYSTEM

Aluminum—99% min and greater	1xxx
Copper	2xxx
Manganese	3xxx
Silicon	4xxx
Magnesium	5xxx
Magnesium and silicon	6xxx
Zinc	7xxx
Other elements	8xxx
Unused series	9xxx

The temper of an aluminum alloy is one of the major factors governing its strength, hardness, and ductility. Some alloys are hardened and strengthened by cold work such as cold rolling, drawing, or stretching. Other alloys are heat treatable and their properties can be improved appreciably by appropriate thermal treatments. The temper designation is usually specified by a suffix to the alloy number, as indicated by Table 14-19.

Table 14-19   TEMPER DESIGNATIONS FOR ALUMINUM

Temper	Process
-F	As-cast condition
-O	Annealed temper of wrought materials
-H14	Cold worked to tensile strength midway between tempers O and H18
-H18	Cold worked to full hard temper
-H34	Cold worked beyond temper range and stabilized
-H38	Same as H34 but with higher strength properties
-T3	Solution heat treated followed by strain hardening
-T4	Solution heat treated followed by natural aging at room temperature
-T6	Solution heat treated followed by artificial aging
-T8	Solution heat treated, strain hardened, and artificial aging

The theory for the increased strength and hardness of heat-treated aluminum alloys differs in several respects from that for steel. After aluminum is quenched, the material is a supersaturated solid solution of the alloying elements. At this time it is soft and ductile. If allowed to stand at room temperature the alloying elements are precipitated as hard particles which are effective in preventing movement within the crystal structure. Since considerable time may be required for the greatest strength to be reached, the process is called aging. For some alloys the process can be hastened by heating to a temperature that is known to give the optimum particle size for the maximum strength and hardness. After precipitation of the hardening constituents, the material is said to be in the T condition.

29. Composition and Mechanical Properties

The composition and mechanical properties of several typical wrought aluminum alloys are given in Table 14-20. Alloys in the O-condition are suit-

Table 14-20 COMPOSITION AND MECHANICAL PROPERTIES OF WROUGHT ALUMINUM ALLOYS*

Alloy	Nominal Composition Balance: Aluminum Plus Inpurities								Type
	Si	Fe	Cu	Mn	Mg	Cr	Zn	Ti	
Sheets and Plates									*Non-heat Treatable*
1100	1.0		0.20	0.05			0.10		Commercially pure aluminum
3003	0.6	0.7	0.20	1.0–1.5			0.10		Manganese-aluminum alloy
5005	0.4	0.7	0.20	0.20	0.5–1.1	0.10	0.25		Magnesium-aluminum alloy
Bars and Rods									*Heat Treatable*
2011	0.4	0.7	5.0–6.0				0.30		Copper-aluminum alloy
6061	0.4–0.8	0.7	0.15–0.4	0.15	0.8–1.2	0.15–0.35	0.25	0.15	Magnesium-silicon alloy
6063	0.2–0.6	0.35	0,10	0,10	0,45–0.9	0.10	0.10	0.10	Magnesium silicon alloy
7075	0.5	0.7	1.2–2.0	0.30	2.1–2.9	0.18–0.40	5.1–6.1	0.20	Zinc-aluminum alloy

Alloy and Temper	Tensile Strength, psi	Yield Strength, psi	Elongation 2 in. %		Brinell	Shearing Strength, psi	Endurance Limit, psi
			1/16 in. sheet	1/2 in. round			
1100-O	13,000	5,000	35	40	23	9,000	5,000
1100-H14	18,000	17,000	9	20	32	11,000	7,000
1100-H18	24,000	22,000	5	15	44	13,000	9,000
3003-O	16,000	6,000	30	40	28	11,000	7,000
3003-H14	22,000	21,000	8	16	40	14,000	9,000
3003-H18	29,000	27,000	4	10	55	16,000	10,000
5005-O	18,000	6,000	30		30	11,000	
5005-H18	29,000	28,000	4		51	16,000	
5005-H38	29,000	27,000	5		51	16,000	
2011-T3	55,000	43,000		15	95	32,000	18,000
2011-T8	59,000	45,000		12	100	35,000	18,000
6061-O	18,000	8,000	25	30	30	12,000	9,000
6061-T4	35,000	21,000	22	25	65	24,000	13,000
6061-T6	45,000	40,000	12	17	95	30,000	14,000
6063-O	13,000	7,000			25	10,000	8,000
6063-T4	25,000	13,000	22			16,000	
6063-T6	35,000	31,000	12	18	74	22,000	10,000
6063-T835	48,000	43,000	8		105	30,000	
7075-O	33,000	15,000	17	16	60	22,000	17,000
7075-T6	83,000	73,000	11	11	150	48,000	22,000
7075-T73	73,000	63,000		13			

*Taken mainly from Alcoa Aluminum literature. Properties are typical; variations must be expected in practice.

able for forming, spinning, and deep drawing. Alloys in the work-hardened or fully heat-treated condition are not so workable. Suitable radii must be provided on the forming tools, depending on the composition, the thickness of the stock, and the temper. Table 14-21 indicates some of the fabricating qualities and principal applications for these alloys.

The composition and mechanical properties of a number of widely used aluminum sand casting alloys are given in Table 14-22.

Table 14-21 PRINCIPAL CHARACTERISTICS AND USE OF WROUGHT ALUMINUM ALLOYS

Alloy	Characteristics and Uses
1100	Spinnings, deep-drawn shapes, hand-hammered giftware.
3003	Excellent formability with moderate strength. Food utensils.
5005	Good formability. Fine grain size. Panels, ductwork, decorative trim.
2011	Highest machinability. Most widely used screw machine stock.
6061	Best heat-treated alloy for brazing and welding. Boats, furniture.
6063	Good formability. Furniture tube, moldings, doors, windows, lights.
7075	Forming requires careful attention. Aircraft structures and skins, lock washers, fuel tanks, skis.

Table 14-22 COMPOSITION AND MECHANICAL PROPERTIES OF ALUMINIUM SAND CASTING ALLOYS*

Alloy	Composition†			Applications
	Cu	Si	Mg	
43		12.0		General purpose casting alloy. Cooking utensils, architectural and marine applications, pipe fittings.
214			3.8	Food, chemical, marine and architectural applications.
195	4.5	0.8		Machinery and aircraft structural members.
220			10.0	High-strength and shock-resistant castings. Aircraft structural members.
356		7.0	0.3	Most popular casting alloy. Automotive and railroad fittings, hardware, valves.

Alloy and Temper	Tensile Strength, psi	Yield Strength, psi	Elong. 2 in. %	Compressive Yield Str., psi	Brinell	Shearing Strength, psi	End. Limit
43-F	19,000	8,000	8.0	9,000	40	14,000	8,000
214-F	25,000	12,000	9.0	12,000	50	20,000	7,000
195-T6	36,000	24,000	5.0	25,000	75	30,000	7,500
220-T4	48,000	26,000	16.0	27,000	75	34,000	8,000
356-T6	33,000	24,000	3.5	25,000	70	26,000	8,500

*Taken from *Alcoa Aluminum Handbook*. Values shown are typical; variations must be expected in practice.

†Percentage of alloying elements. Aluminum and normal impurities constitute remainder.

30. Magnesium Alloys

Magnesium, in its various cast and wrought forms, provides another medium for achieving lightweight designs. As shown by Table 2-1 the weight per cubic inch and the modulus are lower than for aluminum. Magnesium is less resistant to corrosion than aluminum alloys, and should be protected against a salt atmosphere and against constant contact with water. Proper provision for drainage or ventilation is necessary when moisture is present.

Magnesium has relatively low hardness, and is very easy to machine. No special composition is required for use in automatic screw machines. Magnesium can be hot forged with hammers or mechanical and hydraulic

Table 14-23 COMPOSITION AND MECHANICAL PROPERTIES OF MAGNESIUM ALLOYS*

Alloy	Percentage Composition—Balance Magnesium and Minor Impurities					
	Al	Mn, min	Zn	Si, max	Cu, max	Ni, max
AZ 31B	2.5–3.5	0.20	0.7–1.3	0.30	0.05	0.005
AZ 31C	2.5–3.5	0.20	0.6–1.4	0.30	0.10	0.03
AZ 63A	5.3–6.7	0.15	2.5–3.5	0.30	0.10	0.01
AZ 81A	7.0–8.1	0.13	0.4–1.0	0.30	0.10	0.01
AZ 92A	8.3–9.7	0.10	1.6–2.4	0.30	0.10	0.01
HK 31A			Zirconium 0.50–1.00		Thorium 2.5–4.0	

Alloy and Temper	ASTM	Tensile Strength, psi	Yield Strength, psi	Comp. Yield Strength, psi	Shear Strength, psi	Elong. in 2 in. %	Brinell
Sand castings†							
AZ 63A-F	B80–63	29,000	14,000	14,000	18,000	6	50
AZ 63A-T4		40,000	14,000	14,000	18,000	12	55
AZ 63A-T6		40,000	19,000	19,000	21,000	5	73
AZ 81A-T4		40,000	14,000	14,000	18,000	12	55
AZ 92A-F		24,000	14,000	14,000	19,000	2	65
AZ 92A-T4		40,000	14,000	14,000	19,000	10	63
AZ 92A-T6		40,000	21,000	21,000	22,000	2	84
Sheets and plates†‡							
AZ 31B-H24	B90–65	42,000	32,000	26,000	23,000	15	73
AZ 31B-O		37,000	22,000	16,000	21,000	21	56
HK 31A-H24		37,000	29,000	25,000	21,000	8	57
Rods, bars, shapes†							
AZ 31B-F ⎱	B107–65	38,000	29,000	14,000	19,000	15	49
AZ 31C-F ⎰							

*Data courtesy Dow Chemical Co.
†These values are typical. Minimum values are lower.
‡Strength values are for minimum sections. Values are somewhat less for larger sizes.

presses. It can be cold formed or drawn, but only with difficulty. The rapid work hardening can be relieved by annealing. Care should be exercised to avoid stress concentration such as caused by notches or sharp fillets. A sharp 60° V-groove will reduce the endurance limit to one-fourth or one-third the value for a standard specimen. When abrasion conditions are present, the soft magnesium must be protected by inserts such as plates, liners, sleeves, shoes, or bushings. Magnesium exhibits a high damping capacity which is effective in preventing vibrations from building up into large amplitudes.

When magnesium is welded with a torch, butt welds are preferable because the removal of the flux is usually easier than for lap welds, and the possibility of corrosion is reduced. All types of joints may be made with the helium arc process, which requires no flux. Resistance welding is satisfactory for magnesium, and may be applied to wrought products or thin-walled castings. Magnesium cannot be used in the form of rivets. Rivets of aluminum alloys 6053 and 6061 have been found to give satisfactory service. Brass and iron rivets should be avoided because of the possibility of galvanic action.

The composition and physical properties for a number of typical magnesium alloys are given in Table 14-23.

31. Copper Alloys

Copper is a very ductile metal, and may be spun, stamped, rolled into sheets, and drawn into wire and tubing. Mechanical working increases the hardness and strength, but decreases the ductility. The effects of cold work can be removed by annealing. Copper can be fabricated by soldering, welding, or brazing. Copper has excellent heat-conducting qualities, and is used in refrigerators, radiators, water heaters, and air-conditioning equipment. It is widely used as an electrical conductor. It cannot be hardened by quenching unless alloyed with beryllium. Copper resists certain types of corrosion.

The most important copper-base alloys are brass and bronze. Brass is a copper-zinc alloy, and bronze is composed principally of copper and tin. Zinc is cheaper than tin, so the cost of brass is less than the cost of bronze. Bronze, however, is usually considered the superior metal. It is commercial practice to designate certain reddish-colored copper-zinc combinations as "bronze" even though no tin is present.

Brass and bronze are used in both the cast and wrought forms. The strength of brass increases with the zinc content. Most of the copper alloys can be work hardened to higher strengths than pure copper. In general, the machinability of the alloys is satisfactory. The free-cutting qualities are improved by the addition of small amounts of lead. Brass is about equal to copper in corrosion resistance, but bronze is superior to both. There are a great many copper alloys with a wide variety of commercial designations.

A selected list of the wrought alloys, giving composition and mechanical properties, are shown in Tables 14-24 and 14-25. Several of the most commonly used casting alloys are listed in Table 14-26.

Table 14-24 COMPOSITION OF BRASS, BRONZE, AND MISCELLANEOUS WROUGHT COPPER ALLOYS*

No.	ASTM spec.	Alloy	Cu, %	Sn, %	Zn, %	Other
1	B133	Copper				
2	B36 Alloy 3	Plates, sheets	85.0		15.0	
3	B36 Alloy 6	Plates, sheets	70.0		30.0	
4	B36 Alloy 8	Plates, sheets	65.0		35.0	
5	B134 Alloy 2	Brass wire	90.0		10.0	
6	B134 Alloy 7	Brass wire	65.0		35.0	
7	B16	FC brass	60.0–63.0		remainder	Pb 2.5–3.7
8	B21 Alloy A & B	Naval brass	59.0–62.0	0.5–1.0	remainder†	
9	B159 Alloy A	Phos. bronze wire	remainder	4.2–5.8	0.30	P 0.03–0.35
10	B139 Alloy B2	Phos. bronze rod, shapes	remainder	3.5–4.5	1.5–4.5	P 0.01–0.50‡
11	B138 Alloy A	Manganese bronze	58.5	1.0	39.0	Fe 1.4, Mn 0.1
12	B171	Muntz metal	58.0–61.0	0.25 max	remainder	Pb 0.4–0.9
13	B150 Alloy 1	Aluminum bronze	80.0–93.0	0.60 max	1.0	Al 6.5–11.0§
14	B196	Beryllium copper	remainder			Be 1.8–2.05
15	B151 Alloy B	18% Ni Silver	53.5–56.5	Ni 16.5–19.5	remainder	Mn 0.5 max
16	B127 (Monel)	Nickel-copper alloy	remainder	Ni 63.0–70.0	Fe 2.5	Mn 1.25

*Some alloys have minor amounts of impurities. See specifications for exact compositions. FC, free cutting.

†Pb, Alloy A, 0.20 max; Alloy B, 0.40–1.0.

‡Pb 3.5–4.5.

§Fe 4.0 max, Ni 1.0 max, Si 2.2, Mn 1.5

Table 14-25 BRASS, BRONZE, AND MISCELLANEOUS WROUGHT COPPER ALLOYS

No.	Alloy	Condition	Tensile Strength,* psi	Rockwell, min.	Elong., %
1	Copper	Bars, all sizes	A 37,000		25
		Bars 0.188–0.375 in.	H 42,000		12
		Bars 0.50–2.0 in.	H 33,000		15
2	Brass, alloy 3	Plates, sheets, strips	1/2 H 51–61,000,	56B	

No.	Material	Form	Strength*	Rockwell B	Elong.
			H 63–72,000	72B	
			Sp 78–86,000	82B	
3	Brass, alloy 6	Plates, sheets, strips	1/2 H 57–67,000,	60B	
			H 71–81,000	79B	
			Sp 91–100,000	89B	
4	Brass, alloy 8	Plates, sheets, strips	1/2 H 55–65,000,	57B	
			H 68–78,000	76B	
			Sp 86–95,000	87B	
5	Brass wire, alloy 2	Round, hexagonal,	1/2 H 56–67,000,		
			H 70–79,000		
		octagonal and square	Sp 84,000		
6	Brass wire, alloy 7	Do	1/2 H 79–94,000,		
			H 92–117,000		
			Sp 120,000		
7	Free-cutting brass	Bars, 1 in. and under	A 48,000		
			YS 20,000		15
		Bars, 1 in.–2 in.	A 44,000		
			YS 18,000		20
		Bars, 0.5 in. and under	1/2 H 57,000		
			YS 25,000		7
		Bars, 0.5 in. and 1 in.	1/2 H 55,000		
			YS 25,000		10
8	Naval brass, A, B	Bars, 1 in. and under	A 54,000		
			YS 20,000		30, 25
		Bars, 1 in.–2 in.	A 52,000		
			YS 20,000		30, 25
		Bars, 1 in. and under	1/2 H 60,000		
			YS 27,000		25, 18
		Bars, 1 in.–2 in.	1/2 H 58,000		
			YS 26,000		25, 20
9	Phos. bronze wire, A	Wire, 0.025 in. and under	Sp 145,000		
		Wire, 0.025–0.0625 in.	Sp 135,000		
		Wire, 0.125–0.250 in.	Sp 125,000		
10	Phos. bronze rod, shapes, B2	Rounds, 0.25–0.5 in.	H 60,000		10
		Rounds, 0.5–1.0 in.	H 55,000		12
11	Manganese bronze, A	Bars, 1.0–2.5 in.	1/2 H 70,000		
			YS 35,000		15
		Bar, 1.0–1.5 in.	H 76,000		
			YS 52,000		10
12	Muntz metal	Tube plates under 2 in.	50,000		
			YS 20,000		35
13	Aluminum bronze, 1	Bars, 0.5 in. and under	80,000		
			YS 40,000		9
		Bars, over 1 in.	72,000		
			YS 35,000		12
14	Beryllium copper	Rods, bars	A 60–85,000	46B–85B	
		Rods, up to 0.375 in.	H 95–130,000	92B–103B	
15	18% Ni silver, alloy B	Rods, 0.02–0.25 in.	H 90–110,000		
		Rods, 0.50–1.0 in.	H 75–95,000		
16	Nickel-copper alloy	Plate, sheet, strip	A 70–85,000		
			YS 28,000	73B	35

*A, annealed; 1/2H, half hard; H, hard; Sp., spring; YS, yield strength.

Table 14-26 TYPICAL BRASS AND BRONZE CASTING ALLOYS

SAE	ASTM	Nominal Composition				Commercial Designation
		Cu	Sn	Pb	Zn	
40	B 62–63	85	5	5	5	Composition bronze or ounce metal
41	B 146–52 6B	67	1	3	29	Leaded yellow brass for general purposes. No. 1
43	B 147–63 8A	58	Fe 1.25	Al 1.25	39.25	Mn 0.25. High-strength yellow brass. Manganese "bronze"
62	B 143–61 1A	88	10		2	Tin bronze. 88–10–2 or "G" bronze
64	B 144–52 3A	80	10	10		High leaded tin bronze. 80–10–10
68	B 148–65 9B	89				Al 10, Fe 1.0 Aluminum bronze. Grade B

ASTM	Tensile Strength psi	Yield Strength, psi	Elong. in 2 in., %	General Properties and Applications
B 62–63	30,000	14,000	20	Good casting and finishing properties. Free cutting
B 146–52 6B	30,000	11,000	20	Used where cheapness is main consideration
B 147–63 8A	65,000	25,000	20	Makes tough, strong castings
B 143–61 1A	40,000	18,000	20	Withstands heavy gear and bearing pressures
B 144–52 3A	25,000	12,000	8	Has low friction. Withstands heavy loads
B 148–65 9B	65,000	25,000	20	BHN 110. For heavy service such as worms, gears, and similar applications

32. Alloys for Die Castings

Die castings are formed by forcing a molten alloy into metal molds or dies under high pressure. Only alloys of relatively low melting points are suitable for this process. Die castings are made principally from zinc, aluminum, and magnesium, and to a lesser extent from brass.

Die casting is strictly a machine operation and lends itself to very rapid production of parts. The chief disadvantage is the high cost of dies, which usually limits the use of the method to parts that are to be made in large quantities. Close tolerances on the finished product can be maintained. Consequently, very little or no machining is required. The die casting process is suitable for parts containing very thin sections or intricate shapes. Inserts such as studs or bushings can be included as a part of the casting. Die castings usually have a very smooth surface suitable for direct application of platings or organic finishes. Dimensions and weights, and the composition and

Table 14-27   APPROXIMATE DIMENSIONAL AND WEIGHT LIMITS FOR DIE CASTINGS

	Zinc	Aluminum	Magnesium
Maximum weight of casting (lb)	30	25	10
Minimum wall thickness, large castings (in.)	0.050	0.080	0.080
Minimum wall thickness, small castings (in.)	0.020	0.050	0.050
Minimum variation from drawing dimensions per inch of diameter or length (in.)	0.001*	0.002*	0.002*
Cast threads (max no. per in. external)	24	12	12
Cast threads (max no. per in. internal)	24†	none	none
Minimum draft on cores (in. per in. of length or diameter)	0.003	0.010	0.010
Minimum draft on side walls (in. per in. of length)	0.005	0.010	0.008
Cored holes, min diameter (in.)	0.030	0.100	0.100

*Depends on conditions.
†Where cheaper than tapping.

mechanical properties of some of the better-known die casting alloys are listed in Table 14-27 and 14-28.

Zinc die castings have gained extensive use because of their high strength, long die life, and moderate casting temperatures. The light weight of aluminum die castings is an important advantage in many applications. The casting temperature is higher than zinc, and die life is shorter. Magnesium castings also possess the advantage of low weight. Casting temperature is the same as for aluminum. The surface of magnesium castings is usually treated to inhibit corrosion and to serve as a base for the finish. Brass is cast at the highest temperature used for die casting, and die life is accordingly the shortest. High strength and hardness are advantages possessed by brass die castings.

The designer must give care and attention to all the details of a part to be made by the die-casting process. The thickness of the sections should be as uniform as possible throughout the casting. Solidification and cooling will thus take place uniformly, and the casting will be free from "hot spots" resulting from the slow cooling of large volumes of metal. Residual stresses are frequently found in castings of unequal sections. The parting line for the dies should be located at a point that permits easy removal of the flash resulting from the imperfect matching of the dies. If a bead can be located on the parting line, the flash can be very easily sheared off. Undercuts that require the use of slides in the die should be avoided whenever possible. The additional die expense for complex castings with undercuts, if made in sufficiently large quantities, is sometimes justified. Fillets should be used at reentrant angles to increase the strength and prevent localization of cooling stresses. Threaded holes can sometimes be cored, and a saving in machining costs

Table 14-28 NOMINAL COMPOSITION AND MECHANICAL PROPERTIES OF DIE CASTING ALLOYS (PER CENT BY WEIGHT)

Alloy	Al	Cu	Mg	Mn	Si	Zn	Pb	Sn
Zinc: ASTM B86								
AG40A (XXIII), or SAE 903, or Zamak* 3	3.5–4.3	0.25 max	0.02–0.05			remainder	0.005	0.003
AC41A (XXV), or SAE 925, or Zamak 5	3.5–4.3	0.75–1.25	0.03–0.08			remainder	0.007	0.005
Aluminum: ASTM B85–60								
S12B, or Alcoa† 13	remainder	0.6	0.10	0.35	11.0–13.0	0.50		0.15
SG100B, or Alcoa 360	remainder	0.6	0.4–0.6	0.35	9.0–10.0	0.50		0.15
SC84B, or Alcoa 380	remainder	3.0–4.0	0.10	0.50	7.5–9.5	3.0		0.35
Magnesium: ASTM B94–58								
AZ91A, or SAE 501, or Dow-metal‡ R, or AM263§	8.3–9.7	0.10		0.13	0.50	0.35–1.0		
Brass: ASTM B176–62								
Z30A	0.25	57.0 min		0.25	0.25 max	30.0 min	1.50	1.50
Z331A	0.15	63.0–67.0		0.15	0.75–1.25	remainder	0.25	0.25

Alloy	Tensile Strength, psi	Yield Strength, psi	Elong. in 2 in., %	Comp. Strength, psi	Shear Strength, psi	Endurance Limit, psi	Brinell Hardness	Melting Point, °F
Zinc								
AG40A (XXIII)	41,000		10	60,000	31,000		82	728
AC41A (XXV)	47,000		7	87,000	38,000		91	727
Aluminum:	43,000	21,000	2.5		25,000	19,000		1065–1080
	44,000	25,000	2.5		28,000	20,000		1035–1105
	46,000	23,000	2.5		28,000	20,000		1000–1100
Magnesium:								
AZ91A	33,000	22,000	3	22,000**	20,000		60	1,105
Brass: Z30A	45,000	25,000	10				120–130	1,650
Z331A	58,000	30,000	15				120–130	1,550

*New Jersey Zinc Co.
†Aluminum Co. of America.
‡Dow Chemical Co.
§American Magnesium Corp.
**Compressive yield strength.
Values of normal impurities not shown. Mechanical properties are typical; variations must be expected in practice.

thereby effected. The designer should consult with a die casting firm on the design of the part in order to receive the guidance of specialists in this branch of engineering. Sometimes minor revisions can be made in the design which will result in a lowering of the cost of the dies and the production parts.[15]

Permanent mold castings are made by filling metal molds under gravity head. Both iron and aluminum castings are made by this process. Iron castings are usually annealed to improve the machining properties. Sand cores are used in the so-called "semipermanent mold method."

REFERENCES

1. Bullens, D. K., *Steel and Its Heat Treatment*, Vol. I, New York: John Wiley & Sons, 1948.

2. Bullens, D. K., *Steel and Its Heat Treatment*, Vol. III, New York: John Wiley & Sons, 1949.

3. Dieter, G. E., *Mechanical Metallurgy*, New York: McGraw-Hill Book Company, 1961.

4. Evans, U. R., *Corrosion and Oxidation of Metals*, London: Edward Arnold, 1960.

5. Fontana, M. G., and N. D. Greene, *Corrosion Engineering*, New York: McGraw-Hill Book Company, 1967.

6. Heyer, R. H., *Engineering Physical Metallurgy*, New York: Van Nostrand Reinhold Co., 1939.

7. Holloman, J. H., and L. D. Jaffe, *Ferrous Metallurgical Design*, New York: John Wiley & Sons, 1947.

8. "Metals," Reference Issue, *Machine Design*.

9. *Metals Handbook, Vol. 1, Properties and Selection of Metals*, Metals Park, Ohio: American Society for Metals, 1961.

10. *Metals Handbook, Vol. 2, Heat Treating, Cleaning, and Finishing*, Metals Park, Ohio: American Society for Metals, 1964.

11. *Metals Handbook, Vol. 3, Machining*, Metals Park, Ohio: American Society for Metals, 1967.

12. *Metals Handbook, Vol. 4, Forming*, Metals Park, Ohio: American Society for Metals, 1969.

13. Sachs, G., and K. R. Van Horn, *Practical Metallurgy*, Metals Park, Ohio: American Society for Metals, 1940.

14. "Selection of Steel for Automobile Parts," *J. Soc. Auto. Eng.*, **57**, Aug. 1949 to **58**, Jan. 1950 (six articles).

15. Williams, G. T., *What Steel Shall I Use*, Metals Park, Ohio: American Society for Metals, 1941.

16. Wulff, J., H. F. Taylor, and A. J. Shaler, *Metallurgy for Engineers*, New York: John Wiley & Sons, 1952.

[15]See the author's *Design Engineering Projects*, Englewood Cliffs, N. J.: Prentice-Hall, Inc., 1968, Chapter 4. See also Barton, H. K., "Die Casting," *Eng. Digest*, **21**, Dec. 1960, p. 77, and Chase, H., *Handbook on Designing for Quantity Production*, New York: McGraw-Hill Book Company, 1944, Chapter 1.

Index